33rd Symposium of Malaysian Chemical Engineers (SOMChE 2022)

IOP Conference Series: Materials Science and Engineering
Volume 1257

Online
8 – 9 August 2022

ISBN: 978-1-7138-6420-2
ISSN: 1757-8981

Printed from e-media with permission by:

Curran Associates, Inc.
57 Morehouse Lane
Red Hook, NY 12571

Some format issues inherent in the e-media version may also appear in this print version.

Institute of Physics
Dirac House, Temple Back
Bristol BS1 6BE UK

Phone: 44 1 17 929 7481
Fax: 44 1 17 920 0979

techtracking@iop.org

Additional copies of this publication are available from:

Curran Associates, Inc.
57 Morehouse Lane
Red Hook, NY 12571 USA
Phone: 845-758-0400
Fax: 845-758-2633
Email: curran@proceedings.com
Web: www.proceedings.com

TABLE OF CONTENTS

Preface

Peer Review Statement

A Critical Review on the Influence of Humidity for Plant Growth Forecasting 1
S Y Chia, M W Lim

Process Simulation of Anaerobic Digestion for Methane Production Using Aspen Plus 7
G Ramasamy, A H Goodman, H M Lahuri, S S Md Shah, K M Sabil

Influence of Calcium Macronutrient on Tip Burn Occurrence in Hydroponically Cultivated Lettuces 13
M Y Yap, M W Lim, L K W Brendan

Comparison of Different Methods for the Extraction of Essential Oil from the Rhizome Part of Kaempferia Galanga Linn 19
C R Nurhaslina, A N Mustapa, C Y Azizi, H Alwi

Promoting Circular Economy Between Palm Oil Sector with Multiple Industries 25
J Z H Hwang, V Andiappan, D K S Ng

Molecular Dynamics Study of CO2/H2 and CH4/H2 Gas Separation by Nanoporous Graphene 31
R Rivero, Y Ren

Terephthalic Acid (TPA) Recovery from Plastic Bottle of Polyethylene Terephthalate (PET) Through Hydrolysis Treatment 37
S Alias, N A Abdul-Hakim

Biofuel Production Via Catalytic Cracking of Waste Cooking Oil Using Sulfonated Activated Carbon Supported La/Ce Bimetallic Catalysts 43
S Z Naji, C T Tye, A R Mohamed

Decision Making on the Renewable Energy Penetration in Malaysia Via SWOT-PESTLE Analysis: Hydrogen Fuel Cell and Solar Photovoltaic Thermal Technology 49
M A Mohd Mokhtar, W N S Wan Mohamed Zuri, N I Mohd Rasid, A Ahmad Asmadi, N Abdul Manaf

Response Surface Optimisation of Biodiesel Synthesis Using Biomass Derived Green Heterogeneous Catalyst 55
W-Y Wong, S Lim, Y-L Pang, C-H Lim, F-L Pua, G Pua

Synthesis of Bio-Hydrogen Supply Network Via Graph-Theoretic Approach Coupled with Monte Carlo Simulation Model 61
S L Y Lo, K G H Kong, M F Tang, J Sunarso, B S How

Experimental Studies on Transesterification of Mango (Mangifera Indica) Seed Kernel Butter for Biodiesel Production 67
G Narayanasamy, TNP Padmesh, S Krishnan

Sustainable Methane Emissions Management 73
H H Hashim, N A Abdul Talip, M H Muhamad Pikri, N A Abu Bakar, A Wan Zakaria

Process Integration and Optimization for Sustainable and Circular Hydrogen Production 79
M Lim Moy Hian, C H Lim, B S How, W P Q Ng, H L Lam

Production of Monoethylene Glycol from Lignocellulosic Biomass Via Catalytic Hydrogenation: A Review 85
A H Norhanifah, A R Norliza, J Rafidah

Kinetic Analysis for Catalytic Co-Pyrolysis of Palm Kernel Shell and Plastic Waste Mixtures with Bifunctional HZSM-5 and Mussel Shell Catalyst 91
A N Kumar, B L F Chin, A L K Chee

Selective Epoxidation of the Unsaturated Fatty Acids 97
M. B Mahadi, I.M. Rasib, N.Z Abas, M.A. Rahman, I. SH Azmi, M.J. Jalil

Potential of Integrating Solid Oxide Fuel Cell Based on Biomass in Power Generation in Malaysia: A Feasibility Study 102
R Ravi, N Shieela Kalib, A Muchtar, M Rao Somalu

Thermal Investigation of Solid Oxide Fuel Cell Ni-YSZ Anode Supported with Cooling System: A FEM Approach 108
N Shieela Kalib, A Muchtar, V Zhen Yuan, M Rao Somalu, A Kamal Ariffin Mohd Ihsan

Hydrogen Co-Firing Characteristics in a Single Swirl Burner: A Numerical Analysis 114
Mohammad Nurizat Rahman, Norshakina Shahril, Suzana Yusup, Ismail Shariff

Decomposition and Decoupling Analyses of CO2 Emissions and Economic Growth: A Comparative Study Between Malaysia and Vietnam 121
B Kiwan, N S Kalib, A Muchtar

Performance Analysis of Neural Network Architecture in Developing Real-Time Malaysian River Water Quality Model 127
D H Djarum, Z Ahmad, J Zhang

Predictive Solubility of Amino Acids in Deep Eutectic Solvents for Chicken Feather Waste Valorisation 133
MZM Salleh, MK Hadj-Kali, AAI Luthfi, MA Hashim

Accelerated Weathering of Limestone Bicarbonate Induced CaCO3 Precipitation in a Mixture of Sand, Fly Ash and CaO 139
S Y W Chai, L H Ngu

Synthesis of Silver Nanoparticles from Microalgae for Potential Photocatalytic Dye Removal 145
H Y Teh, M K Lam, U Suparmaniam, Y H Chai, J W Lim, P L Kiew, I S Tan, B L F Chin, J S Y Lau

Oil Palm Residue as Precursors for Low Transition Temperature Mixture Extracted Lignin 151
J L Wong, D S Khaerudini, R K Hocking, J J Chew, J Sunarso

Design of Control System for Water Quality Monitoring System for Hydroponics Application 157
M R Safira, M W Lim, W S Chua

Synthesis of Integrated Flower Waste Biorefinery: Optimisation with Economic Consideration 163
EHY Chong, DKS Ng, V Andiappan, LY Ng, P Shivaprasad

Design of Lubricants in Electric and Hybrid Vehicles Via Computer-Aided Molecular Design (CAMD) Technique 169
K W Wong, L Y Ng, C H Lim, D K S Ng, V Andiappan

Formulation of Chitosan-Saponin Microbeads Using Microfluidic Technology.......... 175
J Yong, K S Koh, V L Wong, S S Lim

The Effect of Printing Temperature on Bonding Quality and Tensile Properties of Fused Deposition Modelling 3d-Printed Parts.......... 182
R Pang, M K Lai, K I Ismail, T C Yap

Surface Modification of PVDF Membrane Via Graft Polymerization of Acetic and Acrylic Acid.......... 188
S M Saleh, P C Oh, A S Zulkifli

Simulation of Acid Gas Removal Unit Using DIPA+TEA Amine Solvent.......... 194
A S Farooqi, R M Ramli, S S M Lock, N Hussein, A S Farooqi, S M Wajahat

Quality Prediction for Polypropylene Extrusion Based on Neural Networks.......... 200
C H Tan, K M Yusof, S R W Alwi

ANN Based Multi Model Predictive Control for pH-Control.......... 206
ZY Pua, AW Hermansson, CH Lim

Model Predictive Control – Sliding Mode Control of a pH System.......... 212
W X Ooi, A W Hermansson, C H Lim

A New Way of Working: Intelligent Remote Engineering Via Hybrid First Principle Modelling.......... 218
A A M K Hing, A R Norman, A F H A Mubin, M H Mansor

Integration of Neural Network-Based Prediction for Enhanced Process Controllability: Application in Hydrogen Iodide Decomposer for Hydrogen Production Via Water Splitting Process.......... 223
SA Sirajul Adly, N Mohd

Life Cycle Optimisation for Bio-Energy Systems with Carbon Capture and Storage in Electricity Generation.......... 229
X H Cheng, Denny K S Ng, L Y Ng, V A Murugappan

Synthesis of Zinc Oxide Derived from Wastewater of Glove Manufacturing Via Adsorption-Desorption-Precipitation Path.......... 235
C Y Pang, I Gulnaziya, Y L Pang, M C Wong

Investigation of Glyphosate Removal from Aqueous Solution Using Calcium Peroxide.......... 241
F Li, T S Y Choong, L C Abdullah, S N A Md Jamil

Optimization of Aniline Removal Using Graphite Assisted by Response Surface Methodology and Box-Behnken Design.......... 247
C X Tan, V-L Wong, S P Yeap

Foaming Tendency in Anaerobic Digestion of Wastewater Inoculated with Anaerobic Sludge and Cow Manure: Effect of Protein Concentration.......... 253
S N Syed Sabeer Ali, R Omar, H Che Man, A I Md Idris, H A Tajuddin

Investigation of Fluid Capillary Rise Using 3D Printed Microstructure for Solar Desalination.......... 259
R B Ng, C L Hoo, V L Wong

Characteristics of Alginate and Soy Protein Isolates Composite Aerogels as Oil Absorbent.......... 265
H S Roslan, A N Mustapa, S H Hanipah, A B Alias, F M Fuad, Á Martín, M J Cocero

Synergistic Adsorption-Photocatalytic Degradation of Rhodamine B Dye Using a 3D-Printed Photoreactor with g-CN Homojunction-Based Thermosets Coating: Response Surface Methodology and Adsorption Modelling 271
EY Lee, SY Wong, VL Wong, SJ Phang, KH Cheah

Comparative of Alginate/Graphene Oxide and Alginate/Magnetite Graphene Oxide Aerogel as Adsorbents 277
A S Marzuki, A N Mustapa, N H Othman, S H Hanipah, Á Martin, M J Cocero

Fluorine-Free Superhydrophobic Cotton Fabric for Highly-Efficient Gravity-Driven Oil/Water Separation 283
WX Voo, WC Chong, HC Teoh, AW Mohammad

Aspen Adsorption Simulation on Biosorption Between Water Hyacinth (Eichhornia Crassipes) and Pb (II) Ions in Packed Bed Column 289
C Y Boon, TVN Padmesh

Evaluation of the Structure Characterisation of Hydroxypropyl Bispalmitamide MEA and Ceramide 3 296
C H Chong, T-J Khoo

A Sensitivity Study of Salt Concentration in Brine to the Solubility of CO2 in Saline Aquifers 302
A S Setiawan, M A Hamid, E J Mackay

Baking Kinetics and Characteristics of Cake Mixture Formulated with Mealworm Powder 308
C L Hii, N A R S Kumaran, C H Cheah, K L Tan, B Mistry, C L Chiang

Author Index

IOP Conf. Series: Materials Science and Engineering 1257 (2022) 011001 doi:10.1088/1757-899X/1257/1/011001

2022 SOMCHE 33rd Symposium of Malaysian Chemical Engineers 2022
HERIOT WATT UNIVERSITY MALAYSIA
IChemE

33rd Symposium of Malaysian Chemical Engineers (SOMChE)

School of Engineering and Physical Sciences

Heriot-Watt University Malaysia

Wilayah Persekutuan Putrajaya

8th – 9th August 2022

Over the two days of August 8-9 2022, the 33rd Symposium of Malaysian Chemical Engineers (SOMChE) 2022 was successful held attracting more than 100 researchers, academics and industry experts that joined the two-day event. Due to the COVID-19 situation with its uncertain situation and travel restrictions for some countries, SOMChE was held in virtual format via ZOOM.

The theme of SOMCHE 2022 was "Empowering and Building Resilience in Achieving Sustainable Development Goals". Academic and industry experts covered a wide range of topics relevant to cutting-edge research, new technologies, and emerging growth areas in chemical engineering. The symposium was jointly organised by Heriot-Watt University Malaysia and Institution of Chemical Engineers (IChemE), Malaysia, with support from the Academic Research on Palm Oil Sustainability (ARPOS) Network and IChemE's Palm Oil Processing Special Interest Group (POPSIG).

The authors were invited to submit original and unpublished papers in the area of:

[a] Water and Wastewater Treatment
[b] Clean and Sustainable Energy
[c] Green Manufacturing and Technologies
[d] Biotechnology and Bioprocessing
[e] Circular Economy and Sustainable Finance
[f] Polymer Engineering and Science
[g] Smart/Intelligent Material
[h] Environmental Protection and Management
[i] Process Safety and Risk Management
[j] Process Systems Engineering
[k] Ageing Societies and
[l] Other Related Areas.

For More Information, Please Visit Our Website: https://intake.hw.edu.my/somche2022/

The two-day conference was supported by a scientific committee and an advisory board, with a combined number of 45 members. They supported the plenary session, six keynote speeches, two panel discussions and the seven parallel sharing sessions.

The symposium featured seven renowned speakers including Prof. Datuk Dr. Ahmad Fauzi Bin Ismail from Universiti Teknologi Malaysia (UTM); Prof. Dr. Kathleen B. Aviso from De La Salle University; Prof. Ir. Dr. Abdul Aziz Bin Abdul Raman from University of Malaya (UM); Prof. Dr. Mohamed Kheireddine Aroua from Sunway University Malaysia; Assoc. Prof. Dr. Nishanth Chemmangattuvalappil from The University of Nottingham Malaysia; Assoc. Prof. Dr Yamuna Munusamy from Universiti Tunku Abdul Rahman; and, Assoc. Prof. Dr. Wee-Jun Ong from Xiamen University Malaysia. The plenary session was delivered during a 1 hour presentation using ZOOM which was followed by 5 minutes for Q&A session. The keynote sessions, delivered across the two days, was 30 minutes long consisting of three parts: (1) a 5-minute announcement and keynote-speaker introduction, (2) a 20-minute keynote presentation, and (3) a 5-minute moderated Q&A session. The parallel sessions was delivered across the two days, with three sessions in the first day and four in the second day. Each presentation was in the form of a 10 minute pre-recorded video presentation followed by a 5 minutes live Q&A session with the presenter.

The event also included two panel discussions - “Sustainability in Today’s Palm Oil” supported by POPSIG and “Can We Achieve Carbon Neutral by 2050?”.

The first panel consisted of Prof. Ir. Dr. Dominic Foo Chwan Yee from The University of Nottingham Malaysia; Dr. Steven Lim from Universiti Tunku Abdul Rahman; Dr. Carmen Goh Kar Mun from Eureka Synergy Sdn Bhd, and Ir. Dr. Calvin Chok from KL-Kepong Oleomas Sdn. Bhd with Ir. Prasath Ramani as the moderator.

The second panel was moderated by Prof. Ir. Dr. Denny Ng Kok Sum from Heriot-Watt University Malaysia with the following panellists; Prof. Dr. Agamutu Pariatamby from Sunway University Malaysia; Richard James Brice from IChemE Sustainability Hub; Ir. Dr. Suzana Yusup from Tenaga Nasional Berhad (TNB) Research Sdn. Bhd., Malaysia; Kasturi Nathan from KPMG, Malaysia; and Dr. Norfaizah Ab Manan from Petroliam Nasional Berhad (PETRONAS). Each panel

IOP Conf. Series: Materials Science and Engineering 1257 (2022) 011001 doi:10.1088/1757-899X/1257/1/011001

2022 SOMCHE 33rd Symposium of Malaysian Chemical Engineers 2022

HERIOT WATT UNIVERSITY MALAYSIA

IChemE

forum session was one hour long consisting of three parts: (1) a 5-minute announcement and introduction the panellists, (2) a 45-minute forum discussion, and (3) a 10-minute moderated Q&A session with curated questions.

The organising committee of SOMChE 2022 also expressed the deepest gratitude to JJ-Lurgi Engineering Sdn Bhd and Novozymes Malaysia Sdn Bhd for their steadfast support and generosity in sponsoring the four (4) best paper awards (Water, Sustainable Energy, Sustainable Processes, and Sustainable Materials) with monetary prizes and two (2) best presentation awards with monetary prizes.

https://intake.hw.edu.my/somche2022/

For More Information, Please Visit Our Website: https://intake.hw.edu.my/somche2022/

IOP Conf. Series: Materials Science and Engineering 1257 (2022) 011001 doi:10.1088/1757-899X/1257/1/011001

33rd Symposium of Malaysian Chemical Engineers (SOMChE)

ORGANIZING COMMITTEE

Organiser :	School of Engineering and Physical Sciences, Heriot-Watt University Malaysia
Co-organiser :	The Institution of Chemical Engineers
Chairman :	Prof. Ir. Dr. Denny Ng Kok Sum
Deputy Chairman :	Dr. Wong Voon Loong
Secretariat :	Dr. Jecksin Ooi Ms. Syakirah binti Mohamad Talib
Webpage and Publicity :	Dr. Wong Voon Loong
Sponsorship :	Ir. Dr. Edwin Lim Prof. Ir. Dr. Denny Ng Kok Sum
Event Management :	Dr. Koh Kai Seng Dr. Billie Yan Zhang Hiew Ir. Dr. Edwin Lim Dr. Dalila Alias
Publication :	Dr. Ali Mohamed Ali Hamid Dr. Ayman William Hermansson Dr. Choy Sook Yan Dr. Jecksin Ooi
Registration :	Dr. Wong Voon Loong Dr. Billie Yan Zhang Hiew

For More Information, Please Visit Our Website: https://intake.hw.edu.my/somche2022/

IOP Conf. Series: Materials Science and Engineering 1257 (2022) 011001 doi:10.1088/1757-899X/1257/1/011001

33rd Symposium of Malaysian Chemical Engineers (SOMChE)

ADVISORY AND SCIENTIFIC COMMITTEE BOARD

For More Information, Please Visit Our Website: https://intake.hw.edu.my/somche2022/

IOP Conf. Series: Materials Science and Engineering 1257 (2022) 011001 doi:10.1088/1757-899X/1257/1/011001

2022 SOMCHE 33rd Symposium of Malaysian Chemical Engineers 2022 — HERIOT WATT UNIVERSITY MALAYSIA — IChemE

TECHNICAL COMMITTEE

- Prof. Ir. Ts. Dr Show Pau Loke (University of Nottingham Malaysia, Malaysia)
- Prof. Dr. Su Shiung Lam (Universiti Malaysia Terengganu, Malaysia)
- Prof. Ir. Dr. Suzana Yusup (TNB Research Sdn. Bhd., Malaysia)
- Prof. Ir. Dr. Wan Azlina Wan Ab Karim Ghani (University Putra Malaysia, Malaysia)
- Prof. Dr. Lim Jit Kang (Universiti Sains Malaysia, Malaysia)
- Prof. Ir. Dr. Sharifah Rafidah Datu Wan Alwi (Universiti Teknologi Malaysia, Malaysia)
- Assoc. Prof. Dr. Lim Siew Shee (University of Nottingham Malaysia, Malaysia)
- Assoc. Prof. Dr. Ku Zilati binti Ku Shaari (Universiti Teknologi Petronas, Malaysia)
- Assoc. Prof. Dr. Nishanth Chemmangattuvalappil (University of Nottingham Malaysia, Malaysia)
- Assoc. Prof Dr. Hisham Khaled Ben Mahmud (Curtin University Malaysia, Malaysia)
- Assoc. Prof. Dr. Wee-Jun Ong (Xiamen University Malaysia, Malaysia)
- Assoc. Prof. Ir. Dr. Mimi Haryani Binti Hassim (Universiti Teknologi Malaysia, Malaysia)
- Assoc. Prof. Ts. Dr. Ho Wai Shin (Universiti Teknologi Malaysia, Malaysia)
- Dr. How Bing Shen (Swinburne University of Technology Sarawak Campus, Malaysia)
- Dr. Chew Jiuan Jing (Swinburne University of Technology Sarawak Campus, Malaysia)
- Dr. Irene Chew (Monash University Malaysia, Malaysia)
- Dr. Deng Chun (China University of Petroleum, China)
- Dr. Wendy Ng Pei Qin (University Teknologi Brunei, Brunei)
- Dr. Agus Saptoro (Curtin University Malaysia, Malaysia)
- Dr. Pang Yean Ling (Universiti Tunku Abdul Rahman, Malaysia)
- Dr. Steven Lim (Universiti Tunku Abdul Rahman, Malaysia)
- Dr. Shuit Siew Hoong (Universiti Tunku Abdul Rahman, Malaysia)

For More Information, Please Visit Our Website: https://intake.hw.edu.my/somche2022/

2022 SOMCHE 33rd Symposium of Malaysian Chemical Engineers 2022

HERIOT WATT UNIVERSITY MALAYSIA

IChemE

- Dr. Bridgid Chin Lai Fui (Curtin University Malaysia, Malaysia)
- Dr. Rafeqah Raslan (Universiti Teknologi MARA, Malaysia)
- Dr. Nasir Mohammed Amro Al Lagtah (Heriot-Watt University, Dubai)
- Dr. Khalik Mohamad Sabil (PETRONAS Research Sdn. Bhd., Malaysia)

https://intake.hw.edu.my/somche2022/

For More Information, Please Visit Our Website: https://intake.hw.edu.my/somche2022/

IOP Conf. Series: Materials Science and Engineering 1257 (2022) 011001 doi:10.1088/1757-899X/1257/1/011001

2022 SOMCHE 33rd Symposium of Malaysian Chemical Engineers 2022

HERIOT WATT UNIVERSITY MALAYSIA

IChemE

33rd Symposium of Malaysian Chemical Engineers (SOMChE)

SPONSORS AND PROFESSIONAL PARTNERSHIP

SPONSORS

JJ-Lurgi Engineered for you

novozymes

PROFESSIONAL PARTNERSHIP

Academic Research on Palm Oil Sustainability (ARPOS)

energy institute — EI Student Chapter | Heriot-Watt University Malaysia

IChemE Student Chapter — Heriot Watt University Malaysia

ENGINEERS WITHOUT BORDERS
HERIOT-WATT UNIVERSITY MALAYSIA STUDENT CHAPTER

IEM STUDENT SECTION
THE INSTITUTION OF ENGINEERS, MALAYSIA
HERIOT-WATT UNIVERSITY MALAYSIA

IChemE
Palm Oil Processing
Special Interest Group

https://intake.hw.edu.my/somche2022/

For More Information, Please Visit Our Website: https://intake.hw.edu.my/somche2022/

IOP Conf. Series: Materials Science and Engineering 1257 (2022) 011002 doi:10.1088/1757-899X/1257/1/011002

Peer Review Statement

All papers published in this volume have been reviewed through processes administered by the Editors. Reviews were conducted by expert referees to the professional and scientific standards expected of a proceedings journal published by IOP Publishing Publishing.

- **Type of peer review:** Open Review
- **Conference submission management system:** Morressier
- **Number of submissions received:** 76
- **Number of submissions sent for review:** 60
- **Number of submissions accepted:** 52
- **Acceptance Rate (Submissions Accepted / Submissions Received × 100):** 68.4
- **Average number of reviews per paper:** 2
- **Total number of reviewers involved:** 38
- **Contact person for queries:**
  Name: Ayman William Hermansson
  Email: a.hermansson@hw.ac.uk
  Affiliation: Heriot-Watt University - EPS

IOP Conf. Series: Materials Science and Engineering 1257 (2022) 012001 doi:10.1088/1757-899X/1257/1/012001

A critical review on the influence of humidity for plant growth forecasting

S Y Chia and M W Lim

Centre for Water Research, Faculty of Engineering, Built Environment & Information Technology, SEGi University, 47810 Kota Damansara, Selangor Darul Ehsan, Malaysia

*E-mail: limmeewei@segi.edu.my

Abstract. Relative humidity of ambient air is a critical parameter for crop production as it influences the water balance and photosynthesis process in the plants. A comprehensive review on leafy, fruiting, flowering plants and grains were conducted to determine the effects of varying humidity levels on plant growth. Data from published works concluded that the plant growth improved with increasing humidity, as higher humidity conditions help to keep the stomata open to maintain the photosynthesis process and minimize evaporation process of the plants. By using regression analysis, it is forecasted that the increase in the air relative humidity by 10% would increase the leaf length, width and aerial dry weight for *Lactuca Sativa* by 10±3%, 2±5% and 28±3% respectively. The data trend suggests that the optimal relative humidity that could improve plant growth is 85±2%, as low humidity below optimum level will increase stomatal resistance, which leads to a reduction of carbon dioxide uptake and photosynthesis rate. Inversely, relative humidity above the optimal levels will adversely reduce the transpiration rate, which negatively affects the plant growth and leaf development due to impaired nutrient assimilation. The outcome of this study provides an insight for the development of plant growth forecasting model.

1. Introduction

Plants are typically grown using soil as the growing medium. The soil provides nutrients, water, anchorage, and air for plant growth. However, soil culture has a few limitations for the plants to grow such as unsuitable soil reaction, erosion issues leading to soil degradation, presence of disease-causing organisms and nematodes, poor drainage and the soil compaction is unfavorable. This leads to the application of soil-less agriculture, also known as hydroponics which is a culture technique for plant growth in nutrient rich water solution [1]. This technique can manage the agriculture system more efficiently because it has much higher crop yields than traditional agriculture and the growth of plants in a hydroponic system is significantly faster as the environmental conditions under hydroponic systems can be well controlled, including the optimization of the nutrient elements supplied to the plants [2]. Deep-water culture is the simplest active hydroponic system which is usually a choice for water-loving plants which are fast-growing such as leaf lettuce, cucumber, and tomato [3].

In a closed chamber, the growth of plants would be affected by several environmental factors, including relative humidity. The relative humidity (RH) of the ambient air is known as the ratio of the water vapor density to the saturation water vapor density and it has a different percentage due to evaporative forcing at different air temperatures [4]. The RH value is a critical environmental parameter for crop production since it affects the leaf transpiration rate and can influence the water balance in crops. Generally, too high relative humidity which is above the optimal levels will affect the leaf surface area's development by causing reduction of transpiration rates and lowering growth and nutrient assimilation. In contrast, low relative humidity raises water flow into plants and increases transpiration with extreme problems in species with decreased ability to control stomatal opening. Therefore, the photosynthesis rate reduces as well [5].

According to Shibata, Iwao, and Takano (1995), low relative air humidity and high air velocity can cause water stress and stomatal closure. Based on their works, the best relative air humidity level to maximize the rate of photosynthetic for lettuce plants grown in a plant factory was approximately 80 % of relative humidity [6]. This was in good agreement with works reported by Tibbitts and Bottenberg (1976), whereby the authors reported that the plants grown under 85% of relative humidity had a faster growth rate than those grown under 50% RH [7]. In addition, the increase in the relative humidity to 76% could also reduce lettuce tip burn occurrence as reported by Goto and Takakura (1992) [8]. Similar recent works by various authors also indicated that the optimal relative humidity for maximum plant growth lies within 60-70% RH [9-14].

Typically, the relative air humidity is also affected by the air temperature as changes in the ambient air temperature will affect the moisture content in the atmosphere leading to varying relative humidity. In addition, the ambient temperature influences the speed at which the plants process the energy. The changes in the relative humidity and temperature have a direct impact on the photosynthesis process; thus, it will influence the growth and development of plants. If these conditions are maintained within the desired range, the photosynthesis capability of the plant is not compromised [15]. Therefore, it is important to control the environmental parameters such as relative humidity and air temperature to improve plant growth. As optimal humidity and temperature are essential for transpiration and photosynthesis of the plants in hydroponics applications, it is crucial to observe and monitor these two parameters for plant growth by using the hydroponics system in a plant growth chamber.

The objective of this study was to conduct a thorough review on the effects of various relative humidity levels on lettuce plant growth in terms of leaf length, leaf width, and dry weight of the leaves under a controlled environment. Subsequently, the effects of relative humidity ranging from 55% to 85% on lettuce (*Lactuca sativa* L.) plant growth were predicted under constant air temperature (18±2°C) in an enclosed plant growth chamber using a regression analysis method.

2. Leafy plants growth trend according to the humidity: A review

In the early years, a preliminary study by Tibbitts and Bottenberg (1976) studied the growth of lettuce (*Lactuca sativa* L. cv. Meikoningen, butterhead type) under controlled humidity levels. In this experiment, the light intensity was maintained at 21.5±5% kilolux with a light duration of 16 hr. The air temperature was set at 20 ± 1°C, while the relative humidity for the first and second chamber were kept consistent at 85±2% and 50±2% respectively. From table 1, the leaf length and leaf width of the lettuce are 16.30% and 18.79% higher under 85% RH than under 50% RH conditions. On top of that, the fresh weight and dry weight of the lettuce plant are 73.63% and 56.63% greater for the lettuce plants growing under the higher relative humidity. Moreover, the relative growth rate of the lettuce plant's growth under higher humidity was determined around 2 to 3 days faster than the one growing up under lower humidity. It can be concluded that the lettuce plants are more suitable and will have a higher growth performance under higher humidity levels [7].

The authors, Tibbitts and Collier (1984) conducted a detailed follow-up study to Tibbitts and Bottenberg (1976), to research in detail the effect of relative humidity on lettuce plants [7,16]. The plant growth such as plant dry weight, length, and width of leaf number 12 was also recorded in the results. In this study, a different type of lettuce was selected which is romaine lettuce (*Lactuca sativa* L. cv. Lobjoits Green Cos) and grown under light intensity of 320±10 μmol s^{-1} m^{-2} with a light duration of 16 hr. The temperature was maintained at 26/12.5 ±1.8°C (day/night) and the relative humidity treatment with different humidity levels was carried out for 14 days. From table 2, the leaf length of the twelfth lettuce leaf is increased by 13 to 22% when the relative humidity increases from 51% to 74% while the leaf width of the twelfth lettuce leaf grown under 74% of relative humidity is approximately 2% and 3% greater than the lettuce plants grown under 51% and 62% of relative humidity treatment. Besides, the lettuce plant dry weight of the lettuce plants under 74% RH treatment is significantly higher than the others. It can be concluded that even different types of lettuce plants are also more suitable and will have a higher growth performance under higher humidity levels [16].

Table 1. Review on growth responses of lettuce plants at varying RH levels for 26 days [7].

Growth Responses	Relative Humidity (% RH)	
	50±2	85±2
Leaf number (quantity)	20	20
Leaf length (cm)	9.20	10.70
Leaf width (cm)	7.45	8.85
Length/width ratio	1.23	1.20
Plant fresh weight (g/plant)	12.06	20.94
Plant dry weight (g/plant)	0.83	1.30
Growth rate (r/day)	1.28	1.31

Table 2. Review on growth responses of lettuce plants at varying RH levels during light period [16].

Growth Responses	Relative Humidity (% RH)		
	51	62	74
Leaf length of leaf 12 (cm)	12.64	14.32	15.38
Leaf width of leaf 12 (cm)	9.88	10.06	10.31
Plant top part dry weight (g)	2.84	3.66	4.61

Subsequently, Bradbury and Ahmad (1996) studied the effect of humidity on the growth of lettuce (*Lactuca sativa* Var. Great Lakes). The relative humidity was set at two different levels which low humidity is 62/81 % while high humidity is 92/100 % in which the relative humidity is night/day. The air temperature was maintained at 23/17°C (day/night) and the light intensity was 0.15 mmol s^{-1} m^{-2} with 11 hr of light duration. The plants were harvested after 70 days. From the results in table 3, the whole plant dry weight of the lettuce plants grown at high humidity was more than double of that grown under low humidity. In addition, the lettuce plants grown under higher humidity had obtained approximately three times greater leaf area than the lettuce plants grown under lower humidity. Besides, the value of leaf area ratio (LAR), leaf weight ratio (LWR), and specific leaf area (SLA) of the lettuce plants grown at higher humidity were also 34%, 9%, and 23% greater than the one grown at lower humidity conditions. Hence, it also can be concluded that the growth of the lettuce is highly affected by the relative humidity and higher growth performance can be obtained under higher humidity conditions [17].

Table 3. Review on growth responses of lettuce plants at different relative humidity levels [17].

Growth Responses	Relative Humidity (% RH, night/day)	
	62/81	92/100
Whole plant dry weight (g)	0.48	1.05
Leaf area (cm^2)	150	450
Shoot/root ratio	3.9 ± 2.7	6.6 ± 2.5
Leaf area ratio (LAR)	3.23 ± 0.22	4.33 ± 0.16
Leaf weight ratio (LWR)	0.79 ± 0.01	0.86 ± 0.01
Specific leaf area (SLA)	4.09 ± 0.26	5.04 ± 0.16

According to the published literature summarized in Section 2, it can be concluded that high humidity conditions are more favourable for leafy plants and can increase plant growth at the same time when compared to low humidity circumstances.

3. Leafy plant growth trend according to humidity: predicted growth

The plant growth trend for lettuces (*Lactuca Sativa*) under varying humidity levels were subsequently predicted using trends from published literature, and actual plant growth data obtained from previous unpublished experimental work. The atmospheric conditions for the plant growth experiment were around 55±2% of RH and a temperature of 25±2°C. The lettuce plants were grown in microbubbles-aerated deep water culture hydroponic system for 2 weeks, and the data for leaf length, leaf width, and aerial dry weight grown under these conditions are summarized in table 4.

Table 4. Actual data for individual lettuce plant growth at 55 ± 2%.

Parameter \ Lettuce	1	2	3	4	5	6	Average
Leaf Length (cm)	16.18	14.83	14.58	17.77	15.73	15.62	15.79 ± 1.14
Leaf Width (cm)	7.32	7.22	7.58	7.48	7.47	7.13	7.37 ± 0.17
Aerial Dry Weight (g)	0.37	0.30	0.40	0.57	0.33	0.41	0.40 ± 0.09

The actual data taken from the unpublished results were used to predict the plant growth with respect to increasing humidity according to the increasing percentage trend from the results in the study of Tibbitts and Collier (1984) using linear regression analysis [16]. The data were comparable due to the similar atmospheric conditions in which the temperature in the chamber was maintained at 26°C and the plants were maintained under the humidity treatment conditions for 14 days. In the Tibbitts and Collier (1984) study, the plants were grown hydroponically under different relative humidity levels which are 51%, 62%, and 74%. The leaf length, leaf width, and the dry weight of the top part of the plant were increased by 10±3%, 2±1%, and 28±2% respectively with increasing humidity levels from 51% of relative humidity to 85% of relative humidity [16]. As the conditions of the control study mimic the study of Tibbitts and Collier, the increasing percentage trend can be used in order to predict the data of individual lettuce leaf length growth with respect to leaf length, leaf width, and the dry weight of the top part of the plant respectively. Table 5 summarizes the predicted data of varying humidity levels from 55-85 %RH from initial actual plant growth conditions at 55±2% of relative humidity.

From table 5, the predicted values showed the growth of six lettuce leaves lengths and widths which were grown under 4 different ranges of relative humidity levels. The growth of the lettuce plants is increasing significantly with increasing relative humidity from 55±2% to 85±2%. The lettuce leaves length at 85±2% of RH is 10.35%, 15.51%, and 34.26% longer than the lettuce leaves length at 75±2%, 65±2%, and 55±2% respectively. On top of that, the increasing percentage for the lettuce leaves width was observed when the humidity increased. The lettuce leaf length and width of six different lettuce grown under the same conditions is increasing while the relative humidity is increasing at the same time. This means that the lettuce plants grown under low humidity will have a smaller size of leaves compared to those grown under high humidity levels. Thus, it can be justified that the lettuce plants that grow under higher humidity conditions will have a greater weight compared to those grown under lower humidity.

From table 5, the dry weight of the lettuce leaves at a relative humidity of 85±2% is approximately 1.3-, 1.6-, and 2.1-fold higher than those grown under 75±2%, 65±2%, and 55±2% respectively. This was attributed to the low humidity condition which will increase the stomatal resistance. This leads to the reduction of the carbon dioxide uptake and decreasing photosynthesis rate. The plant growth would then be significantly affected as the increase of stomatal resistance will also cause a large reduction in dry weight accumulation [18]. Therefore, this can be concluded that high humidity conditions are more suitable for lettuce plants or leafy plants to obtain higher growth performances. This was in good agreement with the published works reviewed in Section 2 which demonstrated that the growth of the plants is favourable under higher humidity conditions when compared to low humidity circumstances.

Table 5. Predicted data for individual lettuce leaf length growth, leaf width growth and aerial dry weight according to varying relative humidity

Relative Humidity Levels (% RH)	Growth Responses of Lettuce Leaf Length (cm)					
	Lettuce 1	Lettuce 2	Lettuce 3	Lettuce 4	Lettuce 5	Lettuce 6
Growth Responses of Lettuce Leaf Length (cm)						
55±2 (Actual)	16.18	14.83	14.58	17.77	15.73	15.62
65±2 (Predict)	18.33	16.80	16.52	20.13	17.82	17.70
75±2 (Predict)	19.69	18.04	17.74	21.62	19.14	19.14
85±2 (Predict)	21.72	19.91	19.58	23.86	21.12	20.97
Growth Responses of Lettuce Leaf Width (cm)						
55±2 (Actual)	7.32	7.22	7.58	7.48	7.47	7.13
65±2 (Predict)	7.45	7.35	7.72	7.62	7.61	7.26
75±2 (Predict)	7.64	7.53	7.91	7.81	7.80	7.44
85±2 (Predict)	7.80	7.70	8.08	7.97	7.96	7.60
Growth Responses of Plant's Dry Weight (g)						
55±2 (Actual)	0.37	0.30	0.40	0.57	0.33	0.41
65±2 (Predict)	0.48	0.39	0.52	0.73	0.43	0.53
75±2 (Predict)	0.60	0.49	0.65	0.93	0.54	0.67
85±2 (Predict)	0.77	0.62	0.83	1.18	0.68	0.85

Humidity is required for photosynthesis to take place. The plant can only absorb a limited amount of humidity and hence has less water evaporation than other plants when the humidity levels in and around the plants are maintained at an appropriate level. Therefore, the plants will be able to keep their stomata open. The carbon dioxide may then be absorbed, and the plant temperature can be controlled by evaporation. Photosynthesis will be feasible even with low light levels if the stomata remain open. It is essential to reduce plant transpiration in the presence of greater irradiation to keep the stomata open. Nevertheless, evaporation can be minimized if the humidity level is kept high. In summary, the plants transpire moisture more quickly in dry air than in humid air. Water in the leaves evaporates rapidly into the air, resulting in the plant losing moisture quickly. When the leaves begin to lose water quicker than the roots can absorb it, it could cause detrimental effects to plant growth due to insufficient water. To avoid losing additional water to the atmosphere, the plant will almost fully close its leaf pores. This not only decreases the flow of moisture from the plant to the atmosphere but also limits the absorption of carbon dioxide and leads to a reduction in photosynthesis. Therefore, the growth of the plant is affected, and the rate of growth decreases. Thus, optimal parameters of the humidity levels are essential to enhance plant growth. From the predicted data in the tables above, the lettuce plants that grown under 85±2% RH have obtained the highest value of leaf length, leaf width, and dry weight of lettuce leaves. As a result, the relative humidity at 85±2% of relative humidity is the most optimum condition to enhance the lettuce plant growth compared to the others.

4. Conclusion

The physical effects of various relative humidity levels on the lettuce plant growth in terms of leaf length, leaf width, and leaves dry weight under a controlled environment were investigated based on published works. From literature, it can be concluded that the physical plant growth improved when the relative humidity was increased. The predicted data showed that the plant growth in terms of leaf length, leaf width, and aerial dry weight increases 10±3%, 2±1%, and 28±2% respectively when the relative humidity increases. The plant growth can be increased by maintaining the air moisture level in the plant (high humidity) to keep the stomata open in order to maintain the photosynthesis process and minimize the evaporation process of the plants. In conclusion, 85±2% of relative humidity is the most optimum condition that can be improved the lettuce plant growth based on the research and results.

References

[1] Sharma N, Acharya S, Kumar K, Singh N and Chaurasia O P 2019 Hydroponics as an advanced technique for vegetable production : An overview *J. Soil Water Conserv.* **17** 364–71

[2] Pandey R, Jain V and Singh K P 2009 *Hydroponics Agriculture : Its Status, Scope and Limitations. Division of Plant Physiology* (New Delhi: ResearchGate) pp 20–29

[3] Domingues D S, Takahashi H W, Camara C A P and Nixdorf S L 2012 *Comput. Electron. Agric.* **84** 53–61

[4] Bramer I, Anderson B J, Bennie J, Bladon A J, De Frenne P, Hemming D, Hill R A, Kearney M R., Körner C, Korstjens A H, Lenoir J, Maclean I M D, Marsh C D, Morecroft M D, Ohlemüller R, Slater H D, Suggitt A J, Zellweger F and Gillingham P K 2018 Advances in Monitoring and Modelling Climate at Ecologically Relevant Scales *Adv.Ecol. Res.* **58** 101

[5] Jolliet D O 1993 Modelling of Water Uptake, Transpiration and Humidity in Greenhouse, and of Their Effects on Crops *Acta Hortic.* **328** 69–78

[6] Shibata T, Iwao K and Takano T 1995 Effect of Vertical Air Flowing on Lettuce Growing in a Plant Factory *Acta Hortic.* **399** 175–82

[7] Tibbitts T W and Bottenberg G 1976 Growth of lettuce (Lactuca sativa) under controlled humidity levels *J. Am. Soc. Hortic. Sci.* **101** 70–3

[8] Goto E and Takakura T 1992 Prevention of Lettuce Tipburn by Supplying Air To Inner Leaves *ASAE.* **35** 641–45

[9] Krisna B, Tarwaca E, Putra S, Rogomulyo R and Kastono D 2017 The Effects of Oxygen and Calcium Enrichment on the Root Growth and Yield of Curly Lettuce (Lactuca sativa L.) that Grow under Floating Raft Hydroponics *Vegetalika* **6** 14–27

[10] Kang W H, Park J S, Park K S and Son J E 2016 Leaf photosynthetic rate, growth, and morphology of lettuce under different fractions of red, blue, and green light from light-emitting diodes (LEDs) *Hortic. Environ. Biotechnol.* **57** 573–79

[11] Sago Y 2016 Effects of light intensity and growth rate on tipburn development and leaf calcium concentration in butterhead lettuce *HortScience* **51** 1087–91

[12] Kim H H, Goins G D, Wheeler R M and Sager J C 2004 Stomatal conductance of lettuce grown under or exposed to different light qualities *Ann. Bot.* **94** 691–97

[13] Lee H I and Kim Y H 2013 Utilization Efficiencies of Electric Energy and Photosynthetically Active Radiation of Lettuce Grown under Red LED, Blue LED and Fluorescent Lamps with Different Photoperiods *J. Biosyst. Eng.* **38** 279–86

[14] Zhang X, He D, Niu G, Yan Z and Song J 2018 Effects of environment lighting on the growth, photosynthesis, and quality of hydroponic lettuce in a plant factory *Int. J. Agric. Biol. Eng.* **11** 33–40

[15] Thazin T, Aung Z L and Win T T 2019 *IoT based Hydroponic Temperature and Humidity Control System using Fuzzy Logic* (Myanmar: ResearchGate)

[16] Tibbitts, Theodore W and Collier G F 1984 Effects of relative humidity and root temperature on calcium concentration and tipburn development in lettuce *J. Am. Soc. Hortic. Sci.* **109** 128–31

[17] Bradbury M and Ahmad R 1996 Effect of Humidity on Growth of Lettuce (Lactuca Sativa, Var. Great Lakes) Under Saline Condition *Pak. J. Bot.* **28** 97–102

[18] Ahmed H A, Tong Y X and Yang Q C 2020 Optimal control of environmental conditions affecting lettuce plant growth in a controlled environment with artificial lighting: A review *S. Afr. J. Bot.* **130** 75–89

Process simulation of anaerobic digestion for methane production using aspen plus

G Ramasamy[1], A H Goodman[2] , H M Lahuri[2], S S Md Shah[2], K M Sabil1[2]
[1]Chemical Engineering Department,Universiti Teknologi Petronas, 32610 Seri Iskandar Perak Malaysia [2]CCUS, Group Research and Technology, Petronas Research Sdn Bhd, Jalan Ayer Hitam, 43000 Kajang, Selangor, Malaysia.

Contact author: abigail.goodman@petronas.com.my

Abstract: A process simulation model was developed for biogas production via an anaerobic digestion process using Advanced System for Process Engineering (Aspen) Plus software. A total of 46 reactions were included in the model and were simulated with appropriate kinetics. For anaerobic digestion of Palm Oil Mill Effluent (POME), few studies have been reported with regards to simulation, and this biogas model has not been validated for use with POME. These works validate the ability of the developed model to predict biogas production using established data on biogas production from literature namely (O-thong et al.,2016; Mamimin et al., 2015). Average Absolute Deviation (AAD) was computed, and a deviation calculated less than 5% was obtained for both case studies of biogas production from POME with the model. The developed model was also used to study the effects of adding an external supply of H_2 and CO_2 at different ratios (2:1, 1:1). The results showed that the ratio of H_2 and CO_2, 2:1, has higher methane production compared to an equal ratio of H_2 and CO_2, 1:1. The developed model is flexible and able to estimate the enhancement of methanation via the input of an external supply of H_2 and CO_2 quantitatively.

1. Introduction

Biogas is a vital renewable resource that can be generated by the anaerobic digestion process (AD). AD is a complicated biological process in which biomass is converted to biogas, which consists of methane (0.50–0.75), carbon dioxide (0.30–0.40), and other components (hydrogen sulfide, hydrogen, etc.), by a microbial community in the absence of oxygen [1]. Methane (CH_4) is generated via the process, which can be a clean alternative biofuel [2]. One of the advantages of biogas is that it can be used as a power supply for biofuels, lighting, refrigeration, cooking, and power generation.

Hydrolysis, acidogenesis, acetogenesis and methanogenesis are the key stages in the biogas generation process. At the first stage, hydrolytic fermentative bacteria hydrolyze large polymers (biomass) like fats, proteins, and carbohydrates into small molecules such as fatty acids, amino acids and simple sugars. The acidogenic bacteria transform the smaller molecules into volatile fatty acids (VFAs), including acetic acid, propionic acid, butyric acid, iso-butyric acid, valeric acid, and iso-valeric acid. In the third stage, acetogenic bacteria consume and convert the VFAs into hydrogen, carbon dioxide and acetic acid. In the final stage of AD, methanogens are responsible for converting the reaction's products into methane and carbon dioxide [3-5].

Carbon Dioxide (CO_2) gas is one of the dominant greenhouse gases (GHG), which can severely threaten the environment with global warming. Biogas production can reduce CO_2 emissions from this anthropogenic activity, which will help maintain natural resources [6]. High purity CO_2 captured from any industrial process can be utilized to produce CH_4 via a hydrogenotrophic methanogenesis process

[7]. Hydrogenotrophic methanogenesis is performed by methanogenic archaea, a specialized group of microbes that consume organic matter rich in nutrients and convert CO_2 to CH_4 in the presence of hydrogen (H_2) [8].

Researchers and industrialists employ process simulations to forecast and improve the overall performance of anaerobic systems. [9]. The Advanced System for Process Engineering (Aspen) Plus software/simulator Version 11 is employed to illustrate and optimize the performance of a process that incorporates chemical reactions. It enables the application of fundamental engineering principles in scenarios involving mass and energy balances, phase equilibrium, and chemical equilibrium and can thus imitate an existing system [10]. Several simulation studies on AD have been published using the software. Nduse et al. (2016) evaluated the amount of biogas produced from co-digestion of cow dung and food waste using Aspen Plus software [11]. Al-Rubaye et al. (2019) applied Aspen Plus to develop the process model for AD with animal manure and distillery wastewater [12]. Rajendran et al. (2014) developed a novel anaerobic digestion model, a process simulation model (PSM), which can estimate the biogas production from any substrate under initial process conditions [13]. However, little attention has been given to palm oil mill effluent (POME) as feedstock, and limited investigation has been conducted into process parameters which significantly affect the CH_4 composition in biogas production.

POME is a highly contaminated effluent produced from three major processes, including the sterilization, condensation, and hydro cyclone of palm oil in mills. For biogas generation, POME is a nutrient-rich organic compound regarded as a renewable energy source. [14]. This study aimed to develop and implement the PSM for biogas production from POME using Aspen Plus software version 11. This paper evaluates the validation of the model with two experimental case studies of biogas production using POME [15-16]. The model is used to study the effects of the external supply of H2 and CO_2 at different ratios (2:1, 1:1) using Aspen Plus software version 11. The research aims to study the feasibility of CO_2 gas being incorporated into hydrogenotrophic methanogenesis reaction and the biogas production from the additional H_2 and CO_2 supply .

2. Simulation model development

The PSM is constructed primarily based on the four main AD stages such as hydrolysis, acidogenesis, acetogenesis and methanogenesis as shown in figure 1. These four stages illustrate on how the large polymers such as fats, proteins and carbohydrates are degraded into smaller molecules and finally to CH_4 and CO_2.

The PSM categorizes anaerobic digestion reactions into 2 sets of reaction: 1) The hydrolysis reactions functioning based on the extent of reaction from a scale of 0.0 to 1.0. fats, proteins and carbohydrates are included in the reaction as shown in table 1. Fats were added as triolein, palmito linolein, palmito-olein, and tripalmate. Proteins were incorporated as soluble and insoluble proteins. Carbohydrates were included as starch, hemicelluloses and cellulose. 2) Consists of acidogenic, acetogenic, and methanogenic reactions operate according to kinetic constant basis as shown in Table 2. Glycerol, valeric acid, butyric acid, propionic acid, linoleic acid, amino acids, sugars, palmitic acid, oleic acid, methanogenesis, and hydrogen using reactions are among the ten sub-sets of reactions that are considered in the calculation of reaction kinetics.

All the components required for the simulation were included in Aspen Plus. Non-Random Two-Liquid (NRTL) model selected as the property package as liquid and gas phases involved in biogas production. It also evaluates the mass fractions and activity coefficient of components. Reactors added after all the compounds are added and the properties verified. Stoichiometric reactor is chosen for the hydrolysis reaction as it is dependent on extent of reaction [20]. Continuous stirred tank reactor (CSTR) chosen for the rest of the reactions (acidogenic, acetogenic and methanogenic) as it is function of kinetics.

There are thirteen and thirty- three reactions altogether in the stoichiometric and CSTR reactor respectively. Monomers are broken down from the hydrolysis reaction undergo blocks of calculator in series at the CSTR. Each reaction in CSTR has a FORTRAN program which calculate products from the reaction. For example, the dextrose produced by the carbohydrates in the hydrolysis reaction, pass through the dextrose calculator block where it is degraded into different (VFA) components. These components pass through several VFA blocks of calculator such as the valeric acid and the propionic acid blocks, then through the methanogenesis block to quantify the amount of biogas that have been generated.

Table 1: Extent of reaction for the hydrolysis reactions involved in model [13]

No	Compound	Hydrolysis Reaction	Extent of reaction
1	Starch	$(C_6H_{12}O_6)n + H_2O \rightarrow n\ C_6H_{12}O_6$	0.6 ± 0.2
2	Cellulose	$(C_6H_{12}O_6)n + H_2O \rightarrow n\ C_6H_{12}O_6$	0.4 ± 0.1
3	Hemicellulose	$C_5H_8O_4 + H_2O \rightarrow 2.5\ C_2H_4O_2$	0.5 ± 0.2
4	Hemicellulose	C5H8O4 + H2O → C5H10O5	0.6 ± 0.0
5	Xylose	$C_5H_{10}O_5 \rightarrow C_5H_4O_2 + 3H_2O$	0.6 ± 0.0
6	Cellulose	$C_6H_{12}O_6 + H_2O \rightarrow 2\ C_2H_6O + 2\ CO_2$	0.4 ± 0.1
7	Ethanol	$2\ C_2H_6O + CO_2 \rightarrow 2\ C_2H_4O_2 + CH_4$	0.6 ± 0.1
8	Soluble protein	$C_{13}H_{25}O_7N_3S + 6H_2O \rightarrow 6.5\ CO_2 + 6.5\ CH_4 + 3H_3N + H_2S$	0.5 ± 0.2
9	Insoluble protein (I.P)	I.P $+ 0.3337\ H_2O \rightarrow 0.045\ C_6H_{14}N_4O_2 + 0.048\ C_4H_7NO_4 + 0.047\ C_4H_9NO_3 + 0.172\ C_3H_7NO_3 + 0.074\ C_5H_9NO_4 + 0.111\ C_5H_9NO_2 + 0.25\ C_2H_5NO_2 + 0.047\ C_3H_7NO_2 + 0.067\ C_5H_{11}NO_2S + 0.074\ C_6H_{11}NO_2 + 0.07\ C_9H_{11}NO_2 + 0.046\ C_6H_{13}NO_2 + 0.036\ C_8H_{11}NO$	0.6 ± 0.1
10	Triolein	$C_{57}H_{104}O_6 + 3H_2O \rightarrow C_3H_8O_3 + 3C_{18}H_{34}O_2$	0.5 ± 0.2
11	Tripalmate	$C_{51}H_{98}O_6 + 8.436\ H_2O \rightarrow 4\ C_3H_8O_3 + 2.43\ C_{16}H_{34}O$	0.5 ± 0.3
12	Palmito-olein	$C_{37}H_{70}O_5 + 4.1\ H_2O \rightarrow 2.1\ C_3H_8O_3 + 0.9\ C_{16}H_{34}O + 0.9\ C_{18}H_{34}O_2$	0.6 ± 0.2
13	Palmito-linolein	$C_{37}H_{68}O_5 + 4.3\ H_2O \rightarrow 2.2\ C_3H_8O_3 + 0.9\ C_{16}H_{34}O + 0.9\ C_{18}H_{32}O_2$	0.6 ± 0.2

Table 2: Kinetic constants for the amino acids, acidogenic, acetogenic, and methanogenic reactions which involved in model [13]

No	Compound	Chemical Reaction	Kinetic constant
Amino acid degradation reactions			
1	Glycine	$C_2H_5NO_2 + H_2 \rightarrow C_2H_4O_2 + H_3N$	$1.28*10^{-02}$
2	Threonine	$C_4H_9NO_3 + H_2 \rightarrow C_2H_4O_2 + 0.5\ C_4H_8O_2 + H_3N$	$1.28*10^{-02}$
3	Histidine	$C_6H_9N_3O_2 + 4H_2O + 0.5\ H_2 \rightarrow CH_3NO + C_2H_4O_2 + 0.5\ C_4H_8O_2 + 2H_3N + CO_2$	$1.28*10^{-02}$
4	Arginine	$C_6H_{14}N_4O + 3H_2O + H_2 \rightarrow 0.5\ C_2H_4O_2 + 0.5\ C_3H_6O_2 + 0.5\ C_5H_{10}O_2 + 4H_3N + CO_2$	$1.28*10^{-02}$
5	Proline	$C_5H_9NO_2 + H_2O + H_2 \rightarrow 0.5\ C_2H_4O + 0.5\ C_3H_6O_2 + 0.5\ C_5H_{10}O_2 + H_3N$	$1.28*10^{-02}$
6	Methionine	$C_5H_{11}NO_2S + 2H_2O \rightarrow C_3H_6O_2 + CO_2 + H_3N + H_2 + CH_4S$	$1.28*10^{-02}$
7	Serine	$C_3H_7NO_3 + H_2O \rightarrow C_2H_4O_2 + H_3N + CO_2 + H_2$	$1.28*10^{-02}$
8	Threonine	$C_4H_9NO_3 + H_2O \rightarrow C_3H_6O_2 + H_3N + H_2 + CO_2$	$1.28*10^{-02}$
9	Aspartic acid	$C_4H_7NO_4 + 2H_2O \rightarrow C_2H_4O_2 + H_3N + 2\ CO_2 + 2H_2$	$1.28*10^{-02}$
10	Glutamic acid	$C_5H_9NO_4 + H_2O \rightarrow C_2H_4O_2 + 0.5\ C_4H_8O_2 + H_3N + CO_2$	$1.28*10^{-02}$
11	Glutamic acid	$C_5H_9NO_4 + 2H_2O \rightarrow 2\ C_2H_4O_2 + H_3N + CO_2 + H_2$	$1.28*10^{-02}$
12	Histidine	$C_6H_9N_3O_2 + 5H_2O \rightarrow CH_3NO + 2\ C_2H_4O_2 + 2H_3N + CO_2 + 0.5\ H_2$	$1.28*10^{-02}$
13	Arginine	$C_6H_{14}N_4O_2 + 6H_2O \rightarrow 2\ C_2H_4O_2 + 4H_3N + 2\ CO_2 + 3H_2$	$1.28*10^{-02}$
14	Lysine	$C_6H_{14}N_2O_2 + 2H_2O \rightarrow C_2H_4O_2 + C_4H_8O_2 + 2H_3N$	$1.28*10^{-02}$
15	Leucine	$C_6H_{13}NO_2 + 2H_2O \rightarrow C_5H_{10}O_2 + H_3N + CO_2 + 2H_2$	$1.28*10^{-02}$
16	Isoleucine	$C_6H_{13}NO_2 + 2H_2O \rightarrow C_5H_{10}O_2 + H_3N + CO_2 + 2H_2$	$1.28*10^{-02}$

17	Valine	$C_5H_{11}NO_2 + 2H_2O \rightarrow C_4H_8O_2 + H_3N + CO_2 + 2H_2$	$1.28*10^{-02}$
18	Phenyalanine	$C_9H_{11}NO_2 + 2H_2O \rightarrow C_6H_6 + C_2H_4O_2 + H_3N + CO_2 + H_2$	$1.28*10^{-02}$
19	Tyrosine	$C_9H_{11}NO_3 + 2H_2O \rightarrow C_6H_6O + C_2H_4O_2 + H_3N + CO_2 + H_2$	$1.28*10^{-02}$
20	Typtophan	$C_{11}H_{12}N_2O_2 + 2H_2O \rightarrow C_8H_7N + C_2H_4O_2 + H_3N + CO_2 + H_2$	$1.28*10^{-02}$
21	Glucine	$C_2H_5NO_2 + 0.5\ H_2O \rightarrow 0.75\ C_2H_4O_2 + H_3N + 0.5\ CO_2$	$1.28*10^{-02}$
22	Alanine	$C_3H_7NO_2 + 2H_2O \rightarrow C_2H_4O_2 + H_3N + CO_2 + 2H_2$	$1.28*10^{-02}$
23	Cysteine	$C_3H_7NO_2S + 2H_2O \rightarrow C_2H_4O_2 + H_3N + CO_2 + 0.5\ H_2 + H_2S$	$1.28*10^{-02}$
Acidogenic reactions			
24	Dextrose	$C_6H_{12}O_6 + 0.1115\ H_3N \rightarrow 0.1115\ C_5H_7NO_2 + 0.744\ C_2H_4O_2 + 0.5\ C_3H_6O_2 + 0.4409\ C_4H_8O_2 + 0.6909\ CO_2 + 1.0254\ H_2O$	$9.54*10^{-03}$
25	Glycerol	$C_3H_8O_3 + 0.4071\ H_3N + 0.0291\ CO_2 + 0.0005\ H_2 \rightarrow 0.04071\ C_5H_7NO_2 +$	$1.01*10^{-02}$
Acetogenic reactions			
26	Oleic acid	$C_{18}H_{34}O_2 + 15.2396\ H_2O + 0.2501\ CO_2 + 0.1701\ H_3N \rightarrow 0.1701\ C_5H_7NO_2 + 8.6998\ C_2H_4O_2 + 14.4978\ H_2$	$3.64*10^{-12}$
27	Propionic acid	$C_3H_6O_2 + 0.06198\ H_3N + 0.314336\ H_2O \rightarrow 0.06198\ C_5H_7NO_2 + 0.9345\ C_2H_4O_2 + 0.660412\ CH_4 + 0.160688\ CO_2 + 0.00055\ H_2$	$1.95*10^{-07}$
28	Isobutyric acid	$C_4H_8O_2 + 0.0653\ H_3N + 0.8038\ H_2O + 0.0006\ H_2 + 0.5543\ CO_2 \rightarrow 0.0653\ C_5H_7NO_2 + 1.8909\ C_2H_4O_2 + 0.446\ CH_4$	$5.88*10^{-06}$
29	Isovaleric acid	$C_5H_{10}O_2 + 0.0653\ H_3N + 0.5543\ CO_2 + 0.8044\ H_2O \rightarrow 0.0653\ C_5H_7NO_2 + 0.8912\ C_2H_4O_2 + C_3H_6O_2 + 0.4454\ CH_4 + 0.0006\ H_2$	$3.01*10^{-08}$
30	Linoleic acid	$C_{18}H_{32}O_2 + 15.356\ H_2O + 0.482\ CO_2 + 0.1701\ H_3N \rightarrow 0.1701\ C_5H_7NO_2 + 9.02\ C_2H_4O_2 + 10.0723\ H_2$	$3.64*10^{-12}$
31	Palmitic acid	$C_{16}H_{34}O + 15.253\ H_2O + 0.482\ CO_2 + 0.1701\ H_3N \rightarrow 0.1701\ C_5H_7NO_2 + 8.4402\ C_2H_4O_2 + 14.9748\ H_2$	$3.64*10^{-12}$
Methanogenic reactions			
32	Acetic acid	$C_2H_4O_2 + 0.022\ H_3N \rightarrow 0.022\ C_5H_7NO_2 + 0.945\ CH_4 + 0.066\ H_2O + 0.945\ CO_2$	$2.39*10^{-03}$
33	Hydrogen	$14.4976\ H_2 + 3.8334\ CO_2 + 0.0836\ H_3N \rightarrow 0.0836\ C_5H_7NO_2 + 3.4154\ CH_4 + 7.4996\ H_2O$	$2.39*10^{-03}$

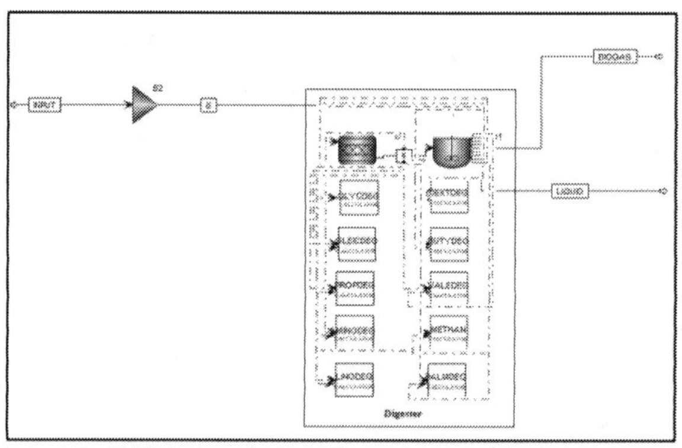

Figure 1: PSM simulation flowsheet of AD process

3. Model validation

Model validation can be performed by differentiating the simulated with literature results (experimental work) under fixed process conditions. Two scenarios were employed in this validation, and the findings from the model were compared with experimental data to determine the model's validity. It is crucial to note that this model is unable to predict the microbial activity involved in the AD process.

For validation, the two case studies from literature considered were: Case study (1): POME consists of total solids (TS), volatile solids (VS), and volatile suspended solids with hydraulic retention time of 15 days (Mamimin et al., 2016). Case (2): Different set up from case 1 which includes the effects recirculation of effluents to the reactors. HRT of POME is fixed at 15 days in both case studies.

According to case study 1, biogas production rate was 4.4 L biogas L-1 d-1 with containing of 51% methane, 35% carbon dioxide and 14% hydrogen [16]. Meanwhile, O-thong et al. (2016) demonstrated effect of effluent recirculation for biogas production, and it resulted to higher biogas production rate which consists of 54.4% of methane, 32.2% carbon dioxide and 13.3% of hydrogen [15].

For validation purpose, absolute average deviation (AAD) calculation performed to evaluate the percentage of CH4 generation from both experiment and simulated results. The AAD was calculated using eq.1 and maximum error tolerance is fixed at 5% for validation purpose.

$$AAD\,\% = \left|\frac{Methane\ production\ (Experiment) - Methane\ production\ (Simulation)}{Methane\ production\ (Experiment)}\right| x\ 100\ \% \quad (1)$$

4. Results and discussion

Using the identical process flowsheet, the generated simulation model was validated for the two situations. eq. 1 was used to determine the difference percentage or error values (1). For the synthesis of bio-methane from POME, the derived models differed from cases 1 and 2 by 1.9 % and 4.4 %, respectively. The permitted range was equal to or less than 5%. The model was validated and might serve as a foundation for further improvement because all the differences were less than 5%.

5. Effects of additional hydrogen and carbon dioxide

Specific amount of hydrogen and carbon dioxide gases is added in AD could increase the CH_4 mole fraction in biogas as it favors the hydrogenotrophic methanogenesis reaction pathway. The PSM used to study gas supply of H_2: CO_2 at two different ratios, 2:1 and 1:1.

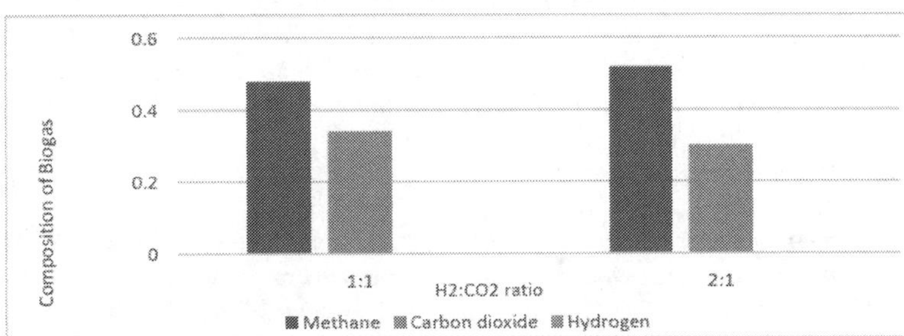

Figure 2: Effect of hydrogen and carbon dioxide addition on methane composition at different H_2: CO_2 ratio

Figure 2 illustrates that external gas supply of H_2: CO_2 influence the composition of biogas at constant flowrate of POME. Based on the results generated from the simulation , hydrogen has been completely utilized at both ratios as the composition of hydrogen nearly zero. Methane production recorded highest at 2:1 ratio compared to 1:1 ratio. The maximum composition of methane obtained was 52% at with 2:1 ratio. The results justified that it works on hydrogenotrophic methanogenesis reaction as shown in the eq. 2. This stochiometric reaction it shows that at least 4 mol of H_2 and 1 mol of CO_2 is needed to produce 1 mol of methane.

$$4H_2 + CO_2 \rightarrow CH_4 + 2H_2O, \qquad \Delta H = 165.0\ kJ/mol \quad (2)$$

In biomethane production from POME, there are two stages of reactions; where first is on biohydrogen production (eq.3) and secondly on in-situ biomethane production which may occur in two pathways namely via CO_2 reduction (eq. 2) and methyl trophic or aceticlastic methanogenesis pathway (eq 4). Adding external CO_2 may increase the methanogenesis towards CO_2 reduction pathway, the effect is less sensitive provided that H_2 is not the limiting or else the acetogenesis way in much in favourable.

$$C_6H_{12}O_6 + 2H_2O \rightarrow 4H_2 + 2CH_3COOH + 2CO_2 \quad (3)$$
$$2CH_3COOH \rightarrow 2CH_4 + 2CO_2 \quad (4)$$

6. Conclusion

The process of anaerobic digestion has been modelled using Aspen Plus. The four phases that make up the anaerobic digestion process are intended to be represented by its four stages. Thirteen (13) reactions were included, and a stoichiometric reactor was utilised to symbolise the hydrolysis process. 33 chemical reactions were defined in a kinetic reactor that represented the remaining stages of anaerobic digestion. The developed model was validated with two experimental data and obtained AAD <5%. After performing the validation, the model was further simulated to investigate the effects of adding external hydrogen and carbon dioxide the bio-methane yield. From the model, it can be shown that adding extra CO_2 has a reversed effect on the production of methane as compared to hydrogen. Methane production is lower at gas ratio H_2: CO_2 1:1 as compared to more H_2 is added to the ratio. Hydrogen has more predominant effect towards this biomethane production.

References

[1] Borja R 2011 2 55-*Biogas Production Comprehensive Biotechnology (Second Edition)* 785-798
[2] Mohan S and Jagadeesan K 2013 Production of biogas by using food waste *International journal of engineering research and application* **3(4)** 390-394
[3] Mata-Alvarez J (Ed) 2002 *Biomethanization of the organic fraction of municipal solid wastes* IWA publishing
[4] Nijaguna B T 2006 *Biogas technology* New Age International
[5] Suhan M B K Shuchi S B Abir R A R and Hasan R 2009 Innovative Multi-Stage Anaerobic Modelling of a 10 MW Biogas Plant Using Aspen HYSYS In ICPE 2019 25
[6] Harun N Hassan Z Zainol N Ibrahim W H and Hashim H 2019 Anaerobic digestion process of food waste for biogas production: a simulation approach *Chemical Engineering & Technology* **42(9)** 1834-1839
[7] Alitalo A Niskanen M and Aura E 2015 *Biocatalytic methanation of hydrogen and carbon dioxide in a fixed bed bioreactor Bioresource technology* 600-605
[8] Thauer R K 1998 *Biochemistry of methanogenesis*: a tribute to Marjory Stephenson: 1998 Marjory Stephenson prize lecture *Microbiology* **144(9)** 2377-2406
[9] Donoso-Bravo A Mailier J Martin C Rodríguez J Aceves-Lara C A and Wouwer A V 2011 *Model selection identification and validation in anaerobic digestion: a review Water research,* **45(17)** 5347-5364
[10] Haydary J 2019 Chemical process design and simulation: Aspen Plus and Aspen Hysys applications John Wiley & Sons
[11] Nduse R and Oladiran T M 2016 Simulation of a co-digester plant using aspen plus In Proceedings of the 5th IASTED International Conference on Power and Energy Systems, *AfricaPES* 325-332
[12] Al-Rubaye H Karambelkar S Shivashankaraiah M M and Smith J D 2019 Process simulation of two-stage anaerobic digestion for methane production *Biofuels* **10(2)** 181-191
[13] Rajendran K Kankanala H R Lundin M and Taherzadeh M J 2014 A novel process simulation model (PSM) for anaerobic digestion using Aspen Plus. *Bioresource technology* **168** 7-13
[14] Rajani A Santosa A Saepudin A Gobikrishnan S and Andriani D 2019 Review on biogas from palm oil mill effluent (POME): challenges and opportunities in Indonesia. *In IOP Conference Series: Earth and Environmental Science* **293** 012004 IOP Publishing
[15] Kongjan P O-Thong S and Angelidaki I 2013 Hydrogen and methane production from desugared molasses using a two-stage thermophilic anaerobic process *Engineering in Life Sciences* **13(2)** 118-125
[16] Mamimin C Singkhala A Kongjan P Suraraks B Prasertsan P Imai T and Sompong O 2015 Two-stage thermophilic fermentation and mesophilic methanogen process for biohythane production from palm oil mill effluent *International Journal of Hydrogen Energy* **40(19)** 6319-6328
[17] Angelidaki I Ellegaard L and Ahring B K 1993 A mathematical model for dynamic simulation of anaerobic digestion of complex substrates: focusing on ammonia inhibition *Biotechnology and bioengineering* **42(2)** 159-166
[18] Angelidaki I Ellegaard L and Ahring B K 1999 A comprehensive model of anaerobic bioconversion of complex substrates to biogas *Biotechnology and bioengineering* **63(3)** 363-372
[19] Peris R S 2011 Biogas process simulation using aspen plus Department of Chemical Engineering, *Biotechnology and Environmental Technology* **88**
[20] Luo G and Angelidaki I 2013 Co-digestion of manure and whey for in situ biogas upgrading by the addition of H2: process performance and microbial insights *Applied microbiology and biotechnology* **97(3)** 1373-1381

IOP Conf. Series: Materials Science and Engineering 1257 (2022) 012003 doi:10.1088/1757-899X/1257/1/012003

Influence of calcium macronutrient on tip burn occurrence in hydroponically cultivated lettuces

M Y Yap[1], M W Lim[1] and L K W Brendan[2]

[1]Centre for Water Research, Faculty of Engineering, Built Environment & Information Technology, SEGi University, 47810 Kota Damansara, Selangor Darul Ehsan, Malaysia
[2]Omnicare Farm Sdn Bhd, Jalan Tandang, Seksyen 51, 46050 Petaling Jaya, Selangor Darul Ehsan, Malaysia

*E-mail: limmeewei@segi.edu.my

Abstract. Tip burn in hydroponics lettuce is an undesirable phenomenon identified by necrotic breakdown of marginal tissue of leaves. A deep-water culture hydroponics system was designed to determine the effect of increasing nutritional calcium concentration and varying nutritional calcium sources in alleviating the tip burn development for the hydroponics growth of lettuce (*Lactuca sativa*). Lettuces was cultivated separately in an untreated fish wastewater (150mg/L) and four treated fish wastewaters with calcium addition by calcium hydroxide and calcium carbonate to concentrations of 250 and 350mg/L respectively. Increasing calcium concentration from 150 to 350mg/L demonstrated not only an increase in tip burn incidence by 33.33% (calcium carbonate) and 50% (calcium hydroxide), but a slower plant growth in comparison with lettuces cultivated at 150mg/L. Besides, lettuces treated by calcium carbonate has greater leaf calcium concentrations by 48.17% at constant calcium concentration, attributed to its high solubility to release more Ca^{2+} ions in fish wastewater for plant adsorption. Results suggest that the addition of calcium carbonate to fish wastewater for hydroponics cultivation is more suitable for reducing the tip burn occurrence and growth inhibition at concentrations up to 150mg/L.

1. Introduction

Hydroponics lettuce (*Lactuca sativa*) is a cool season vegetable, which is highly susceptible to a physiological disorder, also known as tip burn [1]. This disorder is characterized by the presence of necrotic spots on the margins of young, rapidly-expanding leaves, which causes irreversible tissue damage [2]. Severe tip burn development may be subjected to substantial economic yield losses by up to 50% [3]. Tip burn is a symptom of localised calcium deficiency that appears to depend on the supply of calcium. It develops when the calcium uptake by plants is inadequate for re-translocation into the leaves, which results in water-soaked tissue that can lead to necrosis formation [4].

Since tip burn is associated with the calcium uptake, the calcium supplied in nutrient solution is considered as an important aspect to mitigate the tip burn development. Kuronuma et al. (2020) had reported that the optimum calcium concentration in nutrient solution for *Azuma-no-Kaori* and *Voyage Yellow* are 0.12 and 0.08 g/L respectively with negligible tip burn development, while 100% of tip burn incidence with severity of 10% was observed in *Celeb Wine* for all calcium treatments [5]. Yang and Kim (2020) had observed that the tip burn was only developed on lettuce even though the supply of calcium (140-150mg/L) to tomato, basil and lettuce are the same [6]. For strawberry (*cv. Nicolla*),

Coosemans (1989) had reported that the percentage of leaves without tip burn are highest when grown under the nutrient solution with 0.25% Ca $(NO_3)_2$ [7]. Thus, for non-leafy vegetables such as tomato and strawberry, calcium concentration of 150 mg Ca/L is sufficient to prevent the tip burn from occurring. However, a higher calcium concentration in nutrient solution is required for hydroponics lettuce due to its high sensitivity to the tip burn [6]. Since tip burns is still occurred at calcium concentration of 150mg/L, the effect of increasing calcium level in eradicating the tip burn development of hydroponics lettuce is significantly important as plants' responses to the concentrated regime have not been widely studied.

Calcium is normally supplied to nutrient solution in various forms, such as calcium nitrate, calcium carbonate, calcium sulphate, calcium hydroxide and calcium chloride [8]. However, to date, there are no concrete conclusions on the effects of calcium nutrition sources on tip burn for hydroponics lettuces. In addition, the effect of calcium hydroxide and calcium carbonate in preventing hydroponics tip burn are poorly explored, despite being common chemicals that are used to increase the nutrient availability and pH adjustment for fish ponds [9]. As the agricultural trend moves towards aquaponics which integrates hydroponics with aquaculture, there is a need to study the effect of the addition of calcium hydroxide and calcium carbonate on plant growth and its correlation with tip burn incidence.

Therefore, this study aims to determine the effect of increasing nutritional calcium concentration in fish wastewater on hydroponics tip burn by characterizing the tip burn severity and incidence, lettuce growth and calcium acquisition in leaf tissue using varying calcium sources: calcium hydroxide and calcium carbonate in fish wastewater in preventing the tip burn occurrence. The outcome of this study will provide a future reference on lettuce growth in an aquaponics cycle.

2. Materials and Methods

2.1. Materials

Lettuce, *Lactuca sativa* seeds of Batavia type (Evergreen) were commercially purchased. Fish wastewater was obtained from Omnicare Farm, Malaysia as the nutrient solution for this study to increase the sustainability of hydroponics as aquaculture is typically used in symbiotic relationship with hydroponics. The pH of the solution was varied using 25% of potassium hydroxide solution and 50% of phosphoric acid solution (City Farm, Malaysia). The calcium level of the fish wastewaters was varied by adding the powder of calcium hydroxide (R&M Chemicals) or the granular calcium carbonate (OMYA) respectively. The foliar calcium concentration of leaf samples was determined by an atomic absorption spectrometer (Agilent Technologies, Agilent 280FS).

2.2. Hydroponics plant growth in deep-water culture (DWC) hydroponics system

A plastic container (590 mm L × 430 mm W × 370 mm H) is used as the reservoir to contain the fish wastewater. Four air stones with size of 40 mm were connected to the air pump to oxygenate the fish wastewater for the lettuce growth. A polystyrene with six holes (Ø 71mm) in the arrangement of 2 × 3, separated 6 cm horizontally and 12 cm vertically was used as a supporting medium for the lettuces. Two full spectrum light-emitting diode (LED) grow lights (LED T8, City Farm) with photosynthetic photon flux (PPFD) ranging from 53-63 $\mu mol/m^2s$ were hung at the distance of 30 cm from the plants.

2.3. Seed germination and plant growth

At the start of the experiment, seeds of *Lactuca sativa* were first germinated using growth foams in water, placed under two full spectrum LED grow lights with a photoperiod of 16 hours day, 8 hours night. On Day 5, the growth foam with lettuces were transplanted to the net pots with pebbles. Each six lettuces with the net pot were placed over to the holes of polystyrene to grow with either untreated fish wastewater (150 mg Ca/L) or treated fish wastewater with calcium concentration of 250 and 350 mg/L as illustrated in table 1. The environmental parameters for the lettuce growth are kept constant at average ambient temperatures of 25°C, pH level of 6.6-7.0, and dissolved oxygen level between 4-8ppm at the same lighting conditions as per the seed germination.

Table 1. Calcium concentration for each set of lettuces.

Sets	Set Concentration (mg Ca/L)	Nutrition Calcium Sources
1 (Control Set)	150[a]	-
2	250	$Ca(OH)_2$
3	350	$Ca(OH)_2$
4	250	$CaCO_3$
5	350	$CaCO_3$

[a] Control set concentration was determined through calcium analysis by Permulab Sdn. Bhd., Malaysia and is based on the existing calcium concentration in the sample.

2.4. *Growth and tip burn analysis*
After 30 days of cultivation, the lettuces were harvested and its total, shoot and root fresh weight, leaf length and leaf number were recorded. The number of tip-burned leaves were counted and the tip burn of each leaf was scored based on severity index as defined in Kuronuma et al (2020) with ranking from 0 to 1 (0, asymptomatic; 0.2, deformed leaf margins; 0.5, leaf-tip chlorosis; 1, leaf-tip necrosis) [5]. The tip burn severity in each lettuce was defined as Equation 1 and the incidence of tip burn was expressed as the percentage of lettuces exhibiting tip burn [5].

$$\text{Tip burn severity} = \Sigma\ [(\text{severity index} \times \text{leaf number})/\text{total number of leaves}] \times 100\% \quad (1)$$

2.5. *Leaf tissue analysis*
The leaf samples for calcium analysis were washed, dried at 65°C for 48 hours and grounded. 0.15g samples were digested in the acid mixtures which consist of 1.5ml of 70% HNO_3, 0.45ml of 60% $HClO_4$ and 2.1ml of 33% H_2O_2 at 200°C until the presence of white fumes were observed. The digested solution was filtered to 25ml volumetric flask and diluted by deionised water up to 25ml solution, followed by addition of 1% (w/v) lanthanum oxide to the final solution. The calcium of the resulting solutions was analysed through atomic absorption spectroscopy as outlined by Perkin (1996) [10].

3. Results and Discussion

3.1. *Effect of calcium concentration on tip burn occurrence and lettuce growth*
Several studies have shown that higher calcium concentration ranging between 150 and 250 mg/L is required for lettuces, due to its high susceptibility to the tip burn occurrence [5,6]. However, results showed an inverse trend whereby lettuce receiving fish wastewater with high calcium concentration had significantly exhibited more severe tip burn when compared to low calcium treatment. 100% of the lettuces that grew under calcium hydroxide treatment exhibited tip burn and the tip burn severity increased at higher calcium concentrations. For calcium carbonate treatment, 83.3% of the lettuces developed tip burn and the tip burn severity increased gradually from 4.7% to 19.4% with the increase in calcium concentration. These results indicate the increase in calcium concentration leads to more severe tip burns. Figure 1 show the symptoms of tip burn on lettuce at varying calcium concentrations.

Table 2 showed that the increase in calcium concentration in fish wastewater to 250 and 350 mg Ca/L would inhibit the growth of lettuce due to the nutrient antagonism effect. Based on recent studies, magnesium deficiency symptoms could appear when there is an excessive calcium in nutrient solution. Deficiency in magnesium uptake by plants would result in slower growth and interveinal chlorosis, which affect the yield rate indirectly [8,11]. At high nutritional calcium concentration of 350 mg Ca/L, the growth of the lettuces is the slowest for calcium hydroxide and calcium carbonate treatment as high calcium level in fish wastewater results in higher depressive effect on Mg uptake. In contrast, normal growth was observed for lettuces receiving fish wastewater with 150 mg Ca/L. This is in agreement with the work done by Uno et al (2016) who had also recorded that addition of 142 mg Ca/L to nutrient solution did not result in any negative growth effect [3].

Figure 1. Tip burn at varying calcium concentrations for (a) calcium hydroxide and (b) calcium carbonate treatment.

Table 2. Growth data of lettuce at different calcium concentration (n=6)

Parameters	Control Set	Calcium Hydroxide		Calcium Carbonate	
		250 mg Ca/L	350 mg Ca/L	250 mg Ca/L	350 mg Ca/L
Total fresh weight (g)	1.84 ±1.57	1.09 ±0.65	0.93 ±0.42	0.57 ±0.26	0.48 ±0.20
Shoot fresh weight (g)	1.66 ±1.35	1.00 ±0.60	0.83 ±0.38	0.53 ±0.25	0.45 ±0.18
Root fresh weight (g)	0.18 ±0.23	0.10 ±0.07	0.10 ±0.04	0.04 ±0.02	0.04 ±0.02
Leaf length (cm)	5.92 ±3.16	4.04 ±2.21	4.16 ±1.17	4.03 ±1.63	4.23 ±1.77
Total leaf number	8.83 ±1.17	8.17 ±4.07	7.00 ±1.00	6.33 ±1.03	5.67 ±0.82
Number of leaves without tip burn	0.83 ±1.17	2.50 ±1.05	4.00 ±0.00	1.67 ±1.03	1.67 ±1.21

Tip burn is typically associated with rapid growing tissues due to its high demand on calcium uptake [12]. In this study, results showed a lesser number of tip-burned leaves obtained at low calcium treatments, although the total leaves numbers are higher when compared to high calcium treatments. Increasing calcium concentration in fish wastewater to more than 150 mg Ca/L is not feasible to mitigate tip burn, but instead demonstrates an inhibition in lettuce growth. This indicates that 150 mg Ca/L is the critical concentration to prevent the tip burn development while minimising lettuce growth inhibition. This is in good agreement with Jones (2013) and Smith (1999), whom reported that the maximum calcium levels in hydroponic nutrient solution should be less than 200 mg/L to ensure better plant growth with reduced tip burn [13].

3.2. Effect of nutritional calcium sources on tip burn occurrence and lettuce growth

From the results, lettuces receiving the fish wastewater with calcium addition by calcium carbonate and calcium hydroxide had significant differences in tip burn severity and incidence especially when the calcium levels increase to 350 mg Ca/L (table 3). The lettuces cultivated at 350 mg Ca/L of calcium hydroxide are 22.6% more severe than that of calcium carbonate treatments. However, the severity of tip burns for lettuces grown at 250 mg Ca/L did not display a significant difference between calcium hydroxide and calcium carbonate with a tip burn severity of 20.5% and 19.4% respectively. Besides, tip burn incidence of 100% was observed in lettuces grown under calcium hydroxide treatment, which is significantly higher than calcium carbonate treatment (83.3%) at different calcium concentrations.

Table 3. Growth data of lettuce at different calcium concentration and sources(n=6)

Calcium Concentration	Tip Burn Severity (%)		Tip Burn Incidence (%)	
	Calcium Hydroxide	Calcium Carbonate	Calcium Hydroxide	Calcium Carbonate
250 mg/L	20.51 ±8.62	19.36 ±11.68	100	83.3
350 mg/L	36.94 ±13.34	14.31 ±9.78	100	83.3

For leaf tissue analysis, the calcium concentration in lettuce leaves that cultivated under calcium carbonate treatments are approximately 42.3% and 48.2% higher than that of calcium hydroxide treatments at 250 and 350 mg Ca/L respectively as demonstrated in table 4. These results confirmed that there is insufficient calcium transport from root to leaves in plants grown in calcium hydroxide in comparison with plants grown in calcium carbonate, although the calcium supplied in fish wastewater are the same for both treatments. This could be attributed to the lower solubility of calcium hydroxide as compared to solubility of calcium carbonate in water, which resulted in less Ca^{2+} ions contained in fish wastewater for the plant uptake [14,15].

Table 4. Calcium concentration (mg/g) in leaf tissue at the end of experiments

Control Set	Calcium Hydroxide		Calcium Carbonate	
	250 mg Ca/L	350 mg Ca/L	250 mg Ca/L	350 mg Ca/L
14.42	7.6	6.9	13.2	13.4

The effect of varying calcium sources is apparent in the plant growth yield. Although more leaves were exhibiting tip burn when the lettuce grew under calcium hydroxide treatments, the overall growth of lettuces were significantly faster than calcium carbonate treatments in terms of total weight, shoot fresh weight, root fresh weight and leaf number as demonstrated in table 2 at both 250 mg Ca/L and 350 mg Ca/L concentrations. This could be attributed to the lower calcium concentration in the leaf tissue grown under calcium hydroxide treated water at 6.9-7.6 mg Ca/g as opposed to plants grown in calcium carbonate treated water which contained 13.2-13.4 mg Ca/g. Hence, the antagonistic effect of calcium on magnesium uptake by lettuce is minimized, resulting in a higher uptake of magnesium. This prevents the effect of magnesium deficiency, which eventually lead to the faster growth of lettuces [9,12]. The effect of varying calcium sources using both calcium hydroxide and calcium carbonate on the average leaf length, however, did not demonstrate a significant effect.

In terms of the effect of varying calcium sources on the tip burn occurrence, it was observed that a higher percentage of tip burn severity and incidence (table 3) were observed from plant grown in calcium hydroxide treatment. As the hydroponic plants grown in calcium hydroxide treated water had a higher growth rate, therefore the demand of calcium increases as well [16]. Hence, even though the growth of lettuces under calcium hydroxide treatments is faster, poor calcium uptake caused by low solubility of calcium hydroxide in fish wastewater cannot fulfil the high demand of calcium. Due to the poorer calcium uptake, this led to an increased in the tip burn severity and incidence. In contrast, the leaves of lettuces that grew under calcium carbonate treatment had 13 to 14 mg/g of calcium which is almost doubled when compared with calcium hydroxide. This suggested that the higher calcium concentration recorded in the leaf tissue was able to alleviate the tip burn severity and incidence.

These results indicates that calcium hydroxide is not suitable to be used as nutritional calcium sources for hydroponic lettuce in alleviating the tip burn occurrence. Yanagi (1980) also recorded that calcium carbonate is more effective than calcium hydroxide in reducing tip burn development for soil-based cultivars [17]. However, the lettuce growth was inhibited under calcium carbonate treatments which can affect the production yield indirectly. This growth retardation issue can be solved by reducing the calcium concentration of calcium carbonate to 150 mg/L as discussed. Furthermore, hydroxide and carbonate excess disorder were not observed during the experiment, indicating that the difference in anions of calcium sources in fish wastewater had little effect on plant growth.

4. Conclusion

This study concludes that increasing calcium concentration in fish wastewater to concentration levels more than 150 mg Ca/L did not alleviate the tip burn as the calcium transport from roots to shoots is reduced by the concentrated regime resulting from the high calcium in fish wastewater. The low Ca^{2+} ions contained in lettuce leaves suggests that there is a poor calcium uptake by plants, which is the main reason to cause tip burn development. In addition, nutrient antagonism occurred as high calcium levels in fish wastewater results in higher depressive effect on magnesium uptake, and thus the retardation on lettuce growth was demonstrated. Therefore, the calcium concentration in fish wastewater should not be greater than 150 mg Ca/L for reducing tip burn occurrence. In addition, calcium carbonate is more suitable to be used as a nutritional calcium source to solve the hydroponics tip burn due to its high solubility in fish wastewater. A higher concentration of Ca^{2+} ions from the solubility of calcium carbonate in the fish wastewater leads to sufficient calcium uptake during the lettuce growth, resulting in lower tip burn incidence and severity. However, it is to be noted that a high solubility of calcium carbonate also causes more Ca^{2+} ions available in fish wastewater, which inhibits the magnesium uptake by lettuce and is subject to growth retardation. This issue can be solved by reducing the calcium concentration of calcium carbonate to 150 mg/L.

Acknowledgement

The authors gratefully acknowledge the funding support of SEGi University, Malaysia (SEGiIRF/2021-03/FOEBEIT-04/109) and Omnicare Farm Sdn Bhd for providing the research facilities.

References

[1] Sydney C H, Daniel E W, Jeremy M P and Joseph M K 2019 *Horticulturae* **5** 50
[2] De Swaef T, Vermeulen K, Vergote N, Van Lommel J, Van Labeke C, Bleyaert P and Steppe K 2015 *Acta Hortic.* **1099** 63-70
[3] Uno Y, Okubo H, Itoh H and Koyama R 2016 *Sci. Hortic.* **210** 14-18
[4] Bangerth F 1979 *Annu. Rev. Phytopathol.* **17** 97-122
[5] Kuronuma T, Ando M and Watanabe H 2020 *Agronomy* **10** 216-25
[6] Yang T and Kim H J 2020 *Water (Switz.)* **12** 1259-86
[7] Coosemans J 1989 *Acta Hortic.* **265** 489-96
[8] Jones J B 2013 *Complete guide for growing plants* hydroponically (Boca Raton: CRC Press)
[9] Sa M V C and Boyd C E 2017 *Aquaculture* **469** 102-5
[10] Perkin E 1996 *Analytical methods for Atomic Absorption Spectroscopy* (Waltham: The Perkin-Elmer Corporation)
[11] Ranade-Melvi U 2011 *Karnataka J. Agric. Sci.* **24** 106–9
[12] Birlanga V, Acosta-Motos J R and Perez-Perez J M 2021 *Agronomy* **11** 616-36
[13] Smith R 1999 *Growing Edge* **11** 14-6
[14] Robert N 2013 *Focus on calcium: Its role in crop production* (Autralia: GRDC)
[15] Anna M M K, Aneta S K and Tadeusz M 2017 *Intech* **5** 93-110
[16] Frantz J M, Ritchie G, Cometti N N, Robinson J and Bugbee B 2004 *J. Am. Soc. Hortic. Sci.* **129** 331-8
[17] Yanagi A A 1980 *Phytopathology* **68** 1744-53

Comparison of different methods for the extraction of essential oil from the rhizome part of Kaempferia galanga linn

C R Nurhaslina[1], A N Mustapa[1,2*], C Y Azizi[3], and H Alwi[1]

[1]School of Chemical Engineering, College of Engineering, Universiti Teknologi MARA, 40450 Shah Alam, Selangor, Malaysia
[2]Integrated Separation Technology Research Group, School of Chemical Engineering, College of Engineering, Universiti Teknologi MARA, 40450 Shah Alam, Selangor, Malaysia
[3]School of Chemical & Energy Engineering, Faculty of Engineering, Universiti Teknologi Malaysia, 81310 UTM Johor Bahru, Johor, Malaysia.

anajwa@uitm.edu.my

Abstract. *Kaempferia galanga linn* (KGL) is a well-known condiment in Southeast Asia, consisting of huge advantages for health sustainability. This research aims to compare the essential oils (EOs) yield of the KGL rhizome utilising modern technologies – solvent-free microwave extraction (SFME) and microwave-assisted extraction (MAE) to the traditional method of hydrodistillation (HD). The influence of process parameters on the oil yield is determined and the composition of the extracts is characterised using Gas Chromatography-Mass Spectrometer (GCMS). The SFME has been demonstrated as the most excellent method in producing substantially high yield in a shorter time (40 min, 2.54 %) in comparison to the MAE (40 min, 1.03%) and HD (3 h, 0.14%). Ethyl p-methoxycinnamate from the class of cinnamic acid ester appears as the main volatile constituent with a percent content of 11.6%, 10%, and 13.4% obtained by SFME, MAE, and HD, respectively. The SEM analysis of the KGL reveals that microwave technology caused a significant plant cell wall rupture, thus promoting faster oil extraction than the traditional extraction approach. This study demonstrates that SFME is a promising alternative to producing essential oil as natural products without solvent, safer for pharmaceutical and food industries.

1. Introduction

KGL is one of the important *Zingiberaceae* species, including about 60 plant species found in tropical countries. Known as Cekur in Malaysia, and Kencur in Indonesia. Other names of this plant are called as aromatic ginger and sand [1,2] and are popular condiments in Chinese cuisine. Several applications of KGL include fever[3], antidote snake venoms, and vertigo [4]. The mature rhizome is used in the traditional method and boiled in coconut oil or mixed with vaseline before being applied directly to the abdomen [5]. The rhizome was taken orally to treat indigestion, cold, pectoral, abdominal pain, headache, and menstrual pain [6].

Several studies have applied traditional methods such as solvent extraction [2], maceration [7], and Soxhlet [9] to extract the KGL essential oils (EOs). These techniques utilize organic solvents such as hexane, DMSO and ethanol have created safety issues due to their hazardous, flammable, and potentially toxic for food and pharmaceutical applications. Even though the process is considered as inexpensive, however, it is always associated with significant drawbacks such as the use of a large amount of solvent which ultimately requires downstream separation of oil/solvent. Meanwhile, the hydrodistillation (HD) method which has been reported successfully discovered ethyl p-methoxycinnamate [1] faced the main limitation of an extensively longer extraction time and consumes enormous energy.

New approaches such as microwave-assisted extraction [8] and solvent-free microwave extraction [9] have greatly improved the oil yield while lowering the cost. Microwave extraction was the fastest approach compared to other extraction methods. SFME has been known as a unique device that combines microwave heating idea with dry distillation technology at atmospheric pressure. This technique was designed for use in the extraction of EOs from various aromatic plants by exposing plant material to microwave irradiation without adding any solvent or water. The exposure has caused cell distension, gland and oleiferous receptacle rupture, and cell enlargement inside the "fresh" plant material, and ultimately releases EOs evaporated by plant water in-situ.

At present, no published work reports on the comparison of different methods of extraction for KGL rhizomes. Therefore, in this study, extraction methods such as SFME, MAE, and HD were employed to compare their performance, in terms of the extraction yield, extract compositions, and the microstructural changes in the surface area of samples.

2. Methodology

2.1. Plant Material
The KGL rhizomes were bought from the Klang market. At first, the fresh rhizome was washed, peel the skin, cut into pieces, and then dried in an oven at 40 °C for 24 hours to obtain its moisture content in the range of 21-22%. The powdered rhizome was then stored in a freezer at -16°C in an air-tight container until further use for the experiment. To maintain consistency in the extraction outcome, the sample is held for a maximum of three months. Plant deterioration may cause the content of active compounds in the sample to decrease over time [8].

2.2. Extraction method
2.2.1. Solvent-free microwave extraction. The SFME setup was built as shown in figure 1(a) using a microwave oven (model Samsung (MS28F303TFK/SM) operates at 2450 MHz with a maximum power of 1000 W. The inside cavity of the oven is 33 x 22 x 32 cm. A 100 g dried KGL rhizome was soaked in 500 mL distilled water for 1 hour at room temperature (25 °C). The excess water was then drained. This soaking step was performed to moisturise the plant sample and carry out the SFME without any addition of water or solvent. The sample was then placed in a flask with a flat bottom and attached to a Clevenger apparatus. The glass column is outfitted with a condenser that is linked to a water circulation system, allowing the formed vapour to be condensed to a liquid. In principle, the distillate of the KGL is comprised of EOs and hydrosol. Therefore, the distillate of KGL was dried using freeze-drying (ST-FD-10B-MPT Serial no 8737) to remove hydrosol from the EOs – hydrosol mixture, otherwise, the moisture (hydrosol) may cause spoilage to the EOs during storage. A similar step was applied for the KGL distillate obtained from MAE and HD methods. Finally, the weight of the sample was measured and stored in amber vials until GCMS analysis. The oil yield of the KGL rhizome was determined using equation (1):

$$Yield\left(\%\frac{w}{w}\right) = \frac{Mass\ of\ extracted\ Kaempferia\ galanga\ linn\ oil}{Mass\ of\ dried\ Kaempferia\ galanga\ linn\ rhizome}\ x\ 100\% \quad (1)$$

2.2.2. Microwave-assisted extraction. MAE was used similarly to the SFME set-up as shown in figure 1(a). The KGL sample was soaked for 1 hour at room temperature. However, for the MAE method, the water used during the soaking step was kept in the flask and used in the extraction process as the extracting solvent. A 100 g of dried KGL rhizome and 500 mL of distilled water (F/S ratio of 0.20 g/mL) were placed in a 1000 mL flask with a Clevenger apparatus. The distillate was collected in vials,

IOP Conf. Series: Materials Science and Engineering 1257 (2022) 012004 doi:10.1088/1757-899X/1257/1/012004

subjected to freeze-drying to remove the hydrosol, and kept in a freezer at -16 °C until used for GCMS analysis. The freeze-dryer operated at -66 °C and 1 Pa.

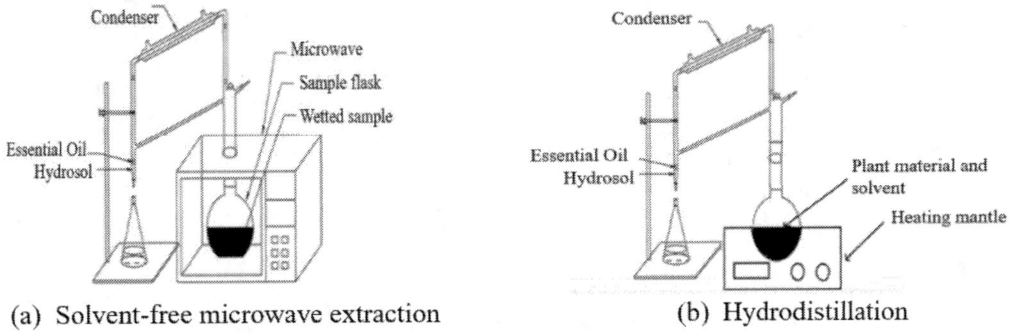

Figure 1. The extraction method used in the study

2.2.3. Hydrodistillation.
A 100 g of dried KGL rhizome was loaded to HD as shown in figure 1(b) and extracted with 1 L of water for 180 min (until no more distillate was extracted). The distillate was collected in amber-colored vials, dehydrated using freeze-drying, keep in freezer at -16°C before being analysed.

2.3. Chemical composition analysis
The EOs was analysed by gas chromatography coupled to mass spectrometry (GCMS) (Hewlett-Packard computerised system comprising a 450 gas chromatograph coupled to a 240-mass spectrometer) using a fused-silica-capillary column with a polar stationary phase SLB®-5ms (30m × 0.25 mm × 0.25 μm film thickness). The GCMS were obtained using the following conditions: carrier gas He; flow rate 1.1 mL/min; injection volume 0.1 L; injection temperature 250 °C. The column temperature was kept at 80°C for 5 min, heated to 150 °C with a 6 °C/min rate and hold for 5 min, and finally kept increased to 220 °C with a 3 °C/min rate and keep constant at this temperature for 5 min.

2.4. Scanning Electron Micrograph
The microstructure analysis was performed on the KGL samples before and after the extraction process, by scanning electron microscope (SEM) (Hitachi SU3500) to analyse the effect of different extraction methods on the pore size of dried KGL. SEM analysis is an important characterisation technique to determine the synergistic effect of microwave heating on morphological and surface area characteristics. The analysis was carried out at a magnification of 1000x and a 15 kV accelerating voltage. A sample size of 4 mm x 4 mm was directly placed on an aluminium holder with a diameter of 12.5 mm using a two-sided adhesive carbon model. A Polaron SC 7620 sputter coater was used to coat the sample with platinum. The prepared samples were introduced into a microscope chamber using a low vacuum.

3. Results and Discussion

3.1 Composition of EOs through SFME, MAE, and HD
Figure 2 shows the analysis result by GCMS for 3 different methods mainly SFME (figure 2(a)), MAE (figure 2(b)), and HD (figure 2(c)). The result shows that ethyl p-methoxycinnamate (EPMC) has been identified as the most abundant component of KGL rhizome extracted by SFME (11.6%), MAE (10%), and HD (13.4%), followed by ethyl cinnamate, consistent with previous studies [10].

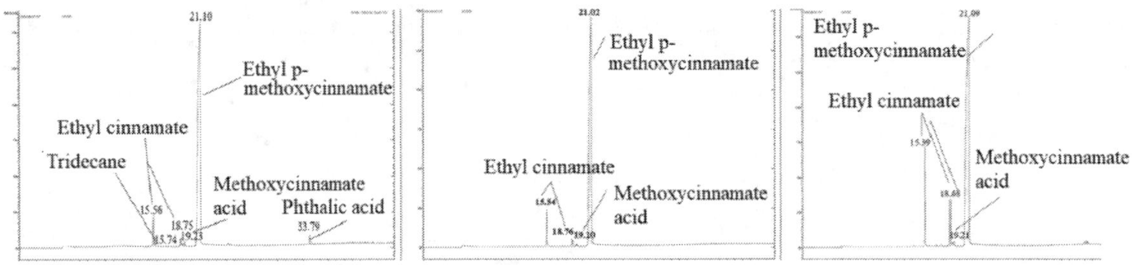

(a) SFME (b) MAE (c) HD

Figure 2. The GC–MS analysis of the KGL EOs from different extraction methods

Figure 3(a) represents the yield of EOs for SFME, MAE, and HD against extraction time. As can be seen, the extraction rate of the SFME was faster than the MAE, and two-fold faster in comparison to the HD method. This indicates that microwave irradiation has facilitated the rate of the extraction, hence improving the yield of EOs. It is important to note that the hydrosol was first recovered from the KGL at the beginning of the process before the extraction of the EOs. The results demonstrated that the extraction of the hydrosol has been accelerated by SFME and MAE methods within 30 minutes compared to the HD which requires more than 55 minutes to obtain the hydrosol.

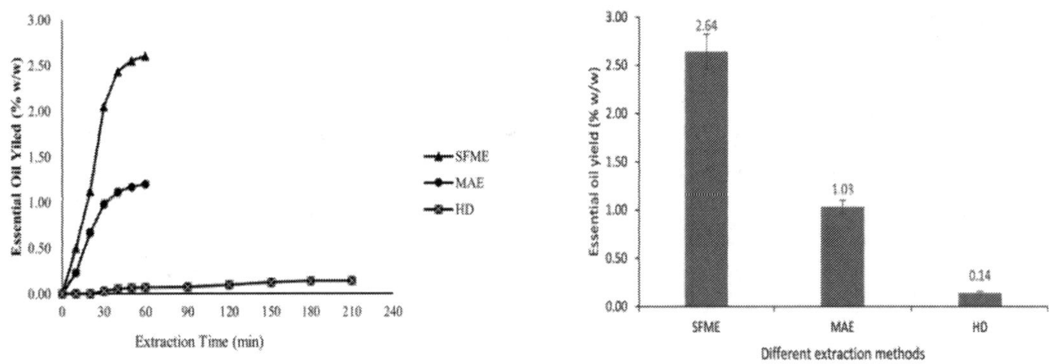

(a) (b)

Figure 3. EOs yield percentage vs (a) Extraction time and (b) Different extraction methods based on SFME, MAE, and HD from KGL rhizome

Figure 3 (b) shows that the EOs yield produced from SFME was much higher and in a shorter time compared to the MAE and HD techniques when the asymptotic yield can be achieved in 40 minutes. A similar finding was reported by [8]. In comparison, the HD appeared to the less efficient when it required more than 3 hours to obtain as low as $0.14 \pm 0.02\%$ yield. In conventional HD, heat transport was limited to conduction and convection. In SFME, however, microwave irradiation could result in a fast increase in temperature due to the absorption of energy by the in-situ water of the KGL rhizome. As a result, the water undergoes a phase transition and becomes steam, and the pressure within the plant cells rises, causing cell enlargement and severe disintegration of cell walls. Similar occurrences were found by [11] and [10] after isolating EOs from *Schisandra Chinensis* and *Mentha* leaves using SFME. The SFME approach was faster and more successful than MAE and conventional extraction methods in separating KGL rhizome EOs.

3.2 Microstructural changes before and after extractions with different extraction methods

The physical alterations of KGL rhizome treated with HD, MAE, and SFME were investigated using SEM and compared to the intact glandular micrograph of the untreated sample (as control). The cells in the untreated samples were smooth and undamaged, as seen in figure 4 (a). Figure 4(b) demonstrates the advantages of soaking in aiding the wall plant to burst. During the soaking, the oil glands in the powder have a strong tendency to swell and enlarge. This would facilitate the easy and quick extraction of oil from the material. Meanwhile, for the non-soaking samples, a greater amount of energy is required to elevate the internal pressure of the oil glands to rupture the cell wall and release oil. Therefore, it is proposed to treat the plant materials with a soaking technique prior to EOs extraction, as it can reduce induction time and overall energy usage [12].

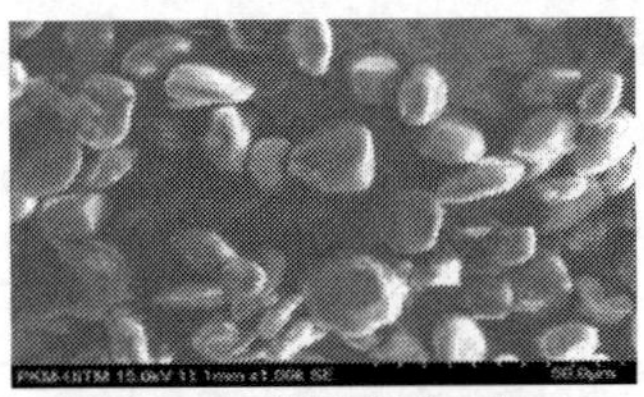

(a) Dried KGL before extraction,

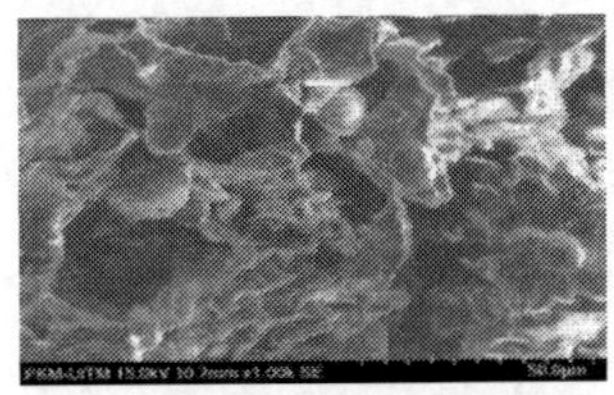

(b) Dried KGL after 60 min soaking

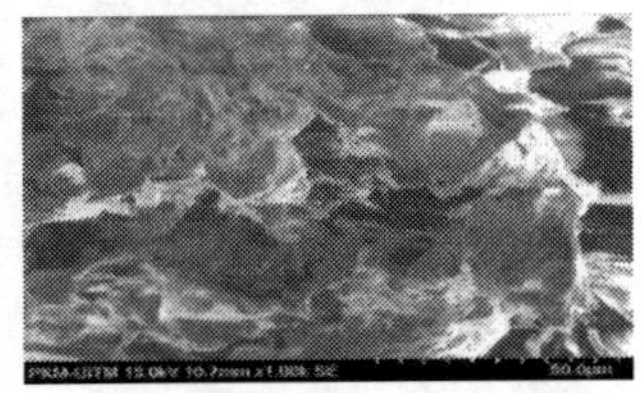

(c) Dried KGL after HD extraction

(d) Dried KGL after MAE extraction at 1000 W

(e) Dried KGL after SFME extraction at 1000 W

Figure 4. Scanning electron micrograph at 1000x and 15 kV

Meanwhile, for the samples after the HD extraction (figure 4(c) for 3 hours, MAE (figure 4(d)), and SFME (figure 4(e) for 40 minutes, various degrees of physical destruction can be observed, which the KGL rhizome became atrophic and wrinkled. However, when the EOs was extracted using the SFME method, the cells of the KGL rhizome were substantially disturbed, wrinkled, and seemed vacuous compared to the MAE and HD procedures. This forceful rupture in plant material aids in EOs extraction, lowering the time and energy required. The form of the heat given during the extraction process can be linked to oil gland disturbance. This result is consistent with [12] and [13]. Microwaves received by water molecules soaking in plant material are transformed into thermal energy, which could cause serious damage to the plant matrix [14]. The SEM results indicate that microwave irradiation accelerates and enhances the breakdown of glandular wall membranes.

4. Conclusion

The composition of KGL extracted using different techniques was investigated. SFME, MAE, and HD methods have successfully obtained ethyl p-methoxycinnamate as the main component, nevertheless, SFME appears as the most promising technique in producing higher yield in a shorter time, followed by the MAE technique. Thus, it is proved that the application of microwave energy to the extraction of natural products enables faster and more successful extraction, less waste, and more energy efficient.

Acknowledgements
The authors are grateful to Universiti Teknologi MARA (UiTM) and the School of Chemical Engineering for providing funds and facilities for this project under the GPK Grant (600-RMC/GPK 5/3(114/2020).

References

[1] Alsalhi M S, Elumalai K, Devanesan S, Govindarajan M, Krishnappa K and Maggi F 2020 The aromatic ginger Kaempferia galanga L. (Zingiberaceae) essential oil and its main compounds are effective larvicidal agents against Aedes vittatus and Anopheles maculatus without toxicity on the non-target aquatic fauna *Ind. Crops Prod.* **158** 113012

[2] Yao F, Huang Y, Wang Y and He X 2018 Anti-inflammatory diarylheptanoids and phenolics from the rhizomes of kencur (Kaempferia galanga L.) *Ind. Crops Prod.* **125** 454–61

[3] Srivastava N, Mishra S, Iqbal H, Chanda D and Shanker K 2021 Standardization of Kaempferia galanga L. rhizome and vasorelaxation efect of its key metabolite ethyl p-methoxycinnamate *J. Ethnopharmacol.* **271** 113911

[4] Elshamy A I, Mohamed T A, Swapana N, Yoneyama T, Noji M, Efferth T, Hegazy M-E F and Umeyama A 2020 Cytotoxic polyoxygenated isopimarane diterpenoids from the edible rhizomes of Kaempferia galanga (kencur) *Ind. Crops Prod.* **158** 112965

[5] Neamsuvan O, Tuwaemaengae T, Bensulong F, Asae A and Mosamae K 2012 A survey of folk remedies for gastrointestinal tract diseases from Thailand's three southern border provinces *J. Ethnopharmacol.* **144** 11–21

[6] Tushar, Basak S, Sarma G C and Rangan L 2010 Ethnomedical uses of Zingiberaceous plants of Northeast India *J. Ethnopharmacol.* **132** 286–96

[7] Tritripmongkol P, Plengsuriyakarn T, Tarasuk M and Na-Bangchang K 2020 In vitro cytotoxic and toxicological activities of ethanolic extract of Kaempferia galanga Linn. and its active component, ethyl-p-methoxycinnamate, against cholangiocarcinoma *J. Integr. Med.* **18** 326–33

[8] Mediani A, Abas F, Tan C P and Khatib A 2014 Effects of different drying methods and storage time on free radical scavenging activity and total phenolic content of cosmos caudatus *Antioxidants* **3** 358–70

[9] Mustapa A N, Martin Á, Mato R B and Cocero M J 2015 Extraction of phytocompounds from the medicinal plant Clinacanthus nutans Lindau by microwave-assisted extraction and supercritical carbon dioxide extraction *Ind. Crops Prod.* **74** 83–94

[10] Singh Chouhan K B, Tandey R, Sen K K, Mehta R and Mandal V 2019 A unique model of gravity assisted solvent free microwave based extraction of essential oil from mentha leaves ensuring biorefinery of leftover waste biomass for extraction of nutraceuticals: Towards cleaner and greener technology *J. Clean. Prod.* **225** 587–98

[11] Yingngam B, Brantner A, Treichler M, Brugger N, Navabhatra A and Nakonrat P 2021 Optimization of the eco-friendly solvent-free microwave extraction of Limnophila aromatica essential oil *Ind. Crops Prod.* **165** 113443

[12] Jeyaratnam N, Nour A H, Kanthasamy R, , Yuvaraj A R and Akindoyo J O 2016 Essential oil from Cinnamomum cassia bark through hydrodistillation and advanced microwave assisted hydrodistillation *Ind. Crops Prod.* **92** 57–66

[13] Modi P I, Parikh J K and Desai M A 2021 Intensified approach towards isolation of cinnamon oil using microwave radiation: Parametric, optimization and comparative studies *Ind. Crops Prod.* **173** 114088

[14] Benmoussa H, Elfalleh W, He S, Romdhane M, Benhamou A and Chawech R 2018 Microwave hydrodiffusion and gravity for rapid extraction of essential oil from Tunisian cumin (Cuminum cyminum L.) seeds: Optimization by response surface methodology *Ind. Crops Prod.* **124** 633–42

IOP Conf. Series: Materials Science and Engineering 1257 (2022) 012005 doi:10.1088/1757-899X/1257/1/012005

PROMOTING CIRCULAR ECONOMY BETWEEN PALM OIL SECTOR WITH MULTIPLE INDUSTRIES

J Z H HWANG[1], V ANDIAPPAN[1,2], D K S NG[1].

[1]School of Engineering and Physical Sciences, Heriot-Watt University, No.1, Jalan Venna P5/2, Precint 5, 62200 Putrajaya, Malaysia.
[2]Research Centre of Sustainable Technologies, Faculty of Engineering, Computing and Science, Swinburne University of Technology, 93350 Kuching, Sarawak, Malaysia.

jasonhwang64@gmail.com, vmurugappan@swinburne.edu.my, denny.ng@hw.ac.uk

Abstract. Malaysia is the second largest producer and exporter of palm oil products in the world, with the increasing production of crude palm oil (CPO) over the recent years. On the contrary, the palm oil sector has also received number of negative perceptions, such as high carbon footprint, unutilised waste and deforestation. The global organization of sustainable palm oil promotes sustainable practices to be being implemented in the palm oil sector to counter the negative image of unsustainable palm oil. One of the important areas to recover the generated waste into value-added products and energy. Therefore, the waste generation in palm oil sector can be viewed as an opportunity rather than a threat. In order to enhance the resource conservation, the generated waste can be reused/recycled within the palm oil sector, as presented in many previous research works. However, there are still large amount of waste generated being disposed. To address such issue, a circular economy approach is proposed around the palm oil industry. The generated waste from palm oil sector can be used as feedstock or input to other industries such as polymer, construction materials, furniture, paper production, etc. In this work, a systematic approach is developed to determine the optimum integrated supply chain network of palm oil sector with other industries Via such approach, materials that are low in cost, weight, renewable and biodegradable can be used as feedstock to create a self-regenerative system amongst the industries.

1. Introduction

The palm oil industry produces various products covering from both non-edible and edible uses. Production of crude palm oil (CPO) has increased substantially over the recent years due to the increasing demand on vegetable oil. The world produced 71 million tonnes of palm oil in 2018 as compared to 58 million tonnes in 2016 [1]. On the contrary, the palm oil industry has also been labelled as non-environmentally friendly industry due to expansion of palm production resulting in clear cutting of old growth rainforests [2]. The palm oil industry contributes to 85.5% of total biomass production in Malaysia [3]. Hence, efforts to minimise the environmental impact of deforestation shall be implemented to ensure the production of palm oil products in a sustainable manner. Through the establishment of biorefineries based of alternative technologies such as physical, chemical, biological, and thermochemical conversion, waste can be utilised as feedstock for palm oil industry and other industries. Note that palm-based biomass is a major sustainable energy and material source that have high recyclability potential. By connecting the output wastes from palm oil industry as the input resources for other industries, a circular economy approach is proposed around the palm oil industry.

Many of the previous research works have focused on the circularity of palm-based oil biomass around palm oil sector, such as feedstock for integrated with biorefineries and combined heat and power (CHP) systems. For example, Ng et al. [5] presented an optimization model to synthesize integration palm oil processing complex (POPC) which integrates palm oil mill, integration biorefinery, combined heat and power system, integrated biogas and wastewater system as a whole. In order to address multiple owner scenario where each of the plants within in POPC is owned by different owners, Ng et al [5] extended the previous model by integrating multi-objective optimization approach (fuzzy optimization) to trade-off the economic and environmental performances among the plants.

Yeo et al [4] presented an optisation approach which evaluate different types of biomass conversion technologies which are biochemical, physical and thermochemical processes to convert palm-based biomass into bioproducts. As reported waste from palm oil mill such EFB, MF, PKS, POME and plantation (trunk and frond) are taken into consideration. A generic superstructure is presented to evaluate all the possible alternatives of energy and material recovery within the network. Based on the optimised result, energy flow of electricity and steam are fed as an input to the palm oil sector. Besides, different products are produced and recovered within palm oil sector including plantations.

It is noted that most of the previous works are centric on palm oil sectors. However, to enhance the circularity of the sector, palm oil waste shall be recovered in other industries. Therefore, in this work, a generic optimization model which is based on superstructure approach, is expanded beyond the palm oil industry to be incorporated with other industries. All the potential utilization palm-based in other industries are to be considered. The concept of industrial symbiosis applied in this work when waste and by-products of the palm oil industry are introduced as raw materials to other industries [7]. Subsequently, utilising the waste as resources from other industries resulting in larger variety of waste to be recycled. This creates new opportunities for production of variety value-added products and reduce the waste generation at the landfill. Hence, creating a self-regenerative system amongst the industries and better sustainability. More parties are able to participate in the circular economy as a joint effort to achieve the objectives of the network. In this work, a case study from Perak, Malaysia is used to illustrate the proposed approach. More detailed discussion on the methodology and case study are presented in the following sections.

2. Problem statement

Given a set of industrial plants owned by different companies are interested in participating in the integrated circular economy framework. The framework poses to share utilities and exchange resources as well as by-products among different processing plants. Total waste is defined by the amount of unutilised waste being sent to the landfill in the network. Economic potential is calculated by revenue generated from selling of products while deducting raw material price. Carbon emission is determined by energy consumption multiplied by constant value of CO2 equivalent per kwh of energy by Tenaga Nasional Berhad [8]. The goal is to optimise the framework by trade-off between three objectives simultaneously via fuzzy optimization.

3. Mathematical optimisation

A number of alternate pathways for each waste are proposed. Each of those pathways are analysed to promote circularity and non-linear flows in the network. By connecting the dots across the biorefineries and multiple industries, a network of can be formed. Possible pathways configurations combined with different biorefineries, and industries are represented in a superstructure of all the available alternative. In the superstructure, details of material flows such as names and their types must be established for the formulation of mathematical model during optimisation. Generic superstructure is shown below:

IOP Conf. Series: Materials Science and Engineering 1257 (2022) 012005 doi:10.1088/1757-899X/1257/1/012005

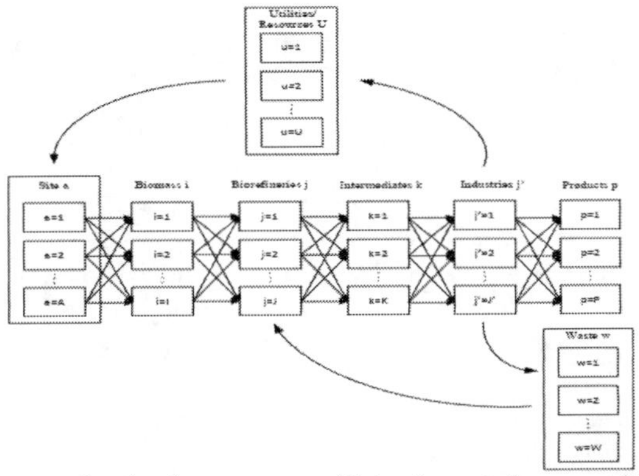

Figure 1: Generic superstructure for circular economy with implementation on multiple industries. Site a (palm oil sector), biomass i, biorefineries j, intermediates k, industries j', products p, waste w, utilities/resources u.

3.1. Model Formulation

Based on the developed superstructure, models are formulated based on a series of formulations below:

Different types of biomass i produced from palm oil sector a through their conversions X_i of fresh fruit bunches *FFB*:

$$F_i^{\text{Biomass}} = F^{\text{FFB}} X_i^{\text{Biomass,a}} \qquad \forall\, i \qquad (1)$$

Each biomass i are sent to multiple type of biomass conversion technologies j:

$$F_i^{\text{Biomass}} = \sum_j F_{ij} \qquad \forall\, i \qquad (2)$$

For each type of biomass conversion technology, j, all of biomass i and waste w are added to be inlet feed flow for biorefinery F_j^{In}:

$$\sum_i F_{ij} = F_j^{\text{In}} \qquad \forall\, j \qquad (3)$$

$$\sum_w F_{wj} = F_j^{\text{In}} \qquad \forall\, j \qquad (4)$$

Process capacities $F_j^{\text{In,cap}}$ are capped at the amount of inlet feed flow for each biorefinery j:

$$F_j^{\text{In}} \leq F_j^{\text{In,cap}} \qquad \forall\, j \qquad (5)$$

Biomass i are converted to intermediates k via biorefinery j through conversions X:

$$F_{j,k}^{\text{Out}} = F_j^{\text{In}} X_{j,k} \qquad \forall\, j\ \forall\, k \qquad (6)$$

Each intermediates k are sent to different industry j':

$$F_{j,k}^{\text{Out}} = \sum_{j'} F_{kj'} \qquad \forall\, j\ \forall\, k \qquad (7)$$

For each type of industry j', all of intermediates k are added to inlet feed flow for the respective industry $F_{j'}^{In}$:

$$\sum_k F_{kj'} = F_{j'}^{In} \qquad \forall j' \tag{8}$$

Intermediates k are converted into product p, utility/resource u and waste w through industry j':

$$F_{j'p}^{Out} = F_{j'}^{In} X_{j'p} \qquad \forall j' \tag{9}$$

$$F_{j'w}^{Out} = F_{j'}^{In} X_{j'w} \qquad \forall j' \tag{10}$$

For industries producing utilities:

$$F_{j'u}^{Out} = F_{j'}^{In} X_{j'u} \qquad \forall j' \tag{11}$$

The revenue R for each individual business in palm oil sector a, biorefineries j and industries $j'$is calculated by multiplying selling price C of the product p, biomass i or intermediates k and subtracting the feedstock selling price and transportation of each material. The total revenue of the network can is denoted as:

$$R^{\mathrm{TOTAL}} = R_a + R_j + R_{j'} + R_{j'} \tag{12}$$

Energy can be generated from utilities u to decrease the energy consumption of palm oil sector. The total production of energy streams from utilities can be determined using the following equation:

$$E_u^{\text{Energy-generation}} = \sum_{j'} \sum_{K} (F_{j'}^{In} X_{j'u}) \tag{13}$$

Energy generation can all be sold to the grid depending on objective functions. Another scenario is consuming within the palm oil sector and having the remaining sold to the grid. Equation is written as below:

$$E_u^{\text{Energy-generation}} = E_u^{\text{Self-consumption}} + E_u^{\text{Grid}} \tag{14}$$

Equation above signifies here E_u^{Grid} have the possibility to be 0 but not a negative value. However, $E_u^{\text{Self-consumption}}$ can be a negative value meaning excess of energy generated. Hence, excess energy of $E_u^{\text{Self-consumption}}$ is brought to left-hand side to be added up with energy generation $E_u^{\text{Energy-generation}}$ and all energy generation will be sold to grid E_u^{Grid}.

For all energy consumption of each processes, there are carbon emission. Carbon emission for palm oil sector, biorefineries, and industries are calculated on by $CO2^{perkwh}$. base on kgCO2 per input.

The total carbon emission of the network can is denoted as:

$$CO2_{total}^{emission} = CO2_a^{emission} + CO2_j^{emission} + CO2_{j'}^{emission} \tag{15}$$

4. Case study

In this work, an agricultural plantation size in Perak, Malaysia is given as with 391,768 HA and FFB yield of 9.81tons/HA as per MPOB in 2020. Assuming operating hour is 8760 a year and the palm oil mill is taking one fourth of FFB, the amount of FFB fed is 110 tons/hr. In order to identify the pathways, each sector and industry are given a label. Referring to figure 1, the description of each pathways are as follow: "A", which is from agriculture to palm oil mill and then to refinery. "B" is the biological refinery pathway which produces biogas to gas engine, bioethanol and biofertilizer up to the combined heat and power system (CHP) excluding the purification system. "C" is the chemical refinery pathway which produces biopolymer and CO_2 adsorbent. "D" being the physical refinery pathway for the production of dried fibre and pellets. "E" is the thermochemical refinery pathway which produces char, tar and biochar. "F" being wastewater treatment to the production of low-pressure steam. "G" is the polymer industry which produces chitosan and protein sludge. "H" is the paper industry producing pulp. "I" for furniture industry produces mattress. "K" is electricity grid. Lastly is "L" for biomethane upgrading from biogas and produces biomethane. A full superstructure with labels at the bottom of the figure is shown below.

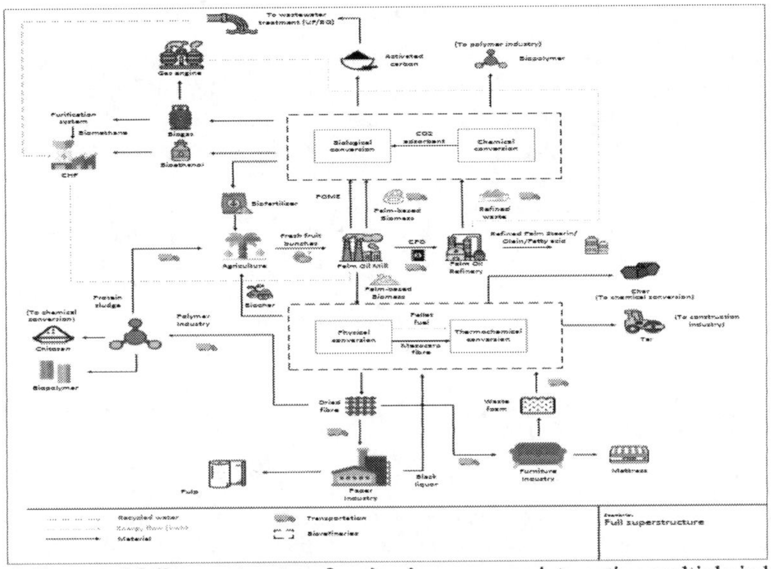

Figure 2: Proposed full superstructure for circular economy integrating multiple industries.

5. Results and discussion

Multi-objective optimisation is carried out to determine the trade-off between waste minimisation, maximum economic potential and carbon emission minimisation. Results of multi-objective optimisation are shown in the table below.

Table 1: Results of multi-objective optimisation.

Pathway	Description	Waste (tons/hr)	Economic potential (RM)	CO2 emission (kgCO2/hr)
A → B → A	Biogas and electricity production sent back to palm oil sector	EFB = 17.3	-9930	20
A → F → B → A	Wastewater treatment for boiler feedwater and low-pressure steam to palm oil sector	POME = 13.9 PKS = 7.5	-4083	31285

A → C → G	Biopolymer production	Sulphur = 11.85	53583	240
A → D → G → A & C	Chitosan production for CO2 adsorbent and protein sludge as biofertilizer	-	7090	-30.32
A → E → A	Biochar for agriculture	-	29309	-1680
Total		50.55	75969	29834.7

First pathway of biogas production and electricity utilised within the palm oil sector significantly reduces the energy consumption of the palm oil sector. Second pathway of wastewater treatment significantly reduces the amount of waste particularly in POME. Third pathway contributes largely to the revenue due for the production of biopolymer from sulphur. The fourth pathway of chitosan production as well as protein sludge are able to generate little revenue while reducing carbon emission through CO_2 adsorbent. Biochar production using pellet fuel can generate high revenue and with large amounts of excess energy produced. The degree of satisfaction, λ, achieved is 0.64 in this multi-objective optimisation scenario.

6. Conclusion

The multi-objective scenario concludes to satisfy three objectives of equal importance using fuzzy optimisation. The entire network is able to achieve high revenue particularly at biopolymer production. Waste and carbon emission has been significantly reduced through recyclability of resources that generates energy. Based on the proposed superstructure, participation of paper and mattress industry not advisable due to low economic potential and little waste usage. However, the work only considers one industrial waste, there are still many waste to be considered in future work. As of now, the optimum network without paper and mattress industry yield the highest degree of satisfaction between three objectives, which are waste minimisation, maximum economic potential and minimum carbon emission. The model can be adapted in different location of case studies and improved upon by accounting for capital expenses of technologies as well as safety assessments.

References

[1] Ritchie H and Roser M 2021 Palm Oil. *Our World in Data.* Available at: https://ourworldindata.org/palm-oil#:~:text=Palm%20oil%20production%20has%20increased,is%20shown%20in%20the%20chart. [Accessed March 30, 2022].

[2] Chang F K 2021. Palm oil: Malaysian economic interests and foreign relations. *Foreign Policy Research Institute.* Available at: https://www.fpri.org/article/2021/04/palm-oil-malaysian-economic-interests-and-foreign-relations/ [Accessed March 30, 2022].

[3] Umar M S, Jennings P J and Urmee T 2013 *Strengthening the palm oil biomass Renewable Energy industry in Malaysia.* Available at: https://www.researchgate.net/publication/236946899_Strengthening_the_palm_oil_biomass_Renewable_Energy_industry_in_Malaysia [Accessed March 30, 2022].

[4] Yeo J Y et al 2020 Synthesis of sustainable circular economy in palm oil industry using graph-theoretic method. *Sustainability*, 12(19), p.8081.

[5] Ng R T L et al 2014 Multi-objective design of industrial symbiosis in palm oil industry. *Proceedings of the 8th International Conference on Foundations of Computer-Aided Process Design*, pp.579–584.

[6] Tay D H et al 2011 Fuzzy optimisation approach for the synthesis of a sustainable integrated biorefinery. *Industrial & Engineering Chemistry Research*, 50(3), pp.1652–1665.

[7] Webgate 2021 *Industrial Symbiosis.* [online] Webgate.ec.europa.eu. Available at: <https://webgate.ec.europa.eu/greencitytool/resources/docs/best_practices/Industrial_Symbiosis_A01.pdf> [Accessed 7 October 2021].

[8] TNB 2020 Climate and energy benchmark - electric utilities. World Benchmarking Alliance. Available at: https://www.worldbenchmarkingalliance.org/publication/electric-utilities/companies/tenaga-nasional/#:~:text=Its%20emissions%20intensity%20decreased%20slightly,new%20coal%20and%20gas%20assets. [Accessed March 31, 2022].

Molecular Dynamics Study of CO_2/H_2 and CH_4/H_2 Gas Separation by Nanoporous Graphene

R Rivero[1], **Y Ren**[1,2,3,4*]

[1]Department of Mechanical, Materials and Manufacturing Engineering, University of Nottingham Ningbo China, Ningbo, China
[2]Research Group for Fluids and Thermal Engineering, University of Nottingham Ningbo China, Ningbo, China
[3]Key Laboratory of Carbonaceous Wastes Processing and Process Intensification Research of Zhejiang Province, University of Nottingham Ningbo China, Ningbo, China
[4]Nottingham Ningbo China Beacons of Excellence Research and Innovation Institute, Ningbo, China
*Email: yong.ren@nottingham.edu.cn

Abstract. The combustion of coal produces CO_2 emissions, rising greenhouse gas levels, and leading to global warming. Hydrogen is one potential clean source of energy. However, current hydrogen production methods produce CO and CO_2 emissions, while pure hydrogen is required for fuel cells. Therefore, efficient methods for capturing greenhouse gases and the extraction of pure hydrogen are required. Membrane gas separation requires low energy and cost. However, current polymer-based membranes struggle with the trade-off between selectivity and permeability. Graphene-based membranes like nanoporous graphene (NPG) are potential candidates due to its monoatomic thickness and high mechanical performance, offering high permeability. Selectivity was altered by controlling the nanopore shape, size, and functionality, and using multi-layered NPG with various interlayer spacing and angle. Molecular dynamics simulations were performed to study the transport phenomena for graphene-based membrane gas separation. CO_2/H_2 and CH_4/H_2 gas mixtures separation were performed on a bottom-up synthesized H-Passivated NPG with Dumbbell-Shaped Nanopores. Results show that 100% H_2 selectivity was achieved for both CH_4/H_2 and CO_2/H_2 gas mixtures, using single layer NPG and tri-layer NPG, respectively, while maintaining high permeability in the order of 10^5-10^6 GPU. Therefore, the H-Passivated NPG with Dumbbell-Shaped Nanopores is an excellent choice of membrane for gas separation.

1. Introduction

It is inevitable to avoid the increasing demand of energy due to the rising population [1]. Coal remains as the main source of energy due to its low cost and high availability [2]. The combustion of coal produces CO_2 emissions – rising levels of greenhouse gases and leading to global warming [1, 2]. The search of renewable sources of energy are also important to minimize the dependence on fossil fuels. Hydrogen has been potentially considered as a clean source of energy due to its high energy density [3] and environmental friendliness. However, current hydrogen generation methods such as methane steam reformation produces CO and CO_2 emissions, while ammonia decomposition produces nitrogen emissions [4]. Alternatively, the electrolysis of water consumes electricity. This means electrolysis is dependent on the cleanliness of electricity production [5]. Therefore, a solution to efficiently remove greenhouse gas emissions, including clean production or extraction of hydrogen is required.

Industrial gas separation methods include cryogenic separation, membrane separation, supersonic separation, and pressure adsorption, but the preferred method is membrane separation due to its low energy requirements, low pollution, and its simplicity [6, 7]. Current well-established polymer-based membranes and the relatively newer materials like metal-organic frameworks struggle with the trade-off between permeability and selectivity [8]. However, graphene-based membranes provide a solution to this trade-off. Graphene is a novel material well known for its single-atomic thickness, high mechanical strength, thermal stability, and chemical inertness. These properties indicate that graphene is an ideal candidate as it satisfies the criteria for effective membrane gas separation [1, 2, 9, 10]. Since graphene is only a single-atom thick, it offers the shortest diffusion paths for gas molecules, offering high permeability. Its highly ordered pore arrays crystal structure also offers high selectivity. However, gas molecules cannot permeate through pristine graphene sheet due to the repulsion force exerted by the aromatic rings [10, 11]. For gas separation, graphene-based derivatives are used instead, such as nanoporous graphene (NPG), including the functionalization of the nanopore with hydrogen, nitrogen, or boron [6, 12-14], or adding oxygen-containing functional groups to form graphene oxide [15].

In 2012, Koenig et al. [12] were the first group to conduct an experimental study on NPG gas separation, they fabricated nanopores in micrometre-sized graphene membranes using oxidative etching induced by ultraviolet light. Liu et al. [13] performed MD simulations to determine the flux and selectivity of various gases (CO_2, H_2, He, CH_4, O_2, and N_2). Their results summarised the trend of the gas flux (permeability) through the 4N4H NPG in the following order, H_2 (easiest), CO_2, Ar, N_2, and CH_4 (hardest). These results are similar to the experimental results provided by Koenig et al. [12].

In 2015, Sun et al. [6] performed MD simulations for NPG gas separation of CH_4/CO_2, CH_4/H_2S, and CH_4/N_2. Their results indicate that CH_4 is impermeable, while all other gases are permeable, where CO_2 is the most permeable and N_2 is the least permeable. Hence, 100% selectivity for CO_2, H_2S, and N_2 were achieved and that the permeability depends on the kinetic diameter of the gas molecules, where CO_2 is 3.3Å (the smallest) and N_2 is 3.64Å (the largest), which also supports the results from Liu et al. [13]. The permeability of CO_2, H_2S, and N_2 were in the order of 10^5-10^6 GPU. In comparison, polymer-based membranes for CO_2 separation only can achieve permeability in the order of 10^2-10^3 GPU [16].

Tian et al. [14] in 2021, performed Density Functional Theory (DFT) and MD simulations for NPG hydrogen separation. The permeability and selectivity of binary gas mixtures of H_2/O_2, H_2/CO_2, H_2/CO, and H_2/CH_4 were studied. H-Passivated Dumbbell-Shaped pores (8.06×3.11Å^2) were used in their setup, which was based on a bottom-up synthesised NPG made by Moreno et al. [17]. Solvik et al. [18] expected ethane to be more permeable than ethane, however the simulation result shows the opposite. Hence, it was concluded that the lower entropic barrier of ethene made ethene more permeable than ethane. In the present paper, MD simulation has been conducted to investigate CO_2/H_2 and CH_4/H_2 gas separation by NPG. While single layer NPG gas separation has been explored by other researchers, this paper explored the use of tri-layer NPG with various interlayer spacing and angle to improve selectivity.

2. Simulation Model

To study the permeability and selectivity of binary gas mixtures (CH_4/H_2 and CO_2/H_2) separation on NPG, MD simulations were performed in Large-scale Atomic/Molecular Massively Parallel Simulator

(LAMMPS) software. For all cases, the size of the simulation box is 3.48×3.01×12 nm^3 in the x-, y-, and z-direction respectively. NPG with six H-Passivated Dumbbell-Shaped pores (8.03×3.05$Å^2$) based on a bottom-up synthesis by Moreno et al. [17] was used. The simulation system and NPG are illustrated in **Figure 1** with the CH_4/H_2 case as an example. Periodic boundary conditions were applied to the x- and y-direction, and reflective walls were applied in the z-direction. The NPG was placed in the midpoint of the z-direction, dividing the system into two chambers with equal volume, the feed chamber and the permeate chamber. The feed chamber initially starts with 30 gas molecules of type A and 30 gas molecules of type B. This setup provides a pressure difference of 1.96 MPa (19.38 atm) between the feed chamber and permeate chamber. To prevent the NPG from translating during simulation, each carbon atom at the corners of the NPG is fixed. Only the corners of the carbon atom were fixed to allow the NPG to vibrate during simulation based on the molecular interactions with other gas molecules.

For every case, the simulation ran for 40×10^6 timesteps, where each timestep is 0.5 fs, resulting in a total simulation time of 20 ns. The NVT (canonical) ensemble was applied at a fixed temperature of 298 K using a Nosé-Hoover thermostat. The NPG, CH_4, and H_2 molecules were characterised by the AIREBO potential as it can accurately describe the C-C and C-H bonding interactions. CO_2 was characterised by the LJ potential and the harmonic potential to accommodate bond stretching and bond angle deformation. The interaction between the NPG and gas molecules were described by the LJ and coulombic potential with a cut-off distance of 12Å to include short-range electrostatic interactions and van der Waals forces. The potential parameters used are presented in **Table 1**, and the potentials between unlike atoms were calculated with Lorentz-Berthelot mixing rule.

Specifically for the CO_2/H_2 case, additional scenarios were performed (with the expectation of improving selectivity), such as using tri-layer NPG (two additional layers of NPG are placed in parallel) with an interlayer spacing of 6Å and 12Å, and an inclination of 0°, 5° and 10° for the second NPG layer.

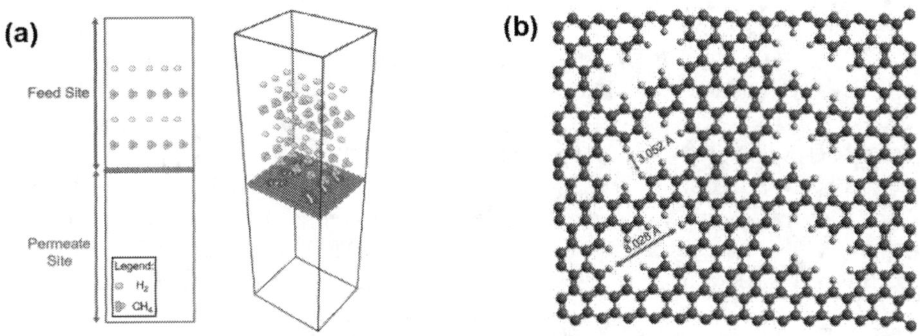

Figure 1. (a) The simulation system and (b) H-Passivated NPG (black atoms are C, grey atoms are H)

Table 1. Potential parameters for CO_2, adapted from Sun et al. [6] and Tian et al. [14]

	ε (eV)	σ (Å)	Charge (e)
C_{CO2}-C_{CO2}	2.424×10^{-3}	2.757	0.6512
O_{CO2}-O_{CO2}	6.938×10^{-3}	3.033	-0.3256
C_{CO2}-O_{CO2}	4.101×10^{-3}	2.895	/
	K_{bond} (eV/$Å^2$)		r_0 (Å)
C-O (CO_2)	6.158		1.160
	K_{angle} (eV/rad^2)		θ_0 (degree)
O-C-O (CO_2)	6.416		180

3. Results and discussion

In general, for all cases, the relationship between the number of permeated molecules against time was non-linear, one factor is because the pressure difference across the feed chamber and permeate chamber decreases with time as the gas molecules permeate to the permeate chamber from the feed chamber. For all simulation cases, the permeability of the permeable gas molecules lies within the order of 10^5-10^6 GPU, hence agrees with the MD simulation results from Sun et al. [6] and Tian et al. [14].

For CH_4/H_2 on single layer NPG, only H_2 could permeate though the NPG (100% H_2 selectivity), which was achieved by molecular size exclusions, and agrees with the MD simulation results from Liu et al. [13] and Tian et al. [14]. The permeability of H_2 is 1.55×10^6 GPU. Since CH_4 has a tetrahedral shape with a kinetic diameter of 3.8Å, it cannot pass through the $3.05\times8.03Å^2$ NPG regardless of its orientation. While H_2 can pass through the NPG at any orientation as its kinetic diameter is only 2.89Å.

For CO_2/H_2 on single layer NPG, both CO_2 and H_2 were able to permeate through the NPG, resulting in 25% CO_2 and 75% H_2 selectivity, with the permeability of 4.30×10^5 GPU for CO_2 and 1.29×10^6 GPU for H_2. Similar to the MD simulation results from Liu et al. [13], a single layer of NPG could not achieve 100% separation for CO_2/H_2 gas mixture. It is challenging to achieve 100% selectivity for the CO_2/H_2 gas mixture as the kinetic diameter of CO_2 and H_2 are 3.3Å and 2.89Å respectively, which are relatively close to each other. Since the shape of CO_2 is linear, despite its kinetic diameter is larger than the shorter side of the nanopore (3.05Å), it is still smaller than the longer side of the nanopore (8.03Å), hence it still can permeate through the $3.05\times8.03Å^2$ NPG but only in the perpendicular orientation to the NPG. Hence the selectivity of CO_2 was lower than H_2. The limited orientation for CO_2 penetration supports the idea of the entropy-driven molecular separation by Solvik et al. [18]. Similar to their ethane/ethene gas separation case, where ethene is more permeable than ethane due to ethene having a lower entropic barrier, H_2 in the CO_2/H_2 gas mixture also has a lower entropic barrier due to having more accessible microstates as it can readily permeate through the nanopore at any possible orientation.

As it was not possible to achieve 100% H_2 selectivity with a single layer NPG, variations to the CO_2/H_2 case were performed, such as using tri-layer NPG, with interlayer spacings of 6Å and 12Å, and the investigation of inclining the second NPG layer. Firstly, consider the tri-layer NPG case with 0° slope. For the case of the 12Å interlayer spacing, 7 CO_2 molecules were able to permeate through the first NPG layer and remain trapped between the first and second NPG layer. This was expected as it had been shown in the CO_2/H_2 single layer NPG case that CO_2 was also permeable. Meanwhile, H_2 was able to permeate through all three layers of NPG. 5 H_2 molecules permeated through all three layers of NPG. This indicates 100% H_2 selectivity, with the permeability of 4.30×10^5 GPU for H_2.

Meanwhile for 0° slope and 6Å interlayer spacing, CO_2 molecules could not permeate through the tri-layer NPG, as there was insufficient space for CO_2 to remain between the interlayers due to the tight interlayer spacing. H_2 permeated through all three layers of NPG as if it only had one layer of NPG. This resulted in 100% H_2 selectivity, with the permeability of 8.59×10^5 GPU for H_2. Interestingly, using tri-layer NPG with either 6Å or 12Å interlayer spacing, 100% H_2 selectivity could be achieved, with a slight reduction in permeability. DFT results from Tian et al. [14] show that CO_2 has a high energy barrier above 1.5 eV, while H_2 has a low energy barrier of 0.24 eV. Hence, CO_2 had difficulties to permeate through the tri-layer NPG. Since the tri-layer NPG with 12Å interlayer spacing entraps the CO_2 molecules between the first and second NPG layer, in a larger scale simulation, CO_2 might block the passage for H_2, hampering the selectivity. Therefore, 6Å interlayer spacing is more favourable.

With 5° slope, for both 12Å and 6Å interlayer, CO_2 molecules could permeate through the first NPG layer but gets trapped between the first and second NPG layer, while H_2 was able to permeate through all three layers of NPG, resulting in 100% H_2 selectivity for both cases, with an increase in permeability compared to the uninclined case, where H_2 permeability is 6.01×10^5 GPU and 1.12×10^6 GPU for 12Å and 6Å interlayer spacing respectively. CO_2 being trapped between the first and second NPG layer was expected, as it was shown in previous cases that CO_2 could permeate through the first NPG layer and remain trapped between the first and second NPG layer if there is enough space for CO_2 to stay within the interlayers. Hence, 5° slope for 6Å interlayer spacing was spacious enough for CO_2 to stay.

Finally, consider the tri-layer NPG case with 10° slope. For 12Å interlayer spacing, CO_2 could permeate through the first and second NPG, where several CO_2 molecules remain between the first and second, as well as the second and third NPG layer, while H_2 could permeate through all three layers of NPG. This resulted in 100% H_2 selectivity with an increase in H_2 permeability to 7.73×10^5 GPU.

For 10° slope and 6Å interlayer spacing, both CO_2 and H_2 could permeate through all three layers of NPG. Where CO_2 molecules occupied the first and second NPG layer as well as the second and third NPG layer. This resulted in 13.3% CO_2 selectivity and 86.7% H_2 selectivity, with the permeability of 1.72×10^5 GPU for CO_2 and 1.12×10^6 GPU for H_2. Due to the 10° slope, the size of the nanopore slightly increases, making CO_2 easier to permeate through. Therefore, CO_2 fills up the interlayer spaces between the first and second NPG layer as well as the second and third NPG layer until there is no more room for CO_2 to stay. Hence, for 6Å interlayer spacing case, CO_2 molecules could permeate through the third NPG layer, as 6Å interlayers were much more occupied that the 12Å interlayers.

The reason why CO_2 could occupy the interlayer spaces but not H_2 can be supported from DFT simulation results from Tian et al. [14], where the adsorption energy of CO_2 and H_2 towards the NPG is 0.151 eV and 0.041 eV respectively. Therefore, CO_2 molecules are much more likely to be adsorbed by the NPG and remain within the NPG. The weak adsorption energy of H_2 allows H_2 molecules to permeate through the NPG due to the pressure difference. Therefore, inclining the second NPG layer for both the 6Å or 12Å interlayers, resulted in increased permeability for both CO_2 and H_2. However, since CO_2 could occupy the interlayer spaces and the nanopore was stretched (easier for CO_2 to permeate), 100% H_2 selectivity might not be achieved, as shown in the 10° slope 6Å interlayer case.

Hence, inclining the second NPG layer is not suitable for the CO_2/H_2 case. The tri-layer NPG with 6Å interlayer spacing at 0° slope remains the best and guaranteed setup for CO_2/H_2 gas separation as it could achieve 100% H_2 selectivity while maintaining very high permeability of 8.59×10^5 GPU.

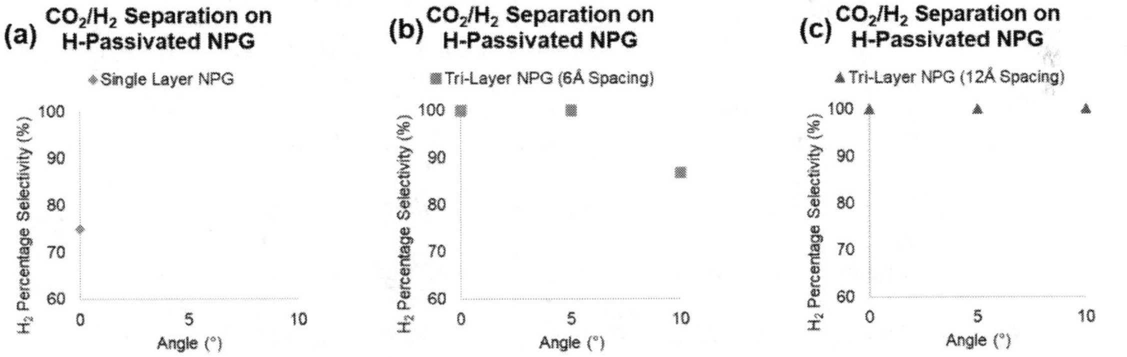

Figure 2. Summary of H_2 selectivity for CO_2/H_2 separation on (a) single layer NPG, (b) tri-layer NPG (6Å spacing), and (c) tri-layer NPG (12Å spacing)

4. Conclusion

MD simulation results demonstrated that at 298 K with 1.96 MPa (19.38 atm) pressure difference (a practical condition in the laboratory or industry), the H-Passivated NPG with Dumbbell-Shaped Nanopores ($8.03\times3.05Å^2$ pore), which can be fabricated through bottom-up synthesis, could achieve 100% H_2 selectivity with high permeability in the order of 10^5-10^6 GPU (significantly higher than polymer-based membranes) for both CH_4/H_2 and CO_2/H_2 gas mixture separation.

For CH_4/H_2, single layer NPG was sufficient to achieve 100% H_2 separation simply due to molecular size exclusions. However, the tri-layer NPG was required for CO_2/H_2 to achieve 100% H_2 separation due to CO_2 and H_2 having similar sizes. For the tri-layer NPG with 6Å interlayer spacing, there was insufficient space for CO_2 to remain between the interlayers, hence the high energy barrier of CO_2 made CO_2 impermeable. For the 12Å interlayer spacing, the high adsorption energy of CO_2 entraps CO_2 between the interlayers. Inclining the second NPG layer was not favourable for CO_2/H_2 as this increases the interlayer spaces and the stretched pore eases the permeation of CO_2. Hence the tri-layer NPG with 6Å interlayer spacing at 0° slope would be the recommended choice.

Therefore, the ideal parameters to achieve 100% selectivity while maintaining high permeability for CH_4/H_2 and CO_2/H_2 gas mixture separation were successfully determined and the H-Passivated NPG with Dumbbell-Shaped Nanopores is an excellent choice of membrane for gas separation.

Future work would include DFT simulations to supplement the MD simulation results. Simulation results would be verified through experimental gas separation by fabricating the H-Passivated NPG with Dumbbell-Shaped Nanopores through bottom-up synthesis suggested by Moreno et al. [17]. Once results are verified, collaboration with industries would be desired to mass produce. Further case would be to explore separation of NOx and SOx.

References

[1] Ali A., Pothu R., Siyal S. H., Phulpoto S., Sajjad M., and Thebo K. H., 2019, Graphene-based membranes for CO2 separation, *Materials Science for Energy Technologies,* vol. 2, no. 1, pp. 83-88

[2] Balasubramanian R. and Chowdhury S., 2015, Recent advances and progress in the development of graphene-based adsorbents for CO2 capture, *Journal of materials chemistry. A, Materials for energy and sustainability,* vol. 3, no. 44, pp. 21968-21989

[3] Wang J., Cui W., Liu Q., Xing Z., Asiri A. M., and Sun X., 2016, Recent progress in cobalt-based heterogeneous catalysts for electrochemical water splitting, *Adv Mater,* vol. 28, no. 2, pp. 215-30

[4] Lang Y., Arnepalli R. R., and Tiwari A., 2011, A review on hydrogen production: methods, materials and nanotechnology, *J Nanosci Nanotechnol,* vol. 11, no. 5, pp. 3719-39

[5] Kothari R., Singh D. P., Tyagi V. V., and Tyagi S. K., 2012, Fermentative hydrogen production – an alternative clean energy source, *Renewable & sustainable energy reviews,* vol. 16, no. 4, pp. 2337-2346

[6] Sun C., Wen B., and Bai B., 2015, Application of nanoporous graphene membranes in natural gas processing: molecular simulations of CH4/CO2, CH4/H2S and CH4/N2 separation, *Chemical engineering science,* vol. 138, pp. 616-621

[7] Freemantle M., 2005, Membranes for gas separation, *Chemical & Engineering News Archive,* vol. 83, no. 40, pp. 49-57

[8] Robeson L. M., 2008, The upper bound revisited, *Journal of Membrane Science,* vol. 320, no. 1, pp. 390-400

[9] Fatemi S. M., Fatemi J., and Abbasi Z., 2020, Gas separation using graphene nanosheet: insights from theory and simulation, *Journal of Molecular Modeling,* vol. 26, p. 322

[10] Jiang D.-e., Cooper V., and Dai S., 2009, Porous graphene as the ultimate membrane for gas separation, *Nano Letters,* vol. 9, pp. 4019-24

[11] Bunch J. S. *et al.*, 2008, Impermeable atomic membranes from graphene sheets, *Nano Letters,* vol. 8, no. 8, pp. 2458-2462

[12] Koenig S. P., Wang L., Pellegrino J., and Bunch J. S., 2012, Selective molecular sieving through porous graphene, *Nat Nanotechnol,* vol. 7, no. 11, pp. 728-32

[13] Liu H., Chen Z., Dai S., and Jiang D.-e., 2015, Selectivity trend of gas separation through nanoporous graphene, *Journal of Solid State Chemistry,* vol. 224, pp. 2-6

[14] Tian L., Duan H., Luo J., Cheng Y., and Shi L., 2021, Density functional theory and molecular dynamics simulations of nanoporous graphene membranes for hydrogen separation, *ACS applied nano materials,* vol. 4, no. 9, pp. 9440-9448

[15] Kim Hyo W. *et al.*, 2013, Selective gas transport through few-layered graphene and graphene oxide membranes, *Science,* vol. 342, no. 6154, pp. 91-95

[16] Wang S. *et al.*, 2016, Advances in high permeability polymer-based membrane materials for CO2 separations, *Energy & environmental science,* vol. 9, no. 6, pp. 1863-189

[17] Moreno C. *et al.*, 2018, Bottom up synthesis of multifunctional nanoporous graphene, *Science,* vol. 360, no. 6385, pp. 199-203

[18] Solvik K., Weaver J. A., Brockway A. M., and Schrier J., 2013, Entropy-driven molecular separations in 2D-nanoporous materials, with application to high-performance paraffin/olefin membrane separations, *The Journal of Physical Chemistry C,* vol. 117, no. 33, pp. 17050-17057

Terephthalic Acid (TPA) Recovery from Plastic Bottle of Polyethylene Terephthalate (PET) through Hydrolysis Treatment

S Alias[1*] and N A Abdul-Hakim[1]

[1]Centre for Civil Engineering Studies, Universiti Teknologi MARA, Cawangan Pulau Pinang, Permatang Pauh Campus, 13500 Seberang Prai, Pulau Pinang, Malaysia.

[*]Corresponding author: salina346@uitm.edu.my

Abstract. Recovery of plastic waste has a positive influence on solid waste management as it can bring new life to plastic waste. Hydrolysis treatment is one of the approaches for plastic waste recovery, in which plastic monomer can be recovered through depolymerisation. This study explored the recovery of polyethylene terephthalate (PET) plastic in nitric acid and leachate. The weight changes were analysed in PET when exposed to different hydrolysis agents (nitric acid and leachate) at different reaction times and temperatures. The monomer recovery of PET, known as terephthalic acid (TPA), was also evaluated. It was found that high temperatures and reaction times promoted to the high recovery of TPA. The PET residue measured as weight difference was found in line with the trend of TPA yield. The hydrolysis using nitric acid can recover more TPA than the leachate system. The knowledge obtained from this study would be beneficial in upcycling plastic waste into high-value end products.

Keywords: Hydrolysis; Plastic Waste; Polyethylene Terephthalate (PET); Terephthalic Acid (TPA); Waste Recovery

1. Introduction

The COVID-19 pandemic has resulted in an exponential increase in plastic waste globally from e-commerce plastic packaging, food packaging, and the rise in personal protective equipment [1,2]. Because plastics are claimed to be no better than other materials in terms of durability, convenience, and affordability, their greatest advantages become the primary issue in reducing plastic waste. As a result, the environmental threat issue arises in tandem with the health crisis. As a result, increasing plastic waste generation necessitates the implementation of an efficient and long-term waste management system.

PET stands for polyethylene terephthalate and is used to make plastic bottles, polyester fibres, food packaging, containers, and films. Its versatility has increased because of its lightweight qualities, great chemical stability, and outstanding barrier properties. Thus, PET consumption has gradually increased every year, owing to continued economic growth [3]. Due to the high demand for PET plastic items and their short life cycles as single-use items, there is a significant amount of discarded PET plastic in the solid waste stream. If not properly treated, this will harm the environment and waste resources [4]. Recycling waste PET plastics could be seen as a viable option for extending their life cycle, reducing nonrenewable resource use, preventing plastic pollution, and protecting the environment in this light [5].

Mechanical recycling is the most common process for waste PET recycling, which involves grinding and melting PET garbage to obtain recycled PET particles [6,7]. However, the necessity that waste plastics be cleaned of contaminants before being processed limits this method. In comparison to mechanical approaches, chemical recycling, biochemical enzymatic recycling, or molecular recycling technologies can achieve permanent recycling of waste PET through the recovery of waste monomer [8]. This is more in line with the present requirement for sustainable development and provides a number of benefits and opportunities [9]. Depolymerization of PET through hydrolysis, the most basic method for PET recovery, yields terephthalic acid (TPA) and ethylene glycol (EG). The basic components of PET polymerization are these monomers.

The cheapest method of PET recovery is natural hydrolysis with water or steam as a reagent. However, the recovered product is less pure, necessitating further depolymerization and purification [6]. As a result, the entire procedure takes longer and costs more money [10]. Nitric acid and sodium hydroxide are often studied as acid and base catalysts that enhance PET hydrolysis with a high yield of product recovery. However, the approach still requires high temperature (220 to 400 °C), pressure and significant range of time (1 min – 3 hours). Furthermore, the reactor setup and handling are considerably complicated and costly [11,12]. Therefore, the limitation could be overcome by using low-temperature conditions (80°C and 150°C), a simple closed reactor and different reagent solutions, namely, leachate and nitric acid.

2. Materials and methods

2.1. *Sample Preparation*

PET bottles were found in a garbage near Universiti Teknologi MARA in Cawangan Pulau Pinang. Only a clear PET bottle was collected to maintain material homogeneity. Prior to use, the PET bottles were rinsed in distilled water and air-dried. PET bottles were cut into small, equal-sized pieces of 1cm by 1cm with scissors, as shown in figure 1. To ensure that the hydrolysis reaction occurs on the same surface area for all samples, the area to be cut must be a flat region. The PET bottles were then rinsed in deionized water and dried at room temperature once again.

2.2. *Hydrolysis Treatment*

Different hydrolysis agent, namely nitric acid and leachate were used in the hydrolysis process. The sliced PET plastic was weighed at 5 g before being immersed in 7 mL 4 M nitric acid and 7 mL leachate in a test tube. After that, the test tube was capped and placed in a reactor (COD reactor) at 80°C and 150°C, respectively as adapted from [13]. The treatment was let react in 1, 3, and 7 days. Then, the reacted solution was filtered by using filter paper to separate the precipitated impurities and unreacted PET waste. The procedures were repeated three times and the results reported as an average value. Standard deviations were shown in each result. Figure 2 shows the process flow of the PET hydrolysis treatment.

2.3. *PET Weight Different*

After the washing and drying operation, the filtered residue was weighed. Equation 1 was used to calculate the weight difference between the final PET bottles and reacted PET as adapted from [12].

$$PET\ Weight\ Different = \{\frac{(W_{(PET,0)} - W_{PET,f})}{W_{PET,0}}\} \times 100 \quad (1)$$

Where $W_{PET,0}$ is the initial weight of PET (g); and $W_{PET,f}$ is the residual weight of PET (g).

2.4. *TPA Yield*

The filtered solution was being acidified with 1.5 M of hydrochloric acid to obtain TPA precipitate. It was then well-washed with distilled water, dried and weight. Next, the yield of solid TPA and non-degraded PET will be calculated by Equation 2 as adapted from [12,14].

$$\text{TPA yield} = \frac{W_{TPA}}{W_{TPA\,theory}}\, x\ 100 \quad (2)$$

Where W_{TPA} is the weight of TPA obtained from hydrolysis (g) after the acid washed while $W_{TPA\,theory}$ is theoretical value of TPA from a complete PET depolymerization to TPA. The maximum value for TPA is 86.4% as a result from the different molecular weight ratio between TPA and PET repeating unit (166 g/mol and 192 g/mol) respectively.

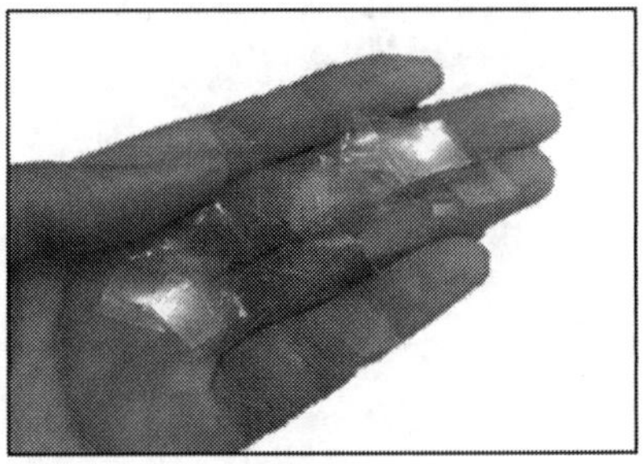

Figure 1. The PET bottle sample

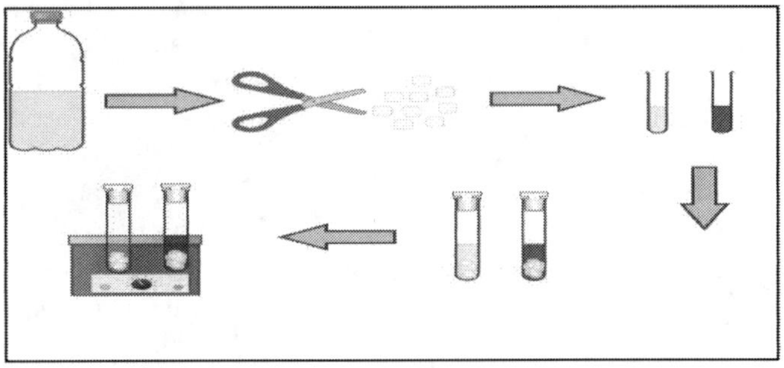

Figure 2. Process Flow of PET Hydrolysis Treatment

3. Results and Discussion

3.1. Effect of Temperature in Weight Different and TPA Yield of PET Plastic

Figure 3 shows the effect of temperature, 80 and 150° C, on weight different (%) of PET plastic in 4 M of nitric acid and leachate solution hydrolysis treatment at day 7. It was observed that by increasing the reactor temperature from 80 to 150°C, the PET weight different (%) that represent the percentage different in the weight of PET before and after hydrolysis process for both solutions increase linearly. The weight different at temperature of 80°C and 150°C were 67% and 77% in nitric acid solution respectively. Almost the same PET weight different is observed in hydrolysis treatment using leachate in which 62% and 73% are observed at temperature of 80°C and 150°C. These indicated that more PET has broken down or depolymerised into its monomer, terephthalic acid (TPA), ethylene glycol or other by-products when subjected to high temperature. As a result, just a minimal amount of unreacted PET will remain in the hydrolysis solution.

Figure 4 depicts the slow dissolving of PET plastic in nitric acid, resulting in a yellow solution with insoluble unreacted PET remaining. The insoluble result, referred to as "residue," appeared on the solution as a yellowish-brown crystal-like substance. The residue was filtered, and the results were expressed as a difference in percentage weight PET. On the surface of PET, cracks and holes can be visible, which is backed up by the weight loss of PET plastic film. A further increase in the reaction was predicted to breakdown PET and make it completely disappear in the solution [13].

It was also observed that the trend of residual PET weight different is also consistent with the trend of TPA yield (%) as shown in figure 5. Similar to the trend of PET weight different, the temperature of 150°C gave the higher value of TPA yield as compared to temperature of 80°C. At day 7, more TPA were observed which 19% yield at 150°C while only 8% at temperature of 80°C in nitric acid solution. The TPA yield also found slightly higher in leachate when subjected to temperature of 150°C as compared to 80°C.

The TPA yield in this study is far behind the results obtained in previous finding. Cˇolnik *et al.* [12] found that nearly 50% and 90% of TPA can be recovered from hydrolysis treatment of PET using sub-critical water of 250°C within 1 and 30 minutes reaction time, respectively. In another study, Yang *et al.* [14] observed that the TPA produced at least 3% and more than 90% at temperatures of 180°C and 210°C, respectively, when subjected to PET hydrolysis using water at 3 hours reaction time and TPA as co-catalyst. The difference could be due the effect of pressure and type of reactor that were used in hydrolysis treatment. Theoretically, at temperature 150°C, the pressure is around 5 bars in which it become double when the pressure reaches 180°C. Meanwhile, at temperature of 250°C the vapor pressure reaches 50 bars. High pressure induces activation of reaction in which the pressure could control products' composition [12,15].

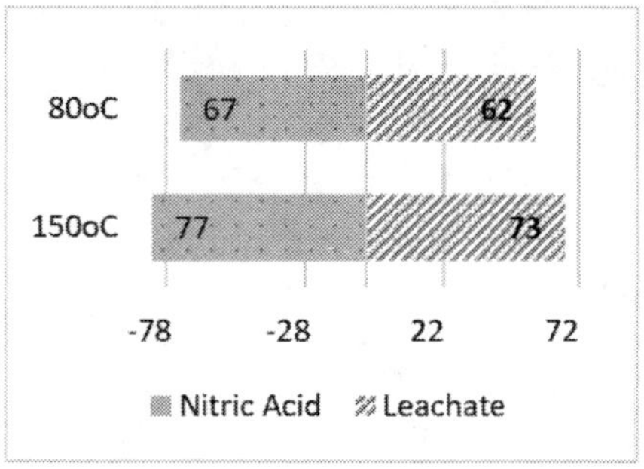

Figure 3. PET Weight Different (%) at Different Temperature

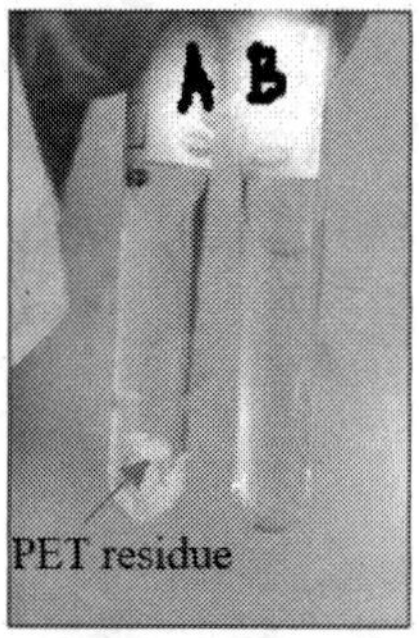

Figure 4. (A) PET residue in acid nitric (B) Nitric acid prior to the PET hydrolysis

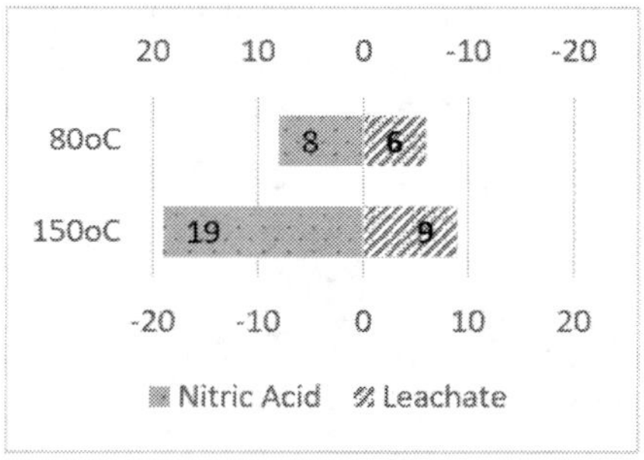

Figure 5. TPA Yield (%) at Different Temperature

Furthermore, the findings show that the type of reactant is an obvious element in the successful recovery of monomers from PET via hydrolysis treatment. The two-fold increase in TPA yield in nitric acid solution compared to leachate at 150°C indicates that other factors other than temperature are important for the effectiveness of TPA recovery from PET hydrolysis. As an acid hydrolysate, nitric acid acts as a catalyst to enhance the breakage of polyester chains, yielding two macromolecules: terephthalic acid (TPA) and ethylene glycol (EG) [10]. It was found in Yoshioka *et al.* [16] study, about 91.3% of PET was degraded by using 13 M nitric acid, reaction temperature of 100°C and 24 hours reaction time. In comparison to the current study, only 67-77% of PET was degraded when nitric acid of 4 M was used at temperature of 80°C and 150°C. It may be deduced from this that the varying concentrations of nitric acid employed during hydrolysis treatment are an essential element.

3.2. Effect of Reaction Time in Weight Different and TPA Yield of PET Plastic

The effect of reaction time at Day 1, 3, and 7 on PET weight different (%) in 4 M of nitric acid and leachate solution for the hydrolysis process of PET plastic are shown in figure 6. The rate of PET weight different (%) for both solutions varied when the reaction time changed from 1 to 3 to 7 days. Generally, more than half of PET (≥ 50%) was degraded in all hydrolysis solution after day 1. Then the changes in PET weigh increase linearly as time which only left PET at around 40% in all experimental condition except not more than 30% in nitric acid at applied temperature of 150°C. Yet, study discovered that the value of the PET weight difference increases virtually for both options. For both solutions, the effect of reaction time on the weight difference of PET plastic was higher for temperature of day 7 compared to day 1 and 3 for both solutions with linear increment.

The effect of reaction time in hydrolysis treatment using nitric acid and leachate as catalyst at temperature of 80◦C and 150◦C is shown in figure 7. It is obviously can be seen that the TPA yield increase linearly in all experimental conditions except in nitric acid 150◦C. In general, the same pattern can be seen for TPA yield (%) of PET plastic, where the reaction time of day 7 generated a greater value of TPA yield percentage for both solutions than day 1 and 3. By referring to the maximum rate of the PET weight difference and TPA yield percentage obtained from the experiment, it can be determined that the reaction day of 7 was the most significant duration for the breakdown of PET plastic employing the hydrolysis approach.

A similar trend was also found by Yang *et al.* [14] where the rate of PET degradation and TPA yield produced under the optimized conditions of 220◦C of reaction temperature, and 180 min of reaction time, was entirely degraded, and the produced TPA yield reached 95.5% respectively. The comparison with this previous study of hydrolysis process with the recent study was it was using concentrated acid and requires high reaction temperature and shorter reaction time. The requirement in terms of acid concentration and reaction temperature can be implemented in the future study for achieving better hydrolysis performance.

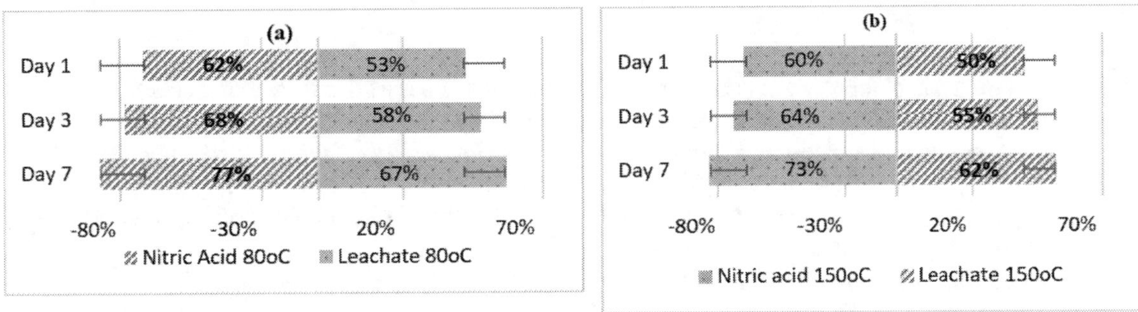

Figure 6. Changes in PET Weight (%) at Different Reaction Time in (a) Nitric Acid and leachate at 80°C (b) Nitric Acid and leachate at 150°C

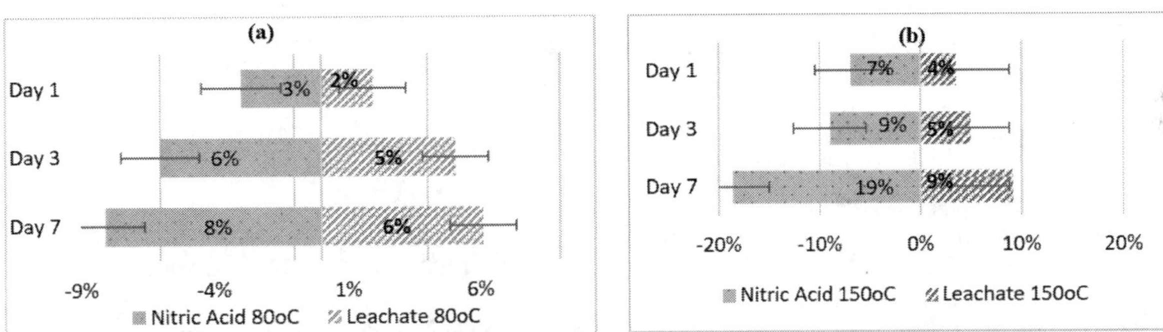

Figure 7. TPA Yield (%) at Different Reaction Time at (a) Nitric acid and Leachate, temperature 80°C and (b) Nitric acid and Leachate, temperature 150°C

4. Conclusion

Hydrolysis is promising technology in monomer recovering from plastic waste through breakdown the plastic waste in depolymerization process. The current studies show temperature of 150°C can recover TPA, a monomer of PET, better than the 80°C. Furthermore, the efficiency of TPA recovery varies depending on the type of hydrolysis solution or catalyst used. Nitric acid showed a better recovery of TPA over the leachate. However, leachate also plays significant advantages and effectiveness in monomer recovery from PET. High temperature, longer reaction time and type of reactant plays important product recovery percentages. However, further research is highly recommended to increase the product recovery through the incorporation of different plastic loading, pH and enzyme. Besides, the recovery performance using different types of acids are also recommended to be explored.

Acknowledgments
The authors thank to Unversiti Teknologi MARA, Cawangan Pulau Pinang on the facilities, financial support and staff that are involved in conducting this research.

References

[1] Patrício S A L, Prata J C, Walker T R, Duarte A C, Ouyang W, Barcelò D and Rocha-Santos T 2021 Increased plastic pollution due to COVID-19 pandemic: Challenges and recommendations *Chemical Engineering Journal* **405** Article 126683

[2] Olatayo K I, Mativenga P T and Marnewick A L 2021 COVID-19 PPE plastic material flows and waste management: Quantification and implications for South Africa *Science of the Total Environment* **790** Article 148190

[3] Furukawa M, Kawakami N, Oda K and Miyamoto K 2018 Acceleration of enzymatic degradation of poly(ethylene terephthalate) by surface coating with anionic surfactants *ChemSusChem* **11(23)** 4018–4025

[4] Gibb B C 2019 Plastics are forever *Nature Chemistry* **11(5)** 394–395

[5] Raheem A B, Noor Z Z, Hassan A, Abd Hamid M K, Samsudin S A and Sabeen A H 2019 Current developments in chemical recycling of post-consumer polyethylene terephthalate wastes for new materials production: A review *Journal of Cleaner Production* **225** 1052–1064

[6] Ragaert K, Delva L and Van Geem K 2017 Mechanical and chemical recycling of solid plastic waste *Waste Management* **69** 24–58

[7] Alias S, Aziz H A and Hung Y 2022 Recovery of Plastic Waste In Lawrence K W, Mu-Hao S W, Yung-Tse H *Solid Waste Engineering and Management Handbook of Environmental Engineering* 217–303

[8] Tournier V, Topham C M, Gilles A, David B, Folgoas C, Moya-Leclair E, Kamionka E, Desrousseaux M L, Texier H, Gavalda S, Cot M, Guémard E, Dalibey M, Nomme J, Cioci G, Barbe S, Chateau M, André I, Duquesne S and Marty A 2020 An engineered PET depolymerase to break down and recycle plastic bottles *Nature* **580(7802)** 216–219

[9] Coates G W and Getzler Y D Y L 2020 Chemical recycling to monomer for an ideal, circular polymer economy *Nature Reviews Materials* **5(7)** 501–516

[10] Siddiqui M N, Redhwi H H, Al-Arfaj A A and Achilias D S 2021 Chemical recycling of pet in the presence of the bio-based polymers, pla, phb and pef: A review *Sustainability (Switzerland)* **13(19)** 1–26

[11] Ügdüler S, Geem K M, Van D R, Roosen M, Mys N, Ragaert K and Meester S D 2020 Towards closed-loop recycling of multilayer and coloured PET plastic waste by alkaline hydrolysis *Green Chemistry* **22** 5376–5394

[12] Cˇolnik M, Knez Z and Škerget M 2021 Sub- and supercritical water for chemical recycling of polyethylene terephthalate waste *Chemical Engineering Science* **233** Article 116389

[13] Hanaoka T, Arao Y, Kayaki Y and Kuwata S 2021 Analysis of nitric acid decomposition of epoxy resin network structures for chemical recycling *Polymer Degradation and Stability* **186** Article 109537

[14] Yang W, Liu R, Li C, Song Y and Hu C 2021 Hydrolysis of waste polyethylene terephthalate catalyzed by easily recyclable terephthalic acid *Waste Management* **135** 267–274

[15] Matsumoto T, Makimoto S and Taniguchi Y 1997 Effect of pressure on the mechanism of hydrolysis of maltotetraose, maltopentaose and maltohexose catalyzed by porcine pancreatic a -amylase *Biochimica Et Biophysica Acta* **1343** 243–250

[16] Yoshioka T, Motoki T and Okuwaki A 2001 Kinetics of hydrolysis of poly(ethylene terephthalate) powder in sulfuric acid by a modified shrinking-core model *Industrial and Engineering Chemistry Research* **40(1)** 75–79

Biofuel production via catalytic cracking of waste cooking oil using sulfonated activated carbon supported La/Ce bimetallic catalysts

S Z Naji, C T Tye* and A R Mohamed

School of Chemical Engineering, Universiti Sains Malaysia, 14300 Nibong Tebal, Pulau Pinang, Malaysia.

***Corresponding author:** Ching Thian Tye, chcttye@usm.my

Abstract. The use of waste cooking oil (WCO) in the production of biofuel as a renewable, clean, and environmentally friendly fuel is of great interest because it reduces the WCO related pollution issues and promote circular economy. In this project, biofuel material production via catalytic cracking of WCO over sulfonated activated carbon supported La/Ce bimetallic catalysts has been investigated. Details of the reaction results and catalyst used are analysed and discussed. The results showed that pre-treatment of the activated carbon with 20% sulfuric acid improved the hydrocarbon content in the product stream by 34% in comparison with the pristine activated carbon catalyst. The conversion has been further improved by using a sulfonated activated carbon-supported bimetallic catalyst.
Keywords. Sulfonated activated carbon; Acid pre-treatment; Bimetallic catalysts; Waste cooking oil conversion.

1. Introduction

The sustainable and renewable energy sector attracts the attention of policy makers and experts owing to the depletion of crude oil and the high demand for natural gas. These energy sources can contribute to reducing greenhouse gas emissions. The conversion of biomass-derived feedstock, including waste cooking oil and inedible oil, into biofuels is a win-win situation in terms of the introduction of alternative sources of energy. Many conversion technologies have been proposed and applied for biofuel production from waste/inedible oils, such as catalytic cracking, sugar fermentation, transesterification, and pyrolysis [1]. The catalytic cracking technology is employed for the production of biofuel with various biofuel fractions such as gasoline, kerosene, and diesel over various types of catalysts [2]. Catalyst surface modifications with transition metals can improve the functional groups that enhance the deoxygenation and decarboxylation reactions rates. Various metals, such as Ni, Cu, Co, Fe, Ag, and some noble metals (Pd, Pt, and Ru), have been investigated in waste/inedible oil upgrading through catalytic cracking [3–6]. Acid modification is another viable strategy for increasing surface acidity that can enhance the deoxygenation reactions and boost the cracking performance of the catalysts [7]. Biomass-derived activated carbon (AC), has demonstrated viability as catalyst support with great thermal stability, and a large surface area with tunned pore size [8]. The aim of this study is to test the performance of a bimetallic catalyst of 5% La-10% Ce over sulfonated activated carbon in the upgrading of WCO to biofuel. There is a lack of data on the preparation and efficacy of metal impregnation over AC supported catalyst via continuous reaction. The performance of the catalyst has been studied in terms of hydrocarbon fractions in the product stream and liquid yield.

2. Experimental

2.1. Materials

The waste cooking oil in this work was provided from the cafeteria in the campus of Universiti Sains Malaysia. Coconut shell-based carbon (AC) was acquired from Kekwa Indah Carbon Solutions. Sulphuric acid was purchased from Merck with 95–98 % purity. Lanthanum (III) nitrate hydrate was supplied by Merck and cerium (III) nitrate hexahydrate was collected from Sigma. The N_2 gas was provided by the Well Gas company with 99.99% purity. GC grade n-hexane (>98% purity) utilized for dilution was supplied by Merck. The major components of WCO used were oleic (40.2677%) and palmitic acids (22.2200%). These results agreed with those reported earlier [9], the major composition was oleic acid followed by palmitic acid.

2.2. Catalyst preparation

AC was dried in an oven at 110 ◦C for 3 h to reduce the humidity. Then, the AC powder was immersed in 100 mL of 20% concentrated sulphuric acid with continuous stirring. The synthesis was done in a reflux system at 100-150 ◦C for 5-6 h. Afterward, the mixture was left at room temperature for cooling followed by filtration. The material was washed with distilled water until it reached naturality. Then, the solid material was dried at 110 ◦C overnight. The sulfonated activated carbon, 20SAC was then impregnated with 5 wt.% of La and 10 wt.% of Ce under continuous stirring for 6 h. After that, the sample was left in an oven overnight at 110°C. Then, the material was calcined at 600 °C under N_2 flow for 4 h. The synthesized samples were named AC, 20SAC, and 5La-10Ce/20SAC.

2.3. Catalyst characterization

The Brunauer–Emmett–Teller and Barrett–Joyner–Halenda methods were conducted to measure the specific surface area, pore volume, and pore distribution of the catalysts. Fourier transform infrared transmission spectra (FTIR) was carried out to determine the surface functional groups on the catalysts. The analysis was directed carried out using a Thermo Nicolet iS1O FTIR Spectroscopy at room temperature. Spectrometer resolution was 4 cm^{-1} and wavelength range were 400 to 4000 cm^{-1}. The scanning electron microscopy (SEM) - EDX was also employed to investigate the surface morphology and composition of the sample.

2.4. Catalyst activity test

The catalytic cracking reaction of WCO was conducted in a fixed-bed reactor. One gram of catalyst was loaded into the reactor on quartz wool as a bed of catalyst. Nitrogen gas (99.99%) was used as carrier gas. A tube furnace (Nabertherm B170) was served for heating the reactor. Once the reaction reached the steady state at 450°C, WCO was fed into the reactor at WHSV of 8 h^{-1} by a high-performance metering pump. A receiver vessel was used to collect the liquid products. The liquid product was diluted with the solvent (n-hexane) before being analysed by the GC–MS that equipped with an Elite-5ms II column (30 m, 250 μm, 0.25 μm) with a flame ionization detector set up at 300 °C. The oven temperature started from 40°C for 4 min then ramping to 280°C at 10°C/min and held for 5 min. A 1 μL volume of the liquid product was injected into the column at 250 °C, where the Helium (99.999%) was as a carrier gas. Gaseous products collected were analysed by a GC fitted with a thermal conductivity detector (TCD) and a Carboxen PLOT 1010 column. The temperature of the oven was set up at 50 °C for 2 mins then up to 200 °C at 10 °C/min with 14 min holding, where the Helium (99.999%) gas was supplied at a rate of 3 ml/min. The weight of the formed coke was measured after the reaction. The liquid product yield was calculated as follows:

$$Yield\ of\ liquid\ product\ (wt.\%) = \frac{weight\ of\ the\ liquid\ products\ (g)}{weight\ of\ feed\ oil\ (g)} \times 100 \quad (1)$$

The yield of hydrocarbon fractions (alkane, alkene, alkyne, aromatic, etc.), and other components such as ester, carboxylic acid, ketones, alcohol, and aldehyde were evaluated as follows:

IOP Conf. Series: Materials Science and Engineering 1257 (2022) 012008 doi:10.1088/1757-899X/1257/1/012008

$$Yield\ of\ desired\ liquid\ compound\ (\%) = \frac{area\ of\ the\ desired\ liquid\ compound}{\Sigma area\ of\ liquid\ products} \times 100 \quad (2)$$

3. Results and discussion

3.1. Characterization of the catalysts

Figure. 1 (a-c) and Table 1 show the surface morphology and elemental compositions of the pristine AC and modified AC catalysts. The SEM images show significant changes in the structure of pristine AC after modification with sulphuric acid and binary metal. Pristine AC (Figure. 1 a) shows a honeycomb shape, while more pores were appeared on the SAC catalyst (Figure. 1 b). The AC morphology changed considerably after the introduction of both La and Ce metal species. As shown in Figure.1 c, 5La-10Ce/SAC catalyst exhibited a compact aggregation of metal particles on the sulfonated AC support. There is a relatively smaller spherical pores structure with a high number of tiny, aggregated particles on the support surface. Doping of metals on the SAC support material results in the collapse of some pore walls. Meanwhile, the EDX spectroscopy had confirmed the presence of sulphur and metal species of La, and Ce for modified AC catalysts. Table 1 shows that carbon was the most abundant (>85%), as expected, followed by oxygen (<10%). The metal species (La and Ce) were detected in relatively low amounts (1.47 and 1.26 wt.%, respectively), which were lower than the theoretical amounts. This could be due to the metal particles being internally embedded in the sulfonated AC and not completely dispersed on the surface [10].

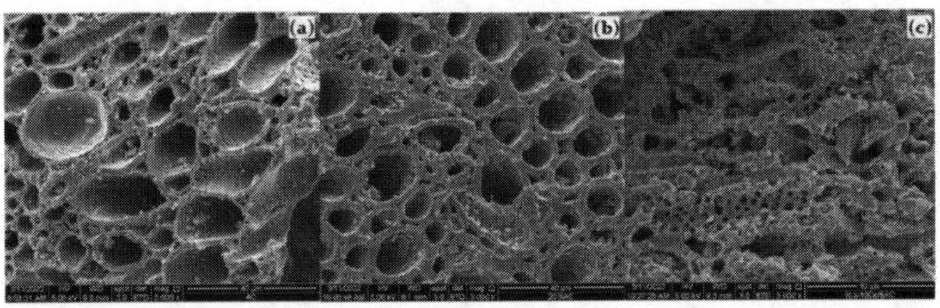

Figure 1. SEM images for the (a) pristine AC, (b) sulfonated AC, and (c) 5La-10Ce/SAC.

Table 1. EDX analysis of catalyst element composition.

	Element composition (%)							
	C	O	S	La	Ce	P	Si	K
AC	91.64	6.25	0.42	-	-	0.60	0.46	0.64
20% SAC	89.98	8.97	1.14	-	-	-	-	-
5La-10Ce/SAC	91.93	5.22	0.13	1.47	1.26	-	-	-

Table 2 summarises the surface area, pore volume, and pore width of the catalysts investigated. The results show that the surface area of the AC raised from 988.37 m^2/g to 1022.38 m^2/g after being treated with sulfuric acid. This may be due to the $-SO_3H$ groups found on the surface of activated carbon (FTIR results in Figure 3). The catalyst surface area is further increased to 1031.01 m^2/g with bimetal incorporation as seen for La(5wt.%)-Ce(10wt.%)/SAC. The increase after bimetallic impregnation could be due to increase the surface roughness, without plugging the pores. Furthermore, the large pore volume of 5La-10Ce/SAC (0.532 cm^3/g) with a higher surface area indicated an outstanding morphological change caused by the introduction of active metals, La and Ce (see SEM image Figure 1. c).

Table 2. Physical properties of the prepared catalysts.

Catalysts	Surface Area (m^2/g)	Microporous Area (m^2/g)	Pore Volume (cm^3/g)[a]	Micropore Volume cm^3/g	Pore diameter (nm)
Pristine AC	988.37	623.575	0.5230	0.3085	2.1166
20% SAC	1022.38	663.380	0.5303	0.3260	2.0748
5La-10Ce/SAC	1031.01	743.351	0.5328	0.3582	2.0674

[a] Total pore volume at p/po = 0.984.

Figure 2 presents the isotherms of N2-adsorption-desorption and pore size distribution of the catalysts investigated. The figure shows that all catalysts exhibit a type-I isotherm at low values of P/P0 according to the IUPAC system [11]. The prepared catalysts had large micropores supplemented with minor mesopores. The pore size of the catalysts reduces moderately from 2.12 to 2.067 nm after acid modification and metal impregnation.

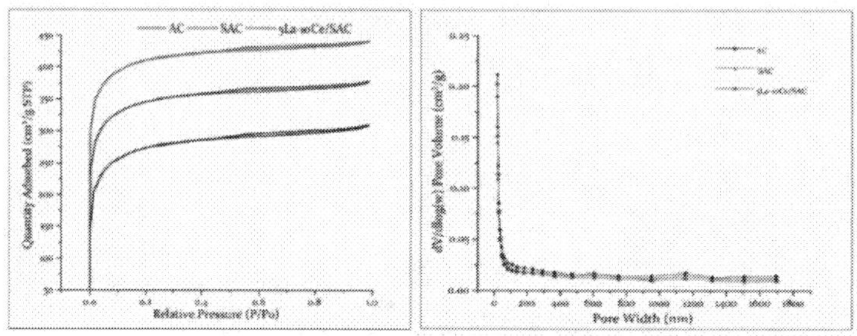

Figure 2. Isotherms of N2 adsorption-desorption and pore size distribution of the AC, SAC, and 5La-10Ce/SAC catalysts.

Figure 3 depicts the FTIR spectra for AC, 20SAC, and sulfonated AC supported bimetallic La and Ce oxide catalysts. The spectrum of pristine AC has fewer peaks than SAC and 5La-10Ce/SAC, indicating that the number of hydrophilic functional groups (carboxyl, and hydroxyl groups) on the AC surface increased after modification with sulfuric acid and binary metal oxide. The absorption bands in AC at 1406.11 cm^{-1}, 1570.06cm^{-1}, and 3425.58 cm^{-1} were referred to the C=C aromatic rings, C=O, and O-H functional groups, respectively [12]. The similar absorption bands of these functional groups with slight shifts at 1431.18 cm^{-1}, 1579.7 cm^{-1}, and 3425.58 cm^{-1}, respectively, were also observed over SAC. Besides, two additional bands were observed: clear asymmetric stretching modes of $-SO_3H$ groups and C–H at 1116.78 cm^{-1} and 2933.73 cm^{-1} over SAC catalyst, respectively [13]. This can be ascribed to the role of acid modification using sulphuric acid, which results in the addition of $-SO_3H$ groups on the AC surface. FTIR spectra of 5La-10Ce/SAC catalyst reported slight shifts in the absorption bands owing to the charge of transition metal ions on the SAC surface. The intense peak at 1103.28 cm^{-1} belongs to the $-SO_3H$ functional group owing to the sulphate ions presented in the catalyst framework which appears after acid modification. The intense peak at 1595.13 cm^{-1} is associated with the C=O (carboxyl) groups on the catalyst surface. Another stretching vibration band at 3466.08 cm^{-1} was related to the O-H stretching mode of phenolic groups. The peak of aromatic rings has shifted to 943.19 cm^{-1} after loading of La and Ce onto the SAC surface. Furthermore, the new peak in the range of 400–600 cm^{-1} was referred to as La–O, and Ce-O bending vibration [14–16].

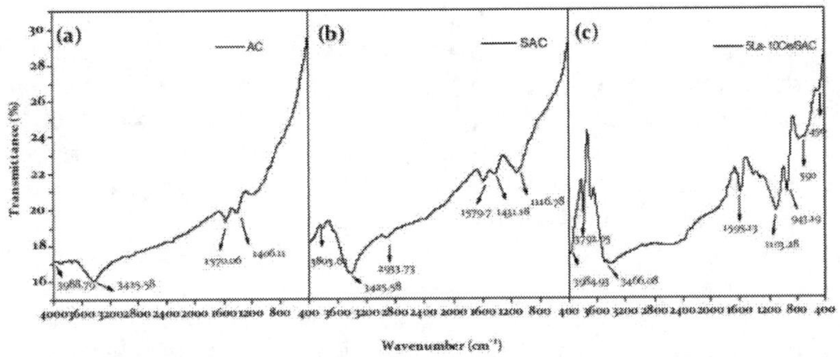

Figure 3. FTIR absorption spectra of the pristine AC, sulfonated AC, and 5La-10Ce/SAC catalysts.

3.2. Catalyst activity

Figure 4a illustrates the liquid, gas, and coke yields from the WCO catalytic cracking over the pristine AC, 20SAC, and 5La-10Ce/20SAC catalysts at 450 °C and 8 h^{-1}. 20SAC catalyst recorded a higher liquid yield (79.41%) than pristine AC catalyst (63.65%) under the same cracking conditions. Over the pristine AC catalyst, the lowest liquid yield (63.65%) with a high amount of coke formation (12.76%) was obtained. Coke deposition blocked the pore-mouth and covered the active sites, making it difficult for reactants to reach the core pore network. The highest gas yield of 23.59% was attained with the pristine AC. The liquid yield increased in the order of the pristine AC< 20SAC < 5La-10Ce/20SAC. The highest liquid yield (84.98%) was observed over 5La-10Ce/20SAC, which could be ascribed to the synergistic incorporation of the acid and basic sites coming from metal molecules. The lowest yield of coke was obtained over 5La- 10Ce/20SAC catalyst at 4.63%. As a result, the inclusion of coke inhibitor species, such as La and Ce into a mixed system could limit the rate of catalyst deactivation.

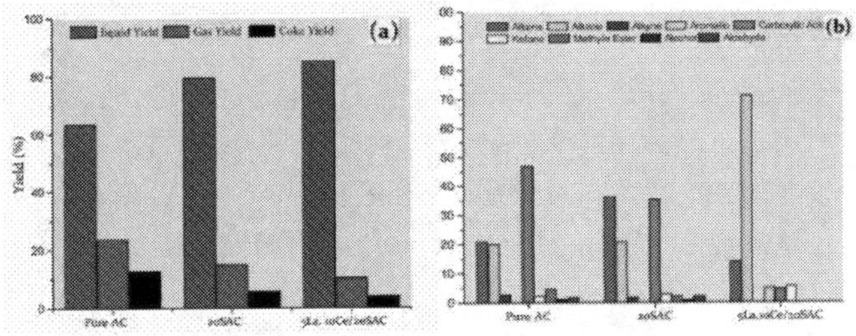

Figure 4. (a) yield of liquid, gas, and coke from the WCO catalytic cracking process over pristine AC, 20SAC, and 5La-10Ce/SAC catalysts, and (b) the liquid product compositions yield over catalysts.

Besides, the liquid compositions in the product stream over the three catalysts (e.g., AC, 20SAC, 5La-10Ce/SAC) under cracking conditions of 450°C and 8hr^{-1} WHSV were analysed. The distribution of hydrocarbon components (alkene, alkane, alkyne, and aromatic) and by products (alcohol, aldehyde, ketone, ester, and carboxylic acids) were analysed, as shown in the figure. 4b. The highest hydrocarbon yield was over 5La- 10Ce/SAC catalyst at 90.04% with the lowest oxygenated components content at 9.96%. This outcome has demonstrated the significance of using mixed metal oxide supported by AC in promoting the cleavage of C-O and C-C bonds via deoxygenation reactions and enhancing hydrocarbon production. While the low hydrocarbon yield over pristine AC was mainly due to its poor active sites. Hence, high coke formation (12.76%) had led to catalyst deactivation, which suppresses the cracking reactions and results in a lower hydrocarbon yield (43.58%). The moderate catalytic activity of pristine AC was also justified by the high yield of oxygenated components (56.43%) such as alcohol, ester, aldehyde, ketone, and carboxylic acids. Notably, 5La-10Ce/SAC catalyst gave the highest yield of n-paraffin (alkane) at 71.12%. This indicated that the major route of deoxygenation reaction over 5La-10Ce/SAC catalyst is decarboxylation, which also yielded a high concentration of CO2 (59.79%) in the gas product stream. Overall, 5La-10Ce/SAC was shown to be an effective catalyst, achieving a remarkable hydrocarbon yield with a high percentage of n-paraffin. Meanwhile, pristine AC, and sulfonated activated carbon catalysts produced 20.02% and 20.81% of alkane, respectively. The highest olefin yield (alkene) was recorded over 20SAC at 36.05%. This suggest that 20SAC catalyst facilitates the cracking of oxygenated compounds of WCO to straight-chain alkene fractions.

4. Conclusion

The catalytic cracking of WCO over modified activated carbon, via acid pre-treatment followed by binary metal introduction, was studied. The results showed that acid pre-treatment was an effective step in boosting the catalyst cracking performance by increasing the surface acidity and producing 58.43% hydrocarbon. The incorporation of bimetallic 5La-10Ce on the acid treated AC surface further improved both liquid yield and hydrocarbon yield to 84.98% and 90.04%, respectively, in the product stream. In general, 5La-10Ce/SAC showed excellent catalyst performance in waste cooking oil cracking. Additional study is needed to determine the recyclability of the catalyst and the impact of coke deposition on the catalyst performance.

Acknowledgement

This study was supported by Ministry of Higher Education Malaysia under FRGS grant scheme with Project Code: FRGS/1/2019/TK02/USM/02/5 (A/C: 203. PJKIMIA. 6071445).

References

[1] Naji A Z, Tye C T and Abd A A 2021 *Process Biochem.* **109** 148–68.

[2] Ooi Y S, Zakaria R, Mohamed A R and Bhatia S 2004 *Biomass and Bioenergy.* **27** 477–84.

[3] Sadeek S A, Mohammed E A, Shaban M, Abou Kana M T H, and Negm N A 2020 *J. Mol. Liq.* **306** Article 112749.

[4] Jiang W, Cao J P, Yang Z, Xie J X, Zhao L, Zhu C, Zhang C, Zhao X Y, Zhao Y P and Zhang J L 2022 *Appl. Catal. A Gen.* **633** Article 118516.

[5] Fan S, Zhang M and Li H 2022 *Fuel.* **315** Article 123060.

[6] Kaewpengkrow P, Atong D and Sricharoenchaikul V 2014 *Renew. Energy.* **65** 92–101.

[7] Naji S Z and Tye C T 2022 *Energy Convers. Manag. X.* **13** Article 100152.

[8] Asikin-Mijan N, Rosman N A, AbdulKareem-Alsultan G, Mastuli M S, Lee H V, Nabihah-Fauzi N, Lokman I M, Alharthi F A, Alghamdi A A, Aisyahi A A and Taufiq-Yap Y H 2020 *Process Saf. Environ. Prot.* **142** 336–49.

[9] Thangadurai T, and Tye C T 2021 *Period. Polytech. Chem. Eng.* **65** 350–60

[10] Ang B C, Yaacob I I and Nurdin I 2013 *J. Nanomater.* **2013** 1-6.

[11] Thommes M, Kaneko K, Neimark A V, Olivier J P, Rodriguez-Reinoso F, Rouquerol J and Sing K S W 2015 *Pure Appl. Chem.* **87** 1051–69.

[12] Iwanow M, Gärtner T, Sieber V and König B 2020 *Beilstein J. Org. Chem.* **16** 1188–1202.

[13] Wong W Y, Lim S, Pang Y L, Shuit S H, Chen W H and Lee K T 2020 *Sci. Total Environ.* **727** Article 138534.

[14] Ngaosuwan K, Chaiyariyakul W, Inthong O, Kiatkittipong W, Wongsawaeng D and Assabumrungrat S 2021 *Catal. Commun.* **149** Article 106247.

[15] Ali M M, Mahdi H S, Parveen A and Azam A 2018 *AIP Conf. Proc.* **1953** Article 030044.

[16] El-Deen L M S, Salhi M S A and Elkholy M M 2008 *J. Alloys Compd.* **465** 333–39.

IOP Conf. Series: Materials Science and Engineering 1257 (2022) 012009 doi:10.1088/1757-899X/1257/1/012009

Decision Making on the Renewable Energy Penetration in Malaysia via SWOT-PESTLE Analysis: Hydrogen Fuel Cell and Solar Photovoltaic Thermal Technology

M A Mohd Mokhtar[1,2], W N S Wan Mohamed Zuri[1,2], N I Mohd Rasid[1,2], A Ahmad Asmadi[1,2] and N Abdul Manaf[1*]

[1]Department of Chemical and Environmental Engineering, Malaysia-Japan International Institute of Technology, Universiti Teknologi Malaysia, Jalan Sultan Yahya Petra 54100 Kuala Lumpur, Malaysia
[2]Graduate School of Science and Technology, University of Tsukuba, 1 Chome 1-1 Tennodai, Tsukuba, Ibaraki 305-8577, Japan

*norhuda.kl@utm.my

Abstract. Malaysia is actively working to move towards renewable energy (RE) generation for sustainable growth. Hydrogen fuel cell (HFC) and solar photovoltaic thermal (PVT) feature emerging RE technologies that can supersede conventional power generation performance. HFC and PVT utilize renewable resources, which are hydrogen, oxygen and sunlight, to produce electricity without compromising the environment. However, no in-depth evaluation has been conducted to assist the decision-making of Malaysia's HFC and PVT technology penetration. This study provides decisions on the feasibility and viability of HFC and PVT based on the Strength, Weakness, Opportunities, Threat (SWOT) and Political, Economic, Social, Technological, Legal, Environment (PESTLE) analyses. Based on the SWOT-PESTLE analysis, PVT exhibits great potential in Malaysia's RE portfolio compared to HFC. Technology readiness and social acceptance are the merits of PVT diffusion in Malaysia. Nevertheless, this preliminary decision entails validation from the industries and experts to underpin the shorthand assessment of the present work

1. Introduction

Excessive amounts of CO_2 have significantly impacted aquatic life, human life and activities, which remain the major challenge to the global [1, 2]. Thus, transitioning the "dirty energy" (such as coal power) to renewable energy (RE), which offers zero amount of CO_2, is a new game changer in the energy sector. Malaysia, one of the countries dominated by fossil fuels, is no exception. The CO_2 emission from the energy industry contributed more than half compared to the automobile, construction and manufacturing industries [3]. According to International Energy Agency (IEA), Malaysia has been highly dependent on fossil fuels since decades ago, whereas coal has conquered Malaysia energy source up until 2019 [4]. This scenario demands proactive action from the power sector to adapt RE subsequently to minimize pollution and mitigate environmental impacts. RE capacity for large hydroelectric, solar photovoltaic (PV), biogas, biomass and small hydro contributes 5692 MW, 1534 MW, 123 MW, 594 MW and 507 MW respectively [5]. Even so, emerging RE technology such as hydrogen fuel cell (HFC) and solar photovoltaic thermal (PVT) is not being applied in power generation in Malaysia despite many advantages compared to the existing REs.

HFC is one of the rapidly grown technologies, as it has become a way to reduce the dependence on fossil fuels that creates an adverse environmental impact. Essentially, HFC generates electricity by combining hydrogen and oxygen atoms. The hydrogen reacts with oxygen across an electrochemical cell to produce electricity, water, and small amounts of heat, as illustrated in figure 1(a). China has started to focus on hydrogen energy at policy-making levels by announcing hydrogen energy as a strategic emerging industry in the 14th Five-Year Plan for National Economic and Social Development [6]. On the other hand, PVT offers a combined utilization of solar energy to produce electricity and thermal [7]. In this hybrid system, heat recovery fluid is used to extract the heat energy and reduce the temperature of PV panels, as illustrated in figure 1(b). Considering both HFC and PVT are still not being accessible with large-scale electricity generation in Malaysia, an assessment tool of Strength, Weakness, Opportunities, Threat (SWOT) and Political, Economic, Social, Technological, Legal, Environment (PESTLE) can be used to give a quick insight on the feasibility and viability of both technologies in Malaysia. Subsequently, provide preliminary decision-making of HFC/PVT technology readiness/penetration based on the contemporary situation in Malaysia. Many studies have been done on assessing RE technologies using SWOT and PESTLE analyses, as illustrated in table 1. This trend shows a gap in decision-making assessment between developing and developed countries, whereas developing countries are more cautious about adapting RE technology. This is due to the high investment cost of RE which caused the main hindrance to RE diffusion [8-10]. In Malaysia, palm based biomass for biogas, biodiesel and biofuel is the main RE technology being evaluated using SWOT-PESTLE analysis mainly due to the abundance of biomass resources and its technology maturity in Malaysia. It was found that RE technology was hindered by the limited subsidy provided by the government to RE investors and the technical incapability (i.e., low power output) of the technology itself. Contrariwise, RE application in Malaysia possesses opportunity in terms of geographical area but the capital cost remains the biggest threat.

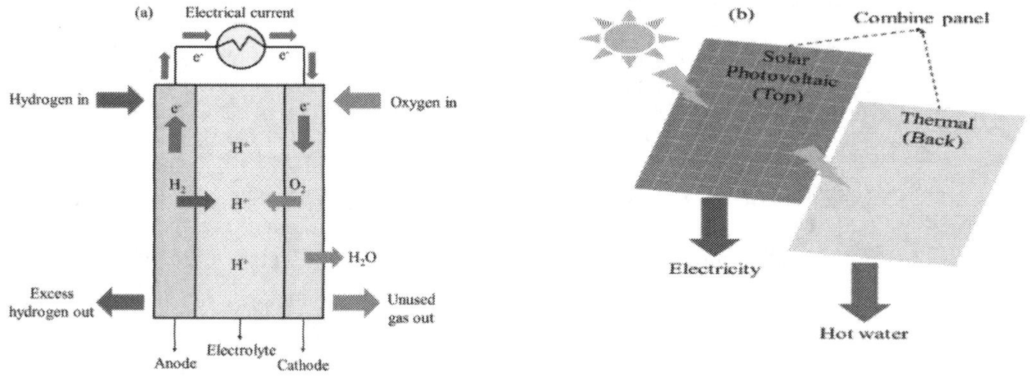

Figure 1. Typical system (a) HFC and (b) solar PVT.

Table 1. SWOT-PESTLE analysis for various RE technologies in different countries.

SWOT-PESTLE	Country	RE	Discussion	Ref.
SWOT	Pakistan	Hydropower	1. Huge availability of hydro resources. 2. 93% of hydro resources are still unexploited.	[11]
SWOT	South Korea	Fuel cell hydrogen	1. Fuel cell hydrogen vehicle has low public acceptance.	[12]
SWOT	Latvia	Waste to energy	1. High research output from renowned institution.	[14]

PESTLE	Rwanda	Solar	1. High demand of solar home system for household and small enterprise.	[15]
PESTLE	Malawi	General	1. Abundance of RE resources 2. Inadequate human resource for large scale energy project.	[16]
SWOT-PESTLE	Malaysia	General	1. Expensive capital cost and unstable supply chain for consumer used.	[17]
SWOT	Malaysia	Biomass	1. Malaysia rich with biomass resources and providing job vacancy to increase income. 2. Biomass has low power efficiency.	[18]
SWOT	Malaysia	General	1. Prioritize subsidy for fossil fuels.	[19]
SWOT	Malaysia	Biofuel	1. Weak access for electricity export using grid.	[20]

2. Methodology

This study employs the SWOT-PESTLE technique to provide preliminary decision-making on Malaysia's HFC and PVT technology penetration. No specific location has been selected for the HFC/PVT installation by assuming that the climatic condition of every states/regime in Malaysia are similar/consistent. The analysis is conducted based on the holistic perspectives of HFC/PVT technology penetration among the public, academics, businesses and governments/non-government agencies. Each factor in SWOT-PESTLE is analysed by referring to existing works of literature (qualitative data) available in Science Direct, Scopus, Web of Science, Wiley Online Library and Non-Scopus Index journals. Grey document such as government reports, policy and statistical reports, and government/non-government organization websites are also being referred to underpin the information obtained from the study. Finally, both technology is compared and identified as the potential RE technology that can be commercialized in Malaysia.

3. Result and Discussion

3.1 SWOT-PESTLE analysis of HFC

To fulfil 21st Conference of the Parties (COP 21) commitment to reduce 45% emissions by 2030, Malaysia needs to develop a hydrogen energy system and fuel cell in the transportation and electricity sectors [14]. Tables 2 and 3 summarize the SWOT-PESTLE analysis of HFC based on a plausible scenario in Malaysia.

Table 2. SWOT analysis for HFC in Malaysia.

	Helpful	Harmful
	Strength	Weakness
Internal	Support from government. Malaysia government has allocated merely RM40 million for plan and development research of HFC [21].	Exorbitant cost. Extraction of hydrogen from common feedstock such as natural gas imposed substantial operating cost [22, 23].
External	Opportunity	Threat
	Create actions to formulate a successful transition to sustainable energy and potential to develop a sustainable transportation/electricity sector in Malaysia by using HFC [22].	By 2030, hydrogen aimed to be a competitive RE source among existing RE sources such as solar and wind [22, 23].

Table 3. PESTLE analysis for HFC in Malaysia.

PESTLE	Details	References
Political	To promote more RE generation, utility-scale batteries will be installed by 2030 in the Peninsular Malaysia	[23]
Environmental	Malaysia has both resources and capacity to level up energy security and contribute to climate damage alleviation by thickening the policy actions towards HFC	[24]
Social	Low stage of hydrogen economy in Malaysia and awareness of HFC is still low hence public acceptance to its cost-effectiveness and associated risks are most likely to become the barriers	[22]
Technological	Since 8th Malaysia Plan, HFC has been identified as main research for technology development	[22]
Economy	Initiatives has been made by Malaysia government to work with foreign countries such as Japan and USA including Malaysia based companies such as Tenaga Nasional Berhad and Petronas for a better hydrogen market plan.	[25]
Legal	11th Malaysia Plan (Green Growth for Sustainability and Resilience 2015)	[21]

3.2 SWOT-PESTLE analysis of PVT

Malaysia is targeted to increase its RE shares to 31% in the power capacity mix by 2025 and 40% by 2035 [15]. Numerous research and development (R&D) activities are being conducted to ensure the viability of PVT technology in a large-scale model. Tables 4 and 5 summarize the SWOT-PESTLE analysis of HFC based on a plausible scenario in Malaysia.

Table 4. SWOT analysis for solar PVT in Malaysia.

	Helpful	Harmful
	Strength	Weakness
Internal	Receive abundant amount of solar radiation throughout the year due to geographical location. According to SEDA, Malaysia has about 7.8 TWh of solar PV potential [26].	Still in development stage and need further research to improve overall output. Solar PV diffusion and contribution in Malaysia remain low (381 MW) compared to neighbouring countries like Thailand (more than 1000 MW capacity in 2014) [27].
External	Opportunity	Threat
	Various schemes and incentives provided by government. Malaysia is one of the leading countries that conduct collaborative PVT research [26, 27].	Lack of awareness and knowledge on the benefits and financial returns from solar PV investment among the Malaysian business and public communities [27].

Table 5. PESTLE analysis for solar PVT in Malaysia.

PESTLE	Details	References
Political	Introduce various new programs and schemes to attain carbon neutral environment including Net energy metering (NEM), large scale solar (LSS), self-consumption (SELCO) and RE incentives.	[1]
Environmental	Promotion of solar power to help Malaysia achieve long-term goals to lower the carbon intensity to 35% by 2030.	[26]
Social	Organize workshop and awareness programs to encourage public engagement in RE programs.	[27]

Technological	A grid connected PVT with nanofluid has been installed for R&D activities at Green Energy Technology and Innovation Park, National University of Malaysia (UKM).	[28]
Economy	Encouragement of RE and energy efficiency (EE) projects by providing tender worth of RM 3.2 billion.	[26]
Legal	Targets to lower the greenhouse gas emissions to 45% by 2030 and launched the Green Technology Master Plan (2017-2030).	[26]

3.3 Technology comparison based on SWOT-PESTLE analysis
As Malaysia targets to increase RE shares in the power capacity mix, both technologies have high potential and are applicable for investment in Malaysia. However, HFC requires a larger investment, and there is no promising return based on its slow-paced development. Malaysia has been listed as the most influential country in collaborative PVT drive research, showing a good sign in the PVT diffusion from the academic/research perspective [29]. In addition, Feed in Tariff (FiT) concept through Renewable Energy Act 2011 may encourage the PVT technology acceleration and innovation, which aims to align R&D towards commercialisation, wealth generation and economic growth. While both technologies receive various schemes and opportunities from the government, the implementation of PVT is seen as more relevant and receives more attention from the stakeholders and investors. PVT technology also exhibits positive economic values compared to HFC and has been largely accepted by the public.

4. Conclusion
Based on the SWOT-PESTLE analysis, PVT is a viable option for a RE technology transition subjected to the present scenario in Malaysia (underpinning with the existing literature studies). Nonetheless, this preliminary decision-making requires validation from the industries and experts to underpin the present work's shorthand assessment.

Acknowledgement
This research is supported by MRUN Young Researchers Grant Scheme project no. R.K130000.7843.4L909 and R.K130000.7843.4L901.

References
[1] Vaka M, Walvekar R, Rasheed A K and Khalid M 2020. A review on Malaysia's solar energy pathway towards carbon-neutral Malaysia beyond Covid'19 pandemic. *Journal of Cleaner Production* **273** 122834
[2] Rahman M N and Wahid M A 2021. Renewable-based zero-carbon fuels for the use of power generation: A case study in Malaysia supported by updated developments worldwide. *Energy Reports* **7** 1986-2020
[3] Abdul Latif S N, Chiong M S, Rajoo S, Takada A, Chun Y Y, Tahara K and Ikegami Y 2021. The trend and status of energy resources and greenhouse gas emissions in the malaysia power generation mix. *Energies* **14** 2200
[4] International Energy Agency *Electricity generation by source, Malaysia 1990-2019* [Internet]. 2020 [cited 2022 February 1]. Available from: https://www.iea.org/fuels-and technologies/electricity
[5] Sustainable Energy Development Authority Malaysia 2021. Malaysia Renewable Energy Roadmap
[6] Li Y and Taghizadeh-Hesary F 2022. The economic feasibility of green hydrogen and fuel cell electric vehicles for road transport in China. *Energy Policy* **160** 112703
[7] Chandrasekar M and Senthilkumar T 2021. Five decades of evolution of solar photovoltaic thermal (PVT) technology – A critical insight on review articles. *Journal of Cleaner Production* **322** 128997
[8] Qaiser I 2022. A comparison of renewable and sustainable energy sector of the South Asian countries: An application of SWOT methodology. *Renewable Energy* **181** 417-25
[9] Agyekum E B, Ansah M N S and Afornu K B 2020. Nuclear energy for sustainable development:

SWOT analysis on Ghana's nuclear agenda. *Energy Reports* **6** 107-115

[10] Igliński B, Iglińska A, Koziński G, Skrzatek M and Buczkowski R 2016. Wind energy in Poland – History, current state, surveys, Renewable Energy Sources Act, SWOT analysis. *Renewable and Sustainable Energy Reviews* **64** 19-33

[11] Sibtain M, Li X, Bashir H and Azam M I 2021. Hydropower exploitation for Pakistan's sustainable development: A SWOT analysis considering current situation, challenges, and prospects. *Energy Strategy Reviews* **38** 100728

[12] Lee Y, Kim Y J and Lee M C 2021. Improving public acceptance of H2 stations: SWOT-AHP analysis of South Korea. *International Journal of Hydrogen Energy* **46** 17597-17607

[13] Wang Y, Xu L and Solangi Y A 2020. Strategic renewable energy resources selection for Pakistan: Based on SWOT-Fuzzy AHP approach. *Sustainable Cities and Society* **52** 101861

[14] Beloborodko A, Romagnoli F, Rosa M, Disanto C, Salimbeni R, Karlsen E N, Reime M, Schwab T, Mortensen J, Ibarra M and Blumberga D 2015. Swot analysis approach for advancement of waste-to-energy cluster in Latvia. *Energy Procedia* **72** 163-9

[15] Thomas P J M, Sandwell P, Williamson S J and Harper P W 2021. A PESTLE analysis of solar home systems in refugee camps in Rwanda. *Renewable and Sustainable Energy Reviews* **143** 110872

[16] Zalengera C, Blanchard R E, Eames P C, Juma A M, Chitawo M L and Gondwe K T 2014. Overview of the Malawi energy situation and A PESTLE analysis for sustainable development of renewable energy. *Renewable and Sustainable Energy Reviews* **38** 335-47

[17] Shadman S, Chin C M M, Yap E H, Sakundarini N and Velautham S 2021. The role of current and future renewable energy policies in fortifying Malaysia's energy security: PESTLE and SWOT analysis through stakeholder engagement. *Progress in Energy and Environment* **16** 1-17

[18] Chala G T, Guangul F M and Sharma R 2019. Biomass Energy in Malaysia-A SWOT Analysis. 2019 *IEEE Jordan International Joint Conference on Electrical Engineering and Information Technology, JEEIT 2019* - Proceedings 401-6

[19] Sharvini S R, Noor Z Z, Chong C S, Stringer L C and Yusuf R O 2018. Energy consumption trends and their linkages with renewable energy policies in East and Southeast Asian countries: Challenges and opportunities. *Sustainable Environment Research* **28** 257-66

[20] Darshini D, Dwivedi P and Glenk K 2013. Capturing stakeholdersĎ views on oil palm-based biofuel and biomass utilisation in Malaysia. *Energy Policy* **62** 1128-37

[21] Ambrose A F, Al-Amin A Q, Rasiah R, Saidur R and Amin N 2017. Prospects for introducing hydrogen fuel cell vehicles in Malaysia. *International Journal of Hydrogen Energy* **42** 9125-34

[22] Ahmed A, Al-Amin A Q, Ambrose A F and Saidur R 2016. Hydrogen fuel and transport system: A sustainable and environmental future. *International Journal of Hydrogen Energy* **41** 1369-80

[23] Thompson S T, James B D, Huya-Kouadio J M, Houchins C, Desantis D A, Ahluwalia R, Wilson A R, Kleen G and Papageorgopoulos D 2018. Direct hydrogen fuel cell electric vehicle cost analysis: System and high-volume manufacturing description, validation, and outlook. *Journal of Power Sources* **399** 304-13

[24] Mah A X Y, Ho W S, Bong C P C, Hassim M H, Liew P Y, Asli U A, Kamaruddin M J and Chemmangattuvalappil, N G 2019. Review of hydrogen economy in Malaysia and its way forward. *International Journal of Hydrogen Energy* **44** 5661-75

[25] Kardooni R, Yusoff S B, Kari F B and Moeenizadeh L 2018. Public opinion on renewable energy technologies and climate change in Peninsular Malaysia. *Renewable Energy* **116** 659-68

[26] Ismail M I, Yunus N A, Mohamed Kaassim A Z and Hashim H 2022. Pathways and challenges of solar thermal utilisation in the industry: ASEAN and Malaysia scenarios. *Sustainable Energy Technologies and Assessments* **52** 102046

[27] Koerner S, Siew W, Salema A, Balan P, Mekhilef S and Thavamoney N 2022. Energy policies shaping the solar photovoltaics business models in Malaysia with some insights on Covid-19 pandemic effect. *Energy Policy* **164** 112918

[28] Sopian K, Alwaeli A H A, Abdulrasool Hasan H and Al-Shamani A N 2020. Advances in High Efficiency Photovoltaic Thermal Solar Collectors. *Journal of Advanced Research in Fluid Mechanics and Thermal Sciences* **47** 1-7

[29] Azad A K and Parvin S 2022. Bibliometric analysis of photovoltaic thermal (PV/T) system: From citation mapping to research agenda. *Energy Reports* **8** 2699-2711

Response surface optimisation of biodiesel synthesis using biomass derived green heterogeneous catalyst

W-Y Wong[1], S Lim[1,2*], Y-L Pang[1,2], C-H Lim[3] and F-L Pua @ G Pua[4]

[1] Department of Chemical Engineering, Lee Kong Chian Faculty of Engineering and Science, Universiti Tunku Abdul Rahman, 43000 Kajang, Selangor, Malaysia.
[2] Centre for Photonics and Advanced Materials Research, Universiti Tunku Abdul Rahman, 43000 Kajang, Selangor, Malaysia.
[3] School of Engineering and Physical Sciences, Heriot-Watt University Malaysia, Jalan Venna P5/2, Precinct 5, 62200 Putrajaya, Malaysia.
[4] Department of Mechanical Engineering, College of Engineering, Universiti Tenaga Nasional, 43000 Kajang, Selangor, Malaysia.

*Email: stevenlim@utar.edu.my

Abstract. Although homogeneous alkali-catalysed transesterification is the typical process used in biodiesel production, it caused complications in downstream separation processes and an oversupply of glycerol as a by-product. The present work studied the synthesis of a novel sulfonated biomass-derived solid acid catalyst and its application in biodiesel production via interesterification of oleic acid. Solid acid catalysts were prepared by direct sulfonation via thermal treatment with concentrated sulfuric acid. The design of experiments was conducted via four-factors central composite design (CCD) coupled with response surface methodology (RSM) analysis. The parameters considered for optimisation included carbonisation and sulfonation temperatures, catalyst loading and reaction time, each varied at five levels. The maximum yield of fatty acid methyl ester (FAME) was obtained using optimum parameters as carbonisation temperature of 586 °C, sulfonation temperature of 110 °C, catalyst loading of 10.5 wt.% and reaction time of 7 h was 54.3 % based on the theoretical ester formation. A quadratic mathematical model in RSM was successfully established that can make effective predictions about the anticipated biodiesel yield. This study proved that the low-cost heterogeneous catalyst derived from biomass waste with a simple production route could catalyse the interesterification process under moderate process conditions.

1. Introduction

Recently, emerging biodiesel production technologies have been developed by employing alternative acyl acceptors such as methyl acetate, ethyl acetate and dimethyl carbonate to render the process free of glycerol. Among the potential acyl acceptors, the interesterification of triacylglycerides with methyl acetate can generate triacetin as the by-product. Contrary to glycerol, triacetin is a more value-added co-product as it functions as a plasticizer or additive for cosmetic, pharmaceutical and tobacco compounds. Since triacetin is compatible with fatty acid methyl ester (FAME) and soluble in biodiesel, it does not need to be separated from the final product. It is possible to incorporate 20 wt.% of triacetin in biofuel composition without compromising the fuel quality [1]. The addition of triacetin in the biofuel brings about improvement in the cold flow properties, viscosity, cloud point and pour point as well as reduces engine knocking. [2,3]. Its presence as an additive also helps to reduce nitrogen oxide, carbon dioxide, carbon monoxide and particulate matter emissions. The esterification of FFA with methyl acetate produces acetic acid as a co-product which has a catalytic effect in the reaction and thus enhances the overall esters yield.

It should be noted that the non-catalytic supercritical and enzymatic interesterifications had been studied extensively in biodiesel production with their own advantages and disadvantages. In comparison, interesterification routes catalysed by enzymes and under pressurized conditions without any catalysts are not as productive as their chemical counterpart. In alkaline interesterification, homogeneous alkali metal alkoxides such as sodium or potassium methoxide record very high catalytic activity. However, the realization of a pure interesterification reaction is difficult to be attained due to the presence of alcohol solvent which initiates a parallel transesterification reaction. To avoid the intrinsic drawbacks of separability, reusability and regeneration of catalysts, heterogeneous catalysts can be used since they are able to catalyse the reaction under relatively moderate conditions. Several solid acid catalysts that have been studied in interesterification reactions were ferric sulfate [4], niobium phosphate [5,6], niobium oxide, zeolite HY [6], γ-alumina [6–8] and persulfate incorporated zirconium dioxide-titanium dioxide-iron (III) oxide ($S_2O_8^{2-}/ZrO_2$-TiO_2-Fe_3O_4) [9].

The proceeding of interesterification reaction with methyl acetate as acyl acceptor and application of the fatty acid on biodiesel synthesis without the co-production of glycerol from low-grade feedstock are the important considerations to be accentuated. Thereby, the interesterification reaction pathway between methyl acetate and oleic acid with the aid of a solid acid catalyst synthesised using empty fruit bunch (EFB) as the carbon precursor was proposed as a promising alternative to eradicate the principal hurdles of traditional biodiesel production in the present study. The acidic, textural and structural properties of the catalyst will be elucidated with the aid of characterisation studies. Although the interesterification reaction has established a promising biodiesel yield in the literature, the optimisation of catalyst synthesising parameters together with reaction conditions is still very much lacking. To address this knowledge gap, this research work reported the combined optimisation of catalyst synthesising parameters (carbonisation and sulfonation temperatures) and reaction conditions (catalyst loading and reaction time) using the statistical method of response surface methodology (RSM) in Design-Expert software.

2. Experimental

2.1. *Materials and chemicals*

EFB was collected from an oil palm factory in Negeri Sembilan, Malaysia. Methyl heptadecanoate (99 %) was obtained from Sigma-Aldrich as the internal standard of Gas Chromatography (GC) analysis. Oleic acid (85-92 %) was procured from Fisher Chemical. Methyl acetate (99 %), ortho-phosphoric acid (85 %), sulfuric acid (95-97 %) and *n*-hexane (≥99 %) were supplied from Merck. All the acquired chemicals were of analytical grade and were utilised without additional purification.

2.2. *Activation and preparation of activated carbon*

EFB was cleaned to eliminate contaminations in advance of overnight drying at 80 °C in an oven. The biomass size was reduced using a commercial blender to create fibre, which was then sieved to a particle size of no more than 850 μm. Afterwards, ortho-phosphoric acid (30 v/v%) was used to chemically treat the EFB at an impregnation mass ratio of 1:7 for 24 h under ambient conditions, preceding washing with distilled water. In a chamber furnace with a heating profile of 5 °C/min for 3 h, the dried pre-treated biomass was carbonised. A test sieve with a mesh size of 300 μm was employed to administer the distribution of activated carbon (AC) particle size.

2.3. *Synthesis of catalyst*

All the sulfonation processes were carried out by continuously stirring AC at a mass ratio of 1:30 wt/wt with concentrated sulfuric acid in a reflux system for 6 h to anchor the sulfonic acid groups (–

SO_3H) onto the polycyclic aromatic carbon sheets. The following precipitates were filtered and rinsed exhaustively with distilled water to eliminate any possible residues of sulfate ions prior to 24 h drying at 80 °C.

2.4. Interesterification reaction
Catalyst performance was evaluated through a batch interesterification reaction between oleic acid and methyl acetate at a molar ratio of 1:50. The as-synthesised catalyst was deposited into a reflux reaction system at 110 °C with constant stirring. Once the desired reaction time was reached, the reaction medium was instantly quenched to ambient temperature. The catalyst was retrieved via filtration to halt any further reaction and the leftover methyl acetate was removed by means of evaporation.

2.5. Catalyst characterisation
The particulars regarding surface morphology and elemental compositions of fabricated catalysts were acquired non-destructively using Scanning Electron Microscopy furnished with Energy Dispersive X-ray (SEM-EDX). Thermogravimetric analysis (TGA) was conducted to thermally analyse the physical and chemical properties of EFB. X-ray Diffractometer (XRD) was applied to ascertain the presence of the crystalline structure in the as-prepared catalysts. All samples tested were pre-treated prior to characterisation analysis for at least 1 h at 100 °C to get rid of the adsorbed water molecules.

2.6. Analysis of product
The products obtained from the interesterification reaction were quantified in terms of FAME yield using gas chromatography (GC) furnished with a flame ionization detector and a capillary column (fused silica, 30 m X 0.53 mm X 0.5 μm). FAME yield was calculated in accordance with equation (1), where m is the theoretical mass of the methyl oleate produced (g), C is the methyl oleate concentration (g L^{-1}) and V_f is the final product volume (L).

$$FAME\ yield\ (\%) = \frac{C \times V_f}{m} \times 100\% \tag{1}$$

2.7. RSM modelling
Among all the designs available such as Central Composite Design (CCD), Definitive Screen design (DSD) and Box-Behnken design (BBD), CCD was chosen for statistical assessment and regression model building. It has been widely applied in the optimisation of numerous physical and chemical procedures. CCD is a full-factorial design that is able to minimise the required experimental trials, evaluate the effects of each independent variable and describe their corresponding relationships, resulting in a comprehensive apprehension of the procedures [10]. It is also a pertinent approach for sequential experiments to attain relevant particulars for testing lack of fit (LOF) in the absence of abundant design points. CCD offers equal predictability in all directions from the centre. It is capable to predict the response within the limits by changing the parameters simultaneously [11].

3. Results and discussion

3.1. Preliminary study
The data of thermal stability of raw biomass waste of EFB is shown in figure 1. TGA endothermic peak and weight loss that took place at temperatures lower than 100 °C were accredited to water and adsorbed free moisture removal via evaporation. A dissociation zone that was observed ranging from 200 to 350 °C originated from the degradation of hemicelluloses and cellulose [12]. The thermal degradation exhibited flat tailing at higher temperatures. Therefore, the carbonisation temperature range chosen to be studied would be varied from 500 °C onwards where the structure of the biomass was at its thermally stable form.

3.2. Catalyst characterisation

3.2.1. Scanning Electron Microscopy furnished with Energy Dispersive X-ray (SEM-EDX). The SEM micrograph in figure 2(a) unveils the surface morphology of the as-synthesised catalyst. Cracks and defects with uneven surfaces, small particles and pores were found. The high-temperature thermal activation developed carbon with hollow perforations owing to the release of volatile compounds. These minute pores would ameliorate the total specific surface area which serves as sites for the attachment of sulfonic group, enhancing the yield of the interesterification reaction. Elemental analysis of the prepared catalyst validated the existence of sulfur which had a content of 1.46 ± 0.03 at.%.

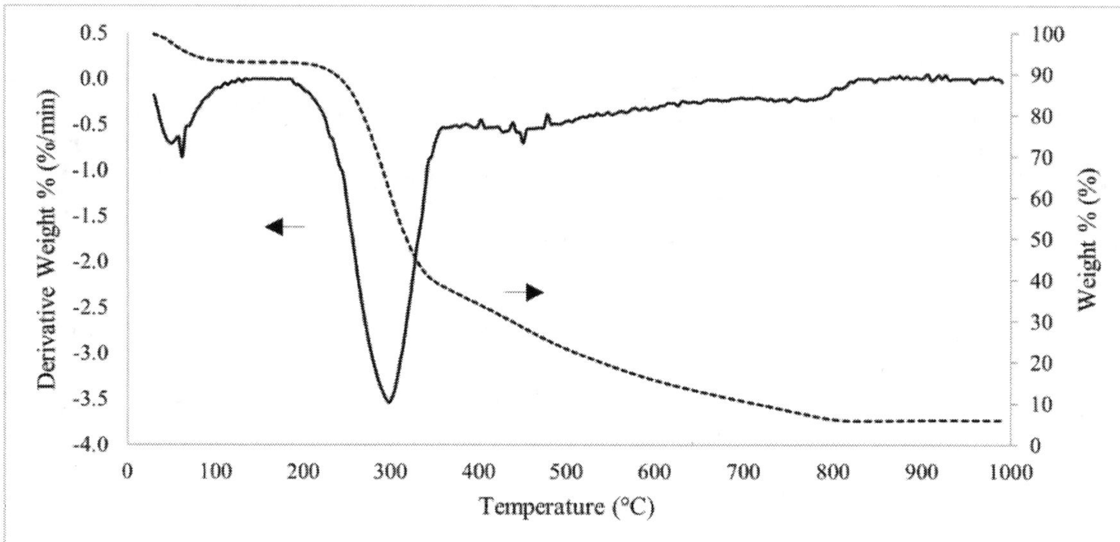

Figure 1. Thermogravimetric analysis (TGA) spectra of the oil palm empty fruit bunch.

3.2.2. X-ray Diffractometer (XRD). Powder X-ray diffractograms in figure 2(b) show evident diffraction peaks from 15° to 35°, representing polycyclic aromatic carbon sheets that made up the randomly arranged amorphous carbon structures. Another broad diffraction peak (40°-50°) depicted the low degree graphitic carbon structures that are favourable for the incorporation of sulfonic acid groups [13]. This showed that the as-prepared catalyst possessed amorphous carbon structures.

3.3. RSM modelling and optimisation of interesterification reaction

There were 30 sets of experiments in total having combinations of different levels of all 4 process parameters including the centre point which has 6 repetitions, 8 axial points and 16 factorial points. As outlined in table 1, the independent factors were coded to two levels, namely +1 and -1 representing the respective high and low levels of the input variables. Alternatively, the axial points were coded as $+\alpha$ and $-\alpha$. The data depicted in table 2 show that quadratic was the most suitable mathematical model compared to linear, two-factor interaction (2FI) and cubic. Based on the analysis of variance (ANOVA), the obtained equation for response in terms of coded factors for the best model was given in equation (2). The value of the coefficient of determination (R^2) for equation (2) was 0.9004. It was shown that the terms in the proposed equation (D, AC, AD, BD, A^2, B^2, C^2 and D^2) which included the linear and quadratic terms were significant terms as suggested by the obtained p-values (probability of error values) of less than 0.05. The optimised interesterification parameters obtained were carbonisation temperature of 586 °C, sulfonation temperature of 110 °C, reaction time of 7 h and catalyst loading of 10.5 wt.% which resulted in the best optimum methyl oleate yield.

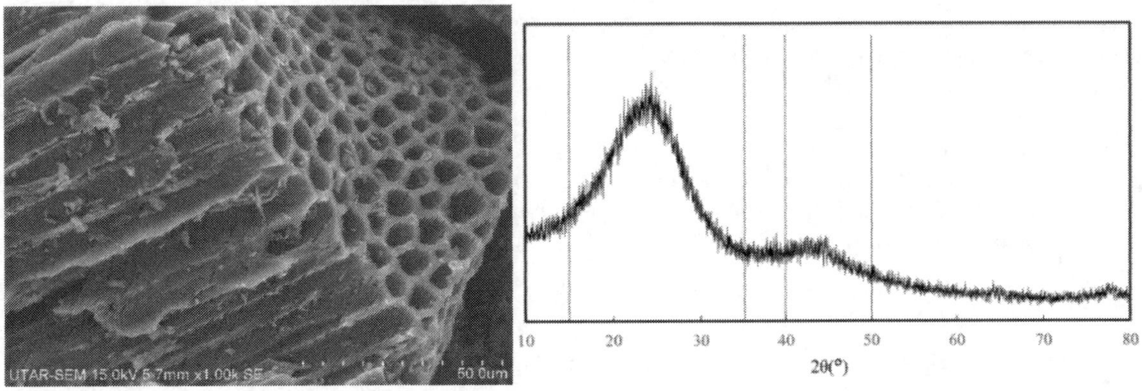

(a) (b)

Figure 2. (a) SEM image (b) XRD patterns of EFB-based solid acid catalyst.

Table 1. Level of independent variables and experimental range used in Central Composite Design.

Variable	Unit	Coded Variable	Experimental Values			
			Low	High	-Alpha	+Alpha
Carbonisation Temperature	°C	A	550	650	500	700
Sulfonation Temperature	°C	B	80	120	60	140
Catalyst Loading	wt.%	C	8	12	6	14
Reaction Time	h	D	5	7	4	8

$$Yield(\%) = 51.44 - 0.3482A + 0.5213B + 0.3107C + 3.80D - 2.16AC - 2.62AD + 3.37BD + 1.04CD - 4.04A^2 - 4.06B^2 - 5.13C^2 - 1.68D^2 \quad (2)$$

Table 2. Analysis of different mathematical models.

Source	Sum of Squares	df	Mean Square	F-value	P-value	Comments
Mean vs Total	25935.78	1	25935.78			
Linear vs Mean	285.71	4	71.43	1.18	0.3423	
2FI vs Linear	207.04	6	34.51	0.5036	0.7980	
Quadratic vs 2F1	1124.45	4	281.11	23.76	<0.0001	Suggested
Cubic vs Quadratic	94.43	8	11.80	0.9949	0.5093	Aliased
Residual	83.05	7	11.86			
Total	27730.47	30	924.35			

4. Conclusion

This research study integrated the advantages of biomass-derived heterogeneous catalysis system and glycerol-free FAME production under moderate reaction conditions. An unconventional EFB-based solid acid catalyst was devised via a consecutive carbonisation-sulfonation method. The amorphous carbon catalyst was furnished with a well-defined porous structure accompanied by a large specific surface area which was shown in the SEM micrograph. Process factors of both catalysts synthesising procedures and interesterification reaction which included the carbonisation temperature, sulfonation temperature, catalyst amount and reaction time were optimised and analysed by CCD of RSM to achieve the best biodiesel yield. The catalyst displayed satisfying catalytic activity in methyl acetate aided interesterification of oleic acid with the biodiesel yield of 54.3 % under optimised carbonisation temperature of 586 °C, sulfonation temperature of 110 °C, reaction time of 7 h and catalyst loading of 10.5 wt.%. RSM presented a stable and reliable model that takes into account the influence of the variable in this study. It can be concluded from the current study that the as-prepared catalyst demonstrated an eco-friendly and auspicious strategy for the production of green fuel and henceforth necessitated supplementary research into this scheme.

Acknowledgements
The authors gratefully acknowledge for the funding given by Fundamental Research Grant Scheme (FRGS/1/2018/TK10/UTAR/02/1) and Universiti Tunku Abdul Rahman (UTARRF) to support this project and the scholarship funding to Ms. Wong Wan Ying.

References
[1] Trentini C P, de Mello B T F, Postaue N, Stevanato N, Cardozo-Filho L and Silva C da 2020 Interesterification of grease trap waste lipids using methyl acetate under supercritical conditions *The Journal of Supercritical Fluids* **164** 104896
[2] Postaue N, Trentini C P, de Mello B T F, Cardozo-Filho L and da Silva C 2019 Continuous catalyst-free interesterification of crambe oil using methyl acetate under pressurized conditions *Energy Conversion and Management* **187** 398–406
[3] Nunes A L B and Castilhos F 2020 Chemical interesterification of soybean oil and methyl acetate to FAME using CaO as catalyst *Fuel* **267** 117264
[4] Tian Y, Xiang J, Verni C C and Soh L 2018 Fatty acid methyl ester production via ferric sulfate catalyzed interesterification *Biomass and Bioenergy* **115** 82–7
[5] Alessio C, Ribeiro J S, Celante D, Brondani L and Castilhos F 2017 Kinetics of methyl esters production with dimethyl carbonate over niobium phosphate *Energy Conversion and Management* **151** 670–80
[6] dos Santos Ribeiro J, Celante D, Simões S S, Bassaco M M, da Silva C and de Castilhos F 2017 Efficiency of heterogeneous catalysts in interesterification reaction from macaw oil (Acrocomia aculeata) and methyl acetate *Fuel* **200** 499–505
[7] Kashyap S S, Gogate P R and Joshi S M 2019 Ultrasound assisted intensified production of biodiesel from sustainable source as karanja oil using interesterification based on heterogeneous catalyst (γ-alumina) *Chemical Engineering and Processing - Process Intensification* **136** 11–6
[8] Visioli L J, de Castilhos F and da Silva C 2019 Use of heterogeneous acid catalyst combined with pressurized conditions for esters production from macauba pulp oil and methyl acetate *The Journal of Supercritical Fluids* **150** 65–74
[9] Wu H, Liu Y, Zhang J and Li G 2014 In situ reactive extraction of cottonseeds with methyl acetate for biodiesel production using magnetic solid acid catalysts *Bioresource Technology* **174** 182–9
[10] Endut A, Abdullah S H Y S, Hanapi N H M, Hamid S H A, Lananan F, Kamarudin M K A, Umar R, Juahir H and Khatoon H 2017 Optimization of biodiesel production by solid acid catalyst derived from coconut shell via response surface methodology *International Biodeterioration & Biodegradation* **124** 250–7
[11] Razack S A and Duraiarasan S 2016 Response surface methodology assisted biodiesel production from waste cooking oil using encapsulated mixed enzyme *Waste Management* **47** 98–104
[12] Ezebor F, Khairuddean M, Abdullah A Z and Boey P L 2014 Oil palm trunk and sugarcane bagasse derived heterogeneous acid catalysts for production of fatty acid methyl esters *Energy* **70** 493–503
[13] Ning Y and Niu S 2017 Preparation and catalytic performance in esterification of a bamboo-based heterogeneous acid catalyst with microwave assistance *Energy Conversion and Management* **153** 446–54

Synthesis of bio-hydrogen supply network *via* graph-theoretic approach coupled with Monte Carlo simulation model

S L Y Lo[1,2], K G H Kong[1,2], M F Tang[1], J Sunarso[1] and *B S How[1]

[1]Research Centre of Sustainable Technologies, Faculty of Engineering, Computing and Science, Swinburne University of Technology, 93350 Kuching, Sarawak, Malaysia.

[2]These authors contribute equally.

*Corresponding author email: bshow@swinburne.edu.my

Abstract. The shift from fossil fuels-derived hydrogen (grey or brown hydrogen) to renewable energy-derived hydrogen (green hydrogen) production is essential to achieve Sustainable Development Goal 7 (SDG7) which aims to ensure the accessibility to affordable, reliable, sustainable, and modern energy. One of the renewable energy sources that have been extensively studied is the biomass-derived energy source. The abundance of palm oil mill effluent (POME) produced annually in Malaysia, poses a unique potential of them being utilized as an alternative renewable feedstock for biohydrogen production. Therefore, this project aims to synthesize an optimal bio-hydrogen supply network *via* the integration of graph-theoretic approach and Monte Carlo simulation model where palm oil mills serve as hydrogen sources, while the ammonia plants serve as hydrogen sinks. Monte Carlo simulation was performed for the top three ranked bio-hydrogen supply networks extracted from the P-graph model. The uncertainties incorporated in the Monte Carlo simulation model are natural gas price and hydrogen price. Using Sarawak as the case study, it was found that the first ranked solution extracted from the P-graph model had the higher mean NPV value of USD 1,202.12 million.

1. Introduction

Recent years, hydrogen plays a key role in the industry as it served as an intermediate in chemical processing such as crude oil refinery, ammonia and urea production. Furthermore, the superiority of hydrogen in environmental reforming has made hydrogen an eye-catching solution to manage sustainable energy without harming the environment [1]. Many countries have shown their interest in investigating the feasibility of hydrogen energy to improve the sustainability of energy. For example, 685 hydrogen refueling stations were in operation worldwide by the end of the year 2021, and this fact itself highlighted the potential of hydrogen energy [2]. As a potential energy source in this era, hydrogen can be generated from diverse resources such as fossil fuel, biomass, and etc. However, only 4% of hydrogen is generated from water electrolysis and other byproduct sources of hydrogen such as heavy oil and naphtha; the remaining 96 % of hydrogen are still produced from non-renewable fossil fuels [3]. This state of affair indicates that hydrogen production is still highly dependent on fossil fuel. Therefore, research on bio-hydrogen production has been widely conducted to enhance the feasibility of green hydrogen production.

The development of bio-hydrogen production from POME treatment has been a huge research interest for the researchers such as the feasibility and profitability in supply chain. The feasibility of bio-hydrogen is highly dependent on supply-chain network design to investigate the deployment of the bio-hydrogen infrastructure [4]. To counter this, an optimal bio-hydrogen production structure with maximum profitability and minimum environmental impact is required. Therefore, in this work, feasible optimal and sub-optimal bio-hydrogen supply networks were determined *via* Process Graph (P-graph). P-graph has been generally utilized in solving Process network synthesis (PNS) problems but has been explored with extended application in supply chain network synthesis. It has been proven that P-graph is capable of providing both optimal and near-optimal solutions for various process synthesis problems, including biomass supply chain [5], water regeneration network [6], and carbon [7]. In fact, P-graph has recently been applied for bio-hydrogen network synthesis. For instance, P-graph had been utilized in Lee et al.'s work in synthesizing the optimal POME-based bio-hydrogen network where geographical factors are incorporated in the model [8].

In order to reflect a more realistic situation in this proposed model, uncertainties need to be incorporated into the model. The effects of fluctuations in hydrogen and natural gas purchase price in bio-hydrogen supply chain are further studied in this work. In this case study, Monte Carlo simulation is adopted to account for the robustness of each structure under uncertain hydrogen and natural purchase price. Top three ranked P-graph networks with the highest degree of satisfaction in terms of both objectives (minimizing carbon emission and maximizing profitability) were extracted and further analyzed. This work aims to provide a significant guideline to evaluate the feasibility of the bio-hydrogen network. The results obtained in this study can assist relevant stakeholders to evaluate the possibility of loss and risk related to a particular investment and thus, providing a guideline to lower the risk involved through possible action plans.

2. Methodology

This work involves P-graph framework synthesis of hydrogen supply chain therefore, essential data information is needed to be input in the P-graph model. Conversion ratios of processes, amount of hydrogen production, hydrogen demand, facilities location and other references are the basic information needed to model the network structure. The methodology for P-graph construction is adopted from the studies of Hemmati et al. [9]. Subsequently, the top three ranked optimal bio-hydrogen were simulated *via* Monte Carlo simulation model to evaluate the financial profitability of the aforementioned supply chain. The two softwares used in this study are P-graph Studio and Microsoft Excel (for Monte Carlo simulation model). P-graph software has the advantage of a visual interface and the capability of generating optimal and sub-optimal solutions simultaneously. Modelling Monte Carlo simulation model using Microsoft Excel is more user-friendly that can be easily implemented by users with minimal computational background.

In this work, Monte Carlo simulation model is adopted to perform stochastic economic evaluation of the top three ranked bio-hydrogen supply chain network. Two uncertainties are integrated into the developed Monte Carlo simulation model, i.e., hydrogen selling price and natural gas price. The economic evaluation indicator utilized in this work is Net Present Value (NPV) (see equation (1)).

$$\mathrm{NPV} = \sum_{n} \frac{(\mathrm{Balance})}{(1+\mathrm{i})^{n}} - \mathrm{IC} \tag{1}$$

$$\mathrm{Balance} = \mathrm{R}^{BH} - \mathrm{C}^{NG} - \mathrm{C}^{\mathrm{OP}} \tag{2}$$

$$\mathrm{R}^{BH} = \mathrm{P}^{BH} \times \mathrm{H}^{\mathrm{Prod}} \tag{3}$$

$$\mathrm{C}^{NG} = \mathrm{P}^{NG} \times \mathrm{W}_{\mathrm{y}}^{NG} \tag{4}$$

where Balance denotes the remaining cash flow for each operating year, n (USD), and i represents the discount rate (%). The breakdown calculation of Balance of each operating year is shown in equation (2) and comprises of three important components that is revenue from the sales of bio-hydrogen, R^{BH}

(USD/y), total cost of purchasing NG, C^{NG} (USD/y), and fixed operating cost of the processing plant, C^{OP} (USD/y). The revenue from the sales of bio-hydrogen can be calculated via multiplication of the selling price of hydrogen, P^{BH} (USD/tonne) with the annual bio-hydrogen produced, H^{Prod} (tonne/y) (see equation (3)). The equation for calculation of total cost of purchasing NG is shown in equation (4) where the price of NG, P^{NG}(USD/tonne) is multiplied by the amount of NG required, W_y^{NG} (tonne/y). It is worthy to note that the annual bio-hydrogen produced, H^{Prod} and the amount of NG required, W_y^{NG} (tonne/y) is extracted from the P-graph results. The Monte Carlo simulation model is developed via Microsoft Excel software and 10,000 iterative runs were performed (note with such large arbitrary sample size, the reliability of the obtained results can be assured and is more reliable to serve as a guidance for decision-makers). The randomized function, RAND (), and Norm.Inv () in Microsoft Excel are used to randomize the input uncertainty variables based on the mean and standard deviation of the uncertainty (assuming the uncertainty is expressed in normal distribution). The 10,000 iterative runs can be simultaneously calculated and generated using the Data Table function. The uncertainties to be randomized in this work are the selling price of hydrogen, P^{BH} and purchasing price of natural gas, P^{NG}.

3. Illustrative Case Study

A Sarawak case study is adopted in this work to demonstrate the applicability of the proposed methodology. A total of 18 palm oil mills were considered as biomass source points whereas two ammonia plants were treated as hydrogen demand points. To reduce the complexity of hydrogen network structure, palm oil mill which has POME capacity lower than 150,000 tonnes per year will be excluded in P-graph synthesis. The proposed compressor substations are located along with Pan Borneo Highway Sarawak. In this case study, hydrogen sources and sinks are connected by pipelines. To overcome the shortage in pressure along the pipelines, compressor substations are constructed to maintain flow pressure. However, certain regions are not suitable for constructing compressor substations while considering the feasibility, such as mountain areas and residential areas. This screening step acts as a crucial step to avoid outcomes that are infeasible in real life. Figure 1 shows the P-graph model developed *via* anaerobic digestion process. Subsequently, the results extracted from P-graph software can undergo stochastic optimization *via* Monte Carlo simulation model. The uncertainties integrated into the Monte Carlo simulation model are tabulated in table 1.

Table 1: Fuzzy Ranking of Process Network Pathway Extracted from P-graph Framework.

Uncertainty	Units	Lower bound	Upper bound	Mean	Standard deviation
Hydrogen selling price	USD/tonne	2,200 [10]	9,660 [10]	5,930	3,730
NG purchasing price	USD/m^3	0.292 [11]	0.140 [11]	0.228	0.048

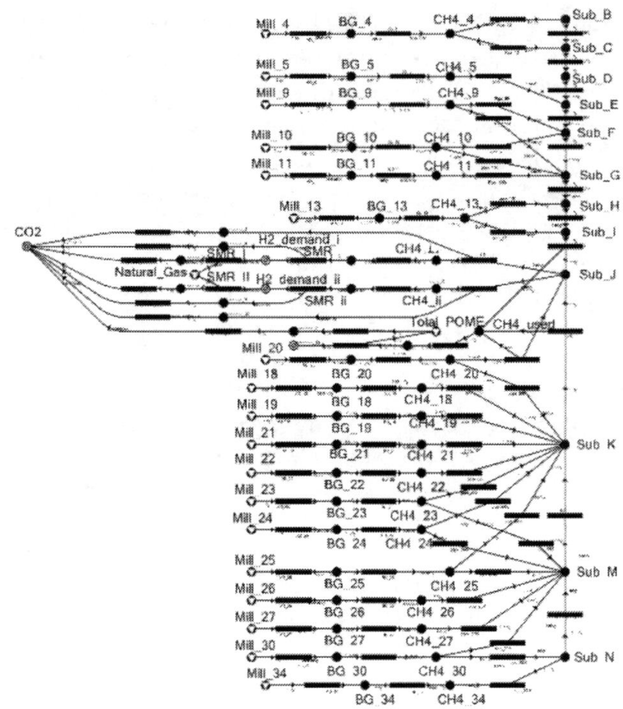

Figure 1: Maximum structure of hydrogen supply network using anaerobic digestion.

4. Results and Discussion

Table 2 summarized the results parameters extracted from the P-graph model that will serves as input to the Monte Carlo simulation model.

Table 2: Ranking of Process Network Pathway Extracted from P-graph Framework.

Rank	W_y^{NG} (m^3/y)	H^{Prod} (tonne/y)	Capital cost (USD/t POME)	Operating cost (USD/y)
1	118,021,000.00	37,170.00	101.05	10.10
2	118,021,000.00	37,170.00	101.06	10.10
3	118,021,000.00	37,170.00	101.10	10.10

Figure 2 presents the structural results extracted from the P-graph model for the first three ranked network. The P-graph model optimized the network with the aim of maximizing profit. Figure 2(a) and figure 2(b) have a similar structural network where the main difference is one of the pipe connections. One different pipe connection between the two structures had led to the difference in capital cost by USD 0.05 per tonne of POME. After inputting the required parameters from table 1 into the Monte Carlo simulation model, the probability density curve for the three ranked networks can be plotted as shown in figure 3. The mean NPV for the first, second, and third-ranked network pathway is USD 1,202.12 million, USD 1,198.90 million, and USD 1,197.21 million whereas the standard deviation for the first, second, and third-ranked network pathway is USD 298.46 million, USD 301.24 million, and USD 294.28 million. The main reason which contributed to the difference in mean NPV value is the capital investment cost for the three ranked pathways. The operating cost, amount of natural gas required, and the annual hydrogen production for the three ranked network are the same. The large standard deviation calculated for the three pathways is due to the large standard deviation of the input uncertainty variable, hydrogen selling price, where the standard deviation is USD 3,370 per tonne of hydrogen. This will significantly impact the randomized values of hydrogen selling price for the calculation of NPV for 10,000 iterations. It is noteworthy that the results obtained are dependent on the values of input parameters, and variables available and will be subjected to changes when the input parameters and variables are changed.

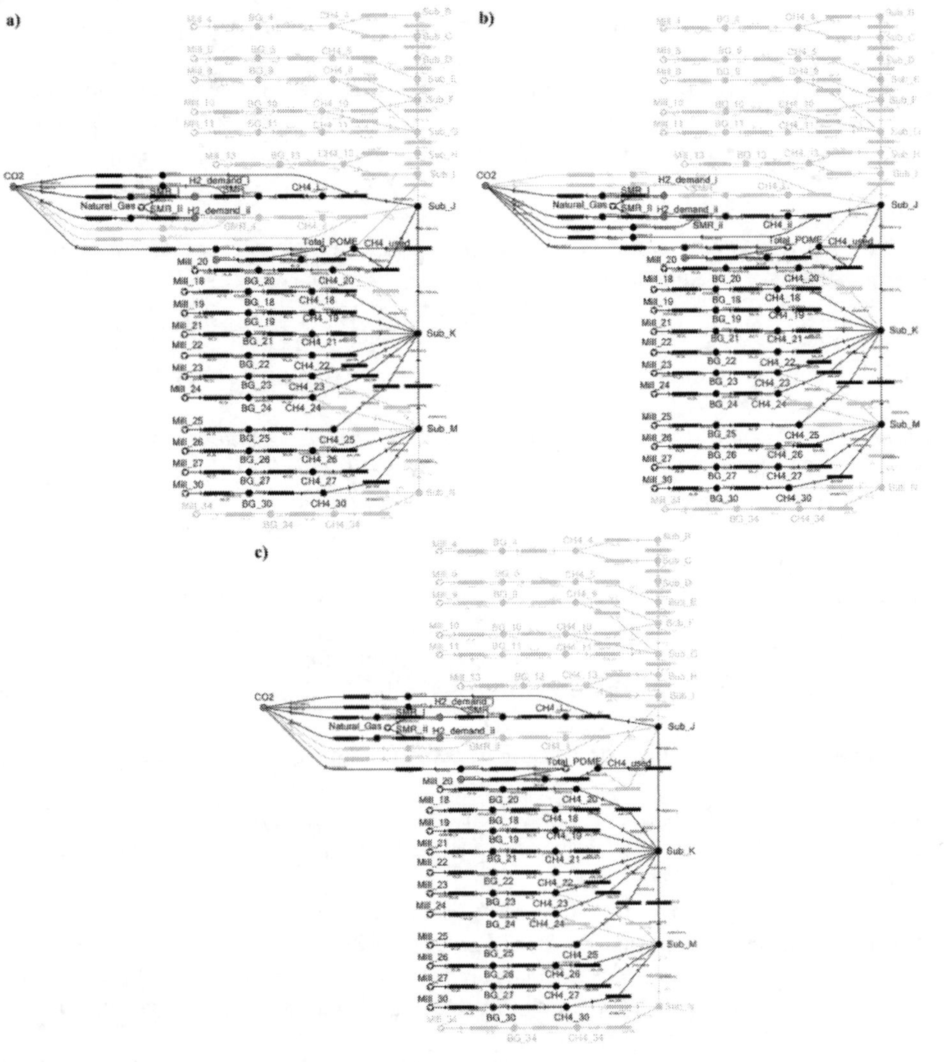

Figure 2: Network structure for (a) first ranked, (b) second ranked, and (c) third ranked.

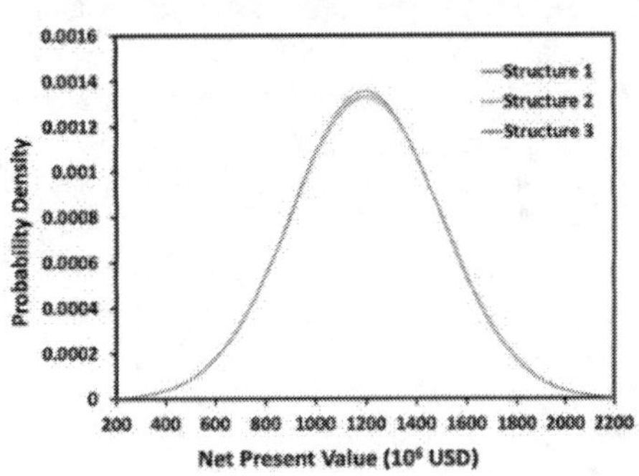

Figure 3: Probability density of NPV (top three ranked structures extracted from the P-graph model).

5. Conclusion and Future Work

This study has proposed optimization of bio-hydrogen supply chain network *via* P-graph framework integrated with Monte Carlo simulation model to generate a financial probabilistic result. The findings of this study had found that the first three ranked pathways had a similar probability distribution profile due to the same input parameter values with the exception of the capital investment cost. The lower capital investment cost for the first ranked pathway had resulted in the first ranked pathway being selected with a mean NPV value of USD 1,202.12 million. Possible extensions to the current study can include the integration of stochastic optimization to compare the economic and environmental probability profile for the selected network structure. This can present a more comprehensive analysis to evaluate the potential deviation from the economic and environmental outcome calculated in P-graph.

Acknowledgement

Authors would like to acknowledge the financial support by Ministry of Higher Education under Fundamental Research Grant Scheme [grant number: FRGS/1/2020/TK0/SWIN/03/3].

References

[1] Kumar R, Kumar A and Pal A 2020 An overview of conventional and nonconventional hydrogen production methods *Mater. Today: Proc* **46**

[2] *Another Record Number of Newly Opened Hydrogen Refuelling Stations.* Fuel Cells Works. 2021 [cited 2022 May 10] Available from: https://fuelcellsworks.com/news/another-record-number-of-newly-opened-hydrogen-refuelling-stations-in 2021/#:~:text=In%202021%2C%20142%20hydrogen%20refuelling,and%2013%20in%20North%20America.>

[3] Vidas L and Castro R 2021 Recent Developments on Hydrogen Production Techologies: State-of-the-Art Review with a Focus on Green-Electrolysis *Applied Sciences* **11** 11363

[4] Li L, Manier H and Manier M 2019 Hydrogen supply chain network: An optimization-oriented review *Renew. Sustain. Energy Rev.* **103** 342-360

[5] How BS, Yeoh TT, Tan TK, Chong KH, Ganga D and Lam HL 2018 Debottlenecking of sustainability performance for integrated biomass supply chain: P-graph approach *J. Clean. Prod.* **193** 720-733

[6] Foo DCY, Chew LX, Chong TL, Teh WH, Lam HL, Affery AP and Pang CH 2021 A Note for the Extended P-Graph Model for the Synthesis of Batch Water Network *Process Integr. Optim. Sustain.* **5** 675-686

[7] Tan RR, Aviso K and Walmsley T 2021 P-graph Approach to Solving a Class of Stackelberg Games in Carbon Management *Chem. Eng.Trans.* **89** 463-468

[8] Lee CH, Chong DYL, Hemmati S, Elnegihi MM, Foo DCY, How BS and Yoo C 2020 A P-graph approach for the synthesis of national-wide bio-hydrogen network from palm oil mill effluent, *Hydrog. Energy* **45(35)** 17200-17219

[9] Hemmati S, Elnegihi MM, Lee CH, Chong DYL, Foo DCY, How BS, and Yoo CK 2020 Synthesis of large-scale bio-hydrogen network using waste gas from landfill and anaerobic digestion: A P-graph approach *Processes* **8(5)** 505

[10] Contributed Content. *Hydrogen Prices Skyrocket Over 2021 Amid Tight Power and Gas Supply.* PowerMag. 2021 [cited 2022 May 15] Available from: https://www.powermag.com/blog/hydrogen-prices-skyrocket-over-2021-amid-tight-power-and-gas-supply/

[11] *Natural gas.* Trading Economics. 2022 [cited 2022 May 08] Available from: https://tradingeconomics.com/commodity/natural-gas

IOP Conf. Series: Materials Science and Engineering 1257 (2022) 012012 doi:10.1088/1757-899X/1257/1/012012

Experimental Studies on Transesterification of Mango (Mangifera indica) Seed Kernel Butter for Biodiesel Production

G Narayanasamy[1], TNP Padmesh[2] and S Krishnan[3]

Department of Chemical Engineering, Manipal International University, 71800 Nilai, Negeri Sembilan, Malaysia.
[1]gubera070898@gmail.com
[2]padmesh.tnp@miu.edu.my
[3]soundarajan.krishnan@miu.edu.my

Abstract. This research study involves transesterification of mango seed kernel butter using methanol as solvent and Tetra Hydro Furan (THF) was used as co-solvent in converting biodiesel. High density and higher acid value in biodiesel cause less thermal efficiency in the engine. The percentage yield of biodiesel has been studied under various parameters such as the molar ratio of methanol: oil, type of alkali catalyst and catalyst concentration (%). As the significant results, the highest biodiesel yield% was attained using the molar ratio of methanol to mango seed oil at 7:1 using sodium hydroxide as alkali catalyst with the catalyst concentration of 1.0% and THF co-solvent. The maximum %yield obtained was 90.79%. The produced biodiesel was characterized by Gas Chromatography-Mass Spectrometry (GC-MS). The properties of biodiesel were analysed, such as freezing point, acid value, saponification value, density, and specific gravity with the respective values of 15, 0.85 mg/g, 70.13 mg/g, 788 kg/m^3, and 0.788, 11 fatty acid methyl esters were found in different concentrations. Analysis of Variance (ANOVA) has been used as a statistical method for optimization to validate the results obtained. The biodiesel results met the ASTM-D6751 and EN14214 standards for use as a working fluid.

1. Introduction

Higher demand for non-renewable energy and environmental pollution at its peak due the increase in population and industrialization, making the world under crisis gave rise for the need of sustainable energy resource. Moreover, the energy from conventional resources is not only toxic (emission of carbon monoxide, carbon dioxide, nitrogen dioxide and sulphur) but also comes at a high cost. In order to meet this crisis, this has simulated to search of renewable and sustainable with cheaper alternative energy resource. Biodiesel industry has gained attention as an alternative to petroleum-based fuels to reduce major concerns of global warming that have been raised because of fossil fuel use. It is eco-friendly fuel because of its biodegradability, high flash point and reduced exhaust emission, and compatibility with existing fuel [1].

Mango seed kernel oil was chosen since it has less FFA percentage when compared with soybean and sunflower oil. Mango seed kernel oil has high carotenoid content that helps in biodegradation. Mango seed kernel oil has higher biodiesel yield when it is compared with biodiesel from corn oil. For biodiesel oil, a higher density means more mass of fuel per unit volume, which means more energy is available for work output per unit volume [2].

Mangoes are members of the Anacardiaceae family and belong to the genus Mangifera indica. During mango season, large amounts of seed waste are collected from households, marketplaces, and processing industries and dumped in landfills, leading to biochemical reactions that produce methane and leachate, which contaminate the environment and underground water. Mango seed kernel oil is in yellowish colour and has a melting point of 32- 42 °C. The mango seed kernel contains 70% of carbohydrates, 15% of fats and 5% of proteins and the other 10% consist of phosphorus and calcium [3]. Mango butter is another form of mango seed kernel oil. Mango butter is available in the market now. When the mango butter is processed, the oil fraction obtained from the processed is known as mango oil, mango kernel fat or mango butter. It is in semi-solid form in ambient temperature.

This research aim is to produce biodiesel by transesterification of mango seed kernel oil under the optimized conditions converting to biodiesel, since limited number of papers were available on transesterification using mango seed kernel oil, in Malaysia. The effects of methanol to oil ratio, catalyst type and concentration of catalyst on the yield of biodiesel were studied using different chemical as co-solvent (THF). The biodiesel has also been characterised through GC-MS and the data were analysed using ANOVA software.

2. Materials and Methods

2.1. *Materials*

The mango seed kernel butter was purchased at (Hermit Resource Sdn Bhd), methanol (99.9% purity), sodium hydroxide and potassium hydroxide from R&M Chemicals, tetra hydro furan, distilled water and phenolphthalein. The experiment was performed at Manipal International University.

2.2. *Transesterification*

The mango seed oil was esterified with 1% (w/v) of sulfuric acid with 300 ml of methanol and reaction time at 60°C for an hour. While FFA percentage was less than 2%, transesterification would be conducted. As for transesterification, 100g of mango butter weighed and melted at 35 °C for 20 minutes.

100 mL of melted mango seed kernel butter and 20 mL of tetra hydro furan added into the 250mL round bottom flask. The mango seed kernel oil heated till it reached 60°C at a constant agitation speed of 600 rpm. 1 %w/v of potassium hydroxide pellets weighed mixed into a beaker that contains 600 mL of methanol and stirred till it was dissolved. Potassium methoxide was formed, and it was poured into the preheated mango seed kernel oil. The temperature of the mixture was maintained at 60 °C at agitation speed of 600 rpm for 60 minutes. The mixture temperature was kept at 60 °C at agitation speed of 600 rpm for 60 minutes. To ensure thorough separation, the biodiesel and glycerin was transferred to a separatory funnel and kept for overnight. Transesterification of mango seed oil done with a few optimized parameters. (4:1, 5:1, 6:1 and 7:1) were the molar ratios of methanol to mango seed oil that tested. Base catalysts that used were sodium hydroxide (NaOH) and potassium hydroxide (KOH). Concentration of the catalyst tested using 0.5, 1.0 and 1.25 %(w/v). Reaction time was fixed at 60 minutes and reaction temperature at 60°C. The biodiesel yield was calculated. The lower layer of glycerol from the bottom of funnel removed.

To avoid emulsion, the biodiesel rinsed using distilled water to eliminate undissolved methanol. Anhydrous sodium sulphate was used as adsorbent to remove excess water from fatty acid methyl ester. To remove contaminants and reduce the probability of saponification, the obtained methyl ester washed with 1 gram of sodium chloride was diluted in 250 mL of distilled water and rinsed three times with warm distilled water.

Mango seed oil biodiesel was characterized using (GC-MS) method, freezing point, acid value, density, saponification value and specific gravity. Once the characterization of mango seed oil biodiesel was done, the properties were compared with used vegetable oil biodiesel.

IOP Conf. Series: Materials Science and Engineering 1257 (2022) 012012 doi:10.1088/1757-899X/1257/1/012012

3. Results and Discussions

3.1. Free fatty acid test
Higher content of FFA in mango butter deactivates the occurrence of base catalyst and also leads to the glycerol in gel form during transesterification. Hence, FFA test is necessary before conducting transesterification of mango seed kernel butter. 2 FFA test conducted using same mango kernel oil sample. The obtained free fatty acid value in mango butter sample were 0.82% and 0.93% which were less than 2%. Thus, esterification was not done before conducting transesterification of mango seed kernel butter.

3.2. Transesterification of biodiesel

3.2.1. Effect of alkali catalyst. Catalyst increases the rate of transesterification reaction process and make the biodiesel and glycerol layer separation easier and also in reducing reaction temperature with complete conversion of biodiesel. Even though KOH was highly soluble in alcohol due to its molecular size, it was found that NaOH obtained higher biodiesel yield. Studies had shown that mixing for NaOH and methanol would give more yield when compared with KOH [4]. Figure 1 clearly shows that NaOH as catalyst gave a higher yield when compared to KOH. This This is because when transesterification process started with the attack of methoxide ion on the carbonyl carbon atom of the triglyceride molecule, both KOH and NaOH had higher dissociation rate resulting in lower molecular weight. Hence, available hydroxide ions in NaOH increased the methoxide anion production rate. Thus, this has proved in figure 1 that shows the influence of KOH was less than NaOH on biodiesel yield.

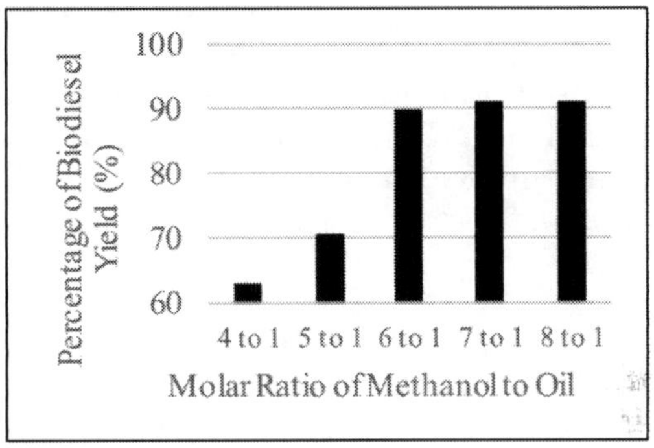

Figure 1. Influence of alkali on biodiesel yield.

3.2.2. Effect of molar ratio of methanol:oil. Though alkali catalysed, the transesterification reaction requires optimum amount of methanol for higher biodiesel conversion. It was found that molar ratio of 7:1 has achieved highest biodiesel yield which was 91.02%. It was observed that lower molar ratio of methanol has higher mass of glycerine. Excess amount of methanol needed for higher conversion of methyl ester. Based on figure 2, the biodiesel yield of methanol to oil ratio of 6:1 was slightly less than ratio of 7:1. Higher ratio of methanol to oil is also needed to complete the reaction faster. Excess alcohol molecules will be converted into ethyl esters in shorter time. This is why there was rise in biodiesel. Hence, 7:1 molar ratio was chosen as optimum parameter for biodiesel yield.

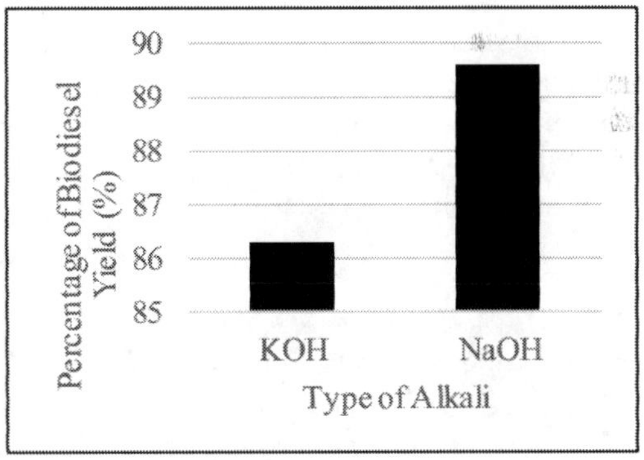

Figure 2. Influence of molar ratio of catalyst on biodiesel yield.

3.2.3. Effect of catalyst concentration. The fatty acid methyl ester obtained at 1.0% of sodium hydroxide catalyst was 91.02%. Usage of 1% of catalyst concentration during transesterification process, has obtained the least glycerol. It was observed and verified during separation of biodiesel and glycerol. Catalyst concentration of 0.5% caused partial conversion of triglycerides into fatty acid esters resulted in lowest biodiesel yield among the 3 different catalyst concentration which was 73.76%. At the catalyst concentration of 1.25%, it has shown reduced biodiesel yield, proving that additional amount of catalyst will react with triglycerides. However, the biodiesel yield was lower though the weight of methyl ester was the highest. It was clearly showed in figure 3 that 1.0% of sodium hydroxide concentration has given the highest yield. Thus, further addition in catalyst concentration impacted negatively in biodiesel yield. This has proved that at the concentration of 1.0% of catalyst, higher fatty acid esters have formed, by a complete reaction.

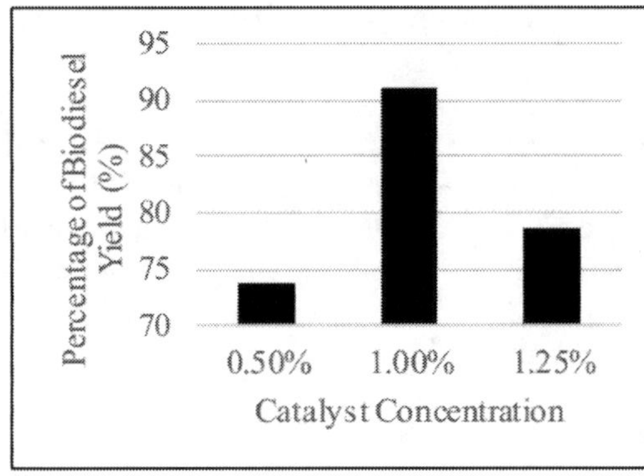

Figure 3. Influence of catalysts on biodiesel yield.

3.2.4. Effect of co-solvent. Short chains alcohol like methanol gives better conversions of biodiesel under the reaction time of an hour. However, co-solvent increases the solubility between methanol and mango seed kernel oil. Biodiesel yield increased once cosolvent was added to the reaction. The biodiesel yield with the addition of co-solvent was 94.11% and the weight of methyl ester obtained was high. This was because co-solvent accelerates the reaction by making mango seed kernel oil more soluble in methanol by improving the contacts between the reactants. Based on figure 4 could clearly notice the yield of biodiesel high with the addition of co-solvent. Co-solvent has improved the rate of reaction by making the mango seed kernel oil more soluble in solvent. At the agitation speed of 600 rpm, the reaction maximizes the surface contact between mango seed kernel oil and methanol resulting in strong mass transfer.

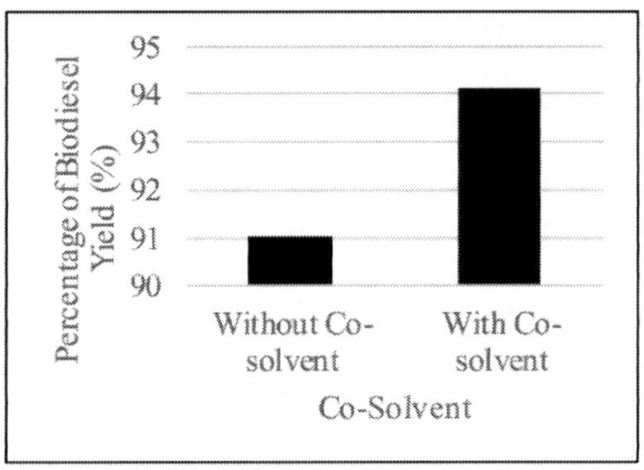

Figure 4. Influence of co-solvent concentration on biodiesel yield.

3.3. Biodiesel characterisation

Table 1 Comparison of the biodiesel produced with other biodiesel standards.

Fuel Property	Standards: Biodiesel ASTM D6751	Biodiesel EN14214	This work	Mango Seed Kernel Oil
Acid Value	≪0.8		0.85	
Density		860-900 kg/m^3	788 kg/m^3	
Specific Gravity	0.88		0.788	
Saponification Value			70.13 mg/g	210 mg/g
Freezing Point			15°C	30°C

Acid value must be checked in biodiesel to find its corrosive nature. If the acid value was high, there was higher amount of water presence in biodiesel. Acid value also indicated the quality of biodiesel. If the acid value was less, the quality of biodiesel was high [5]. Higher acid value in vegetable-based biodiesel was due to its unsaturated bonds in vegetable oil [2]. According to ASTM D6751, acid value in biodiesel should be less or equal to 0.8 [6]. Based on table 1 the acid value of biodiesel from mango seed kernel oil was 0.85 mg of KOH/g of oil. Density of mango seed kernel biodiesel obtained was 788 kg/m^3 at the room temperature. The range of biodiesel under EN 14214 should be in the range of 860-900 kg/m^3 at the temperature of 15°C [4]. The obtained value was less than given standard value but not much of difference. Specific gravity obtained from calculating was 0.788. Under ASTM D6751 standards, the required biodiesel should be in the value of 0.88. The value obtained was nearer the given value. If specific gravity of biodiesel was not considered, it might likely damage the engine. Presence of water in biodiesel also cause formation of soap or in other word, undergoes saponification. Saponification value will be high if the content of free fatty acid is high in biodiesel. The saponification value in mango seed kernel oil should be around 210 mg/g. Based on the table 1, saponification value was 70.13 mg of potassium hydroxide/ g of oil, which is less than the mango seed kernel oil saponification value. The freezing point of biodiesel was 15°C as stated in table 1. It was also clear the different source of vegetable oil-based biodiesel has different freezing point. The freezing point of mango seed kernel oil was 30°C.

3.3.1. Gas chromatography- mass spectrometry. Based on the obtained results from the GC-MS analysis (table 2), 11 Fatty Acid Methyl Esters (FAME) have been identified. FAME names that present in the biodiesel were stated in table 2. Palmitic acid (0.1021%), stearic acid (2.4878%), linoleic acid (0.527% and 0.199%) is present in the biodiesel. It can be seen that the acid percentages were less in biodiesel compared to mango seed kernel oil. The initial yield obtained was 94.11%. It was then reduced with the fatty acids that obtained from GC-MS analysis. Therefore, the final yield was 90.79%.

Table 2 Fatty acid methyl ester composition.

PK	RT	Area Pct	Library/ID	Ref	CAS	Qual
1	13.0489	0.1656	Butanoic acid, hexyl ester	39518	002639-63-6	86
2	24.2619	0.0522	Methyl tetradecanoate	95859	0124-10-7	99
3	27.0125	0.0651	Benzoic acid, 2-hydroxy-, phenylmethyl ester	84374	000118-58-1	81
4	27.8326	6.552	Hexadecanoic acid, methyl ester	119400	000112-39-0	99
5	29.3908	0.0511	Hexadecanoic acid, 14-methyl-, methyl ester	131316	002490-49-5	97
6	30.4696	3.1341	9,12-Octadecadienoic acid (Z,Z)-, methyl ester	139724	000112-63-0	99
7	30.7157	27.1385	cis-13-Octadecenoic acid, methyl ester	141299	1000333-58-3	99
8	31.1446	33.136	Methyl stearate	143126	000112-61-8	99
9	33.5041	0.1361	cis-13-Eicosenoic acid, methyl ester	164512	1000333-52-1	99
10	33.8826	0.4429	Methyl 18-methylnonadecanoate	166215	1000352-20-6	99
11	36.5764	0.0515	Docosanoic acid, methyl ester	186930	000929-77-1	98

3.3.2. Statistical analysis of mango seed biodiesel. The importance of each parameter and their interactions was determined using the ANOVA test (table 3). Due to the low and negligible noise, the chosen model is clearly significant with the highest F value (1.63) and lowest P value (0.3165). Mangifera indica biodiesel production was influenced by catalyst concentration (CC) and methanol to oil molar ratio. The most major significant variable in the synthesis of biodiesel with the greatest F value was CC (0.6809). The effect of reaction temperatures on biodiesel yield was negligible. This could be because the transesterification reaction slowed as the temperature raised. The lack of fit parameters is F value 23.34 and P value 0.1507 in table 3. The more relevant the associated coefficient was, the bigger the magnitude of the t-value and the smaller the P-values. The R^2 coefficient of determination was 76.42%, while the corrected R^2 (Adj R^2) coefficient was 71.92, indicating that the chosen model was accurate.

Table 3 Significant tests and ANOVA results.

Source	Sum of Squares	df	Mean Square	F-value	p-value
Model	478.79	3	159.60	1.63	0.3165
Catalyst Concentration	66.61	1	66.61	0.6809	0.4556
Ratio	310.13	1	310.13	3.17	0.1496
Temperature	0	0			
Reaction	0	0			
Agitation	0	0			
Co-solvent	0.2312	1	0.2312	0.0024	0.9636
Residual	391.33	4	97.83		
Lack of Fit	385.82	3	128.61	23.34	0.1507
Pure Error	5.51	1	5.51		
Cor Total	870.12	7			

4. Conclusion

In this study, mango seed butter is suitable to produce as methyl ester since it has same properties as the other biodiesel. After 8 test runs, it was concluded molar ratio of methanol to oil was 7:1, catalyst concentration of 1.0%, sodium hydroxide type of alkali catalyst and presence of THF as co-solvent were the chosen parameters to obtain 90.79 % yield of biodiesel. Molar ratio of methanol to oil, concentration of alkaline catalyst, type of alkaline catalyst and presence of co-solvent were key parameters in this experiment. GC-MS results stated that there was less percentage of FFA. According to ANOVA analysis that have done for this study, experimental data had satisfied the statistical data. Mango seed kernel oil biodiesel had the same properties to other edible seed oil biodiesel. It also satisfied ASTM-D6751 and the EN14214 standard requirements for biodiesel. Therefore, mango butter biodiesel was a potential replacement for petrol.

Acknowledgments

I would like to thank the Management of Manipal International University, Nilai for providing the facilities to carry out my research work. I am also extremely grateful to my dearest parents for constantly providing me emotional, mental, or financial support and for their sacrifices for me.

References

[1] Ashok B and Nanthagopal K 2019. Eco Friendly Biofuels for CI engine applications. *Advances in Eco-Fuels for a Sustainable Environment* **1** 407–40

[2] Aliyu and Oluwakemi A. A comparative study of biodiesel produced from some vegetable and waste oils in Kaduna

[3] Momoh O R, Audu H I and Binta Z B 2014. Investigating the production of biodiesel from Alphonso mango seed oil. *Nigerian Journal of Technology* **33(4)** 497

[4] Atapour M, Kariminia H R and Moslehabadi P M 2014. Optimization of biodiesel production by alkali-catalyzed transesterification of used frying oil. *Process Safety and Environmental Protection*, **92(2)** 179–85

[5] Musa U, Ibrahim A M, Sadiq M M, Aliyu A M, Suleiman B and Segun T 2014. Production and characterization of biodiesel from Nigerian mango seed oil **1** 2078-0958

[6] Hiwot T 2018. Mango (*Magnifera indica*) seed oil grown in Dilla town as potential raw material for biodiesel production using NaOH-A homogeneous catalyst **4(4)** 198-205

Sustainable Methane Emissions Management

H H Hashim[1,3], **N A Abdul Talip**[1,4], **M H Muhamad Pikri**[1,5], **N A Abu Bakar**[2,6] **and A Wan Zakaria**[1,7]

[1]Group Technical Solutions, Project Delivery and Technology Division, PETRONAS, 50088 Kuala Lumpur, WP, Malaysia
[2]GHG Management, Group HSE, PETRONAS, 50088 Kuala Lumpur, WP, Malaysia

[3]hassaruddin@petronas.com.my, [4]arnida_talip@petronas.com.my, [5]mhafiz_pikri@petronas.com.my, [6]nuratikah@petronas.com.my, [7]aidi_wanzakaria@petronas.com.my

Abstract. Methane is one of the primary and more potent Greenhouse Gas (GHG) with Global Warming Potential (GWP) of 28-36 compared to CO_2. In Malaysia, the Energy Sector including Oil and Gas Industry contributes ~20% of GHG emissions out of total GHG emissions as of 2020 majorly from main sources coming from flaring, venting and combustion. Other methane sources, like fugitive emission is currently quantified by high level estimation while from other small intended sources such as compressor seals and pneumatic devices are usually not quantified and reported. Therefore, for PETRONAS to meet stakeholders' expectations and be on par with other global peers, a comprehensive and sustainable Methane Emissions Management is required to accurately quantify and reduce methane emissions. Methane emissions was calculated based on Scope 1 emissions from gas value chain facilities. In 2020 and 2021, PETRONAS have conducted methane emissions quantification baseline study on LNG, gas processing, gas transmission plants as well as upstream facilities using process information and ground measurement data. Based on the studies, it is proven that about 70-80% methane emissions were from venting, flaring and fugitive emissions. Comprehensive methane emissions will provide a better purview on methane contributions which will facilitate further actions on reductions.

1. Introduction

The crucial part of the effort to tackle climate change is to cut methane emissions through enhanced methane management. There are five areas that encompass Methane Emission Management program: Measurement, Quantification, Verification, Reporting and Reduction. Measurement covers the activities to detect and measure methane emission parameters (e.g. concentration or flowrate) from identified sources. Quantification covers the activities to determine the amount of methane emissions from each common methane sources using measurement parameters obtained from either direct or indirect measurement techniques. Verification covers the activities to verify the quality of emission source data, quantification approach and relevant work processes to ensure credible and transparent methane emission data. Reporting covers disclosing of methane emission data to meet internal and external stakeholders' requirement. Reduction covers the activities to eliminate or reduce methane emission from the quantified sources using available abatement technologies.

Gradual reduction of methane emissions would be able to be achieved with effective GHG emissions monitoring from all relevant sources. PETRONAS has been reducing GHG emissions which includes methane emissions since introduction of PETRONAS Carbon Commitments in 2012. Major methane emissions reductions are from continuous flaring and venting reductions in upstream. These 5 areas need to be sustainable to effectively manage methane emission to reduce global warming.

2. Materials and methods

The key step in reducing methane emissions is to identify and detect sources of emissions. Subsequently, the emissions are then measured, quantified and recorded in inventories, which are a starting point for prioritization of mitigation activities in order to reduce emissions.

Sources of methane emissions comprise of both intended and unintended sources. The intended methane emissions sources refer to the process systems that are designed to emit methane gas within the permissible limit. In some cases, methane emissions from these systems will be routed to common flare/vent header where methane gas will be combusted or vented. Nine (9) typical intended methane emissions sources are as flaring, venting, stationary combustion, compressor seals, pneumatic controls and pumps, liquid unloading from well, hydrocarbon storage tanks, gas dehydration and well test.

Identification and detection of methane emissions sources should also take into account unintended releases, where these emissions are not expected and not designed to emit methane emission. Besides contributing to GHG emissions footprint, these methane emissions also pose threats to cause process safety-related incidents i.e., fire or explosions if not regularly monitored and mitigated. Two (2) typical unintended methane emissions sources are fugitive leaks and loss of primary containment (Tier 3).

2.1. Measurement and Quantification

The oil and gas industries are still making continuous effort in developing regulations of methane emission measurements for effective monitoring and ultimately setting the targets for emission reduction. The choice of methods to be used to estimate emission depends on how the estimates will be used and the degree of accuracy required. Methods using site-specific data from direct measurement are preferred over other methods relative to accuracy and confidence level. Figure 1 shows the five-level approach of measurement and quantification sorted into following declining order with respect to accuracy and confidence level (Ref [5]).

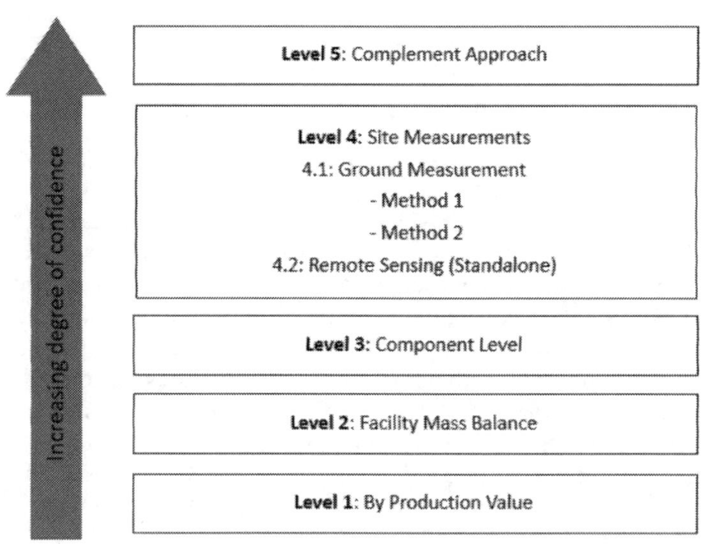

Figure 1. Five-level approach of measurement and quantification.

The amount of methane emissions is to be quantified for each respective source identified in Section 2 and totalled up to yield the amount of methane emission from the facility in tonne per year (tonne/yr) or million standard cubic feet per year (mmscf/yr) or standard cubic meter per year (sm3 /yr). Each facility should have its own quantified methane emissions and the numbers will not be generic. Each facility also should strive to apply the higher level of measurement and quantification for each of the 11 sources such as quantification using measurement from flow measurement device.

2.2. *Verification and Reporting*

In order to provide assurance of the disclosure and the credibility of the methane emissions reported data to be rolled up in the overall GHG emissions (Ref [2]), external verification or/and audit by relevant third-party (i.e. Environmental Resources Management) are performed and undertaken as part of the scope in GHG verification exercise.

All facilities across the gas value chain (i.e. Upstream and Gas Business) should report their methane emissions on annual basis, considering the leak survey exercise is conducted at minimum, annually to measure and quantify methane emission from fugitive emissions (Ref [5]. If the leak survey exercise is conducted more than once in a year, an average number or the most representative methane emissions number will be used. The reported methane emissions should include all methane emissions sources applicable to each facility. For effective monitoring of methane emission, each facility will report the following indicators:

2.2.1. Indicator 1. Absolute Methane Emissions – The amount of methane emission from all applicable sources. The amount is represented in emissions rate. Units can be in tonne/yr or mmscf/yr or sm^3/yr.
2.2.2. Indicator 2. Methane Emissions Intensity- The percentage release of methane emissions from the facility over the total production or throughput. Unit is in %.

2.3. *Reduction*

At any time, the project (both brown or green field) should consider minimising methane emissions in the facilities design and equipment selection at the very early stage of the project (i.e., FEL2 or FEL3 stage). Table 1 shows the list of the recommended technologies and methodologies to reduce methane emissions from 4 major sources and the respective equations used in quantification.

Table 1. Recommended Technologies and Methodologies to Reduce Methane Emissions.

	Methane Emission Sources	Methane Emission Reduction Options	Reflection of Methane Emission Reduction in Quantification Formula (Ref [1])
1	Flaring	Improve flaring efficiency by changing flare tips with single nozzle to tips with multiple nozzles / nozzles of different shape that will increase combustion efficiency due to better mixing. Methane emission from flaring reduction is reflected in lower ***% residual CH4*** in Equation (1).	$\textbf{Emission rate}\left(\frac{\textbf{ton}}{\textbf{yr}}\right) = V \times CH_4\ \textbf{mole fraction} \times \%\ \textbf{residual}\ CH_4 \times \frac{1}{\textbf{molar volume conversion}} \times MW_{CH4} \times AOD\left(\frac{days}{yr}\right)$ (1) *V = Volume flared (scf/d or sm³/d) at 15°C and 1 atm* ***% residual CH₄*** *= Non-combusted fraction of flared stream (default = 0.5% or 2%)* ***Molar volume*** *= Conversion from molar to mass (379.3 scf/lbmole or 23.685 m³/kgmole)* ***MW*** $_{CH4}$ *= CH4 molecular weight, 16.04 kg/kmol* ***AOD*** *= Annual operating days (days/yr)*

2	Venting	When gas utilization is not economically or technically possible, may consider converting existing vent stack/boom to flare stack/boom without energy recovery instead of venting. Methane emission from venting reduction is reflected in lower ***VR*** in Equation (2).	$\boldsymbol{Emission\ rate} = \boldsymbol{VR} \times \boldsymbol{F} \times \frac{\boldsymbol{MW}}{\boldsymbol{molar\ volume\ conversion}}$ (2) ***VR*** = *Annual vent rate in volume at STP conditions (at 15°C and 1 atm) per unit of time (i.e scf/yr or sm³/yr)* ***F*** = *Methane mol fraction (mol/mol)* ***MW*** = *Methane molecular weight (kg/kmol)* **Molar volume conversion** = *Conversion from molar to mass (379.3 scf/lbmole or 23.685 m³/kgmole)*
3	Stationary Combustion	Improve the efficiency of fuel combustion, by reducing incomplete combustion for engines, turbines and fired heaters. May consider installing automated air to fuel ratio controls or replacing cylinder unloaders as unloaders leak methane through o-rings, covers and pressure packing. Methane emission from venting reduction is reflected in lower ***M*** in Equation (3).	$\boldsymbol{Emission\ rate} \left(\frac{ton}{yr}\right) = \boldsymbol{M} \times \boldsymbol{GHV} \times \boldsymbol{EF} \times \boldsymbol{AOD}$ (3) ***M*** = *Fuel consumption quantity (ton/day)* ***GHV*** = *Fuel gross heating value (MMBTU/ton)* ***EF*** = *Specific methane emission factor for a stationary combustion equipment (ton/MMBTU)* ***AOD*** = *Annual operating days (days/yr)*
4	Fugitive Leaks and LOPC (Tier 3)	Threaded connections and flanges are used to connect piping, equipment, and other components such as valves. There is the potential for natural gas to leak from threaded connections and flanges. These connections can sometimes be replaced with welded connections, which are much less likely to leak. Fugitive emission reduction is reflected in lower number of components leaking with ***Screened Value*** in Equation (4).	$\boldsymbol{Emission\ rate} \left(\frac{ton}{yr}\right) = \boldsymbol{A} \times \boldsymbol{Screened\ Value\ (ppmv)^{B}} \times \boldsymbol{MethaneWeight\ Fraction} \left(\frac{wt}{wt}\right) \times \boldsymbol{Annual\ Operating\ Hours} \left(\frac{hrs}{yr}\right) \div \boldsymbol{1000} \left(\frac{kg}{ton}\right)$ (4) ***A*** *and* ***B*** = *Constants developed in the correlation fitting with respect to type of leaked equipment (i.e valves, flanges, connectors, open end lines, pump seals and others)*

3. Sample Case Studies for Selected Facilities

The method for obtaining a holistic methane emission number for a plant is by covering accurate quantification plantwide based on the previous mentioned approaches in Section 2.1 for all applicable methane sources. Table 2 shows an example of capturing methane number for a gas processing plant.

Table 2. Gas Processing Sample Plant Methane Emission Quantification.

No.	Source	Fugitive Emission	LOPC	Pneumatic Device	Stationary Combustion	Flaring	Process Vents	Total Emission (ton/yr)
1	Plant 1	2.5	0.0	-	5.0	1.5	230.0	239.0
2	Plant 2	9.0	0.0	-	3.5	500.0	230.0	742.5
3	Utilities Area	0.3	0.0	-	77.0	0.0	0.0	77.3
4	Terminal	0.3	0.0	0.2	1.5	30.0	0.0	32.0
	Total Emission(ton/yr)	12.1	0.0	0.2	87.0	531.5	460	**1090.8**
	Equivalent GHG Emission(tonCO$_{2e}$/yr)	338.8	0.0	5.6	2436.0	14882.0	12880.0	**30542.4**

Figure 2 depicts the methane emission contribution by sources. Flaring and Process Vents are found to be the main sources of methane emission for this sample gas processing plant and this is due to the significant amount of flared and vented volumes.

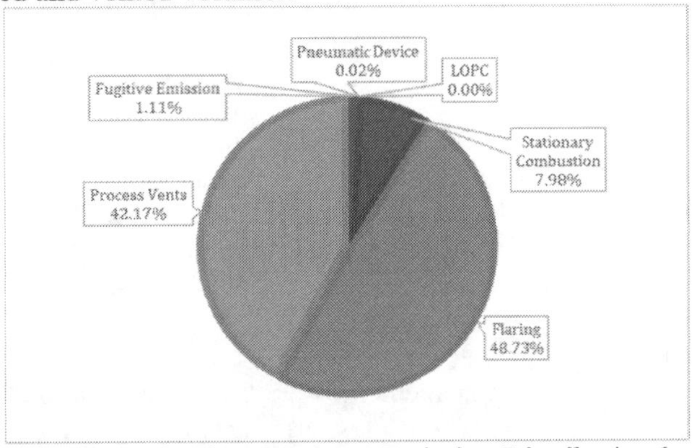

Figure 2. Gas Processing Sample Plant Methane Emission Distribution by Sources (in %).

Another case study is performed for a sample gas transmission and regasification unit for multiple regions. Table 3 shows the typical methane emission quantification for this sample unit. Different region may be associated with different major methane source. From this sample case, the main methane source for Region 1 is Pneumatic Device, for both Region 2 and Region 3 is Process Vents and for Region 4 is Pneumatic Device closely followed by Process Vents.

Table 3. Gas Transmission and Regasification Unit Methane Emission Quantification Result.

No.	Source	Fugitive Emission	LOPC	Pneumatic Device	Stationary Combustion	Flaring	Process Vents	Total Emission (ton/yr)
1	Region 1	27.0	0.4	65.0	0.0	0.0	2.0	94.4
2	Region 2	32.0	0.2	60.0	7.0	0.0	85.0	184.2
3	Region 3	26.0	0.0	56.0	0.0	0.0	150.0	232.0
4	Region 4	30.0	0.0	58.5	5.5	0.0	50.0	144.0
	Total Emission(ton/yr)	115.0	0.6	239.5	12.5	0.0	287.0	**654.6**
	Equivalent GHG Emission(tonCO$_{2e}$/yr)	3220.0	16.8	6706.0	350.0	0.0	8036.0	**18328.8**

Figure 3 shows the methane emission contribution by sources. From this sample gas transmission and regasification unit, the three main methane emission sources in sequence are Process Vents, Pneumatic Device and Fugitive Emission. Process Vents being the highest is due to the high vent volume, meanwhile Pneumatic Device is contributed by the total count of pneumatic devices utilizing instrument gas and Fugitive Emission is contributed by the fugitive leaks identified during on-site measurement.

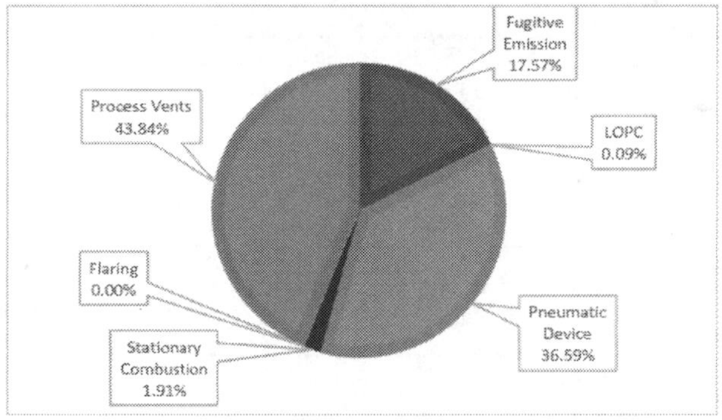

Figure 3. Sample Gas Transmission and Regasification Unit Methane Emission Distribution by Sources (in %).

4. Conclusion

Methane emission management efforts are a critical part of an Oil & Gas industry's strategy to thrive during the global energy transition by providing more and cleaner energy. The driver for effective management of methane emissions for oil and gas plants, would be through improving the accuracy of methane emission measurements and quantification as well as operational excellence (flaring reduction and electrification initiative).

The significant impacts of effective management of methane emission include reduction in potential occurrence of process safety incidents (fire/explosion), demonstrate support on aspiration towards Net Zero Carbon Emissions while continue promoting energy transition.

References

[1] Shires T, Loughran C, Jones S and Hopkins E 2009 *Compendium of greenhouse gas emissions methodologies for the oil and natural gas industry* **Section 3-6** https://www.api.org/~/media/Files/EHS/climate-change/2009_GHG_COMPENDIUM.pdf

[2] US EPA *Stationary Fuel Combustion Sources – Final Rule: Mandatory reporting of greenhouse gases* 2014, February [2022, May, 18] https://www.epa.gov/sites/production/files/2015-02/documents/stationaryfuel_infosheet.pdf

[3] US EPA *Climate Change* 2022 www.epa.gov/climatechange/emissions/ghgrulemaking.html

[4] https://www.ccacoalition.org/sites/default/files/resources/2017_OGMP-TGD1-Pneumatic-controls-and-pumps_CCAC.pdf

[5] Abu Bakar N A, Wan Zakaria A and Abdul Talip N A 2021 *Petronas Technical Standards (PTS) 18.75.03: Methane emissions quantification and reporting*

[6] Muniandi V, Narayanamoorthy T, Othman N.S and Inayat Hussain S H 2021 *Petronas Technical Standards (PTS) 18.06.04: Greenhouse gas (GHG) emissions management*

[7] United Nations Environment Programme (UNEP) 2020 *Oil and gas methane partnership (OGMP) 2.0 reporting framework*

[8] Willems K and Tielemans J 2017 *The Sniffers Manual Conform EPA Method 21 SOCMI*

IOP Conf. Series: Materials Science and Engineering 1257 (2022) 012014 doi:10.1088/1757-899X/1257/1/012014

Process Integration and Optimization for Sustainable and Circular Hydrogen Production

M Lim Moy Hian[1], C H Lim[1,*], B S How[2], W P Q Ng[3] and H L Lam[4]

[1]School of Engineering and Physical Sciences, Heriot-Watt University Malaysia, Jalan Venna P5/2, Precinct 5, 62200 Putrajaya, Malaysia.
[2]Swinburne University of Technology Sarawak, Research Centre for Sustainable Technologies, Faculty of Engineering, Computing and Science, Jalan Simpang Tiga, 93350, Kuching, Sarawak, Malaysia.
[3]Petroleum and Chemical Engineering Programme Area, Universiti Teknologi Brunei, Jalan Tungku Link, Gadong BE1410 Brunei Darussalam.
[4]The University of Nottingham Malaysia Campus, Department of Chemical and Environmental Engineering, Jalan Broga, 43500 Semenyih, Selangor, Malaysia.

*E-mail : l.chun_hsion@hw.ac.uk

Abstract. With omnipresent issues concerning environmental degradation following the extensive use of fossil fuels, there is an urgent need to move towards more sustainable fuels such as hydrogen, which produces only water as by-product upon combustion. However, hydrogen production processes have limitations in terms of raw materials, process parameters and energy input which affects the sustainability of the product. As hydrogen demand is forecasted to increase, process optimization studies are favourable due to limited current work on this topic. In this paper, six hydrogen production processes are integrated into a superstructure with special consideration given to implementation of circular economies. Mathematical modelling of the superstructure then allows optimization of three case studies based on economic and environmental performance. This study shows the possibility of achieving zero net CO_2 emissions through water electrolysis, provided that electricity generated with the produced hydrogen is used to power the plant. Gross profits of 1489.14 RM/hr and 4075.60 RM/hr show the potential of Palm Oil Biomass as feedstock. Overall, implementation of circular economy was found favourable to achieve sustainable production. This framework created has the potential to act as decision- making tool for policy makers to assist them in determining optimum hydrogen production pathways.

1. Introduction

Nowadays, the rapid climate change caused by the emissions of greenhouse gases in the atmosphere is an ever-growing challenge. From 1990 to 2019, the amount of Carbon Dioxide (CO_2) emissions, the main greenhouse gas, has increased by 63 % [1]. The main reason for this is the global reliance on fossil fuels as an energy source. As such, there is an increasing urge to move towards greener, renewable fuels such as hydrogen which has the potential to decrease greenhouse gases emissions as its direct combustion only produces water as by-product. Moreover, it is the most abundant element and has a higher energy density and efficiency than fossil fuels [2]. However, although hydrogen combustion itself produces near net zero greenhouse gas emission, its production is often derived from fossil fuels and involves the release of carbon dioxide which affects the sustainability of the hydrogen produced. Some production methods also require large energy input, making the hydrogen expensive. Hence, with the various pathways for hydrogen production coupled with the aforementioned limitations it is important to find an optimized process pathway in terms of environmental impact and cost minimization.

Several research works have dealt with the optimization of hydrogen networks using mathematical optimization of superstructures and programming [3, 4]. However, although cost optimization and sustainability aspects were considered, they mainly focused on hydrogen refinery networks. Recent work by Ibrahim et al. (2021) dealt with a sustainable framework for hydrogen production but included only two production pathways with emphasis put on carbon sequestration [5]. Moreover, few studies consider the implementation of circular economies where usage of wastes is extended optimally [6], although circularity might be beneficial for more sustainable production. Hence, in this study a mathematically modelled superstructure will be developed, which can act as a decision-making tool for policy makers, to assist them in determining optimum hydrogen production pathways based on economic and environmental performances while considering circular economy concepts.

2. Methodology

The first section involves gathering information on available hydrogen production methods, from which the most mature ones were selected. These processes were then integrated in a superstructure with proper consideration given to waste management to implement circular economies as shown in **figure 1**.

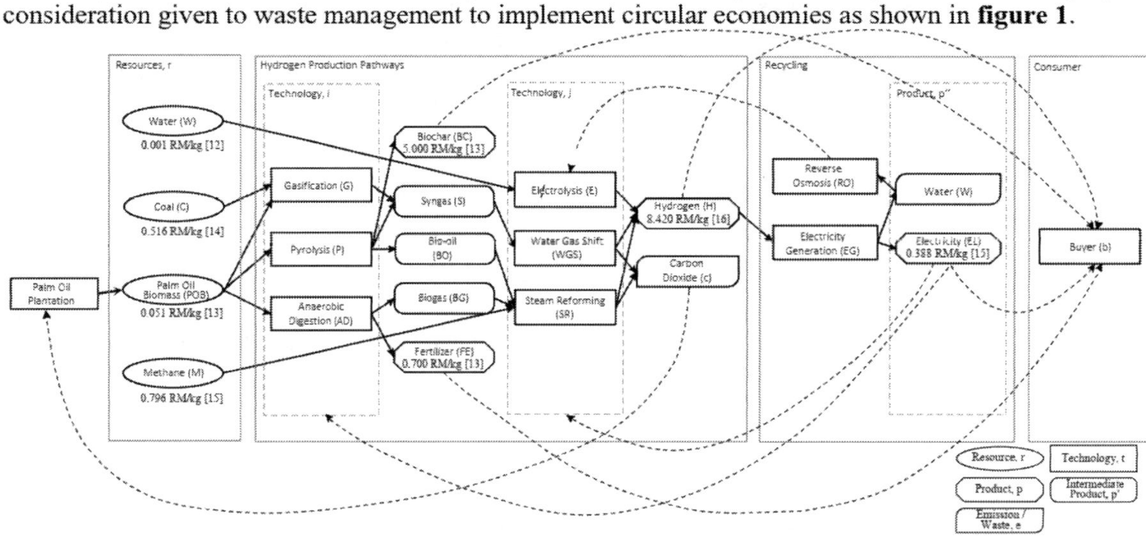

Figure 1. Superstructure for sustainable and circular hydrogen production pathway.

2.1 Modelling Equations and Data

To allow optimization, the superstructure was modelled using the software LINGO. First, hydrogen supplied to the consumer (H_1), was established as the modelling parameter. The flowrates of the different products (p) to the buyer (b) defined as $F_{p,b}$ are set at the desired throughput as illustrated in equation (1). The feedstock flowrates (r) to subsequent technologies (t which can be ti or tj), defined as $F_{r,t}$ were also limited based on their availability as shown in equation (2). Moreover, the flow of Palm Oil Biomass (POB) can be directed to 3 processes as illustrated in equation (3).

$$F_{p,b} \leq F_{p,b}^{demand} \qquad \forall p, b \in P, B \tag{1}$$

$$F_{r,t} \leq F_{r,t}^{availability} \qquad \forall r, t \in P, B \tag{2}$$

$$F_{"POB"} = F_{"POB","G"} + F_{"POB","P"} + F_{"POB","AD"} \tag{3}$$

The output flowrate from the technologies (technologies (i) or intermediate technologies (j)), can be defined in terms of the input flowrate and the corresponding conversion, X, as illustrated by the

IOP Conf. Series: Materials Science and Engineering 1257 (2022) 012014 doi:10.1088/1757-899X/1257/1/012014

following general equation. The input can either be resources (*r*), or intermediate products (*p'*). The outputs can be intermediate products (*p'*), hydrogen (H) or carbon dioxide (C).

$$F_{"input","technology","output"} = F_{"input","technology"} \times X_{"technology","output"} \quad (4)$$

The total flowrate of hydrogen produced ($F_{"H"}$), is then given by the sum of hydrogen flowrates from processing technologies (*tj*) and is defined as $F_{tj,"H"}$ in equation (5). From the total flowrate of hydrogen obtained, part is sold as product to the consumer ($F_{"H1"}$) and the remaining hydrogen is sent for electricity generation ($F_{"H2"}$) as shown in equation (6) below.

$$F_{"H"} = \sum_{tj} F_{tj,"H"} \quad (5)$$

$$F_{"H"} = F_{"H_1"} + F_{"H_2"} \quad (6)$$

For the net CO_2 emissions, the first source considered is the CO_2 produced by the processing technologies ($F_{tj,"c"}$) and the second one is that associated with the energy consumed by the processes Assuming that electricity imported from the grid, often derived from fossil fuels, has an average carbon footprint of 0.386 kg of CO_2 per kWh of energy, the associated CO_2 emissions ($F_{EL,"c"}$) can be obtained from equations (7) and (8). The total electricity required is calculated from equation (9). From H_2, the products, p'' (EL or W), can be obtained from equation (10). Part of the products (EL_1 or W_1) can be sent to the grid or discarded while part (EL_2 or W_2) can be recycled back as shown in equation (11).

$$EL^{import} = EL^{required} - F_{"H2","EG","EL_2"} \quad (7)$$

$$F_{EL,"c"} = EL^{import} \times 0.386 \quad (8)$$

$$EL^{required} = \sum_{r}\sum_{ti}\left(F_{r,ti} \times EL_{ti}^{consumption}\right) + \sum_{tj}\left[\left(\sum_{r}\sum_{p'} F_{r,tj} + F_{p',tj}\right) \times EL_{tj}^{consumption}\right] \quad (9)$$

$$F_{"H2","EG",p''} = F_{"H_2"} \times X_{"EG",p''} \qquad \forall p'' \in P'' \quad (10)$$

$$F_{"H2","EG",p''} = F_{"H2","EG",p_1''} + F_{"H2","EG",p_2''} \qquad \forall p'' \in P'' \quad (11)$$

In the case of palm oil biomass, a certain amount of CO_2 can be recycled back to the palm oil plantation, which accounts for the CO_2 absorbed by the Oil Palms. This value can then be used to offset the carbon emissions, giving the net carbon dioxide emission (c) of the process as shown below:

$$c = \left(\sum_{tj} F_{tj,"c"} + F_{"EL","c"}\right) - \left(F_{"POB"} \times c^{absorbed}\right) \quad (12)$$

For calculation of the gross profit, the cost of the raw material (C^{total}), the total OPEX and CAPEX is subtracted from the total revenue (R^{total}) as shown in the following equations. The input refers to either resources (*r*) or products (*p, p', p''*) to technologies, *t*, which can be intermediate technologies (*i*), main technologies (*j*) or recycling technologies.

$$C^{total} = \sum_{r} F_r \times C_r \quad (13)$$

$$OPEX^{total} = \sum_{"input"}\sum_{t} F_{"input",t} \times OPEX_{"input",t} \quad (14)$$

$$CAPEX^{total} = \sum_{"input"}\sum_{t} F_{"input",t} \times CAPEX_{"input",t} \quad (15)$$

$$R^{total} = \sum_{p} F_p \times R_p \quad (16)$$

$$GROSS\ PROFIT = R^{total} - C^{total} - OPEX^{total} - CAPEX^{total} \quad (17)$$

Data for modelling of the superstructure are tabulated in **table 1**.

Table 1. Modelling data for various technologies.

Tech.	Input (kg)	Output (kg)	X (kg/kg)	Energy Consumption (kWh/kg)	CAPEX (RM/kg)	OPEX (RM/kg)	Ref.
G	C	S	1.683	0.280	0.150	0.180	
	POB	S	2.039				
P	POB	BC	0.040				
		S	0.043	0.180	0.141	0.171	[7, 8]
		BO	0.248				
AD	POB	F	0.897	-	0.202	0.173	
		BG	0.063				
WGS[a]	S[b]	H	0.069				
		c	1.423	0.150	0.056	0.175	[9]
	S[c]	H	0.060				
		c	1.418				
SR	BO	H	0.066				
		c	0.969	0.183	0.416	0.679	[10, 9]
	BG or M	H	0.339				
		c	2.709				
E	W	H	0.111	0.333	0.746	[d]	[11]

a: WGS also takes into consideration pressure swing adsorption for separation of hydrogen from carbon dioxide
b: Syngas derived from coal gasification.
c: Syngas derived from gasification and pyrolysis of palm oil biomass.
d: OPEX is taken as sum of raw material and cost associated with electricity consumption.

2.2 Case Studies

To assess the mathematical model, a set of case studies were conceptualized. **Case 1** assumes only coal, methane, and water as possible feedstock (conventional practice). For **case 2**, alternative feeds (biomass) are considered and finally for **case 3**, it is assumed that with the increasing pressure to achieve net zero emission by 2050 and the rising popularity of hydrogen as fuel, the Malaysian government opens the Feed-in-tariff program for hydrogen fuel cells at a capacity of 20 MW and a FiT rate of 0.400 RM/kWh. In all cases, two objective functions are set which are the minimum net carbon emission pathway (A) and the maximum gross profit (B) pathway. In this study, the country considered is Malaysia and the product outputs are set based on the local market demand. In Malaysia, the use of petroleum, natural gas and coal is still predominant, accounting to a total of 94% of the primary energy consumption in 2019 according to the Energy Information Administration (2021). As of 2022, although hydrogen as a fuel is gaining popularity, Malaysia still lacks infrastructure to accommodate its use. As a result, the demand for hydrogen is set in the lower range of 100 kg/hr. The output of biochar and fertilizer is set based on currently existing plants in Malaysia and are given a maximum limit of 5000 kg/hr. Finally, the output of electricity to be sent to the grid is set as zero since, as of 2022, Malaysia's government Feed-in- Tariff plan does not have any allowance for electricity derived from hydrogen.

3. Result and Discussion

Using the mathematically modelled superstructure, optimization of the cases outlined were carried out. The results obtained are displayed in **table 2** below. Results of the optimization gives the same hydrogen production pathway for cases 1A, 2A and 3A. This pathway involves the electrolysis of water to produce hydrogen. Part is sent to the electricity generation unit where all electricity produced is fed back to the electrolysis process. The water by-product however is discarded. From the results in **table 2**, this pathway produces net zero CO_2 emissions and is favourable for the production of green hydrogen. However, an important aspect in achieving this net zero CO_2 emission is the hydrogen-derived electricity which is recycled back to the process. This circularity allows the energy requirement to be covered by the electricity produced so that no electricity has to be imported from the grid. This is an important aspect since energy imported from the national grid is most often derived from fossil fuels and can thus add to the total CO_2 emission of the process. This is especially significant for electrolysis which itself has no associated CO_2 emission. The circularity thus allows it to truly become a 'green' process.

Table 2. Modelling results for minimization of CO_2 (A), and maximization of gross profit (B).

Variable	Case					
	1A	2A	3A	1B	2B	3B
Input flowrate (kg/hr)	1031.80	1031.80	1031.80	301.38	4791.69	5555.56
Electricity Usage (kWh)	343.59	343.59	343.59	55.71	55.24	522.48
Net CO_2 emissions (kg/hr)	0.00	0.00	0.00	816.44	811.28	7726.84
OPEX (RM/hr)	6.22	6.22	6.22	205.65	1034.94	3271.15
CAPEX (RM/hr)	776.61	776.61	776.61	126.49	1094.61	2721.30
Revenue (RM/hr)	842.00	842.00	842.00	842.00	3860.77	12342.00
Gross Profit (RM/hr)	58.08	58.08	58.08	270.06	1489.14	4075.60

For case 1B, the optimum pathway for the maximum gross profit, is by steam reforming of methane. Part of the hydrogen produced is sold directly to the consumer while part is sent to the electricity generation unit. From there, electricity is recycled back to the process. Implementation of the circularity allows an increase in the gross profit by eliminating utility cost incurred from electricity importation from the national grid. In this instance, implementation of the circular economy is favourable.

For cases 2B and 3B, similar optimized pathway was obtained as shown **figure 2**, which was that of POB anaerobic digestion to produce biogas and fertilizer. The former is then sent to the steam reforming process to produce hydrogen, part of which is sold to the consumer and part is sent for electricity generation, after which the electricity is recycled back to the process thus forming a circular economy model.

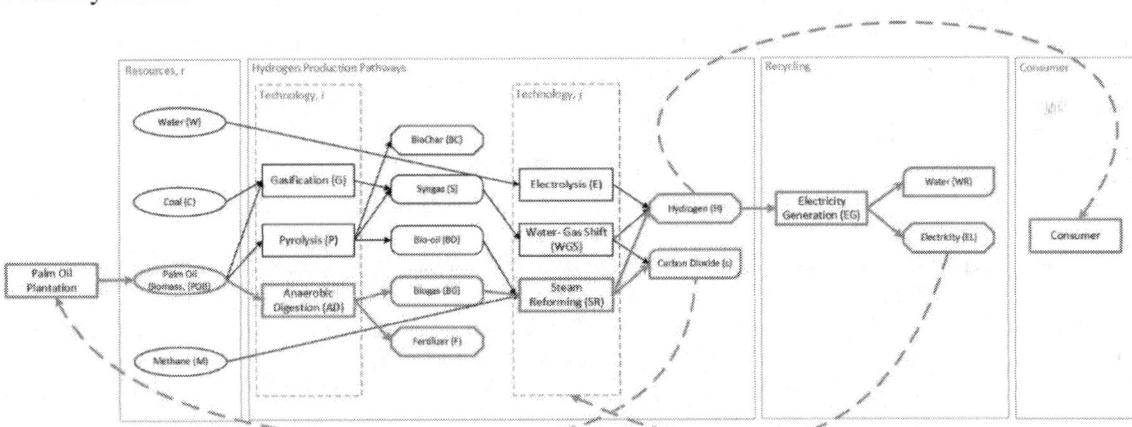

Figure 2. Optimized pathway for case 2B & 3B.

This pathway leads to higher gross profit as compared to the methane steam reforming process outlined in case 1B. This higher profit is accounted for by the fertilizer sold as product. Moreover, the cost of POB is significantly lower than methane. For case 3B, an increase of nearly 3 times can be observed in the gross profit. As the market for hydrogen-based electricity expands, the model encourages the production of hydrogen to be used as fuel for electricity generation to maximise the profit. In both cases, circularity in the pathway allows higher gross profit rates to be achieved.

4. Conclusion

In this paper, different hydrogen production processes were integrated in a superstructure which was modelled using LINGO and used to determine optimum hydrogen production pathways based on two objective functions: minimizing net CO_2 emission and maximizing gross profit. Special focus was also given to the implementation of a circular economy. Through application of the model, it was seen that circular economy has a positive impact on the net CO_2 emission of the electrolysis process. Moreover, when it comes to profit maximization, the model tends to favour the use of the palm oil biomass as it is cheap and abundant. Overall, it was also noted that maximum gross profit is favoured by the circularity While the model presented proves to be efficient, it can be further improved by integrating other technologies such as dark fermentation and autothermal reforming into the superstructure and multi objective functions can be added to the model. Also, modelling data can be further refined, and special attention could be given to economies of scale to account for small-scale and large-scale operations.

Acknowlegment

This work is part of research funded by the Universiti Teknologi Brunei under UTB Seed Fund [UTB/GSR/3/2021(5)].

References

[1] IEA 2021 Global energy review *IEA, Paris*

[2] Oruc O and Dincer I 2021 Assessing the potential of thermo-chemical water splitting cycles: A bridge towards clean and sustainable hydrogen generation *Fuel*, **286**

[3] Liu F and Zhang N 2004 Strategy of purifier selection and integration in hydrogen networks *Chem. Eng. Research and Design* **82** 10 1315-30

[4] Akulinin E, Ishin A, Skvortsov S, Dvoretsky D and Dvoretsky 2017 Mathematical modeling of hydrogen production process by pressure swing adsorption method *Advanced Materials & Technologies* **2** 38-49

[5] Ibrahim Y, Al-Mohannadi D M and Linke P 2021 Modelling and optimization of hydrogen production in an industrial cluster accounting for economic cost and environmental impact *Chemical Engineering Transactions* **88** 439-44

[6] Kirchherr J, Reike D and Hekkert M 2020 Conceptualizing the circular economy: An analysis of 114 definitions *Resources, Conservation and Recycling* **127** 221-32

[7] Li G, Wang L,Wang C, Wu P and Che D 2020 Experimental study on coal gasification in a full-scale two-stage entrained-flow gasifier *Energies* **13**

[8] Yeo J, How B S, Teng S Y, Leong W D, Ng W P Q, Lim C H, Ngan S L and Lam H L 2020 Synthesis of sustainable circular economy in palm oil industry using graph-theoretic method *sustainability* **12**

[9] Lee S, Kim H, Park J, Kang B, Cho C, Lim H and Won W 2021 Scenario-based techno-economic analysis of steam methane reforming process for hydrogen production *Appl. Sci.* **11** 13

[10] Galdamez J R, Garcia L and Bilbao R 2005 Hydrogen production by steam reforming of bio-oil using coprecipitated ni–al catalysts *energy & fuels* **19** 3 1133-42

[11] Rievaj V, Gana J and Synak F 2019 Is hydrogen the fuel of the future? *Tranportation Research Procedia* **40** 469-74

[12] kwb *Water rates* [Internet] 2022 [2022 April 1].

[13] Reeb C, Hays T, Venditti R and Gonzalez R W 2014 Supply chain analysis, delivered cost, and life cycle assessment of oil palm empty fruit bunch biomass for green chemical production in malaysia *Bioresources* **9** 3 5385-16

[14] Coal Agency *What is price of coal in Malaysia today?* [Internet] 2022 April [2022 April 1]

[15] Global Petrol Prices *Malaysia Methane prices* [Internet] 2022 April [2022 April 1]

[16] Azhar K *Charting the growth of the hydrogen economy in Malaysia* [Internet] 2021 November 9 [2022 April 1]

Production of Monoethylene Glycol from Lignocellulosic Biomass via Catalytic Hydrogenation: A Review

A H Norhanifah[1], *A R Norliza[2], and J Rafidah[3]
[1,2]Department of Chemical and Process Engineering, Faculty of Engineering and Built Environment, Universiti Kebangsaan Malaysia, 43600 UKM Bangi, Selangor Darul Ehsan, Malaysia
[3]Forest Research Institute Malaysia, Kepong, Selangor, Malaysia

*Email: norlizajkkp@ukm.edu.my

Abstract. Ethane and ethanol are produced through steam cracking and fermentation into ethylene respectively, which is then hydrolysed into monoethylene glycol (MEG). The disadvantages of both processes included used of easily oxidized substance and large quantities of water in order to minimize by-products such as diethylene glycol and triethylene glycol. Apart from that, MEG can also be produced by catalytic hydrogenation of biomass at extreme temperature and pressure with presence of catalyst. At the same time, this process uses lignocellulosic waste that have a high cellulose content such as residues from the agricultural and food industries. However, lignocellulosic biomass has to be treated to remove lignin content that may lower the rate of hydrogenation activity. In addition, most studies have found that the temperature in range of 240 °C to 280 °C and pressure of 5 MPa to 6 MPa are able to produce 18 wt% to 64 wt% of MEG. Meanwhile, the catalyst that have attract the researchers' attention are nickel and tungsten species which are able to increase the MEG yield by overcoming the activation energy of the hydrogenation process. Factors such as lignocellulose's pre-treatment, operating temperature and pressure, and the presence of catalyst will be discussed further.

1. Introduction

Monoethylene glycol (MEG) is a colourless and non-volatile liquid which commonly used as a mixed substance in polyester fibre, and polyethylene terephthalate [1]. Additionally, it is also used as an antifreeze medium in the heating and cooling system. Its widespread and diverse uses make the demand for MEG growing every year. The production of MEG can be divided into three methods which are conventional, fermentation, and catalytic hydrogenation. MEG production via conventional method uses a diminishing and expensive fossil fuel fuels [2]. Furthermore, conventional techniques use a large water ratio during the hydrolysis process, commonly 20:1 of ethylene oxide to ethylene glycol [3]. In the meantime, Dahmen et al. (2019) estimated 181.5 million of lignocellulosic biomass are produced annually globally. However, only 1.2 million tons of industrial waste are recycled and reused to produce different types of products [4]. Hence, catalytic hydrogenation process which use lignocellulosic biomass is attracting the researchers' attention to substitute the conventional and fermentation methods. However, this process requires high temperature and pressure as well as the presence of catalysts to overcome the activation energy [5], [6]. Thus, this review will discuss on the production of MEG via catalytic hydrogenation and few factors that may affect the process, especially the pre-treatment of lignocellulosic biomass, process temperature, hydrogen pressure, and catalysts.

2. Production of MEG via catalytic hydrogenation

Catalytic biomass hydrogenation method is commonly called as "One Pot Catalytic Cellulose Conversion" as it is able to convert cellulose to MEG only in a tank. Figure 1 show the process flow of catalytic hydrogenation. There are three reactions that occur during the catalytic hydrogenation process. Firstly, hydrolysis reaction breaks the long carbon chain of cellulose into a shorter carbon chain, called oligomers and simple sugars or glucose. Then, both oligomers and glucose will undergo retro-aldol reactions and give aldehyde glycol. During this reaction, the catalyst plays an important role to separate the bond of two carbon [7]. Moreover, catalyst also helps in the hydrogenation of ethylene aldehyde to MEG. At the same time, this hydrogenation process is also capable of producing sides products such as 1,2- propylene glycol, sorbitol, mannitol, and erythrose. In addition, glycerol may also produce from glucose and fructose isomerization during retro-aldol reactions that secrete three carbon sugar, i.e. dihydroxyacetone and glyceraldehyde. These two substances are then hydrogenated and produced glycerol which decrease the percentage of MEG production [6]. This can be avoided by using the optimum parameters to accelerate the hydrogenation process.

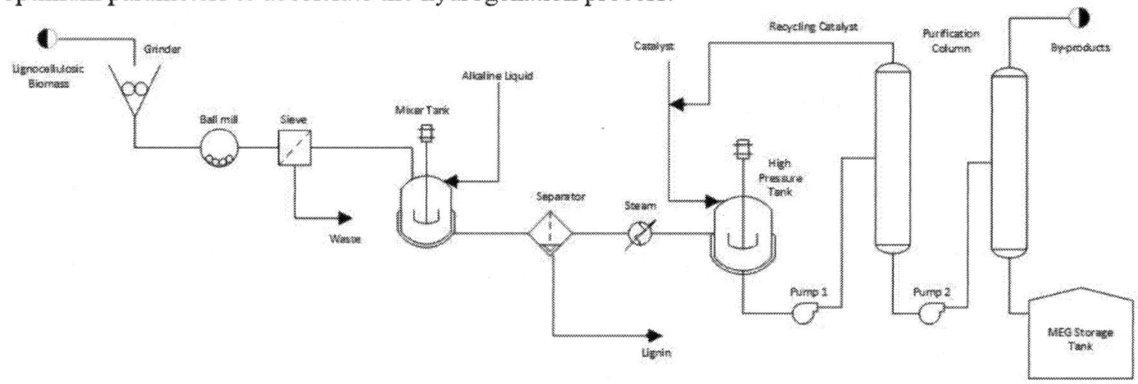

Figure 1. Process flow diagram of monoethylene glycol production via catalytic hydrogenation method.

3. Demand and supply of MEG

Zhang and his colleagues (2019) reported that global demand of MEG has been increased by 18% since 2018 [8]. In the meantime, Malaysia, China, and United States have started to produce MEG using conventional methods in the mid of 2020 with a capacity of 3.2 million tonnes per year. Malaysia's company namely Pengerang Refining and Petrochemical's is capable to produce 740,00 metric tonnes of MEG annually. While China companies of Hengli Petrochemical and Zhejiang Petrochemical are able to produce MEG around 750,000 metric tonnes and 828,000 metric tonnes annually, respectively [8]. Other than that, Nan Ya Plastic Company in Texas, United Stated could produce 828,000 metric tonnes of MEG per year [8]. The production and demand of MEG is expected to increase due to the industrial evolution especially plastic production.

4. Pre-treatment of lignocellulosic biomass

Pre-treatment of lignocellulosic is carried out to remove lignin content that disturb the hydrogenation process. The disadvantage of lignin is proven when Pang and his colleagues (2011) got a low yield of MEG when the corn cob is treated with SC-CO_2 that eliminate hemicellulose while maintaining lignin content compared to another treatment that removed lignin as desired. In addition, his study that mixed microcrystalline cellulose and lignin also showed a decrease in MEG results. However, this is occurred due to the fact that lignin is degraded to another substance during the hydrogenation process which prevents the active site of the catalyst [9], [10]. Moreover, Pang and his colleagues (2019) stated that lignin may form humin when react with glycol aldehyde, which is an intermediate substance of catalytic hydrogenation process before being converted to MEG. Meanwhile, Xiao et al. (2018) stated that lignin will minimize and inhibit the hydrolysis activity of cellulose to glucose [11].

Common pre-treatment method chosen are alkaline and mechanical pre-treatment. Most researchers found that cellulose content and MEG yield increase up to 25% when using alkaline treatment such as sodium hydroxide, butanediol, and hydrogen peroxide [5], [9], [10]. Pan et al. (2016) state that the treated lignocellulose appears rougher and more porous physically than without treatment when observe using SEM machines, making it easier to convert the cellulose to MEG. Nevertheless, this treatment could not remove all lignin which able to form sticky char which will prevent the maximum yield of MEG [9]. While ball mill is one of mechanical pre-treatment method that reduce the size of the lignocellulosic biomass, alter the lignocellulosic structure, increase the surface area, and allow hydrogenation activity by the catalyst [12]. Pang and his colleagues (2019), found that MEG production is doubled after various type of lignocellulosic biomass were treated using the ball mill. However, milling treatment can cause cellulose crystallinity to decrease and more susceptible break into oligosaccharide to forms by-product [12]. Apart from that, there is also a study that combine alkaline and mechanical treatments where Ribeiro et al. (2020) found that the characteristics which represent lignin had disappeared after going through the pre-treatment of ball mill and the lignin removal [13].

5. Operating Condition of Catalytic Hydrogenation

5.1. Temperature and Hydrogen Pressure

Both temperature and hydrogen pressure have a different role in catalytic hydrogenation process, i.e. hydrogen gas pressure affects the hydrogenation reaction while temperature affects the retro-aldol reactivity [5]. A study on catalytic hydrogenation of papaya waste found that the yield of MEG and 1,2-propylene glycol increased when the temperature and pressure of hydrogen gas increased to 245 °C and 6 MPa respectively [14]. This is in accordance with Yu & Wu statement (2010), at a temperature of 200 °C to 370 °C and the pressure at 4 MPa to 20 MPa, the water will produce H3O+ and OH- ions [15]. The production of H3O+ ions will reach a maximum rate of 250 °C and then increase the effectiveness of the hydrogenation process [16], [17]. However, 80% of biomass did not convert into MEG or any other by-products if the temperature is set below 230 °C. This statement is similar to Luo and his colleagues (2007) who stated that the temperature for this reaction should be carried out above 230 °C because the cellulose hydrolysis process requires an acidic condition where by-products such as cellobiose and cellotriose can be reduced [18]. Ooms and his colleagues, (2014) also noted that MEG production increased up to 66 moles % at a temperature of 260 °C, while lower temperatures led to the production of by-products such as sorbitol [6]. This statement is also supported by Wang and his colleagues (2019) who stated that polyol which is MEG and 1,2-propylene glycol are stable when the temperature rises to 250 °C in the water without any reaction or decomposition. Therefore, more extreme conditions are required to activate the substrate [19].

5.2. Catalytic Activity

Catalyst is used generally to lower the activation energy and speed up the reaction. Most researchers used tungsten and nickel species as a catalyst due to its environmentally friendly properties and high selectivity in the process of catalytic biomass hydrogenation [10], [14], [20], [21]. In addition, tungsten species is capable of being an active component in the separation of carbon-carbon bond to form glucose and water-soluble oligosaccharide during the cellulose hydrolysis process [12], [14]. This is proven in a study on the catalyst effectiveness in catalytic hydrogenation process that used a catalyst consisting of several species of tungsten which successfully shows all product results in MEG as the main product over other side products with percentage of less than 20% respectively [14].

Meanwhile, nickel catalysts are known for their cheap prices and play an important role in hydrogenation reaction [24]. However, the use of nickel alone provides a low MEG yield and toxicity [22]. At the same time, Wang & Zhang (2013) stated the tungsten catalyst should be combined with transitional metals because tungsten only acts in retro-aldol reactions instead of hydrogenation [23].

Nevertheless, the percentage of each combined catalyst plays an important role as too much nickel will lower the MEG yields while high quantity of tungsten did not change the yield but lead to waste [10].

5.2.1. Catalyst Supporter. A few studies on catalytic biomass hydrogeneration are done using catalysts with catalyst supporter such as Al2O3, TiO2, SiO2, carbon active, ZnO, ZrO2 and MgO [19], [21], [22]. Catalyst supporters helps to increase the spread of the other catalyst during the process [21]. However, the use of ZnO and MnO increase the yield of 1,2-propylene glycol instead of MEG. Apart from that, MCM-41 and carbon nanotube, CNT are also used as a catalyst supporter as it has porous surface and effective in the catalytic spread [23], [24]. In a study done by Ribeiro and his colleagues (2020) who used the Ru-W/CNT catalyst, found that the acid sites on the CNT and tungsten catalysts helps the hydrolysis process of holocellulose to oligosaccharide and simple sugars [13]. In a meantime, it was also found that OH element presence in catalyst supporter promotes the splitting of hydrogen molecules into hydrates and protons on the reaction surface which helps the hydrolysis process of cellulose to glucose [11]. It is proven when Xiao et al. (2018) observe an increase in MEG percentage from 24.3% to 30.9% when the catalyst supporters were combined with OH element [11].

5.2.2. Recycling Catalyst. Several studies conducted a repetition test on the catalyst to examine its performance. The catalysts were recycled up to four times and it show a decline performance as the yield of MEG also decrease although the type of catalyst used were differ [10], [11], [13], [14]. The reduction of MEG yield could be due to catalyst that has been eliminated by high-pressure hot water during the catalytic hydrogenation process in the autoclave [11].

6. Summary of past study

Table 1 below summarize the condition and the yield of MEG from variety of lignocellulosic sources. It was found that corn cob gives the highest percentage of MEG yield. Most of the optimum temperature used are 240 °C to 250 °C with pressure 5 MPa to 6 MPa. While, nickel and tungsten catalyst are the most selected catalyst for catalytic hydrogenation process.

Table 1. Summary of past studies.

Raw Material (RW)	Temp, °C	Hydrogen Pressure, MPa	Catalyst (C)	Mass, g		MEG yield, %	Source
				RW	C		
Papaya Residues	245	6.0	H_2WO_4–Ru/AC $(NH)_6H_2W_{12}O_{40}$–Ru/AC	0.50	0.37	49.90 48.20	[14]
Australia Tree Oak Tree	205	5.0	W/CNT_0 Ru/CNT_1	0.75	0.30	32.40 29.70	[13]
Empty Palm Oil Fruit Bunches	240	2.2	Ra-Ni/W	2.00	0.12	26.54	[5]
Elephant Grass	245	5.0	5wt % Ru/C,	1.00	0.25	45.60	[12]
Microalgae	250	6.0	5 wt% Mn/Ni/ ZnO	0.25	0.15	16.17	[20]
Corn Cob	250	6.0	5 wt% Ni–15 wt% W/MCM-41.	0.50	0.25	63.60	[10]
	245		2 wt% Ni-W_2C		0.15	18.30	[9]
Hard Wood Pulp Soft Wood Pulp	240	5.0	15 wt% Ni-20 wt% W-SiO_2-OH	0.50	0.20	42.10 37.60	[11]

7. Conclusion

Lignocellulosic biomass has the potential to produce MEG and is always available in large quantities. Thus, conventional process with use of limited raw materials and volatile mid-products could be prevented. However, lignocellulosic biomass required pre-treatment due to lignin which inhibits maximum MEG production. In addition, high temperature and pressure are required in the process of catalytic hydrogenation to break the carbon bonding of polymers and reduce the production of by-products with aid of catalyst. The past studies have successfully achieved 18% wt to 64% wt of MEG by using different sources of lignocellulosic biomass.

8. Recommendation

A wider and deeper research is needed to identify an optimum condition for the process of catalytic hydrogenation of biomass, especially using raw materials available in Malaysia such as oil palm tree residues.

9. Acknowledgement

The authors acknowledge the support by Universiti Kebangsaan Malaysia with fund of TAP-K008291 and TR-UKM-007.

References

[1] National center for biotechnology information, "*1,2-ethanediol*", [internet] 2022, pubchem compound summary [cited 2022 May 17], available from https://pubchem. ncbi.nlm.nih.gov /compound/1,2-Ethanediol

[2] Foster G, 2018, "*Ethylene supply in a fluid context: implications of shale gas and climate change,*" *Energies (basel)*, **11**, 11

[3] Harmsen J and Verkerk M, 2020, "*15 Shell OMEGA only monoethylene glycol advanced process,*" *Process intensification*, 166–175

[4] Dahmen N, Lewandowski I, Zibek S, and Weidtmann A, 2019, "*Integrated lignocellulosic value chains in a growing bioeconomy: Status quo and perspectives,*" *GCB bioenergy*, **11**, 107–117

[5] Law P G, Sebran N H, Zawawi A Z, and Hussain A S, May 2020 "*Optimization study of biomass hydrogenation to ethylene glycol using response surface methodology,*" *Processes*, **8**, 5

[6] Ooms R, Dusselier M, Geboers JA, Beeck B, Verhaeven R, Gobechiya E, Martens J A, Redl A and Sels B F, 2014, "*nickel tungsten carbide in a fed-batch reactor,*" 695–707

[7] Pang J, Zheng M, Wang A, Sun R, Wang H, Jiang Y and Zhang T, 2014, "*Catalytic conversion of concentrated miscanthus in water for ethylene glycol production,*" *AIChE journal*, **60**, 6, 2254–2262

[8] Zhang M, Cambeiro K, and Miguel H, "*Surging monoethylene glycol supply shifts global trade flows,*" [internet] 2019, [cited 2021 Jun 02], available from https://www.spglobal.com/ commodityinsights/en/market-insights/latest-news/petrochemicals/112519-commodities-2020-surging-monoethylene-glycol-supply-shifts-global-trade-flows

[9] Pang J, Zheng M, Wang A, and Zhang T, 2011 "*Catalytic hydrogenation of corn stalk to ethylene glycol and 1,2-propylene glycol,*" *Industrial and engineering chemistry research*, **50**, 11, 6601–08

[10] Pan G Y, Ma Y L, Ma X X, Sun Y G, Lv J M, and Zhang J L, 2016 "*Catalytic hydrogenation of corn stalk into polyol over Ni-W/MCM-41 catalyst,*" *Chemical engineering journal*, **299**, 386–392

[11] Xiao Z, Fan Y, Cheng Y, Zhang Q, Ge Q, Sha R, Ji J, and Mao J, 2018, "*Metal particles supported on SiO2-OH nanosphere: New insight into interactions with metals for cellulose conversion to ethylene glycol,*" *Fuel*, **215**, 406–416

[12] Pang J, Zheng M, Li X, Sebastian J, Jiang Y, Zhao Y, Wang A, and Zhang T, 2019, "*Unlock the compact structure of lignocellulosic biomass by mild ball milling for ethylene glycol production,*" *ACS sustainable chemistry and engineering*, **7**, 1, 679–687

[13] Ribeiro L S, Órfão J J D M, Fernando M and Pereira R, 2020, "*Direct catalytic conversion of agro-forestry biomass wastes into ethylene glycol over CNT supported Ru and W catalysts,*" *Materials today sustainability*, 58

[14] Zhang K, Wu S, Yang H, and Li G, 2017, "*Catalytic conversion of cassava residues to ethylene glycol via a one-pot hydrothermal reaction,*" *Waste and biomass valorization*, **8**, 2443–50

[15] Yu Y and Wu H, 2010, "*Understanding the primary liquid products of cellulose hydrolysis in hot-compressed water at various reaction temperatures*," *Energy and fuels*, **24**, 3, 1963–1971

[16] Kruse A and Dahmen N, 2015, "*Water - a magic solvent for biomass conversion*," *Journal of supercritical fluids*, **96**, 36–45

[17] Pang J, Zheng M, Sun R, Song L, Wang A, Wang X, and Zhang T, 2015, "*Catalytic conversion of cellulosic biomass to ethylene glycol: Effects of inorganic impurities in biomass*," *Bioresource technology*, **175**, 424–429

[18] Luo C, Wang S, and Liu H, 2007, "*Cellulose conversion into polyols catalyzed by reversibly formed acids and supported ruthenium clusters in hot water*," *Angewandte*, **46**, 40, 7636–7639

[19] Wang X, Beine A K, and Palkovits R, 2019, "*1,2-propylene glycol and ethylene glycol production from lignocellulosic biomass*," *Elsevier*, **178**

[20] Tan Z, Shi L, Zan Y, Miao G, Li S, Kong L, Li S and Sun Y, 2018, "*Crucial role of support in glucose selective conversion into 1,2-propanediol and ethylene glycol over Ni-based catalysts: A combined experimental and computational study*," *Applied catalysis a: general*, **560**, 28–36

[21] Miao G, Zan Y, Sun Y, Wang H, Li S, Liu C, Li S, Kong L and Sun Y, 2018, "*Mn-promoted hydrogenation of microalgae (Chlorococcum sp.) to 1,2-propanediol and ethylene glycol over Ni-ZnO catalysts*," *Applied catalysis a: general*, **565**, 34–45

[22] Zan Y, Miao G, Wang H, Kong L, Ding Y, and Sun Y, 2019, "*Revealing the roles of components in glucose selective hydrogenation into 1,2-propanediol and ethylene glycol over Ni-MnOx-ZnO catalysts*," *Journal of energy chemistry*, **38**, 15–19

[23] Wang A and Zhang T, 2013 "*One-pot conversion of cellulose to ethylene glycol with multifunctional tungsten-based catalysts*," *Accounts of chemical research*, **46**, 7, 1377–86

[24] Ribeiro L S, Órfão J, de Melo Órfão J J, and Pereira M F R, 2018, "*Hydrolytic hydrogenation of cellulose to ethylene glycol over carbon nanotubes supported Ru–W bimetallic catalysts*," *Cellulose*, **25**, 4, 2259–72

IOP Conf. Series: Materials Science and Engineering 1257 (2022) 012016 doi:10.1088/1757-899X/1257/1/012016

Kinetic analysis for catalytic co-pyrolysis of palm kernel shell and plastic waste mixtures with bifunctional HZSM-5 and mussel shell catalyst

A N Kumar[1], B L F Chin[1,2*], A L K Chee[1]

[1]Chemical and Energy Engineering Department, Faculty of Engineering and Science, Curtin University Malaysia, 98009 Miri, Sarawak, Malaysia.
[2]Energy and Environmental Research Cluster, Faculty of Engineering and Science, Curtin University Malaysia, CDT 250, 98009 Miri, Sarawak Malaysia.

*e-mail: bridgidchin@curtin.edu.my

Abstract. The present study is dedicated to investigate the kinetic analysis for catalytic co-pyrolysis of palm kernel shell (PKS) and polyethylene waste (HDPE) mixtures with bifunctional HZSM-5 and mussel shell (MS) catalyst. Artificial neural network (ANN) modeling through 17 models based on the functions of reaction mechanism denoted as chemical reactions, diffusion reactions, nucleation and growth reactions, interfacial phase reactions and power law reactions was used in this study to achieve a suitable order of reaction promoting higher rate of accuracy based on the data achieved. It was found that the 2nd order, 3rd order, anti jander, jander and ginsling were selected as the suitable models out of 17 reaction mechanism kinetic models due to the giving a positive values produced. And also, it was observed that 3rd order reaction mechanism model provided higher activation energy (E_A) and A values for all feedstock used. Comparison between experimental data and predicted data using Logsig-Tansig (LT) and Tansig-Tansig (TT) were carried out and it was observed that the predicted results from ANN showed similar trend as experimental data with minimal error of 0.67%, 6.61%, and 2.46% for PKS, HDPE, and mixture of PKS and HDPE with the presence of MS and HZSM-5 catalyst, respectively. From the kinetic analysis, E_A and A value of PKS, HDPE, and of PKS and HDPE with the presence of bifunctional of HZSM-5/MS catalyst are 196.93 kJ/mol, 388.00 kJ/mol, 147.12 kJ/mol, 3.24×1013 s^{-1}, 6.00×1026 s^{-1}, and 1.72×10-1 s^{-1}, respectively.

1. Introduction

Biomass and plastic waste have regained its potential back to be use as the energy source as there is abundant of the waste mentioned all around the world. Biomass is a possible alternative for fossil fuels due to its longevity, an abundance of availability, biodegradability, durability, sizeable volatile matter accumulation and low centimeter content [1]. Biomass can be converted into valuable goods, including biochar, syngas, and bio-oil, through thermochemical transformation [2,3]. World plastics production has substantially increased to double in volume to 460 million tonnes (Mt) from year 2000 to year 2019 [4]. Due to its light weight, easy formation, low cost and compatibility for different purposes such as storage and packaging, there has been a huge increase in plastic use in the recent decades [5].

Pyrolysis is a technique of thermochemical transformation that requires no oxygen to degrade the biomass into the usable solid, liquid fuels, or gaseous products [6]. It is said that this mechanism will help create a circular economy and recycle hydrocarbons in a closed-loop way [7]. Alternatively, co-pyrolysis involving the mixing of two or more feedstock such as plastic waste enables solid, liquid, and gaseous materials to be modified and converted into liquid fuel. The binary mixture of plastic and biomass waste for co-pyrolysis contributed to synergistic results that could lead to lower manufacturing costs, minimisation of environmental impacts, and enhance liquid oil production [8]. The plastic waste has the high hydrogen to carbon ratio and the low ratio of oxygen to carbon that sustain the intrinsically high oxygen to carbon ratio and low hydrogen to carbon ratio solid biomass during co-pyrolysis leads to higher product quality and consistency while reducing deposition from the coke, which is shown by pyrolysis from pure plastic wastes [9]. Co-pyrolysis provides a high potential desirable route for mixed waste by minimizing waste separation requirements [10]. The advantages of using pyrolysis in relation to incineration were stated by Grieco and Baldi [11] such as lower gas flow rate is required for eliminating pollutants and it is easier to wash off Sulphur in the form of H_2S than SO_2. Additionally, pyrolysis occurs in an oxygen-deficient atmosphere in comparison to the incineration process and thus reduced production of pollutants such as dioxins and furans. Aside from that, there are better technologies alternative to incineration process such as gas turbines and internal combustion engines which are able to provide electric electricity, instead of restricting themselves to the Rankine cycle [11].

The novelty of this research is to determine the kinetic parameters from artificial neural network (ANN) modelling from the catalytic co-pyrolysis of palm kernel shell (PKS) and polyethylene waste with bifunctional zeolite HZSM-5 and mussel shell (HZSM/MS) catalyst which had not been conducted before. However, several studies were conducted on the catalytic pyrolysis PKS [12] and catalytic pyrolysis HDPE [13]. Hence, in this work, the thermal degradation study and kinetic analysis focused on the catalytic co-pyrolysis of PKS/HDPE mixture with the presence of bifunctional of HZSM-5/MS catalyst using ANN modelling. The outcomes of the present study could provide the necessary information needed for reactor upscale and design.

2. Experimental materials and methods

2.1. Material and catalyst preparation

The PKS was obtained from Sarawak Oil Palms Berhad (SOPB), Malaysia. The polyethylene waste, specifically the high-density polyethylene (HDPE), was obtained from Shen Foong Plastic Industries Sdn Bhd, Malaysia. The raw materials, PKS and HDPE were pulverised and sieved into a particle size of less than 700 μm. Characterization of PKS and HDPE was done via proximate and ultimate analyses. Proximate analysis allows the ash, moisture, fixed carbon and volatiles to be measured using the ASTM standard methodology and the thermogravimetry analyzer EXSTAR TG/DTA 6300 (Seiko Instrument Inc.). Meanwhile, the ultimate analysis was conducted using the LECO CHNS-932 elemental analyzer which determines the content of carbon, hydrogen, oxygen, nitrogen and sulphur. To investigate the synergistic effect between PKS and HDPE, PKS was blended with HDPE at weight blend ratios of 20% of HDPE in the total mixture of HDPE and PKS mixture (0.2HDPE/0.80PKS), respectively. Thermogravimetric analysis (TGA) of blends and raw material were performed in order to assess interaction effects of HDPE together with PKS. Two types of catalyst were purchased to be used in this work such as HZSM-5 zeolite catalyst from Sigma-Aldrich, Malaysia and mussel shell from Calrock Sdn. Bhd., Malaysia. The mussel shell was milled and sieved into powder form, and was heated at temperature 900 °C for 4 hours to decompose calcium carbonate ($CaCO_3$) to calcium oxide (CaO). To activate the composition of the HZSM-5 zeolite powder, calcination was done in a furnace for 5 hours. Subsequently, the mussel shell powder and HZSM-5 catalyst were placed at a cool down position based on room temperature along with an air-constricted container. The PKS and HDPE was mixed at a homogenous manner towards the catalyst at a ratio of 10:1.

2.2. *Thermogravimetric analysis (TGA) approach*

Co-pyrolysis behavior of PKS, HDPE, and their blends using HZSM-5 and MS as catalysts were investigated by using the thermogravimetry analyzer EXSTAR TG/DTA 6300 (Seiko Instrument Inc.) at four different heating rates of 10, 20, 50 and 100 K/min. For each of the analysis, the TGA experimental configuration were conducted based on previous studies [12-13].

2.3. *Artificial neural network*

Artificial Neural Network (ANN) model has been developed to predict the mass loss of PKS, HDPE and the mixture of PKS and polyethylene waste with the presence of mussel shell and zeolite HZSM-5. ANN is comparable to the human brain in that it examines input from many senses and re-enacts its memory to fully comprehend the association. Since it can map nonlinear data using various approximation functions, ANN has attracted a lot of attention in predictive modelling for nonlinear and complicated systems [14]. Also, ANN model uses 2 inputs and 1 output to map the patterns. For this study, the inputs are heating rate in K/min unit and temperature in K unit whereas the output is weight loss percentage (%). The advantage of using ANN is that research had shown to provide powerful and high accuracy performance on the prediction of the thermogravimetric analysis (TGA) which will benefit in kinetic analysis for this study [13]. According to Naqvi et al. [15], it was mentioned that the number of hidden layers can be chosen between 1 to 3. Since most of the studies were using two hidden layers to keep things simple while still predicting the ANN model's performance and accuracy, this study made use of the same design [16]. Then, the network was fine-tuned by altering the number of neurons from 2 to 20 in order to see how the model's Mean Square Error (MSE) changed for each training session. The ANN model would be trained using 70% of the data points. The remaining 30% of the data would be used for the network validation and testing respectively. Logsig-tansig (LT) and tansig-tansig (TT) transfer functions is used to determine the values of linear regression determination factor (R^2) and mean square error (MSE) between the experimental and predicted data. These values are required to measure the performance of the results of the ANN prediction. A TG curve prediction will be made using the weight loss percentage (wt%) from the ANN predictions of PKS, HDPE, and the mixture of PKS and HDPE with the presence of MS and zeolite HZSM-5. Using the Coats-Redfern method with 17 reaction mechanism kinetic models, kinetic parameters can be calculated from the TG curve prediction.

2.4. *Kinetic model*

Solid biomass undergoing non-isothermal decomposition can be represented by equation (1) in the form of rate equation:

$$\frac{d\alpha}{dT} = \beta \cdot k(T) \cdot f(\alpha) \tag{1}$$

where α, β, $k(T)$ and $f(\alpha)$ denotes the conversion, heating rate, rate constant and reaction model, respectively.

Arrhenius equation is described by equation (2):

$$k(T) = A \cdot \exp\left(-\frac{E_A}{RT}\right) \tag{2}$$

where R and T are gas constant and absolute temperature, respectively.

Upon substitution of rate constant from equation (2) into equation (1), equation (3) is derived:

$$\frac{d\alpha}{dT} = \beta \cdot A\, exp\left(-\frac{E_A}{RT}\right) \cdot f(\alpha) \tag{3}$$

Simplified Coats-Redfern integral solution of equation 3 can be transformed into the following straight-line equation:

$$\ln\left[\frac{g(\alpha)}{T^2}\right] = \ln\frac{AR}{(\beta E_A)}\frac{1-2RT}{E_A} - \frac{E_A}{RT} \tag{4}$$

The most frequent reaction mechanism function $g(\alpha)$ was adopted from the literature and can be found in this reference [15]. An Arrhenius plot can be obtained based on equation (4) where E_A can be determined from the slope of the graph and A from the y-intercept of the graph.

Coats-Redfern through non-isothermal integral process was procured by a researcher to achieve the kinetic parameters, namely the activation energy (E_A) and pre-exponential factor (A) [17-19]. This method is applied in the present study as it allows different models according to different reactions such as chemical, diffusion, nucleation and growth and interfacial phase reactions to be considered when determining the kinetic parameters.

3. Results and discussion

3.1. Thermal decomposition behaviour

The thermal decomposition behaviour of binary mixtures of PKS and HDPE with the presence of zeolite HZSM-5 and mussel shell catalyst are investigated in various heating rates starting from 10 to 100 K/min using a TGA approach. Figure 1 displays on the thermogravimetric (TG) and derivative thermogravimetric (DTG) curves of binary mixtures of PKS and HDPE in catalytic co-pyrolysis process. At Stage I (538K–668 K), it is observed that there are varying number of endothermic peaks for different heating rates. There are four small endothermic peaks obtained at 10 K/min. One medium endothermic peak is observed at 100 K/min. One small and two medium endothermic peak is illustrated at 20 K/min. When 50 K/min heating rate was achieved, there is one small and one medium endothermic peak. It can be observed that there is more fluctuation compared to the pyrolysis of PKS and HDPE individually due to the mixture of the biomass and plastic with the presence of catalyst. Moreover, at Stage II (538K-1018K) it can be illustrated no endothermic peak at all heating rates. During Stage III (1018K-1100K), there is only one small endothermic profile curve observed at 100 K/min which indicates that the remaining residue was evaporated at that specific temperature.

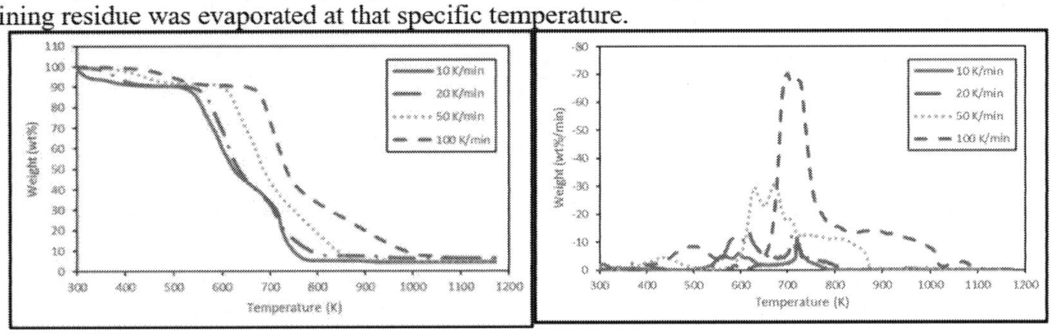

Figure 1: (a) Thermogravimetric (TG) and (b) derivative thermogravimetric (DTG) curves of binary mixtures of palm kernel shell (PKS) and polyethylene waste (HDPE) with the presence of bifunctional zeolite HZSM-5 and mussel shell catalyst.

3.2. Experimental and Artificial Neural Network Predicted Weight Loss

Logsig-tansig (LT) and tansig-tansig (TT) transfer functions is used to determine the values of linear regression determination factor (R^2) and mean square error (MSE) between the experimental and predicted data. Out of 5700 data point, the model would be trained using 70% of the data points. The remaining 30% of the data would be used for the network validation and testing respectively. The weight loss for the catalytic co-pyrolysis of PKS and HDPE mixtures using the bifunctional mussel shell and HZSM-5 catalyst was measured with respect to temperature in TG curve via logsig-tansig (LT) and tansig-tansig (TT) transfer functions and it is shown in figure 2. It was found that the LT and TT transfer function ANN model demonstrated 99.99% similarity when compared to experimental TGA curves. This indicates that the ANN estimation evaluation is accurate when the mean square error is close to zero during the network development.

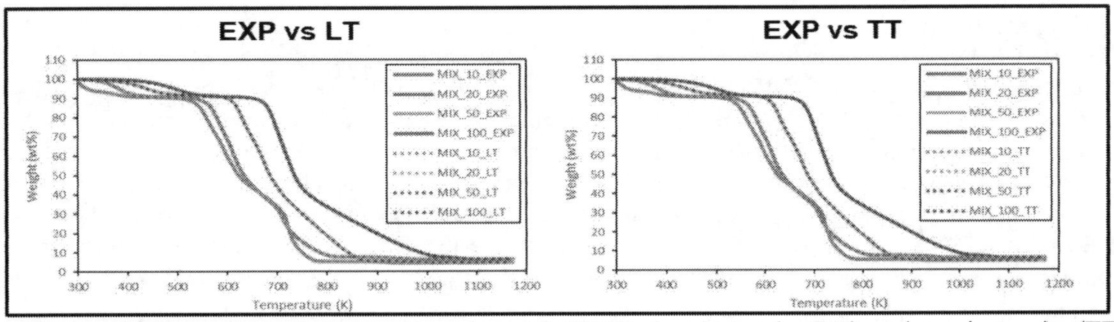

Figure 2: Comparison between Experimental (EXP), logsig-tansig (LT) and tansig-tansig (TT) predicted TG curves on the binary mixtures in catalytic co-pyrolysis process.

3.3. Kinetic study

Tables 1 and 2 show the comparative results on the experimental (EXP) and tansig-tansig (TT) and experimental and logsig-tansig (LT) respectively for various kinetic models. Chemical reactions, nucleation and growth reactions, diffusion reactions, interfacial phase reactions and power law reactions was selected from the 17 models based on the functions of reaction mechanism. The E_A and A is determined based on the mass loss data obtained directly from the thermogravimetric analysis (TGA). The highest experimental E_A was found to be 147.12 kJ/mol for a mixture of PKS and HDPE with MS and zeolite HZSM-5, while the TT predicted E_A was 150.74 kJ/mol (percentage error of 2.46%) and LT predicted E_A was 156.80. The lowest E_A values obtained from the 3rd order reaction mechanism kinetic model was 8.32 kJ/mol (percentage error of 6.58%). This finding is found to be consistent with previous research that used the same kinetic model to predict E_A values [15].

Table 1: The comparative results on the experimental and tangsig-tansig (TT) for various kinetic models.

Kinetic Model	2nd Oder		3rd Order		Anti Jander		Jander		Ginsling	
Source	EXP	TT	EXP	TT	EXP	TT	EXP	TT	EXP	TT
α	E_A (kJ/mol)	E_A (kJ/mol)	E_A (kJ/mol)	E_A (kJ/mol)	E_A (kJ/mol)	E_A (kJ/mol)	E_A (kJ/mol)	E_A (kJ/mol)	E_A (kJ/mol)	E_A (kJ/mol)
0.1	8.44	9.11	8.32	9.08	7.40	7.21	7.35	7.16	7.40	7.22
0.2	7.95	9.09	7.60	8.92	6.33	8.20	6.20	8.13	6.33	8.20
0.3	11.13	11.81	11.25	12.07	11.25	12.22	11.29	12.31	11.25	12.22
0.4	11.29	10.82	11.35	10.75	11.40	10.82	11.43	10.80	11.40	10.82
0.5	9.08	9.57	8.18	8.88	8.90	9.42	8.65	9.21	8.91	9.42
0.6	9.96	9.91	9.01	8.93	10.11	10.07	9.86	9.81	10.12	10.07
0.7	15.50	15.50	17.28	17.27	14.61	14.63	15.02	15.04	14.61	14.63
0.8	17.86	18.07	21.22	21.59	15.53	15.62	16.18	16.31	15.53	15.62
0.9	82.45	84.27	147.12	150.74	23.52	23.62	30.87	31.11	23.48	23.58
Average	**19.29**	**19.80**	**26.82**	**27.58**	**12.12**	**12.42**	**12.98**	**13.32**	**12.11**	**12.42**

Table 2: The comparative results on the experimental and logsig-tansig (LT) for various kinetic models.

Kinetic Model	2nd Oder		3rd Order		Anti Jander		Jander		Ginsling	
Source	EXP	LT	EXP	LT	EXP	LT	EXP	LT	EXP	LT
α	E_A (kJ/mol)	E_A (kJ/mol)	E_A (kJ/mol)	E_A (kJ/mol)	E_A (kJ/mol)	E_A (kJ/mol)	E_A (kJ/mol)	E_A (kJ/mol)	E_A (kJ/mol)	E_A (kJ/mol)
0.1	8.44	8.82	8.32	8.73	7.40	8.10	7.35	8.07	7.40	8.10
0.2	7.95	5.42	7.60	4.73	6.33	2.36	6.20	2.10	6.33	2.37
0.3	11.13	11.21	11.25	11.35	11.25	11.35	11.29	11.40	11.25	11.35
0.4	11.29	10.38	11.35	10.17	11.40	10.26	11.43	10.20	11.40	10.26
0.5	9.08	12.19	8.18	12.57	8.90	12.16	8.65	12.27	8.91	12.16
0.6	9.96	9.71	9.01	8.63	10.11	9.89	9.86	9.61	10.12	9.90
0.7	15.50	15.45	17.28	17.19	14.61	14.60	15.02	15.00	14.61	14.60
0.8	17.86	18.45	21.22	22.27	15.53	15.77	16.18	16.51	15.53	15.77
0.9	82.45	87.35	147.12	156.80	23.52	23.96	30.87	31.77	23.48	23.92
Average	**19.29**	**19.89**	**26.82**	**28.05**	**12.12**	**12.05**	**12.98**	**12.99**	**12.11**	**12.05**

4. Conclusions

This study focused on the kinetic analysis specifically for catalytic co-pyrolysis of PKS and polyethylene waste using the bifunctional HZSM-5 and mussel shell (MS) catalyst. The implementation of the artificial neural network (ANN) modeling had been employed in this study as well. Chemical reactions, nucleation and growth reactions, diffusion reactions, interfacial phase reactions and power law reactions was selected from the 17 models based on the functions of reaction mechanism. The kinetic analysis had been determined using Coats Redfern method via non-isothermal method. It was found that E_A and A value of PKS, HDPE, and of PKS and HDPE involving the bifunctional HZSM-5/MS catalyst are 196.93 kJ/mol, 388.00 kJ/mol, 147.12 kJ/mol, 3.24×1013 s^{-1}, 6.00×1026 s^{-1}, and 1.72×10-1 s^{-1}, respectively.

Acknowledgements

Financial support from Japan-ASEAN Science, Technology, Innovation Platform (JASTIP) 2022 Collaboration Research on the project grant which entitled, "Techno-economics/Environmental/Societal/Governmental (TESG) evaluation of thermochemical conversion of oil palm residues to alternative energies for ASEAN region".

References

[1] Vassilev S V, Vassileva C G and Vassilev V S 2015 *Fuel* **158** 330
[2] Loy A C M, Yusup S, How B S, Yiin C L, Chin B L F, Muhammad M and Gwee Y L 2019 *Bioresour. Technol.* **294** 122089
[3] Chin B L F, Yusup S, Al Shoaibi A, Kannan P, Srinivasakannan C and Sulaiman S A 2014 *Energy Conserv. Manage.* **87** 746
[4] edie, https://www.edie.net/the-facts-and-figures-the-global-state-of-plastics-production-and-pollution/, 2022
[5] Romasanta R R, Slander B O, Gaihre Y K, Alberto M C, Gummert M, Quilty J, Nguyen V H, Casalone A G, Balingbing C, Sandro J, Correa Jr T and Wassman R 2017 *Agri. Ecosys. Environ,* **239** 143
[6] Loy A C M, Yusup S, Lam M K, Chin B L F, Shahbaz M, Yamamoto A and Acda M N 2018 *Energy Conserv. Manage.* **165** 541
[7] Qureshi M S, Oasmaa A, Pihkola H, Deviatkin I, Tenhunen A, Manila J, Minkkinen H, Pohjakallio M and Laine-Ylijoki J L 2020 *J. Anal. Appl. Pyrol.* **152** 104804
[8] Zhang L, Bao Z, Xia S, Lu Q and Walters K B 2018 *Catalysts* **8** 659
[9] Chattopadhyay J, Pathak T S, Srivastava R and Singh A C 2016 *Energy* **103** 513
[10] Johansson A C, Sandström L, Öhrman O G W and Jilvero H 2018 *J Anal. Appl. Pyrol.* **134** 102
[11] Grieco E M and Baldi G
[12] Hii S W, Chin B L F, Yiin C L, Majing F R S, Jawad Z A, Lim H Y, Yusup S, Acda M N, Unrean P and Rianawati E 2021 *IOP Conf. Ser.: Mater. Sci. Eng.* **1195** 012010
[13] Yap T L, Loy A C M, Chin B L F, Lim J Y, Alhamzi H, Chai Y H, Yiin C L, Cheah K W, Wee M X J, Lam M K, Jawad Z A, Yusup S and Lock S S M 2021 *J. Environ. Chem. Eng.* **10** 107391
[14] Xu Y and Chen B 2013 *Bioresour. Tech.* **146** 485
[15] Naqvi S R, Hameed Z, Tariq R, Taqvi S A, Ali I, Niazi M B K, Noor T, Hussain A, Iqbal N and Shahbaz M 2019 *Waste Manage.* **85** 131
[16] Bong J T, Loy A C M, Chin B L F, Lam M K, Tang D K H, Lim H Y, Chai Y H, Yusup 2020 *Energy* **207** 112829
[17] Naqvi S R, Uemura Y, Osman N, and Yusup S 2015 *Research on Chem. Intermediates.* **41** 9713
[18] Naqvi S R, Uemura Y, Osman N, and Yusup S 2015 *BioEnergy Research* **8** 1800
[19] Wang Z H, Ma X Q, Yao Z L, Yu Q H, Wang Z and Lin Y S 2018 *Appl. Therm. Eng.* **128** 662

IOP Conf. Series: Materials Science and Engineering 1257 (2022) 012017 doi:10.1088/1757-899X/1257/1/012017

Selective epoxidation of the unsaturated fatty acids

[1]M. B Mahadi, [1]I.M. Rasib, [1]N.Z Abas, [1]M.A. Rahman , [2]I. SH Azmi and [1]M.J. Jalil*

[1]Centre of Chemical Engineering Studies, College of Engineering, Universiti Teknologi MARA, Cawangan Pulau Pinang, Kampus Permatang Pauh, 13500 Permatang Pauh, Pulau Pinang, MALAYSIA.
[2]Centre of Chemical Engineering, College of Engineering, Universiti Teknologi MARA, Cawangan Johor, Kampus Pasir Gudang, 81750 Masai, Johor, MALAYSIA.
mjumain0686@uitm.edu.my

Abstract. Nowadays, epoxidized vegetable oils are of major significance since they are derived from renewable and sustainable natural resources and are eco-friendly. Epoxide is a crucial chemical precursor to produce alcohols, glycols, and polymers such as polyester and epoxy resin. Oleic acid's long alkyl chain including an unsaturated double bond (C=C) makes it an appealing fatty acid for producing eco-friendly epoxide. The epoxidation of unsaturated fatty acid was accomplished by generating performic acid in situ to yield epoxidized oleic acid. The result of this reaction will be dihydroxystearic acid (DHSA). DHSA is one of the chemical precursors used to manufacture cosmetics. The optimal level of epoxidized unsaturated oleic acid was 82%.

1. Introduction

Vegetable oils and their unsaturated can be converted into epoxidized that act as intermediate product by participant in many reactions due to high oxirane ring reactivity [1]. In many study domains, the quantity of research focusing on these transversal subjects is increasing at an exponential rate. Because both fossil fuels and oils generated by plants were once plants countless years ago, it seems logical that they're chemically similar [2][3]. Today, epoxidized vegetable oil is becoming increasingly popular because it is utilised to replace petroleum-derived sources [4]. Due to the attachment of an oxygen atom to the double bond structure in oleic acid, the initial double bond is transformed into an epoxide ring [5][6]. Double bonds within fatty acid chains are reactive sites that permit vegetable oils to be chemically altered to increase their range of applications. Because the produced carboxylic acid is stronger than the hydrogen-bound peracid, additional ring-opening reactions are possible, particularly in the case of formic acid [7].

In the epoxidation process, formic acids function as oxygen carriers, while hydrogen peroxide acts as an oxygen donor [8]. Compared to hydrogen peroxide, formic acid is recommended as an oxygen carrier due to its strong reactivity and lack of catalyst production requirements [9]. Its greater oleic acid concentration makes it suitable for oil production by epoxidation. Due to the high cost of peracid, it is commonly employed in industry for epoxidation [10]. In-situ preparation of peracid is favoured over other methods since it is the safest[11]. However, there are less references about the formation of DHSA from the ring opening of epoxidized oleic acid at this time. Determining the influence of agitation speed on oxirane ring opening on the rate of crude oleic acid epoxidation is the purpose of this work.

2. Research Methodology

2.1 Materials

The experiment utilized oleic acid (75%), formic acid (85%), hydrogen peroxide (50%), and sulphuric acid as a catalyst. The epoxidation procedure is carried out by preparing all the necessary components, including formic acid, hydrogen peroxide, and oleic acid, in a 1:1:1 molar ratio.

2.2 Experiment

Based on Equation (1) and (2), the relative percentage conversion to oxirane content (RCO) can be computed. While gradually heating the mixture to the appropriate temperature (700C), it was continuously stirred at a consistent pace. Before being examined using the oxirane oxygen content (OOC) analytical method, the samples were cooled and quenched for two minutes. The OOC exp value was calculated with a solution of hydrobromic acid (HBr) in glacial acetic acid using a direct approach (AOCS Cd 9-57).

$$RCO = \frac{OOC\ exp}{OOC\ theo} \times 100\% \quad (1)$$

$$OOCexp = \frac{T \times N_{HBR} \times 1.6}{W} \quad (2)$$

2.3 Experimental Design

To study the effect of various parameters affecting the epoxidation of oleic acid, the range of each reaction condition was tabulated in Table 1. This manuscript discusses agitation speed since a fast-stirring speed can result in a thoroughly mixed solution, yet overmixing can affect the stability of the oxirane rings.

Table 1: Reaction parameters involved for the overall reaction

Reaction parameters	Range	Units
Agitation speed	200-400	rpm
Concentration of aqueous hydrogen peroxide	35	% (w/v)
Catalyst (sulfuric and phosphoric acid)	0.2	gram
Formic acid to oleic acid molar ratio	1.0	-
Hydrogen peroxide to oleic acid molar ratio	1.0	-

3. Results and Discussion

3.1 Influence of agitation speeds

Epoxide, commonly known as oxirane, is a three-membered cyclic ether made up of two carbons and oxygen. When a double bond carbon, such as ethylene (C=C), reacts with the addition of an oxygen atom to produce a cyclic ether, epoxide is formed. The ring is extremely stressed and reactive, with bond angles of 60 degrees. Epoxidation is commonly accomplished by reacting unsaturated hydrocarbon chains with peracid, which is either preformed or generated in situ. The reaction of carboxylic acid with hydrogen peroxide produces peracid. When in situ performic acid is used, formic acid acts as an oxygen transporter, reacting with hydrogen peroxide, which acts as an oxygen donor. The oxirane formation was not substantially affected by stirring speed at the investigated range as proven in Figure 1. Besides that, it was also observed that oxirane formation increases with the increase in agitation speed. As a result, a favorable RCO (%) of 83.0% was produced at 300 rpm, in comparison with 60 % and 65% RCO (%) yield at 200 and 400 rpm respectively. In other words, the effect of stirring speed is vital to ensure that the reaction is free from external mass transfer limitations in the form from diffusion [12]. Based on Figure 1, it is evident that the reaction time required to attain maximum RCO was around 20 minutes. This is because epoxidized oleic acid was stable and breakdown did not occur until 25 minutes later.

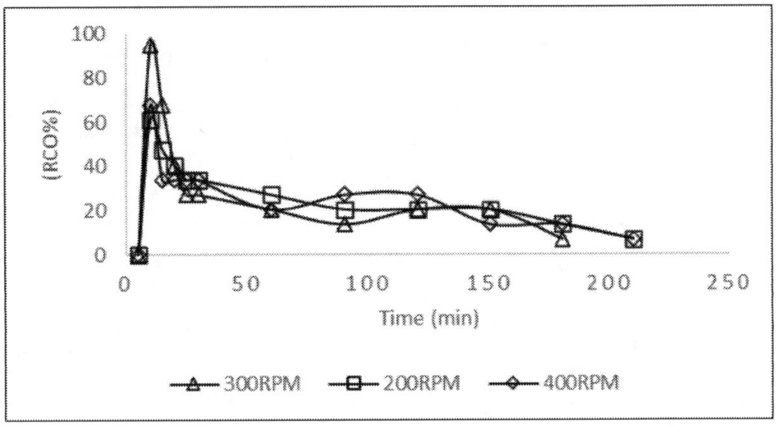

Figure 1: RCO% data based on different agitation speeds.

3.2 Analytical data of FTIR based on agitation speeds

Using Fourier Transform Infrared (FTIR) spectroscopy, the chemical characteristics of both crude oleic acid and epoxidized oleic acid were analysed. The technique enables the identification of C=C double bonds in crude oleic acid as well as the existence of an oxirane ring in the epoxidized form of an unsaturated fatty acid. The FTIR spectrum of DHSA is seen in Figure 2. In general, the existence of an oxirane ring can be identified between 750 cm^{-1} and 880 cm^{-1}.

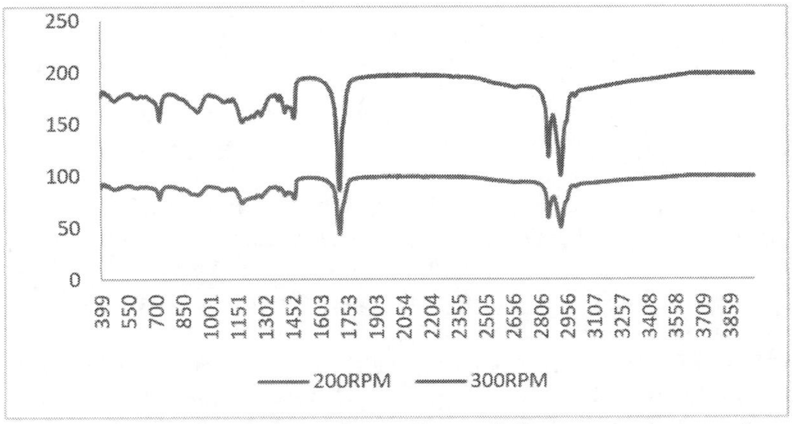

Figure 2: FTIR spectrum for DHSA at different agitation speed

Epoxidized unsaturated fatty acid looks as a whitish semisolid, whereas oleic acid is a transparent liquid. Table 2 displays the DHSA density at various agitation speeds. Peracids are formed when formic acid reacts with hydrogen peroxide to make performic acids. In the epoxidation process, formic acids function as oxygen carriers, while hydrogen peroxide acts as an oxygen donor. Formic acid is favoured as an oxygen carrier over hydrogen peroxide since it is highly reactive and does not require a catalyst to generate. The epoxy groups are then synthesized using palm oil-derived oleic acid. It is suitable for epoxidation oil production due to its increased oleic acid concentration, and it also performs better. Density is an essential notion because it enables us to determine which substances will float and which will sink in a liquid. In general, substances float if their density is less than that of the liquid in which they are immersed.

Table 2: Density of DHSA at different agitation speeds.

Ratios	Speed (rpm)	Density (g/cm^3)
1:1:1	200	0.8917
1:1:1	300	0.8745

4. Conclusions

Palm oil is one of the industrially available vegetable oils, and it is frequently utilised in the production of polymeric materials. The demand for vegetable oil has increased relative to petroleum-based polymer. In the reaction of epoxidation of vegetable oil, oxirane ring-opening or epoxide ring degradation will occur. In this study, the experimental data on the effect of agitation speeds on the epoxidation process and the characterization of the physicochemical features of DHSA produced by hydrolysis of epoxidation are the primary goals. The DHSA was successfully extracted via hydrolysis from the epoxidized palm oil. Based on these findings, palm oil derived from oleic acid has the potential to be a sustainable and environmentally friendly raw material for the synthesis of DHSA.

Acknowledgment

The registration fee is funded by Pembiayaan Yuran Prosiding Berindeks (PYPB), Tabung Dana Kecemerlangan Pendidikan (DKP), Universiti Teknologi MARA (UiTM), Malaysia

References

[1] M.J. Jalil, M.S. Md Zaini, A.F. Mohd Yamin, I.S. Azmi, S.H. Chang, N. Morad and A. Hadi, Synthesis and physicochemical properties of epoxidized oleic acid-based palm oil, *IOP Conf. Ser. Earth Environ. Sci.* 291 (2019).
[2] S.O. Egbuna, U.J. Nwachukwu, C.M. Agu, C.O. Asadu and B. Okolo, Production of biolubricant samples from palm kernel oil using different chemical modification approaches, *Eng. Reports*. 3 (2021) 1–19.
[3] I.S Azmi, M.J Jalil and A. Hadi, Epoxidation of Unsaturated Fatty Acid-based Palm Oil via Peracid Mechanism as an Intermediate Products. *Biomass Conversion and Biorefinery*. 1, 1-9, 2022
[4] M.S.A. Rahman, M.J. Jalil, M. Muain, V. Gloria, I.S. Azmi, A. Hadi, A.F.M. Yamin, and N. Morad, Epoxidation of waste cooking oil using catalytic in situ generated performic acid, *IOP Conf. Ser. Earth Environ. Sci*. 476 (2020) 0–5..
[5] D. Moreno, M. Velasco and D. Malagón-Romero, Production of polyurethanes from used vegetable oil-based polyols, *Chem. Eng. Trans*. 79 (2020) 337–342.

[6] M. J. Jalil, I. S. Azmi, A. Hadi and A.F. Mohd Yamin, In Situ Hydrolysis Of Epoxidized Oleic Acid By Catalytic Epoxidation-Peracids Mechanism, *Journal of Polymer Research*, 29(10), 1-10, 2022.

[7] M.J. Jalil, A. Hadi, I.S. Azmi, Catalytic epoxidation of palm oleic acid using in situ generated performic acid – Optimization and kinetic studies, *Mater. Chem. Phys.* (2021) 124754.

[8] M. J. Jalil, I. S. Azmi, A. Hadi and A.F. Mohd Yamin, In Situ Hydrolysis Of Epoxidized Oleic Acid By Catalytic Epoxidation-Peracids Mechanism, *Journal of Polymer Research*, 29(10), 1-10, 2022.

[9] M.J. Jalil, I.S. Azmi and A. Hadi, Highly Production Of Dihydroxystrearic Acid From Catalytic Epoxidation Process By In Situ Peracid Mechanism , *Environ. Prog. Sustain. Energy*. (2021) 1–7.

[10] R. Tesser, V. Russo, R. Turco, R. Vitiello and M. Di, Bio-lubricants synthesis from the epoxidized oil promoted by clays : Kinetic modelling, *Chem. Eng. Sci.* 214 (2020) 115445.

[11] M.J. Jalil, A. Hadi, N. Mora, S.H. Chang, A.F.M. Yamin and M.S.M. Zaini, Kinetics of epoxidized oleic acid derived from palm oil by in situ generated performic acid , Kem. u Ind. 68 (2019) 181–187.

[12] S.S. Patil and H.M. Jena, Synthesis of Epoxidized Citrullus lanatus Seed Oil: Experimental Investigation and Statistical Optimization, *Arab. J. Sci. Eng*. 44 (2019) 9965–9976.

IOP Conf. Series: Materials Science and Engineering 1257 (2022) 012018 doi:10.1088/1757-899X/1257/1/012018

Potential of integrating Solid Oxide Fuel Cell based on biomass in power generation in Malaysia: A feasibility study

R Ravi[1], N Shieela Kalib[1,2]*, A Muchtar[2,3], M Rao Somalu[2]
[1]School of Engineering & Physical Science-Mechanical, Heriot-Watt University, 62200 Putrajaya, Malaysia
[2]Fuel Cell Institute, Universiti Kembangan Malaysia, 43600 UKM Bangi, Selangor, Malaysia;
[3]Department of Mechanical and Manufacturing Engineering, Faculty of Engineering and Built Environment, Universiti Kebangsaan Malaysia, 43600 UKM, Bangi, Selangor, Malaysia

E-mail: n.kalib@hw.ac.uk

Abstract. Although many initiatives have been launched to achieve the ambitious Net Zero 2050 target, Malaysia is still struggling to control carbon emissions, particularly from electricity and heat generation. Fortunately, the potential to generate electricity from biomass in Malaysia reduces reliance on fossil fuels for power generation. Combining biomass gasification with a solid oxide fuel cell (SOFC) promises sustainable and more efficient performance than the traditional approach. To ensure optimal operation, the biomass-derived fuel should be compatible with the gasification system. A numerical simulation was carried out to gain insight into the feasibility of integrating hybrid gasification (GT) with SOFC based on selected biomass-derived fuels (i) palm oil mill effluent (POME) and (ii) refuse-derived fuel. A detailed anode-based planar SOFC model is developed and adopted into a gasification system that operates with an ideal gas under steady-state conditions. The other components considered in this study are a recuperator, combustor, and high-pressure turbine. A comprehensive energy balance for each stage was prepared to determine the overall performance of the system with higher accuracy. A detailed analysis shows that POME gave better results owing to its higher calorific value, resulting in higher thermal energy production. Reducing the required fuel flow rate has a positive effect on the overall performance compared with increasing the injected air flow rate. Furthermore, reducing the fuel flow rate tends to increase the combined efficiency. The study concludes that the fuel flow rate has a significant impact on the performance of the SOFC-GT hybrid system, particularly on combined efficiency and overall performance.

1 Introduction

In 2005, Malaysia committed to reducing approximately 40% of its carbon emissions by 2020 and then to 45% by 2030, in line with the Paris climate agreement in 2015 [1]. However, as of the end of 2021, according to Our World Data, Malaysia struggles to keep up with this commitment as the trend of CO_2 steadily increases by 15% on average from 2005 to 2021 [2]. Nevertheless, Malaysia further pledged to increase the goal to 45% by 2030. Another analysis found that 49.6% of Malaysia's CO_2 emissions come from electricity and heat generation [3]. Solid oxide fuel cells (SOFCs) are a device that converts chemicals stored as fuel into thermal or electrical energy without a combustion process [4]. Compared with other fuel cells, such as polymer electrolyte membrane fuel cells and direct methanol fuel cells, SOFC is more promising in generating clean electricity [5]. The SOFC has a high efficiency compared with other fuel cells [6].

IOP Conf. Series: Materials Science and Engineering 1257 (2022) 012018 doi:10.1088/1757-899X/1257/1/012018

Malaysia has the potential to use biomass for power generation as abundant biomass-based resources exist. The potential to use biomass for power generation in Malaysia can be explored with SOFC as they have higher fuel flexibility than other fuel cell applications [7]. Malaysia produces approximately 168 million tonnes of biomass annually, including wood, coconut fiber, rice husk, municipal , sugarcane waste, and oil palm waste [8]. Palm oil represents 91% of biomass-derived fuel in Malaysia [9]. In addition, there is the possibility of generating energy from municipal solid waste (MSW) by the end of 2022, as the amount of solid waste collected would increase to up to 14 million tonnes per year. Furthermore, 85% of the collected waste ends up in landfills, and the frequency of incineration can be reduced, decreasing the risk of pollution [10]. Therefore, the use of combustible fuel from municipal waste, also known as refuse derived fuel (RDF), as primary fuel in the gasification plant is beneficial to reduce CO_2 emissions in power generation.

Recently, Shafie et al [11] reported that Malaysia's potential for biogas generation is promising, with an average of 130 MW of electricity generated based on biogas. By converting municipal waste into substitute fuels, the amount of waste can be significantly reduced, and the frequency of incineration can be reduced, thereby decreasing the risk of environmental pollution. The use of SOFC power plants fired by biomass to generate electricity could solve the problem of carbon emissions in Malaysia and help with waste management. In addition, using palm oil mill effluent (POME) and RDF as the primary fuel for the SOFC combined gasification cycle (SOFC-GT) is beneficial. Both fuels operate at high enthalpies and temperatures in the gaseous state, which is beneficial for the overall performance of the system [12].

1.1 Solid Oxide Fuel Cell (SOFC)

SOFC utilizes the movement of electrons, generates electricity, and operates at a range of 600–1000 °C [13] . The electrolytes used are conductors of oxygen ions caused by oxygen being the oxidant for the process. In **Figure 1**, considering hydrocarbon as the main fuel, the processes begin as they are reformed into hydrogen (H_2) and carbon monoxide (CO) before entering the system acting as fuel. This process happens as air/oxygen enters the cathode side of the fuel cell. The oxygen from the air then becomes oxide ions after combining with free electrons, which then travels to the anode through the electrolyte. Next, at the anode, the reaction between the oxide ions and hydrogen forms steam from the water alongside minimal discharge of CO_2 from the fuel cell. The reaction releases free electrons, which travel through the external circuit resulting in electricity.

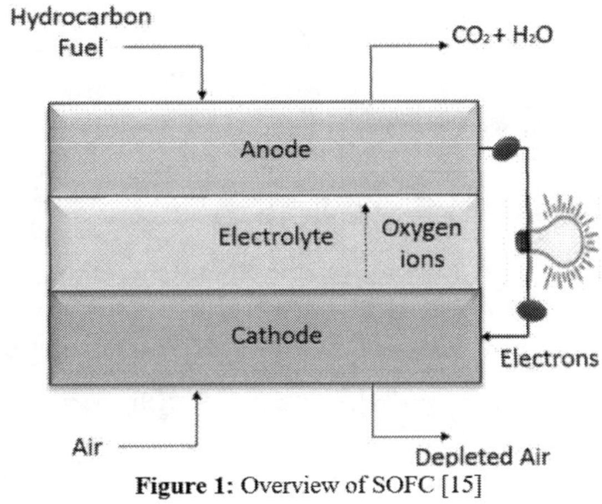

Figure 1: Overview of SOFC [15]

1.2 SOFC – GT Hybrid System

Figure 2 illustrates the SOFC-GT hybrid system: the compressor releases air under pressure, which enters the SOFC. In addition, fuel is fed into the system from the pre-heater, entering the SOFC so that oxidation of the fuel can take place. Then the gas from the SOFC enter the turbine, where it causes a rise in temperature and pressure and generate energy that drives the generator and produces electricity. After the gas leaves the turbine, heat recovery steam generators (HRSG) are added to use the waste heat to preheat the steam that enter the second turbine system. This case reduces the risk of thermal pollution, which also helps to improve the overall performance of the system [14].

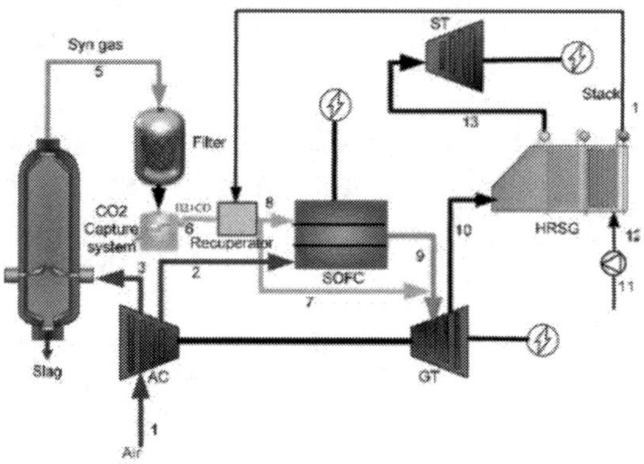

Figure 2: SOFC system integrated into gas turbine system [15]

2 Material and Methodology

Table 1 shows the properties of the SOFC components adopted for the gasification system and the technical specification of the system.

Table 1: Selected parameters and materials

Components	Selection	References
Anode	Yttria stabilized zirconia (Nickel-YSZ)	
Cathode	Strontium doped lanthanum manganite (LSM)	[16]
Electrolyte	Yttria stabilized zirconia (YSZ)	
Interconnect material	Crofer 22 APU	[17]
Compressor and turbines	Efficiencies at 75%	[18]
Fuel	Utilisation at 75%	
Recuperator	Effectiveness at 75%	[19]

2.1 Thermodynamic Analysis

The thermodynamic states in the plant were calculated using the mass and energy Eq. (1) for each equipment as follows. And all these formulas, Eqs. (1)–(3) were adopted from Welaya et al.[19]

$$\dot{Q} + \dot{W} = \sum \dot{m}_{out} h_{out} - \sum \dot{m}_{in} h_{in}. \quad (1)$$

The mass flow rate of the fuel and air entering the SOFC were defined in Eqs. (2) and (3).

$$\dot{m}_{air} = \dot{m}_1 = \frac{3.57 \times \lambda_{air} \times P_{FC}}{10^4 \times V_{cell}}, \quad (2)$$

$$\dot{m}_f = \frac{P_{FC}}{\eta_{FC} \times CV}. \quad (3)$$

3 Results and Discussions

3.1 Effect of Mass Flow Rate on Combined Efficiency

Figure 3 (a) shows the effect of the air supply rate on the performance of the system. The air mass flow rate fed into the system varied by 10%, using 0.31 kg/s as a reference point. This analysis shows that the overall performance of the system increases when the mass flow rate is increased. This case could be caused by more air mass flowing through the turbine and the amount of energy generated tends to increase [20]. This case also showed a similar pattern to previous analyses, where the turbine performed better when the SOFC was fired with POME compared with RDF. Figure 3 (b) shows that an increase in fuel throughput tends to improve cycle efficiency, which has the greatest effect on the overall performance of the system when comparing the differences between all analyses. As a result of increased specific fuel consumption, more fuel is needed to produce a given output.

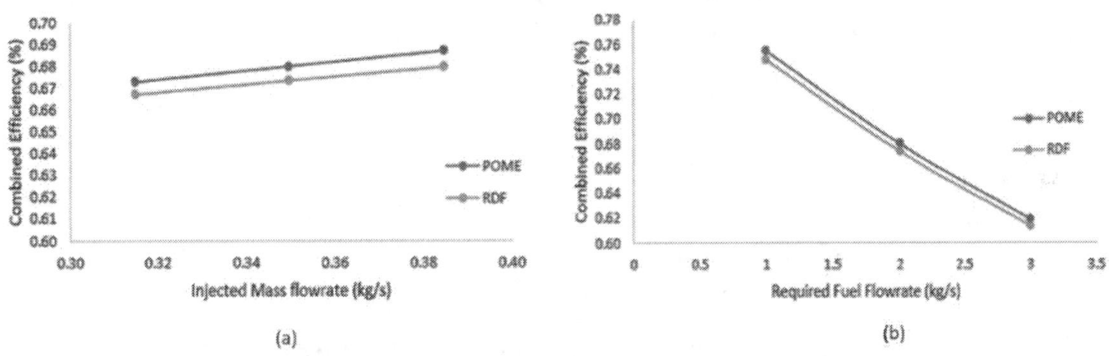

Figure 3: Combined efficiency at different mass flow rate (a) air and (b) fuel

3.2 Performance analysis of SOFC's hybrids GT

Figure 4 shows the comparative analysis between the POME-fired and RDF-fired SOFC-GT hybrids. The appearance of the POME-fired plants provides 49% more power than RDF. The reason is that POME has a higher heating value than RDF. This case may be because a higher calorific value releases more heat energy, so less fuel is needed to run the plant. The biomass used to produce these fuels can be further dried to lower the moisture content and increase the calorific value. From this, it is justified that the calorific value can play a role in fuel consumption. Like how a higher calorific value gives a better fuel efficiency, by ensuring that the calorific value is increased, the required fuel flow rate will drop, resulting in higher overall performance with an increase in the combined efficiency of the system [21]. This makes the combustion process more cost-effective and energy-efficient.

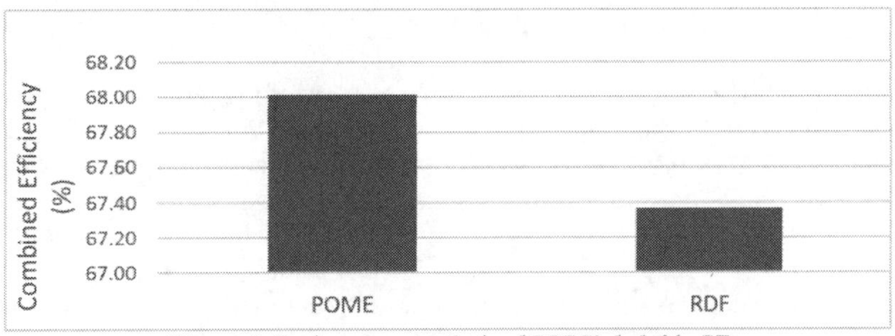

Figure 4: Performance Analysis of SOFC's hybrids GT

4 Conclusion

SOFC-GT hybrid power plant offers a more sustainable and efficient solution than fossil-derived power plant, as carbon emissions from heat and power generation can be reduced by 15%–20%. The biomass-derived fuel must be compatible with the gasification system to ensure optimal performance. In this work, the feasibility of using biomass-derived fuels in SOFC-GT hybrid power plants was investigated from a thermodynamic point of view through numerical analysis.

The feasibility of integrating SOFC hybrid gasification based on selected biomass-derived fuels (i) POME and (ii) RDF was investigated using numerical simulations. This work shows that the mass flow rate of the injected air and fuel significantly affects the performance of the plant. Moreover, the analysis showed that the combined efficiency of POME and RDF was 68% and 67.4%, respectively. The efficiency is high, justifying the feasibility of introducing this technology in Malaysia. POME hasan overall better performance than RDF owing to its higher calorific value, which helps to produce higher thermal energy

Acknowledgement

The authors would like to acknowledge funding from the Ministry of Higher Education Malaysia, through a Transdisciplinary Research Grant Scheme (TRGS) with code TRGS/1/2019/UKM/01/1/1.

References

[1] Bernama, "PM: Malaysia plans to reduce greenhouse gas emission by 45%," *The star*, 2021. [Online]. Available: https://www.thestar.com.my/news/nation/2021/10/11/pm-malaysia-plans-to-reduce-greenhouse-gas-emission-by-45

[2] I. E. Agency, "Total energy supply (TES) by source, Malaysia 1990-2019," *Total energy supply (TES) by source, Malaysia 1990-2019*, 2022. https://www.iea.org/countries/malaysia

[3] Z. Zakaria, S. K. Kamarudin, K. A. Abd Wahid, and S. H. Abu Hassan, "The progress of fuel cell for malaysian residential consumption: Energy status and prospects to introduction as a renewable power generation system," *Renew. Sustain. Energy Rev.*, vol. 144, p. 110984, 2021, doi: https://doi.org/10.1016/j.rser.2021.110984.

[4] M. S. A. Ali, A. Muchtar, J. C. W. Mah, J. Raharjo, and D. S. Khaerudini, "Effect of open pore and pore interconnectivity in the Ni-SDC cermet anode microstructure on the performance of solid oxide fuel cells," *IOP Conf. Ser. Mater. Sci. Eng.*, vol. 1195, no. 1, p. 012042, 2021, doi: 10.1088/1757-899x/1195/1/012042.

[5] A. J. Abd Aziz, N. A. Baharuddin, M. R. Somalu, and A. Muchtar, "Review of composite cathodes for intermediate-temperature solid oxide fuel cell applications," *Ceram. Int.*, vol. 46, no. 15, pp. 23314–23325, 2020, doi: https://doi.org/10.1016/j.ceramint.2020.06.176.

[6] L. S. Mahmud, A. Muchtar, and M. R. Somalu, "Challenges in fabricating planar solid oxide fuel cells: A review," *Renew. Sustain. Energy Rev.*, vol. 72, pp. 105–116, 2017, doi: https://doi.org/10.1016/j.rser.2017.01.019.

[7] N. Khalid *et al.*, "Symmetric and Asymmetric Impacts on Carbon Dioxide Emissions in Palm Oil Industry, Malaysia," pp. 0–30, 2021, [Online]. Available: https://doi.org/10.21203/rs.3.rs-748709/v1

[8] M. Ozturk *et al.*, "Biomass and bioenergy: An overview of the development potential in Turkey and Malaysia," *Renew. Sustain. Energy Rev.*, vol. 79, pp. 1285–1302, 2017.

[9] S. F. Salleh, M. E. Mohd Roslan, A. Abd Rahman, A. H. Shamsuddin, T. A. R. Tuan Abdullah, and B. K. Sovacool, "Transitioning to a sustainable development framework for bioenergy in Malaysia: policy suggestions to catalyse the utilisation of palm oil mill residues," *Energy. Sustain. Soc.*, vol. 10, no. 1, pp. 1–20, 2020, doi: 10.1186/s13705-020-00269-y.

[10] F. P. Paula Costa Rui Neto André,Paula Marques, "Integration of Gasification and Solid Oxide Fuel Cells (SOFCs) for Combined Heat and Power (CHP)," Processes, Basel, 2021.

[11] S. M. Shafie, Z. Othman, N. Hami, S. Omar, and A. H. Nu'man, "Solid Oxide Fuel Cells

Fueled with Biogas in Malaysia: A Review of Potential," in *Journal of Advanced Research in Fluid Mechanics and Thermal Sciences*, Kedah, Malaysia and West Java, Indonesia: Penerbit Akademia Baru, 2021, pp. 18–25.

[12] M. A. Ehyaei and M. A. Rosen, "Optimization of a triple cycle based on a solid oxide fuel cell and gas and steam cycles with a multiobjective genetic algorithm and energy, exergy and economic analyses," *Energy Convers. Manag.*, vol. 180, pp. 689–708, 2019, doi: https://doi.org/10.1016/j.enconman.2018.11.023.

[13] N. A. M. N. Aman, A. Muchtar, M. R. Somalu, M. I. Rosli, N. A. Baharuddin, and N. S. Kalib, "A short review on the modeling of solid-oxide fuel cells by using computational fluid dynamics: Assumptions and boundary conditions," *Int. J. Integr. Eng.*, vol. 10, no. 5, pp. 87–92, 2018, doi: 10.30880/ijie.2018.10.05.014.

[14] P.-C. Kuo, B. Illathukandy, C.-H. Kung, J.-S. Chang, and W. Wu, "Process simulation development of a clean waste-to-energy conversion power plant: Thermodynamic and environmental assessment," *J. Clean. Prod.*, vol. 315, p. 128156, 2021, doi: https://doi.org/10.1016/j.jclepro.2021.128156.

[15] S. Rudra, A. Sayem, S. Biswas, S. Lee, and H. Kim, "Simulation study on a hybrid Solid Oxide Fuel Cell (SOFC) - Heat Recovery Steam Generator (HRSG) - Gas turbine in IGCC power plant," in *Proceedings of the ASME 3rd International Conference on Energy Sustainability 2009, ES2009*, 2009, vol. 2. doi: 10.1115/ES2009-90084.

[16] N. F. Raduwan, A. Muchtar, M. R. Somalu, N. A. Baharuddin, and S. A. Muhammed Ali, "Challenges in fabricating solid oxide fuel cell stacks for portable applications: A short review," *Int. J. Integr. Eng.*, vol. 10, no. 5, pp. 80–86, 2018, doi: 10.30880/ijie.2018.10.05.013.

[17] D. Papurello, D. Canuto, and M. Santarelli, "CFD model for tubular SOFC stack fed directly by biomass," *Int. J. Hydrogen Energy*, vol. 47, no. 10, pp. 6860–6872, 2022, doi: https://doi.org/10.1016/j.ijhydene.2021.12.015.

[18] P. Saisirirat, "The Solid Oxide Fuel Cell (SOFC) and Gas Turbine (GT) Hybrid System Numerical Model," *Energy Procedia*, vol. 1, no. 79, pp. 845–850, 2015.

[19] Y. M. A. Welaya, M. Mosleh, and N. R. Ammar, "Thermodynamic analysis of a combined gas turbine power plant with a solid oxide fuel cell for marine applications," *Int. J. Nav. Archit. Ocean Eng.*, vol. 5, no. 4, pp. 529–545, 2013, doi: 10.2478/IJNAOE-2013-0151.

[20] M. M. Rahman, T. K. Ibrahim, and A. N. Abdalla, "Thermodynamic performance analysis of gas-turbine power-plant," *Int. J. Phys. Sci.*, vol. 6, no. 14, pp. 3539–3550, 2011.

[21] C. J. Mueller, W. J. Cannella, and G. T. Kalghatgi, *Fuels and the Impact of Fuel Composition on Engine Performance*. California: Chevron Corporation, 2012.

IOP Conf. Series: Materials Science and Engineering 1257 (2022) 012019 doi:10.1088/1757-899X/1257/1/012019

Thermal Investigation of Solid Oxide Fuel Cell Ni-YSZ Anode supported with Cooling System: A FEM Approach

N Shieela Kalib[1,3], A Muchtar[1,2*], V Zhen Yuan[3], M Rao Somalu[1], A Kamal Ariffin Mohd Ihsan[2]

[1]Fuel Cell Institute, Universiti Kebangsaan Malaysia, 43600 UKM Bangi, Selangor, Malaysia
[2]Department of Mechanical and Manufacturing Engineering, Faculty of Engineering and Built Environment, Universiti Kebangsaan Malaysia, 43600 UKM, Bangi, Selangor, Malaysia
[3]School of Engineering & Physical Science-Mechanical, Heriot-Watt University, 62200 Putrajaya, Malaysia

*E-mail : muchtar@ukm.edu.my

Abstract. Solid oxide fuel cell (SOFC) is preferred for power generation owing to its high-temperature waste heat recovery and low pollutant emissions. Nevertheless, SOFC operation is highly dependent on an effective cooling system, and the temperature gradient (ΔT) causes the distribution of thermal stress. The trade-off between meeting the SOFC requirement to operate at elevated temperature and minimizing thermal stress by reducing ΔT in the SOFC stack is required. Therefore, a cooling system for the SOFC stack is required to control the temperature homogeneity in the stack. In this study, a 3D finite element method containing heat transfer and energy charge equations was developed and then applied to investigate the effects of the (i) cooling plate and (ii) cooling fin on temperature distribution. The conventional stack orientation (i) parallel flow and (ii) counter flow were considered and compared with the adiabatic model in Ansys thermal analysis. The analysis was carried out under steady-state conditions. Considering that the temperature distribution varies according to the operating temperature and environment, the ambient temperature was set at 800 °C, 750 °C, 700 °C, and 650 °C. The study showed that the steepness of ΔT for a fin with the parallel flow is 1.5% better than for counterflow in the same operating scenario. Notably, the contribution of the cooling plate can minimize ΔT by 10% more than the fin. In addition, cooling measures are required to ensure the long-term stability of the cell during prolonged operation.

1 Introduction

Fuel cells can replace these thermal systems owing to their high energy conversion efficiency, which is twice that of a conventional engine [1]. Among the many types of fuel cells, the solid oxide fuel cell (SOFC) system offers an attractive solution for power generation technology with its high-temperature waste heat recovery, low CO_2 emissions, and high efficiency [2].

The SOFC is an energy conversion system that converts chemical energy directly into electrical energy for efficient power generation. The system uses an ion oxide conductive ceramic material as the electrolyte. In addition, considering the high operating temperature (600 °C–1000 °C), the natural gas fuel can be easily modified within the cell, and electrocatalysis can be carried out quickly with non-noble metals [3]. A key advantage of the SOFC over other fuel cells is, of course, its ability to process a wide range of hydrocarbon fuels, such as natural gas, methanol, and ethanol [4].

The temperature inhomogeneity or also known as temperature gradient (ΔT) often arises from the heat generated during the electrochemical reaction, particularly during ion migration from the cathode to the anode [5]. A large gap in temperature inhomogeneity often leads to performance degradation due to thermal stress [6]. The researchers found that the counterflow design has a lower impact on the ΔT than the parallel flow. Reza et al. [7] performed a CFD analysis assuming that endothermic reactions generate the temperature. In their work, the partial differential equations for the 3D model were derived based on the conservation laws for mass, energy, momentum, and ions. The results show that the uniformity of the current density distribution in the counterflow direction is better than in the parallel direction, which has a steeper ΔT. Therefore, the counterflow direction with uniform temperature distribution is preferred.

However, Iwata et al. [8] found that gas flow arrangements, such as parallel flow and counter flow, tend to significantly affect the temperature profile and thermal gradients of the SOFC. To cope with this situation, a cooling system is essential to remove the excess heat and minimize the ΔT in the SOFC to improve the SOFC performance. Compared with the experiment, the ΔT distribution can be simulated faster as it involves fewer parameters and still provides a quick insight, which is particularly beneficial at the beginning of the design phase.

In this study, the temperature distributions in a SOFC anode with NiO/YSZ support are presented to illustrate the thermal interaction through the temperature distribution on two selected cooling approaches (i) cooling plate and (ii) extended surface (fin). The thermal interactions were analyzed using the finite element method (FEM), and the SOFCs were modeled with a size of 10 cm × 10 cm by three-dimensional modelling. The SOFC stacks were oriented according to (i) parallel flow and (ii) counter flow as both are the most recommended orientations in the industry.

2 Modeling Approach

Figure 1 presents a view of the cross-section of a pair of the flow channel (fuel and air) in an elevated operating temperature of the SOFC fuel cell. The fuel is transported to the anode and the oxidizer to the cathode. The electrolyte conducts oxygen ions formed by the electrochemical reaction of molecular oxygen from the cathode side of the SOFC to the anode side. The fuel then diffuses through the anode to the interface between anode and electrolyte. Here, the fuel reacts catalytically with oxygen ions, releasing electrons that are transmitted through an external circuit to produce electricity.

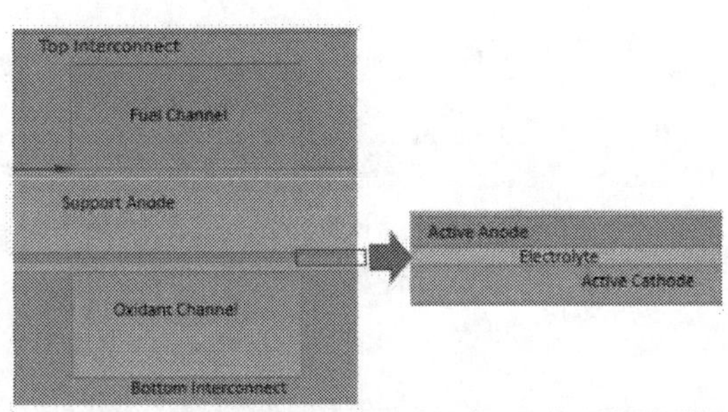

Figure 1: Schematic diagram of the SOFC model

2.1 *Cooling methods*

2.1.1 *Extended surface (Fins)*

Figure 2 illustrates the design of the extended surface, also known as the fin, attached to the top of the interconnect. The length of the SOFC stack is 100 mm and the distance between the fins is 10 mm. The temperature of the fins is regulated by natural convection as cold air is passed through the fins.

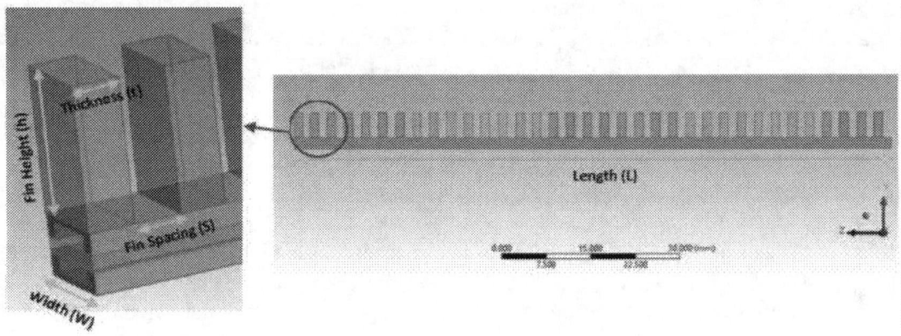

Figure 2: Model for the extruded straight fins

2.1.2 *Cooling Plate*

In the second cooling approach, a cooling plate is attached to both sides of the interconnect. The temperature in the SOFC is controlled by directing cold air onto the surface of the cooling plate. Figure 3 shows the configuration of the cooling plates.

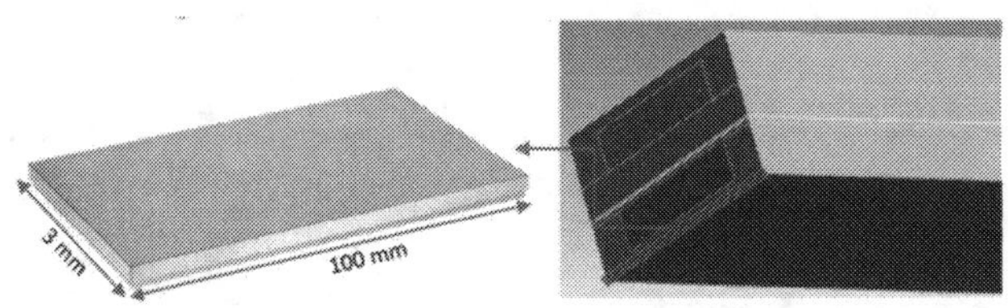

Figure 3: Dimensions of a cooling plate

2.2 *Thermal Analysis*

With the FEM approach, the temperature field in the SOFC stack can be determined according to the principle of heat transfer. The principle of heat transfer is adopted in the Poisson energy, which is expressed as follows:

$$\frac{\partial}{\partial x}\left(k_x \frac{\partial T}{\partial x}\right) + \frac{\partial}{\partial y}\left(k_y \frac{\partial T}{\partial y}\right) + q_i = 0. \tag{1}$$

As all layers are in contact, the internal heat generated by the electrochemical reaction is expressed as a heat flux. In this study, the internal heat flux for each layer of the SOFC components is used. Equation (2) shows the relationship between the heat flux, q_i, and the length of the stack, L.

$$Q_i = \frac{q_i}{L}. \tag{2}$$

The value of the internal heat radiate in the SOFC's components during the electrochemical reaction depends on the electrical/ionic conductivity. The electronic/ionic conductivities, σ, in the anode, cathode, and electrolyte are defined in Equation (3) – Equation (5) [9], respectively

$$\sigma_{anode} = \frac{9.5 \times 10^7}{T} e^{-\frac{1150}{T}}, \tag{3}$$

$$\sigma_{cathode} = \frac{4.2 \times 10^7}{T} e^{-\frac{1200}{T}}, \quad (4)$$

$$\sigma_{electrolyte} = 3.34 \times 10^4 e^{-\frac{10300}{T}}. \quad (5)$$

2.3 *FEM model*

Table 1 shows the technical specifications used in this work. The parameters were determined according to the operating temperature of the SOFC between 850 °C and 860 °C.

Table 1. Working parameters adopted in this work

Operating temperature for SOFC stack		
Fuel inlet		850 °C [10]
Oxidant inlet		860 °C [10]
Cooling methods		
Fins	Ambient temperature:	800 °C [10]
	Film coefficient:	17.3 W/m^2 °C [11]
Cooling plate	Ambient temperature:	350 °C [10]
	Film coefficient:	24.7 W/m^2 °C

3 Results & Discussions

3.1 *With Fins*

The effect of stack alignment between counterflow and parallel flow in the SOFC single stack can be illustrated by the thermal contour distribution shown in figure 4. The temperature distribution for parallel flow is higher at the inlet of the channel than at the outlet. This trend is consistent with the work published by Lee et al. [12]. Although the same operating temperature and internal heat are applied for both orientations of the stack, the temperature distribution at the inlet of the fuel is higher for parallel flow than for counterflow at the same location. However, the uniformity of the temperature distribution in the parallel flow is more linear. The heat was found to increase from point A and decrease over the entire length toward point B. In the counterflow, the thermal contour is higher at points A and B, whereas it is more relaxed in the middle of the stack. These trends indicate that the steepness ΔT is higher in counterflow than in parallel flow.

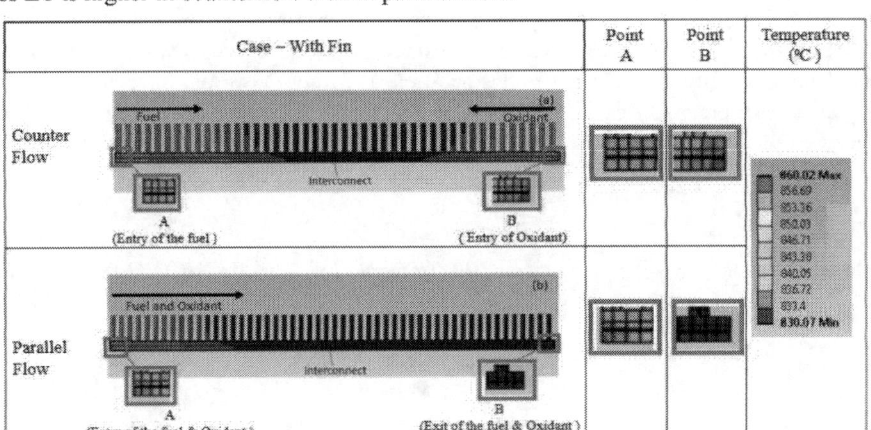

Figure 4: Temperature contour in SOFC model with fins: (a) counterflow and (b) parallel flow

3.2 With cooling plate

Comparing the model with the fins, the heat dissipation capacity of the cooling plate is better, and the minimum temperature drops significantly, as shown in figure 5. Although the maximum temperature is approximately the same for both configurations, a difference can be seen between the two minimum temperatures. The parallel flow has a lower minimum temperature than the counterflow, 547 °C and 628 °C, respectively. This result indicates that the inclusion of the cooling plate in the parallel flow makes heat removal from the stack better than the counterflow. However, with regard to the ΔT distributions, a similar trend as in figure 4 can be seen here; however, the slope of the ΔT in the cooling plate is significantly higher than in the stack with the fin.

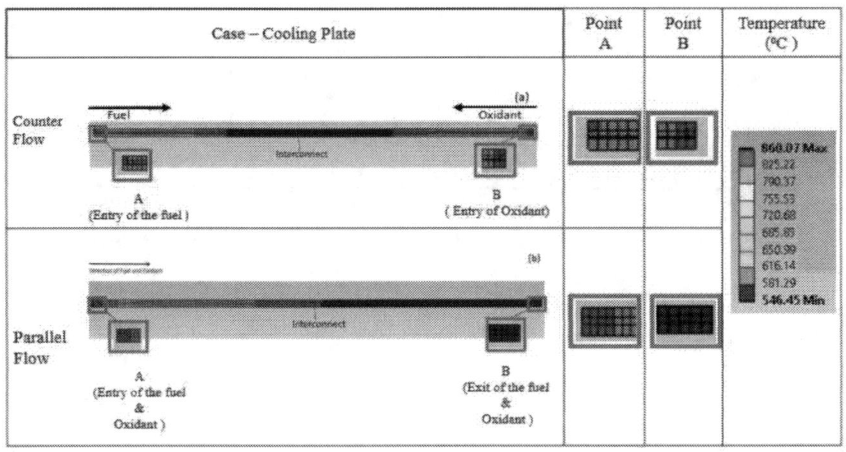

Figure 5: Temperature contour in SOFC model with cooling plate: (a) counterflow and (b) parallel flow

4 Conclusion

SOFC is an alternative for a green energy conversion system that converts chemical energy directly into electrical energy for efficient power generation. Operating at high temperatures is one of the advantages of SOFC as it improves the electrical output of the fuel cell by increasing the kinetic energy. However, the steepness of ΔT distribution across the SOFC stack is larger when operating at higher temperatures, which leads to thermal stress.

In this study, a 3D model FEM was developed for high-temperature SOFC to investigate the best cooling approach to reduce the ΔT. Considering the thermal radiation from the exothermic reaction, the temperature distribution in the stack is shown considering two cooling approaches: (i) fin and cooling plate. In addition, the effects of the stack configuration (i) parallel flow and (ii) counter flow on both cooling approaches were investigated. The results show that both cooling approaches are able to reduce the ΔT in the SOFC stack. The range of reduced ΔT is 36% higher for the cold plate compared with the fin.

Acknowledgement

The authors would like to acknowledge funding from the Ministry of Higher Education Malaysia, through a Transdisciplinary Research Grant Scheme (TRGS) with code TRGS/1/2019/UKM/01/1/1.

IOP Conf. Series: Materials Science and Engineering 1257 (2022) 012019 doi:10.1088/1757-899X/1257/1/012019

5 References

[1] J. C. W. Mah, M. I. Aznam, A. Muchtar, M. R. Somalu and M. J. Ghazali, "Cr Vaporisation and Oxidation Behaviour of SUS430 Stainless Steel in the Simulated Cathodic Environment of Solid Oxide Fuel Cells," *Malaysian Journal of Microscopy,* vol. 16, pp. 124-132, 2020.

[2] A. J. A. Aziz, N. A. Baharuddin, M. R. Somalu and A. Muchtar, "Review of composite cathodes for intermediate-temperature solid oxide fuel cell applications," *Ceramics International,* vol. 46, no. 15, pp. 23314-23325, 2020.

[3] N. S. Kalib, A. Muchtar, M. R. Somalu, A. K. A. M. Ihsan, N. A. M. N. Aman and N. A. Baharuddin, "Factors Influencing Thermal Stress Development in Solid Oxide Fuel Cells," *Journal of Advanced Research in Fluid Mechanics and Thermal Sciences,* vol. 50, pp. 146-152, 2018.

[4] N. A. M. N. Aman, A. Muchtar, M. R. Somalu, M. I. Rosli, N. A. Baharuddin and N. S. Kalib, "A short review on the modeling of solid-oxide fuel cells by using computational fluid dynamics: assumptions and boundary conditions," *International Journal of Integrated Engineering: Special issue 2018: Mechanical Engineering,* vol. 10, pp. 87-92, 2018.

[5] N. A. M. N. Aman, A. Muchtar, M. I. Rosli, N. A. Baharuddin1, M. R. Somalu and N. S. Kalib, "Influence of Thermal Conductivity on the Thermal Behavior of Intermediate-Temperature Solid Oxide Fuel Cells," *Journal of Electrochemical Science and Technology* , vol. 11, pp. 132-139, 2020.

[6] D. Panuh, S. Ali, D. Yulianto, M. F. Shukur and A. Muchtar, "Effect of yttrium-stabilized bismuth bilayer electrolyte thickness on the electrochemical performance of anode-supported solid oxide fuel cells," *Ceramics International,* vol. 47, no. 5, pp. 6310-6317, 2021.

[7] H. R. Amedi, B. Bazooyar and M. R. Pishvaie, "Control of anode supported SOFCs (solid oxide fuel cells): Part I. mathematical modeling and state estimation within one cell," *Energy,* vol. 90, pp. 605-621, 2015.

[8] M. Iwata, T. Hikosaka, M. Morita, T. Iwanari, K. iIto, K. Onda, Y. Esaki, Y. Sakaki and S. Nagata, "Performance analysis of planar-type unit SOFC considering current and temperature distributions," *Solid State Ionics,* vol. 132, no. 3-4, pp. 297-308, 2000.

[9] M. Andersson, H. Paradis, J. Yuan and B. Sundén, "Three dimensional modeling of an solid oxide fuel cell coupling charge transfer phenomena with transport processes and heat generation," *Electrochimica Acta,* vol. 109, pp. 881-893, 2013.

[10] M. Promsen, Y. Komatsu, A. Sciazko, S. Kaneko and N. Shikazono, "Feasibility study on saturated water-cooled solid oxide fuel cell stack," *Applied Energy,* vol. 279, 2020.

[11] H. I. Abu-Mulaweh and J. Donald W. Mueller, "Heat Transfer Coefficient Correlation for Circular Fin Rods," pp. 7-10, 2004.

[12] Anh Dinh Le, Steven B Beale , J.G. Pharoah, "Validation of a Solid Oxide Fuel Cell Model on the International Energy Agency Benchmark Case with Hydrogen Fuel," *Fuel Cells,* 2014.

Hydrogen Co-Firing Characteristics in a Single Swirl Burner: A Numerical Analysis

Mohammad Nurizat Rahman[*,1], **Norshakina Shahril**[2], **Suzana Yusup**[3], **Ismail Shariff**[4]

[1,2,3,4] Fuels and Combustion Section, Generation Unit,
Department of Generation and Environment, TNB Research Sdn Bhd
No 1, Kawasan Institusi Penyelidikan, Jalan Ayer Hitam, Kajang, 43000, Selangor. Malaysia.

*Email: corresponding author: nurizat.rahman@tnb.com.my

Abstract. Hydrogen is gaining traction as an energy carrier in the decarbonisation and net-zero-emissions agenda. Because hydrogen is a clean energy carrier, increasing the percentage of hydrogen in natural gas mixtures aids in the decarbonisation initiatives. Hence, the flame characteristics of the natural gas mixtures, together with hydrogen are explored in the current study through a numerical assessment of a single swirl burner (swirl number, S_N 0.78) since the said burner is widely used in gas turbine (GT) combustors. The baseline CFD and experimental cases referred to natural gas compositions primarily composed of methane (CH_4). The results reveal that the CFD model can effectively represent the swirling component of the flame as seen in the experiment. A 5% hydrogen addition had virtually no effect on the swirling flame structure, as shown by qualitative evaluation of hydroxyl (OH) behaviour and flame temperature in comparison to the baseline methane flame. Despite this, the addition of hydrogen has increased the OH radical pool during combustion, causing a small change in flame temperature. Overall, the novelty of the current research is the opportunity to fire 5% hydrogen in a CH4-dominated GT combustor without any major retrofitting operations, as the study discovered that 5% hydrogen in a pure CH4 stream has a minor affect. However, more research is needed to properly capture the flame structure and strain for assessing transient-related phenomena like flashback and blow off by increasing the hydrogen proportion and using a higher accuracy turbulence model.

1. Introduction

Energy sector in Malaysia is identified as the largest contributor of greenhouse gas (GHG) emissions with 251.7 Mt carbon dioxide (CO_2) corresponding to 79.4% of the total emission[1]. In total, the GHG emissions of Malaysia in 2018, according to World Bank is about 306.7 Mt (0.7% of the world total emission)[2] without Land Use Change and Forestry. Thus, in gearing towards reduction of the GHG emissions, the National Determined Contribution of Malaysia by 2030[3], is to reduce, its carbon intensity against GDP by 45% compared to 2005 level. The Green Technology Master Plan of Malaysia (2017-2030)[4] covers policies and action plans towards a low carbon and resource-efficient economy under the National Green Technology that relates to Energy, Environment, Economy and Social. Among other is improvement in the efficiency in power generation. Nevertheless, emerging technology such as utilising hydrogen in power plant is yet to be considered by Malaysia government due to high cost. The government has pledged to stop building new coal-fired power plants in the future. Effort such as co-firing of hydrogen has gained tractions to reduce CO_2 and to be further explored its potential as part of power plant development. Malaysia Renewable Energy Roadmap (MyRER)[5], highlights decarbonisation of the electricity sector in Malaysia by 2025, focused on exploration of bioenergy such as biomass co-firing. Hydrogen has yet to be highlighted. Nevertheless, hydrogen has gained acknowledgment since the Eight Malaysia Plan (2001-2005)[6-7] through funding on hydrogen fuel cell research & development program. Currently under initiatives of MOSTI[8] led by Nano Malaysia, partnering with Malaysia Green Technology Corporation, Academy Science Malaysia, local universities and Nano Verify, are conducting study to formulate Hydrogen Road Map of Malaysia and is targeted to be completed in 2022. Japan AETI[9] (The Asia Energy Transition Initiatives), have identified potential technologies to support energy transition agenda and among others it includes Conversion to Gas technologies (Coal to Gas, Diesel to Gas) and Co-firing or full combustion of ammonia-hydrogen.

Owing to yet commercial deployment of hydrogen co-firing or full firing at power plant in Malaysia, this paper investigates the potential of employing hydrogen at a certain percentage within natural gas mixtures (primarily methane) to support decarbonisation initiatives using Computational Fluid Dynamics (CFD) modelling from a single swirl burner. The purpose of using a single swirl burner is to emulate the swirler features found in a typical gas turbine (GT) combustor. The single swirl burner's experimental work was carried out using a natural gas mixture, and it was utilised to validate the CFD model. Following validation, a 5% of hydrogen was co-fired in the CFD model of the single swirl burner. In the Results and Discussion section, the essential combustion properties were thoroughly examined using a comparative analysis of the pure natural gas and hydrogen co-firing cases.

2. Methodology

2.1 Experimental setup

To create a natural gas swirl flame, a single annulus, axial swirl GT type burner was used. The axial swirler is made up of eight straight vanes that are angled at 45° from the axial centerline axis. The axial swirler's swirl number (S_N) is estimated to be 0.78, with a vane thickness of 1.5 mm and swirler hub and swirler D_s diameters of 19.2 and 38.4 mm, respectively. The swirler is powerful enough ($S_N > 0.6$) to provide high-intensity recirculation flow, which aids in flame stabilisation. The swirler is positioned concentrically with respect to the burner outlet. A circular quartz tube with a diameter of 100 mm and 180 mm in length was used to form a combustor wall at the burner outlet. The swirl flame burner is shown in Figure 1.

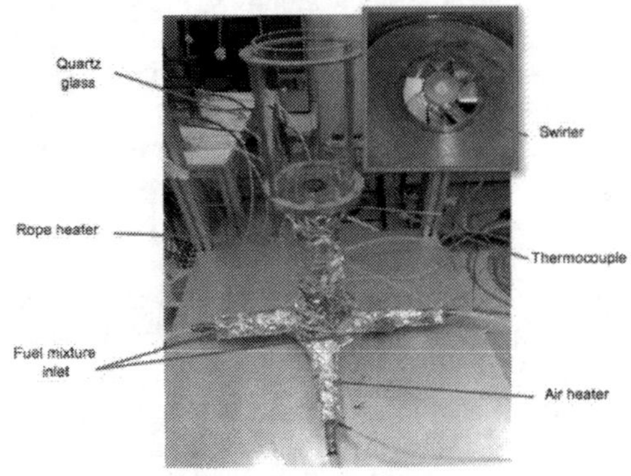

Figure 1: Swirl burner

The supplies of natural gas mixture (Methane (CH_4) = 94 %, Ethane (C_2H_6) = 2%, Propane (C_3H_8) = 0.5%, Nitrogen (N_2) = 1.5%, and CO_2 = 2%) were regulated by mass flow controllers (Sierra; accuracy ±1% full scale). A mass flow controller (Sierra; accuracy ±1% full scale) at the burner plenum blended the modelled natural gas mixture with main airflow at 200 LPM.

2.2 Numerical procedure

This study used as-built dimensions from the swirl burner to develop the three-dimensional (3D) geometry of the burner. The burner geometry was simplified to capture important features, while non-critical components were left out to reduce computational expenses. The computational domain of the burner, as well as the related boundary condition locations, are shown in Figure 2.

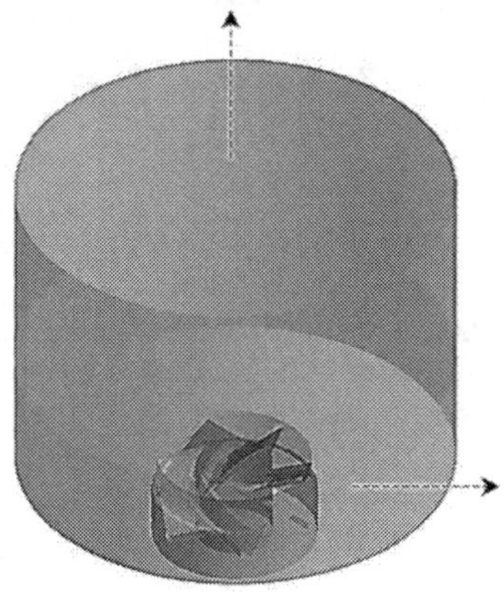

Figure 2: Computational domain of the swirl burner

Using a commercial CFD software package, ANSYS Fluent V.19 R1, the governing equations (steady state and compressible) are transformed into discrete forms using the finite volume approach. All setups and numerical processing are done with ANSYS Fluent V.19 R1. The governing equations are solved using the pressure-based solver. The Semi-Implicit Method for Pressure Linked Equations (SIMPLE) technique was adopted to solve the pressure-velocity coupling. Detail information on the constants and formulations utilized in the SIMPLE algorithm can be found in the study of Ferziger *et al.,*[10]. The Discrete Ordinates (DO) model was employed to resolve the radiative heat transfer from the reacting flow. The Shear Stress Transport (SST) $k-\omega$ model based on the Reynolds-averaged Navier–Stokes (RANS) formulation is employed to resolve the turbulence behaviour.

The probability density function (PDF) and flamelet formulations have previously been demonstrated to be sufficient for modelling the complicated turbulence-chemistry interaction within the combustor[11]. Hence, the current study used the non-adiabatic steady diffusion flamelet model with a detailed chemistry mechanism (GRI Mech 3.0), which computes temperature and species composition using a variable known as the mixture fraction that reflects the local fuel/oxidizer ratio[12]. More information on the constants and formulations utilized in the flamelet approach can be found in Rochaya[11]. Since the diffusion flamelet model is intended for non-premixed reactants, the reactants were blended in a non-premixed manner before entering the swirl burner inlet to ensure premixed reactants were achieved at the burner inlet.

Table 1 shows the case studies for investigating the effect of hydrogen co-firing on combustion parameters in the burner model. A pure CH_4/air mixture is used as the baseline scenario. In comparison to the experiment, the natural gas mixture for the baseline case of CFD simulation is different than in the experiment, where a number of species accounted for the natural gas, yet pure CH_4 is implemented for baseline CFD simulation. This is done to reduce computational complexity by lowering the number of reactants used in the kinetics calculation. Furthermore, more than 90% of the natural gas analysed in the swirl burner experiment contains CH_4. Hence, the use of pure CH_4 in the baseline CFD case can be considered suitable for direct validation with the swirl burner experiment, which used natural gas mixtures.

Table 1. Boundary conditions for each case studies.

Case study	Composition	Flowrate	Equivalence ratio
Baseline	100% CH_4	200 LPM	1.0
Hydrogen co-firing	95% CH_4, 5% H_2	200 LPM	1.0

3. Results and Discussion

3.1 Validation of the CFD Model

One of the first checks in any numerical analysis, prior to the validation phase, is the dependability of the elements (grids) used to construct the CFD model. To investigate the sensitivity of the computational domain to grid resolution in the CFD model, the baseline case simulation was run repeatedly, with the grid system's refinement increasing with each simulation. Grid independence, which indicates numerical result stability, was accomplished in the current research using 1.92 million elements. A validation exercise was then carried out against the generated swirl flame from the experiment. Figure 3 depicts a qualitative comparison of the flames from the CFD baseline and the experiment.

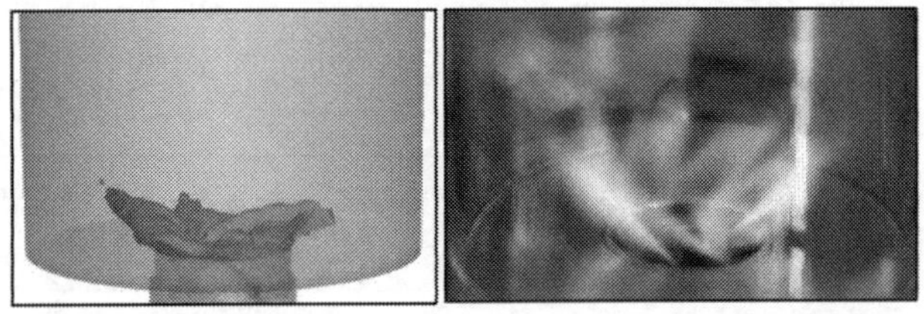

Figure 3: Qualitative comparison of the flames (CFD and experiment)

The iso-surface method is used to depict the flame front in the CFD baseline. More information on the method can be found in our earlier study[13]. The whirling aspect of the flame can be captured appropriately by the CFD model, as seen in Figure 3. The employment of a swirler allows the flame fronts at the burner outlet to be stabilised at the angle of the swirl flow exit[13]. A bluish flame was produced as seen in the experiment data. The bluish flame appears in the heat reaction zone, which is where natural gas chemically reacts with air. The complete reaction of fuel and oxidizer molecules produces water (H_2O) and CO_2, while the flame reaction zone is typically bluish[14]. This type of flame is desirable and has been used in low-emission GT combustor technology, specifically lean-premixing GT technology[14].

3.2 Hydrogen co-firing

Following validation, the simulation for the hydrogen co-firing case was done. The comparison of hydroxyl (OH) radical species and temperature from baseline and hydrogen co-firing simulations is shown in Figure 4. OH is a radical species that dominates chain propagation, chain branching, and chain termination events in a combustion process at various temperatures[13]. The OH is produced in a zone where extremely exothermic chemical processes are occurring. As a result, the chemical kinetics of OH in a combustion event enables researchers to conduct in-depth analyses of flame front behaviour based on OH formation[13]. Table 2 shows the OH generation in terms of mass fraction and flame temperature at three height-above-burner (HAB) locations of 2, 4, and 6 mm. The HAB has a datum just after the burner inlet (above swirler).

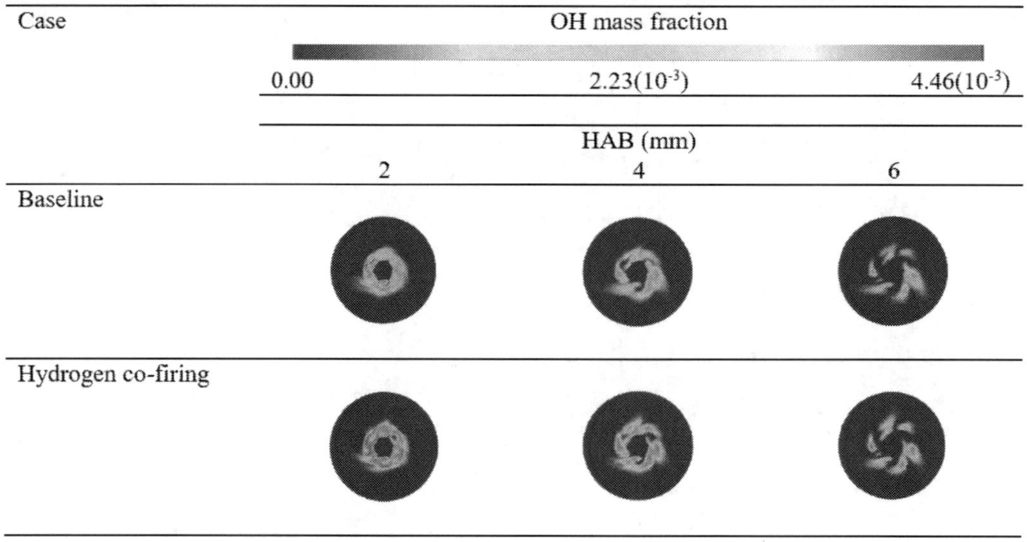

Figure 4: OH radical behaviour in baseline and hydrogen co-firing cases

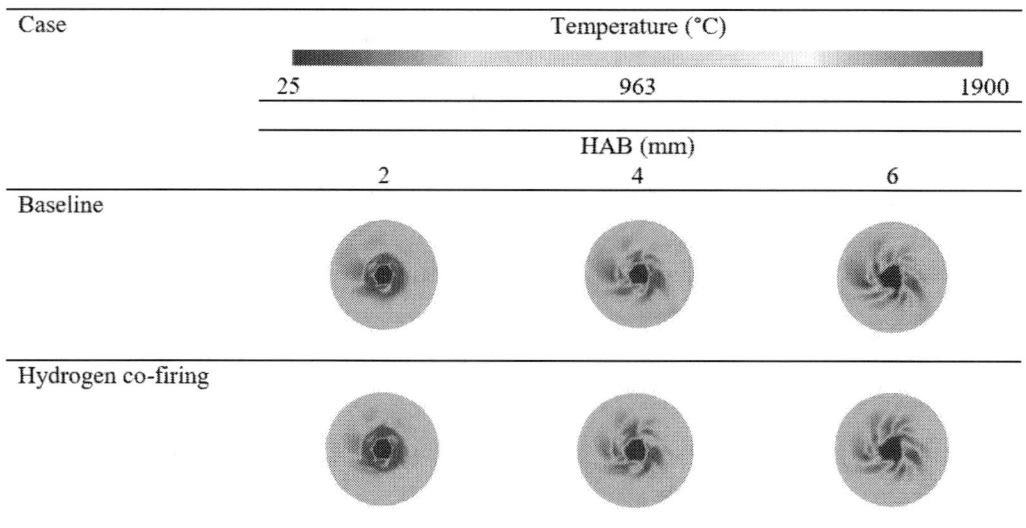

Figure 5: Temperature behaviour in baseline and hydrogen co-firing cases

The flame with hydrogen addition reveals greater OH concentrations, as shown in Figure 4. Furthermore, the OH distributions are more continuous, as observed in HABs 2 mm and 4 mm, where the OH levels along the swirling flame's periphery are higher. These findings reveal that OH levels improve in a hydrogen-enriched flame, implying that the increased radical pools of hydroxyl ions is responsible for a modest change in flame structure with the addition of hydrogen. The OH level is minuscule at the centre of the toroidal swirling flame, implying that there is no flame owing to the fuel-air stream that flows throughout the swirler. In the hydrogen co-firing case, a somewhat greater OH concentration towards the inner wall of the swirl burner domain was predicted at HAB 6 mm. Fluctuations in OH levels represent the swirling flame's movement in response to local flow turbulence.

Despite this, the flame size and shape of the hydrogen-enriched case were almost identical to those produced from the baseline methane flame.

According to the CFD prediction, low levels of hydrogen enrichment resulted in small increases in OH radical concentrations. Higher OH radical concentrations, which boost several critical reaction rates[15], are postulated to be responsible for the slightly improved flame temperature with hydrogen enrichment as seen in Figure 5. Overall, a qualitative assessment of the flame structure with and without 5% hydrogen addition reveals that the flame structure is essentially comparable. As a result of the qualitative comparison of the OH structure and the generated flame temperature, the impact of 5% hydrogen inclusion in the pure methane stream can be deemed minimal. Hence, it reveals that including 5% hydrogen in the natural gas-fired GT combustor with essentially no modification is achievable, providing a significant benefit in terms of retrofitting cost (cheaper) and carbon reduction.

4. Conclusion

We have demonstrated how the effects of hydrogen addition in a methane swirling flame, with the whirling component of the flame exhibited in the experiment, can be reproduced accurately by the CFD model. Based on a qualitative evaluation of OH behaviour and flame temperature with respect to the baseline methane flame, a 5% hydrogen addition had essentially no impact on the whirling flame structure. Nonetheless, the addition of hydrogen has increased the OH radical pool during combustion, which is responsible for a minor variation in flame temperature. A number of hydrogen co-firing cases with increasing hydrogen proportions are required to provide more detailed insights into the impact of hydrogen addition on the natural gas flame. Furthermore, transient-related phenomena such as flashback and blow off should be explored further in future studies using a Large Eddy Simulation (LES) model rather than the RANS model employed in this study. This is done to ensure that the majority of the time and length scales are resolved in order to accurately capture the flame structure and strain.

References

[1] UNFCC:Malaysia's Third Biennil Update Report to the UNFCC 2016, https://unfccc.int/sites/default/files/resource/MALAYSIA_BUR3-UNFCC_Submission.pdf

[2] World Bank: Total greenhouse gas emission: http://data.worldbank.org/indicator/EN.ATM.GHGT.KT.CE?view=chart

[3] UNFCC: Maysia's Update of Its First Nationally Determined Contribution.https://www4.unfcc.int/sites/ndcstaging/PublishedDocuments/Malaysia%20First/Malaysia%20NDCUpdated%20Submission%20to20%UNFCC%July%202021%20final.pdf

[4] Ministry of Energy, Green Technology and Water (KeTTA): Green Technology Master Plan Malaysia : 2017-2030. https://www.pmo.my/ms/2019/07/pelan-hala-tuju-teknologi -hijau-malaysia/

[5] Sustainable Energy Develoment Authority (SEDA), Malaysia: National Renewable Energy policy; http: //www.seda.gov.my/policies/national -renewable-energy-policy -and action plan -2009

[6] Economic Planning Unit, Prime Minister's Deparrtment, Eight Malaysia Plan, 2001-2005. https://www.epu.gov.my/en/economic-developments/development_plans/rmk/eight-malaysia-plan-2001-2005

[7] Rahman MN and Wahid MA (2021). Renewable-based zero-carbon fuels for the use of power generation: A case study in Malaysia supported by updated developments worldwide. *Energy Reports*, 7, 1986-2020. https://doi.org/10.1016/j.egyr.2021.04.005

[8] Ministry of Science, Technology and Innovation (MOSTI), : The Strategic Plan MOSTI 2021-2025. https://www.mosti.gov.my/en/pengummuman /pelan strategik -mosti-2021-2025

[9] Ministry of economy, Trade and Industry (METI) Japan, Minister Kajiyama announced the Asia, Energy Transition Initiative (AETI); https: www.meti.go.jp/english/press/2021/0528_002.html

[10] Ferziger JH, Perić M, and Street, RL. (2002). *Computational methods for fluid dynamics* (Vol. 3, pp. 196-200). Berlin: springer. https://doi.org/10.1007/978-3-642-56026-2

[11] Rochaya D. "Numerical simulation of spray combustion using bio-mass derived liquid fuels." (2007).

[12] Tyliszczak A, Boguslawski A, and Nowak D. (2016). Numerical simulations of combustion

process in a gas turbine with a single and multi-point fuel injection system. *Applied Energy*, *174*, 153-165. https://doi.org/10.1016/j.apenergy.2016.04.106
[13] Rahman MN, Yasin MFM, and Aris MS. (2021). Reacting Flow Characteristics and Multifuel Capabilities of a Multi-Nozzle Dry Low NOx Combustor: A Numerical Analysis. *CFD Letters*, *13*(11), 21-34. https://doi.org/10.37934/cfdl.13.11.2134
[14] Chong CT, Ng JH, Aris MS, Mong GR, Shahril N, Ting ST, and Zulkifli MF (2019). Impact of gas composition variations on flame blowout and spectroscopic characteristics of lean premixed swirl flames. *Process Safety and Environmental Protection*, *128*, 1-13. https://doi.org/10.1016/j.psep.2019.05.015
[15] Schefer RW, Wicksall DM, and Agrawal AK. (2002). Combustion of hydrogen-enriched methane in a lean premixed swirl-stabilized burner. *Proceedings of the combustion institute*, *29*(1), 843-851. https://doi.org/10.1016/S1540-7489(02)80108-0

Acknowledgement

This work is carried out with the financial support from TNB Research (TNBR) Seeding Fund. TNBR Board Technical Committee, which is the fund committee. Stakeholders TNB GENCO, TNB STRATEGIC CORPORATE VENTURE DEPARTMENT. *This work was supported by Grant no. 'TNBR/SF 412/2021"*

Decomposition and decoupling analyses of CO_2 emissions and economic growth: A comparative study between Malaysia and Vietnam

B Kiwan[1], N S Kalib[1] and A Muchtar[2]

[1] School of Engineering & Physical Sciences, Heriot-Watt University Malaysia, No 1 Jalan Venna P5/2, Precinct 5, 62200 Putrajaya, Malaysia.
[2] Fuel Cell Institute, Universiti Kebangsaan Malaysia (UKM), 43600 Bangi, Selangor, Malaysia

* E-mail: n.kalib@hw.ac.uk

Abstract. During the COVID-19 pandemic, Malaysia and Vietnam has managed to show a strong positive growth domestic product (GDP) growth indicator – with both countries recording an average increase of more than 4% annually. However, there was a strong positive correlation between GDP and carbon dioxide (CO_2) emissions worldwide. As the impact of climate change becomes more evident, especially in Southeast Asia, it is crucial to assess the relationship between these two indicators caused by gross electricity generation, especially in the post-pandemic period. In this study, we compared the decoupling performance of CO_2 emissions and GDP growth between Malaysia and Vietnam. The data for Malaysia and Vietnam were quantified by the decoupling method based on the Tapio decoupling model. The data were then decomposed using the Kaya identity into the driving factors affecting decoupling. This was then used to form the decomposition index using the Logarithmic Mean Divisia Index-I (LMDI-I). Data were generated based on three scenarios: i) 2017-2019, i.e. pre-pandemic, ii) 2019-2020, peri-pandemic and iii) 2021-2025, to predict post-pandemic. In terms of the overall decoupling trend, the study shows that: (1) the change in CO_2 emissions in both sovereign states decreased dramatically at the beginning of the COVID-19 pandemic. However, Malaysia experienced negative change in CO_2 emissions, while Vietnam still experienced a positive change. This is due to Vietnam's slightly higher reliance on coal for power generation than Malaysia. (2) A strong decoupling in the projected decoupling between the change in CO2 emissions and GDP in Vietnam, while the two indicators remain coupledfor Malaysia. This is due to Vietnam's more aggressive policy in the adoption of renewable energy - especially through the mass installation of rooftop photovoltaic systems. In contrast, Malaysian policy remains conservative, emphasising reliability at the cost of a huge year-on- year change in CO_2 emissions.

1. Introduction

The ASEAN State of Climate Change Report states that '*Southeast Asia is one of the most at-risk regions in the world to the impacts on climate change*'. However, there is little that has been done by way of a concerted effort in collating, consolidating and synthesising data related to climate change into a coherent report.

One key area of data in this field is looking into the change in CO_2 emissions from electricity generation, which amplifies efforts within ASEAN in making the energy transition from fossil fuels to renewable energy sources [1]. To highlight the disparities in the region in approaches towards adopting renewable energy sources, two ASEAN sovereign states are compared: Malaysia and Vietnam, withthe former taking a more conservative approach, and the latter being more aggressive in its approach – where it had the largest cumulative installed capacity for solar photovoltaic (PV) panels in ASEAN.

Another interesting factor in the data is to examine three periods with relation to the emergence of COVID-19: pre pandemic (2017-2019), peri-pandemic (2019-2020), and post-pandemic (2021-2025). This would then illustrate the differences in the change in CO_2 emissions in periods of economic growth, recession, and recovery – respectively, thereby truly testing to what extent the change in CO_2 emissions is coupled with GDP growth. The data for 2021-2025 are projected from multiple sources.

2. Method

The proposed methodology can be summarised into two components – with the input being processed through two different analyses: (i) decomposition analysis and (ii) decoupling analysis. These two analyses then produce two sets of distinct, yet complementary outputs. This methodology is illustrated in Fig 1 and is further elucidated upon in the subsequent subsections.

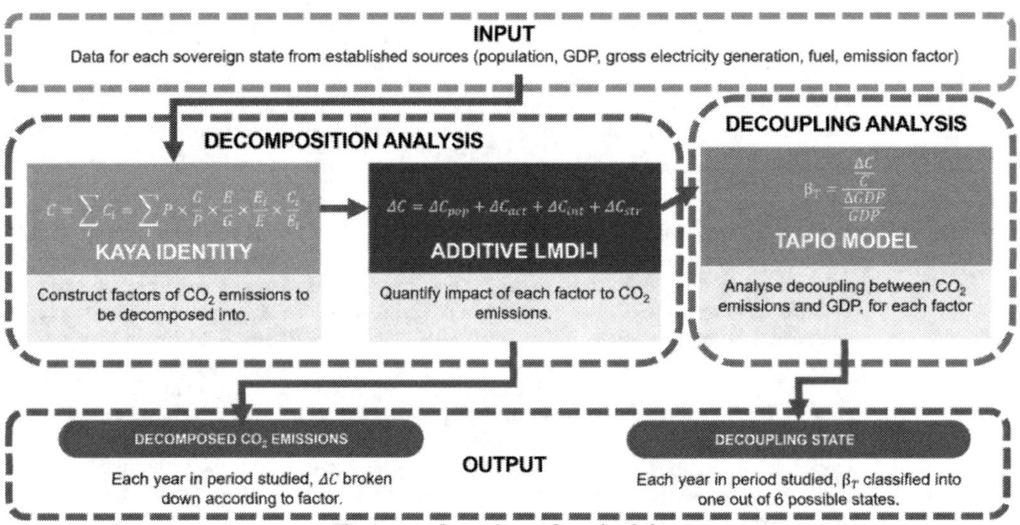

Figure 1: Overview of methodology

2.1. Input

The data sets for each sovereign state to be obtained are tabulated in Table 1 (with their relevant units and abbreviations

Table 1: List of data sets and associated abbreviations.

Data set	Abbreviation
Population	P
GDP (2017 US\$, PPP)	G
Total fuel consumption (TWh)	E
Share in energy mix for fuel type (TWh)	E_i

It must be noted, from here on out, all mentions of US$ would refer to the purchasing-power-parity (PPP) value of the US$ in the year 2017 to account for the effects of inflation and the variation in purchasing power of the currencies of each sovereign state.

2.2. Decomposition analysis

The Kaya identity is first used to decompose the annual CO_2 emissions for each sovereign state for theyears 2017 to 2020 and years 2021 to 2025 into several factors, as shown in Eq. (1):

$$C = \sum_i C_i = \sum_i P \times \frac{G}{P} \times \frac{E}{G} \times \frac{E_i}{E} \times \frac{C_i}{E_i} \\ = \sum_i P \times M \times I \times S_i \times EF_i \tag{1}$$

Using these factors, an additive LMDI-I decomposition model can then be constructed for each sovereign state for each period starting from 201 until 2025, as shown in Eq. (2):

$$\Delta C = \Delta C_{pop} + \Delta C_{act} + \Delta C_{int} + \Delta C_{str} \tag{2}$$

where

pop = population (tied to population growth),
act = activity (tied to GDP growth),
int = intensity (tied to the amount of electricity consumed) a
str = structural (tied to the electricity generation mix).

2.3. Decoupling analysis

For the decoupling analysis, the Tapio decoupling model is utilised, the deoupling elasticity index, β_T,is as seen in Eq. (3):

$$\beta_T = \frac{\frac{\Delta C}{C}}{\frac{\Delta GDP}{GDP}} = \frac{\Delta C \times GDP}{C \times \Delta GDP} \tag{3}$$

where

β_T = Decoupling index from the subsequent year
C = CO_2 emission in the initial year (tonnes)
GDP = GDP in the initial year (US$)
ΔC = Change in CO_2 emissions in the subsequent year from the initial year (tonnes)
ΔGDP = Change in GDP in in the subsequent year from the initial year (US$)

2.4. *Output*

The decomposition analysis finally provides one set of output for the years 2017-2020 and 2021-2025. To provide an apples-to-apples comparison between ASEAN sovereign states, decoupling states were determined for each sovereign state for the periods 2017-2019, 2019-2020 and 2020-2025. This is compare the impact of the emergence of the COVID-19 pandemic on CO_2 emissions and compare it to the periods immediately preceding and following this emergence.

3. Results and Discussion

Therefore, the following results are obtained for each of the selected ASEAN sovereign states, which provides a micro and macro view of CO_2 emissions from electricity generation.

3.1. *Total CO_2 emissions by fuel*

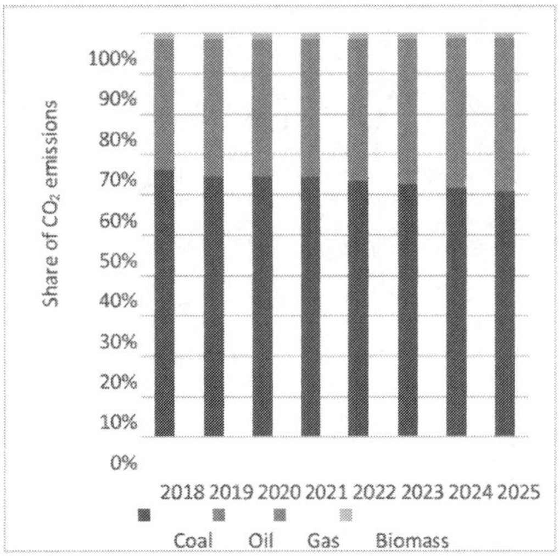

Figure 2: CO_2 emissions breakdown by fuel for Malaysia.

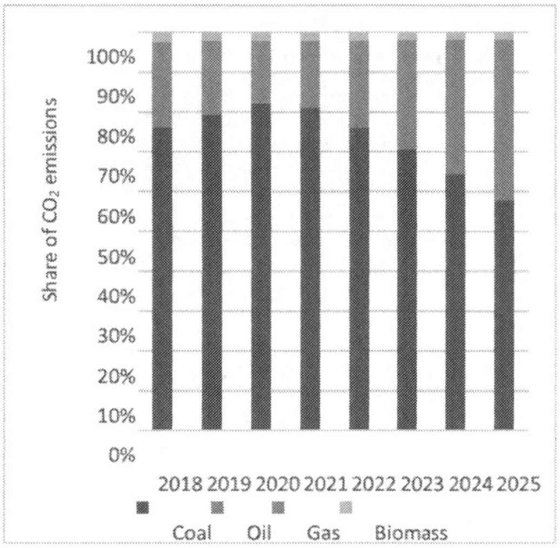

Figure 3: CO_2 emissions breakdown by fuel for Vietnam.

As shown in Fig 2 and Fig 3, both Malaysia and Vietnam, coal contributes to the bulk of CO_2 emissions from electricity generation throughout all three periods. This is due to coal been perceived as reliable and affordable energy source for electricity generation within the short- and middle-term within ASEAN. Natural gas comes in second for these two ASEAN sovereign states. Biomass and oil constitute the smallest sharesin CO_2 emissions, due to their limited use in electricity generation.

3.2. *Decomposition analysis of change in CO_2 emissions*

There is a distinct difference seen between the decomposition analysis of the change in CO2 emissions of Malaysia and Vietnam, for all three periods (Fig 4 and Fig 5). It is important to note that the Malaysian electricity grid is more mature and developed compared to that of Vietnam, hence this would explain the dynamism in the results for Vietnam compared to that of Malaysia

During the pandemic, it is interesting to note that the net change in CO2 emissions for Malaysia is positive, whereas that of Vietnam is negative. Again, the structural factor contributes to this trend mostly. For Malaysia, this mostly counteracted the negative effect of the activity factor, which was roughly as significant as the structural factor. In Vietnam, both factors are smaller in magnitude than that of Malaysia.

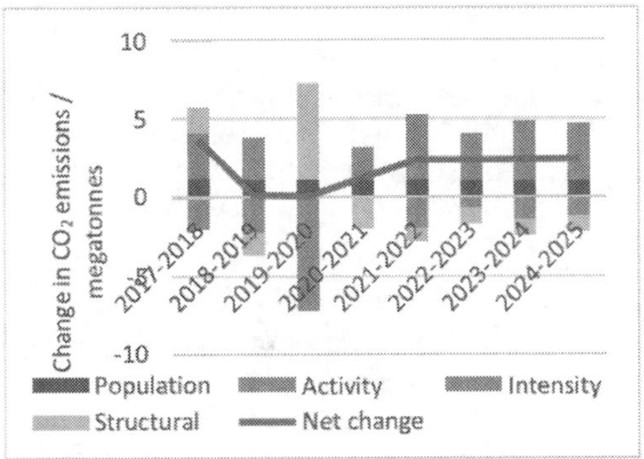

Figure 4: Effect on CO_2 emissions for Malaysia.

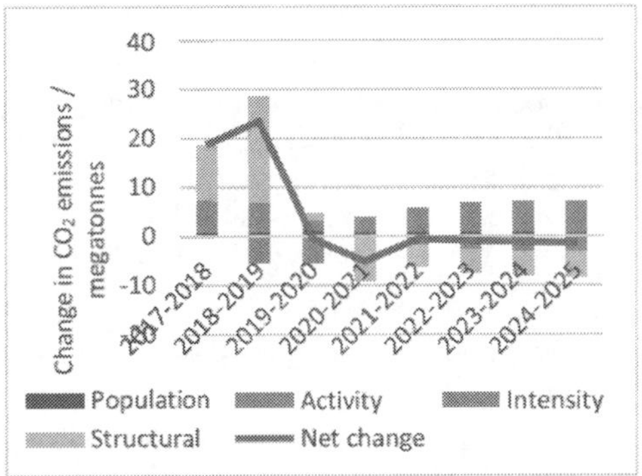

Figure 5: Effect on CO_2 emissions for Vietnam

Post-pandemic, the net change in Malaysia rebounds near to pre-pandemic values, whereas those ofVietnam are negative. This is due to the combined effect of the structural and intensity factors in counteracting the population and activity factors, which is the case for Vietnam (and the opposite is true for Malaysia). Malaysia is relatively more economically developed compared to Vietnam, therefore the magnitude for the activity factor for Malaysia is greater than that for Vietnam. However, when comparing the magnitude of the structural factor, the factor has a greater magnitude in Vietnam compared to Malaysia – which reflects the projected fruits of Vietnam's aggressive policy in adopting renewable energy source as a significant portion of its national electricity mix.

3.3. Decoupling analysis

Table 2: Decoupling type for net decoupling index

Period / Nation	2017-2019	2019-2020	2020-2025
Malaysia	II	VI	II
Vietnam	I	III	III

As shown in table 2, for the pre-pandemic period, both Malaysia and Vietnam can be said to have their CO_2 emissions coupled to GDP growth. During the pandemic, both countries managed to decouple CO_2 emissions from GDP growth, with Vietnam in the desired direction and Malaysia in the opposite direction. This reflects the findings of Conte Grande, which states that decoupling is not the end goal per se, but merely one indicator towards reducing overall CO_2 emissions. In the post-pandemic period, the relationship returns to being coupled, whereas it remains decoupled for Vietnam. From this, CO_2 emissions are seen to consistently increase for Malaysia, whereas the cumulative amount of CO_2 emissions for Vietnam can be seen to trend towards plateauing, before eventually decreasing in the future, after the studied post-pandemic period.

4. Policy Recommendations

Based on the results of the methodology, several policy recommendations can be made to reduce total CO_2 emissions and decouple them from GDP growth. All these policy recommendations are complementary to each other to achieve this goal, as it requires an overhaul of an already existing ecosystem of infrastructures and policies to deal with the disruptive effect of reducing CO_2 emissions with a web of interlocking measures. These recommendations can be divided into four broad areas.

First, reducing CO_2 emissions from fossil fuels and biomass power plants. In the short and medium term, fossil fuels cannot simply be eliminated without a viable alternative capable of continuously supplying electricity to meet demand. One important measure in this area is high-efficiency, low-emission (HELE) coal-fired power plants. This refers to the shift from subcritical to supercritical and ultrasupercritical coal-fired power plants, increasing the efficiency of coal-fired power generation from 31% (subcritical) to 38% (supercritical) and 41% (ultrasupercritical) [2] - as well as the use of fluidised bed combustion.

Furthermore, by implementing carbon capture, utilisation and storage (CCUS) systems in fossil power plants, CO_2 emissions can be captured and potentially stored in geological formations [3]. For example, the CCUS system at the Petra Nova power plant can capture at least 33% of total CO_2 emissions [4]. In the long term, CCUS could also be stand-alone systems to remove CO2 from the atmosphere and thus reduce the CO_2 content in the atmosphere.

Second, facilitate an increase in the share of renewable electricity sources in the electricity mix of sovereign ASEAN countries. It is imperative that the previously described measures to reduce CO2 emissions from fossil fuel power plants be accompanied by a complementary increase in the share of renewable energy sources in the power mix to free ASEAN sovereign states from their heavy dependence on fossil fuels for power generation. ASEAN sovereigns are already implementing several policies that set targets for the power generation mix [5]. However, as BAU projections for 2021-2025 show, these do not go far enough to bring about a decline in CO_2 emissions, let alone achieve net CO_2 emissions. Nevertheless, Vietnam can serve as a model for other sovereign ASEAN countries aiming for a greater share of renewable energy - with the introduction of feed-in tariffs to incentivise the construction of solar and wind power plants in the country. Tax incentives, subsidies and financing schemes can also incentivise the construction of renewable energy infrastructure. In the medium and long term, dependence on fossil fuels would be overcome by energy storage, which solves the problem of intermittency of most renewable energy sources [6].

References

[1] Q. Munir, H. H. Lean, and R. Smyth, "CO2 emissions, energy consumption and economic growth in the ASEAN-5 countries: A cross-sectional dependence approach," *Energy Econ.*, vol. 85, p. 104571, 2020, doi: 10.1016/j.eneco.2019.104571.

[2] World Coal Association, "Clean Coal Technology in ASEAN Introduction," 2021. [Online]. Available: https://www.worldcoal.org/coal-facts/coal-electricity/

[3] P. Newell and A. G. Ilgen, "Chapter 1 - Overview of Geological Carbon Storage (GCS)," in *Science of Carbon Storage in Deep Saline Formations*, P. Newell and A. G. Ilgen, Eds. Elsevier, 2019, pp. 1–13. doi: https://doi.org/10.1016/B978-0-12-812752-0.00001-0.

[4] M. Sucipta *et al.*, "Combined solid oxide fuel cell and gas turbine systems for efficient power and heat generation," *J. Power Sources*, vol. 86, no. 1, pp. 442–448, Oct. 2008, doi: 10.1016/j.jpowsour.2006.01.044.

[5] ASEAN Centre for Energy (ACE), "The 6th ASEAN Energy Outlook (AEO6)," 2017.

[6] S. O. Amrouche, D. Rekioua, T. Rekioua, and S. Bacha, "Overview of energy storage in renewable energy systems," *Int. J. Hydrogen Energy*, vol. 41, no. 45, pp. 20914–20927, Aug. 2016, doi: 10.1016/j.ijhydene.2016.06.243.

IOP Conf. Series: Materials Science and Engineering 1257 (2022) 012022 doi:10.1088/1757-899X/1257/1/012022

Performance Analysis of Neural Network Architecture in Developing Real-Time Malaysian River Water Quality Model

D H Djarum[1], Z Ahmad[1*], J Zhang[2]

[1]School of Chemical Engineering, Engineering Campus, Universiti Sains Malaysia, 14300 Nibong Tebal Penang, Malaysia
[2]School of Engineering, Merz Court, Newcastle University, Newcastle upon Tyne NE1 7RU, United Kingdom.

*e-mail: chzahmad@usm.my

Abstract. According to the world health organization, 485,000 people died each year due to water-related health diseases which are mainly contributed by poor river water quality. As a result, water quality monitoring stations have been deployed across the world. Unfortunately, due to the complex nature of the off-site water quality parameters, the water quality index (WQI) cannot be assessed in real-time. This has led to a significant push for the scientific community to develop an accurate and robust water quality prediction model. The dynamic and nonlinear nature of water quality parameters are major challenges for traditional machine learning algorithms such as multi linear regression to capture. In this study, the water quality index prediction model was developed using the feedforward artificial neural network (FANN) algorithms utilizing only on-site parameters. The performance of different nonlinear activation functions in the hidden neurons was thoroughly analysed which includes rectified linear unit (ReLU), scaled exponential linear unit (SELU), and exponential linear unit (ELU). Additionally, various initialization and optimization algorithms were also evaluated for maximum performance and efficiency. The results shows that FANN-ELU model coupled with Glorot initialization technique and AdaGrad optimizer outperformed other model combinations with an R^2 value of 0.88 and mean squared error (MSE) value of 22.74.

1. Introduction

Approximately two third of the global world population is expected to be that of the urban's area by 2050 [1], a rapid rise in urban population will lead to an exponentially higher demand for river water consumptions. Many studies have shown that urbanization (i.e., Agricultural activities, industrial expansion, power generation, etc.) has a consequential negative impact on river water quality and therefore human's health [2]. In the 2020 Malaysian environmental quality report, despite the country's movement control order (MCO), 195 rivers are still classified as slightly polluted and 34 rivers as polluted, attributing to nearly 34% of the total rivers in Malaysia [3]. Although effort has been made to continuously monitor river water quality, such approach requires tedious measurement of complex offsite parameters namely biochemical oxygen demand (BOD) and chemical oxygen demand (COD), making it impossible to alert the public of the water quality index (WQI) value in real-time. With an increasing environmental concern, there is an urgent need for an effective and robust real-time river water quality forecasting model. Over the years, classical machine learning

(ML) algorithms have gained popularity in predicting river water quality, however due to the highly dynamic and nonlinear characteristics of water quality parameters, traditional ML model might not perform as well compared to the nonlinear neural network model.

In a 2020 study, the development of an artificial neural network (ANN) model to predict dissolved oxygen (DO) of rivers in Kenya were analyzed, the results shows that the ANN model achieved r value of 0.85 which significantly outperforms the multi-linear regression (MLR) model with r value of only 0.62 [4]. In another study, the performance of a novel deep neural network model called deep echo state network (Deep ESN) was compared to gradient boosting tree regression and random forest regression model to predict BOD [5]. They concluded that the Deep ESN model outperform the other models with an R^2 value of 0.79 and RMSE of 0.055. The performance of ANN in predicting water quality index (WQI) using physicochemical parameters was studied by Kadam et al. [6], their results also shows that the ANN model has a better performance compared to the MLR model.

Although many existing works have discussed the advantage of neural network model compared to traditional ML model, very limited studies explored the importance of hyperparameters selection in predicting river water quality. Hyperparameters such as number of hidden layers, number of neurons, learning rate, and activation functions are extremely important in developing high performance neural network model [7]. In this study, combinations of different activation functions, initialization techniques, and optimization algorithms were analyzed and compared to develop an optimized feedforward artificial neural network (FANN) architecture to predict river water quality index (WQI). The model was trained with only onsite parameters (excluding offsite parameters such as BOD and COD) in order to deploy the WQI forecasting model in real-time. The model performance and efficiency were evaluated using R^2, mean squared error (MSE), and training time.

2. Methodology

2.1. Dataset

The area being covered by this study includes two different states in Malaysia (4.2105° N, 101.9758° E) namely Perak (4.5921° N, 101.0901° E) and Penang (5.4164° N, 100.3327° E). There are in total of 2843 samples and 16 different river water quality attributes of non-sequenced measurements being collected from 2014 to 2018 for this work (table 1). Any non-numerical classes were encoded into numeric values to improve the readability of the ML algorithms. The data were also thoroughly analyzed and cleaned to remove any outliers and missing values.

Table 1. Input and output parameters of the water quality dataset.

	Parameters
Input	States, Days, Months, Years, dissolved oxygen, suspended solids, pH, NH3-NL, Conductivity, Turbidity, chlorine, cadmium, chromium, zinc, E coli
Output	Water quality index (WQI)

The varying unit and scale of water quality attributes can cause unwanted bias during model development, as ML algorithms tend to give more weights towards parameters with larger scales. Therefore, feature scaling was performed on the dataset by using the standardization technique where the parameter has a zero-mean value, and it is distributed with a unit of standard deviation. Equation (1) shows how the standardized value of a parameters is calculated where μ is the mean values and σ is the standard deviation of the attributes.

$$X' = \frac{X-\mu}{\sigma} \tag{1}$$

2.2. Feedforward Artificial Neural Network (FANN) Model Architecture

The concept of FANN was first introduced in 1958 by scientist Frank Rosenblatt through his paper on the artificial perceptron [8]. Generally, FANN is composed of three different layers: Input layer, hidden layers, and output layer. The input and output layers as the name suggest are responsible for receiving input from the external environment and generating the predicted output respectively. The hidden layers are composed of fully connected neurons (including the bias term) whereby each neuron has an activation function that is responsible to introduce nonlinearity on the neural network architecture. FANN utilizes backpropagation (BP) algorithms pioneered by David Rumelhart et al. (9), which has been one of the more popular training algorithms since its inception. BP algorithm essentially computes the gradient of the prediction error measured for each training step, then adjust the neuron's connection weight with the objective of minimizing error until the network converges to the global minimum.

In this study, initially the FANN model was developed using Python programming language and undergoes basic hyperparameters analysis which consists of the number of hidden layers, number of hidden neurons, and learning rate. After performing grid-search analysis, it was found that the optimal basic hyperparameters structure for the FANN WQI prediction model has 4 hidden layers, 100 hidden neurons, and a learning rate of 0.1. Figure 1 depicts the overall FANN architecture utilizing onsite water quality parameters as the input.

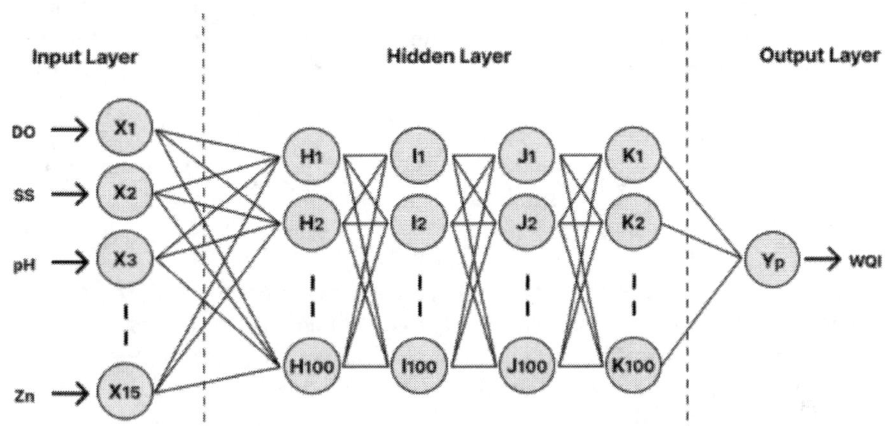

Figure 1. The overall architecture of the FANN WQI prediction model.

2.3. Initialization Technique

In the first training step, the connection weight of each neuron must be randomly initiated. It is important to select the most optimal initialization technique as even a slight variation in the starting weight can cause a significant difference in the network convergence behaviour [10], and the selection is highly dependent on the type of training data set. In this study, two of the more popular techniques were analysed and discussed namely Glorot and He initialization.

The traditional approach of initialization is to set an arbitrary value between two random upper and lower bounds (i.e., -0.1 and 0.1). Although, many researchers admire the simplicity of this method, it does have a very crucial drawback. Since the upper and lower limit are randomly set, the initial weight can cause either an extremely large error leading to exploding gradient or generating negligible derivatives during backpropagation leading to a vanishing gradient. Glorot solves this problem, as the upper and lower limit of the random weight initialization is determined based on the number of input neurons (ni) and hidden neurons (no) as shown in equation (2). This approach improves the consistency and avoids vanishing or exploding gradients [11]. He initialization is very similar to

Glorot initialization but instead of using both number of input neurons and hidden neurons, it considers only the number of input neurons (ni) for the initialization limit as in equation (3) [12].

$$Limit = \sqrt{\frac{6}{ni+no}} \tag{2}$$

$$Limit = \sqrt{\frac{6}{ni}} \tag{3}$$

2.4. *Activation Function*

The activation function is a nonlinear transfer function found in the neurons of the hidden layers and in some cases output layer. It is extremely important when designing a neural network model as it introduces nonlinearity to the network, especially since most real-world problems are nonlinear in nature. In this study, the performance of three different activation functions were analysed and compared namely rectified linear unit (ReLU), exponential linear unit (ELU), and scaled exponential linear unit (SELU).

ReLU has gained a significant amount of interest since it was first introduced in 2010 by Vinod Nair et al. [13]. The main advantage of ReLU is its simplicity, ReLU functions output a linear relationship for any input with a value larger than 0, and output 0 for any input equal to or less than 0. ReLU function can be described as $f(x) = \max(0, x)$, with x as the input. Although ReLU generally has better performance and efficiency compared to traditional sigmoid or tanh activation functions, ReLU still suffers from dying neurons problem during training. In 2015, Djork-Arné Clevert et al. introduced the ELU activation function, which in theory should solve the dying ReLU problem [14]. ELU is essentially identical to the ReLU activation function for any input larger than 0, however for input equal to or less than 0, the output is described as $f(x) = \alpha(e^x - 1)$, where α is a positive constant. SELU as the name suggest is a normalized version of ELU, components of the network such as weight, bias, and activation are all normalized [15].

2.5. *Optimizing Algorithm*

By default, batch gradient descent (BGD) is the algorithms used to optimize the backpropagation process in a neural network architecture. However, BGD calculates the derivatives of the cost function for the entire dataset in every single training step, resulting in an excessively slow training speed. In this work, more efficient optimizing algorithms were explored to improve the efficiency of the backpropagation process, namely stochastic gradient descent (SGD), adaptive gradients (Adagrad), and adaptive delta (Adadelta).

Unlike BGD, stochastic gradient descent (SGD) computes the derivative of the cost function for a randomly selected sample, therefore exponentially reducing the training time by order of magnitude. However, due to the random nature of SGD, it never converges to the most optimal solution as it bounces around the global minimum. This can be solved using the Adagrad optimizer introduced by John Duchi et al. [16], where the learning rate scales based on the historical gradients of each parameter, hence it is more stable compared to SGD, converging much closer to the global optimum and faster. However, as Adagrad optimizers keeps scaling down the learning rate, the network will stop learning at some point when the learning rate approaches 0. Matthew D. Zeiler develop Adadelta algorithm where instead of computing the sum of squared of the whole historical gradients like Adagrad, Adadelta constrained the past gradients into a more recent window. This approach reduces the aggressive nature of Adagrad when scaling down the learning rate [17].

3. Results and Discussion

In this study, the performance of 18 unique hyperparameters combination between activation function, initialization technique, and optimization. algorithms for the FANN model were analysed. The pre-processed water quality dataset was split into 64% training set, 20% test set, and 16% validation set. Grid-search algorithms were utilized to train all 18 unique model combinations for 300 epochs and the performance were evaluated based on the R^2 and MSE loss values. Table 2 shows the three best and the three worst performing hyperparameter combinations for the FANN model on the unseen test dataset.

The best performance could be observed from the FANN-ELU model utilizing Glorot initialization and Adagrad optimizers, this model scores an R^2 values of 0.8806 and MSE of 22.74 with a training time of 12.46s. On the contrary, the worst performing model is the FANN-ReLU model utilizing He initialization and Adagrad optimizers, with an R^2 value of 0.7108 and MSE value that more than double the best performing model at 55.06.

Table 2. Performance comparisons for the WQI FANN model combinations on the unseen test dataset.

Initialization	Activation function	Optimizer	R^2	MSE	t_{Train} (s)
Glorot	ELU	Adagrad	0.8806	22.74	12.46
Glorot	ELU	SGD	0.8579	27.04	11.93
Glorot	SELU	Adagrad	0.8494	28.67	12.67
He	ReLU	SGD	0.7456	48.43	10.33
He	ReLU	Adadelta	0.7273	51.91	11.47
He	ReLU	Adagrad	0.7108	55.06	10.68

The results shows that the top two best performing models have the ELU activation function, which shows that for the WQI dataset, the FANN model with ELU activation perform better compared to ReLU. Since the ReLU activation function always outputs zero for negative input, it can cause the network to stop learning and limit the performance even with large epochs. ELU solves this problem as it has a transfer function that transforms negative input; therefore, the neuron is still activated, and the network can progressively improve with epochs resulting in better performance as shown in table 2. The longer training time of the ELU model is also attributed to the more complex transfer function for $x \leq 0$ compared to ReLU.

It could also be observed that Glorot initialization generally performs better compared to the He initialization technique when predicting WQI, with all three best performing models utilizing Glorot initialization and all three worst performing models utilizing He initialization. The reason could be because the Glorot approach has an additional term being considered when determining the random initialization range which is the number of hidden neurons.

Although, the best choice of activation functions and initialization techniques could be concluded from our analysis, the selection of the optimization algorithms is a little more complicated. From table 2, it could be observed that both the best and worst performing model combinations utilized Adagrad optimizers. This result seems to suggest that when coupled with an optimized activation function and initialization technique, Adagrad optimizers performs well. However, the performance deteriorates rapidly when the other hyperparameters in the network are not optimized.

4. Conclusions

In this study, 18 unique combinations of different hyperparameters were analyzed in the development of the FANN model to predict the water quality index (WQI) of Malaysian Rivers. The results show that the FANN-ELU model utilizing Glorot initialization and Adagrad optimizers have the best performance with an R^2 values of 0.8806 and MSE of 22.74. From the analysis, it was found that for WQI prediction, the ELU activation function has an overall better performance compared to ReLU, which is interesting as ReLU has been the most popular choice of activation function in ANN development. Our findings also shows that Glorot initialization generally performs better than He initialization techniques in WQI prediction. However, the selection of optimizing algorithms cannot be fully concluded since both the best and the worst performing model utilizes Adagrad optimizers. The temporary hypothesis is that Adagrad performance seems to degrade significantly when the other hyperparameter in the network is not optimized. Further investigation on the optimization algorithms is required to fully conclude if our hypothesis for this observation is valid.

References

[1] United Nations 2014 World Urbanization Prospects: the 2014 Revision, Highlights *Department of Economic and Social Affairs,Population Division*
[2] Mateo-Sagasta J, Zadeh SM, and Turral H eds 2018 More people, more food, worse water? *a global review of water pollution from agriculture*
[3] The Department of Environment (DoE) 2020 Malaysian environmental quality report 2020
[4] Ouma Y O, Okuku C O, and Njau E N 2020 Use of artificial neural networks and multiple linear regression model for the prediction of dissolved oxygen in rivers: case study of hydrographic basin of River Nyando, Kenya *Complexity 2020*
[5] Kim S, Alizamir M, Zounemat-Kermani M Kisi O, and Singh V P 2020 Assessing the biochemical oxygen demand using neural networks and ensemble tree approaches in South Korea *Journal of Environmental Management* **270** p 110834
[6] Kadam A K, Wagh V M, Muley A A, Umrikar B N, and Sankhua R N 2019 Prediction of water quality index using artificial neural network and multiple linear regression modelling approach in Shivganga River basin, India *Modeling Earth Systems and Environment* **5(3)** pp 951-62
[7] Madić M J and Radovanović M R 2011 Optimal selection of ANN training and architectural parameters using Taguchi method: A case study *FME Transactions* **39(2)** pp 79-86
[8] Rosenblatt F 1958 The perceptron: a probabilistic model for information storage and organization in the brain *Psychological review* **65(6)** p 386.
[9] Rumelhart D E, Hinton G E, and Williams R J 1985 Learning internal representations by error propagation *California Univ San Diego La Jolla Inst for Cognitive Science*
[10] Kolen J and Pollack J 1990 Back propagation is sensitive to initial conditions *Advances in neural information processing systems* **3**
[11] Glorot X and Bengio Y 2010 Understanding the difficulty of training deep feedforward neural networks *Proceedings of the thirteenth international conference on artificial intelligence and statistics* pp 249-56
[12] He K, Zhang X, Ren S, and Sun J 2015 Delving deep into rectifiers: Surpassing human-level performance on imagenet classification *Proceedings of the IEEE international conference on computer vision* pp 1026-34
[13] Nair V and Hinton G E 2010 Rectified linear units improve restricted boltzmann machines *Icml*
[14] Clevert D A, Unterthiner T, and Hochreiter S 2015 Fast and accurate deep network learning by exponential linear units (elus) *arXiv preprint arXiv:1511.07289*
[15] Klambauer G, Unterthiner T, Mayr A, and Hochreiter S 2017 Self-normalizing neural networks *Advances in neural information processing systems* **30**
[16] Duchi J, Hazan E, and Singer Y 2011 Adaptive subgradient methods for online learning and stochastic optimization *Journal of machine learning research* **12(7)**
[17] Zeiler M D 2012 Adadelta: an adaptive learning rate method *arXiv preprint arXiv:1212.5701*

Acknowledgement
This work was supported by Kementerian Pendidikan Malaysia (KPM) through Fundamental Research Grant Scheme (FRGS) grant number FRGS/1/2018/TK02/USM/02/10. Special gratitude to the Department of Environmental (DOE) Malaysia for providing the water quality data for this study and Universiti Sains Malaysia (USM) for the support.

IOP Conf. Series: Materials Science and Engineering 1257 (2022) 012023 doi:10.1088/1757-899X/1257/1/012023

Predictive solubility of amino acids in deep eutectic solvents for chicken feather waste valorisation

MZM Salleh[1*], MK Hadj-Kali[2], AAI Luthfi[1] and MA Hashim[3]
[1]Department of Chemical and Process Engineering, Faculty of Engineering and Built Environment, Universiti Kebangsaan Malaysia, 43650 Bangi, Selangor, Malaysia
[2]Chemical Engineering Department, King Saud University, P.O Box 800, Riyadh 11421, Saudi Arabia
[3]Department of Chemical Engineering, University of Malaya, Kuala Lumpur, Malaysia

zulhaziman@ukm.edu.my

Abstract. Waste chicken feathers can be potentially utilized by its dissolution in deep eutectic solvents (DESs) to extract keratin. Keratin in chicken feathers is composed by a series of amino acids which can be categorized into negatively charged, positively charged, hydrophobic and hydrophilic natures. As the DESs are also composed by polar and non-polar moieties, their capability to dissolve amino acids needs to be investigated. In this study, the relative solubility of these amino acids in DESs were screened using Conductor-like Screening Method for Real Solvents (COSMO-RS). Five DESs were selected from literature as the targeted solvents, i.e., choline chloride:urea (ChCl:Ur (1:2)), choline chloride:oxalic acid (ChCl:OA (1:2)), choline chloride:ethylene glycol (ChCl:EG (1:2)), sodium acetate:urea (NaAc:Ur (1:2)) and urea:acetate (Ur:Ace (1:2)), to dissolve amino acids in keratin according to the reported compositions. Additionally, the performance of these DESs were benchmarked with other common volatile organic solvents. It was found that the hydrophobic amino acids, particularly cystine, are highly soluble in ChCl:Ur (1:2), ChCl:OA (1:2) and Ur:Ace (1:2). The selected DESs were also predicted to have good dissolution capacity towards the hydrophobic amino acids. It was concluded that the use of DESs in dissolution and extraction of keratin provides more flexible dissolving capability and more environmentally friendly compared to the conventional organic or acidic solvents. This study provides positive insights on the potential of valorising chicken waste feathers using DESs.

1. Introduction

Chicken is one of the famous and preferred sources of protein. As global poultry consumption rises, millions of tons of chicken feather are produced yearly as by-product in the poultry industry [1]. Improper management and treatment of chicken feather waste may degrade the surrounding landscapes and cause disruptions such as bad odour, pests attraction, water pollution and land degradation [2]. Chicken feather waste is traditionally treated by land disposal and incineration [3]. These waste elimination route could impose negative consequences such as high running costs, excessive energy consumption, and severe environmental impacts [4]. Therefore, recycling or converting chicken feather waste into usable products is an interesting alternative. This can be achieved by its sustainable conversion into value added applications.

Keratin in feather was reported to be constituted by more than 90% of protein that comprises of various amino acids such as cysteine, glycine, arginine, and phenylalanine [5]. Keratin can be potentially extracted from chicken feather waste and converted into bioplastic films or biofertilizers [6], vehicle composites [6], papers [7], non-woven insulators [8] and performance enhancer in textile industry [2]. Several methods have been developed to treat feather remnants to add value such as alkaline extraction, reduction, and oxidation [2]. However, these methods are largely expensive, and the process may destroy or change the keratin structure [9].

Deep eutectic solvents (DESs) have attracted extensive attention for many years due to its benign properties [10]. DESs are highly biocompatible, reusable, non-hazardous, low cost, and offering variety of other benefits. DESs are also regarded as the analogue of ionic liquids due to its similar properties with ionic liquids, i.e., negligible vapor pressure and its potential as a designer solvent. DES is formed by complexation of hydrogen bonding receptors (HBA) and hydrogen bond donors (HBD) [11]. When different types of HBAs are matched to different types of HBDs, the dissolution efficiency of DESs varies. It is important to find green solution on valorising the waste chicken feather to protect and conserve the environment. This work provides insights on the potential of valorising chicken waste feathers using DESs by utilizing a computational program known as the Conductor-like Screening Model for Real Solvents (COSMO-RS).

2. Methodology

2.1 Molecular geometry optimisation

All compounds are imported from the C30-1201 COSMO-RS database. In the case of molecules unavailability in the database, the molecular geometry optimisation was performed using the Turbomole programme package. In Turbomole, the 3D chemical structure of the target molecule was firstly drawn. Then, the geometry optimisation was performed at the Hartree–Fock level and 6–31G* basis set. The generation of *.cosmo* file was then conducted through a single-point calculation by using DFT with Becke–Perdew and the Triple-f Zeta Valence Potential (TZVP) basis set. Finally, the *.cosmo* files were exported to the COSMOthermX programme with parameterisation BP_TZVP_C30_1301.ctd.

2.2 DES representation in COSMOtherm

As a single DES is composed of more than one molecule, its representation is performed by adopting the same approach with that of ionic liquids, which can be chosen from among three types: (i) the metafile approach, (ii) the ion pair approach, and (iii) the electroneutral approach [12]. In this study, the electroneutral approach was considered as it is the most appropriate and nearest to the actual nature, where the DES representation is based on the molar composition of their constituents (hydrogen bond acceptors and hydrogen bond donors). The mathematical adaptation has been detailed in our previous works [13, 14].

2.3 Relative solvent screening

The solubility screening was predicted using 'multiple solvents' mode in the COSMO-RS program. Amino acids were set as the solute and its respective solubility in DES or organic solvents was calculated at 25 °C and 1 atm via the relative solvent screening. The relative screening option was selected as it is the fastest method which avoids the commonly large error sources of Gibbs free energy of fusion (ΔG_{fus}) of solids in multiple solvent screening. The result is expressed as a relative molar solubility ranking, where the values differ from absolute values by a simple shift. Although the results are not the absolute solubility values, the qualitative trend and the relevant information about solute-solvent interactions are valid [15].

2.4 Selection of DES and amino acids

Five DESs were selected from literature as the targeted solvents to dissolve amino acids in keratin, i.e. choline chloride:urea (ChCl:Ur (1:2)), choline chloride:oxalic acid (ChCl:OA (1:2)), choline chloride:ethylene glycol (ChCl:EG (1:2)), sodium acetate:urea (NaAc:Ur (1:2)) and urea:acetate

(Ur:Ace (1:2)). On the other hand, the amino acids in keratin were listed and categorized according to the reported compositions literature (table 1).

Table 1. Amino acids in keratin from chicken feather [5]

Functional groups	Amino acids	Contents (%)
Positively charged	Arginine	4.30
Negatively charged	Aspartic acid	6.00
	Glutamine	7.62
Hygroscopic	Theronine	4.00
	Serine	16.00
Hydrophobic	Tyrosine	1.00
	Leucine	2.62
	Isoleucine	3.32
	Valine	1.61
	Cystine	8.85
	Alanine	3.44
	Phenylalaline	0.86
	Methionine	1.02
Others	Proline	12.0
	Asparagine	4.00

3. Results and discussion

The relative solubilities of all amino acids in the five DESs are gathered and depicted in figure 1. For the solute, cystine, which is a hydrophobic amino acid, extremely showed the highest solubility in all DESs studied. The rank of DES to dissolve cystine can be ranked as ChCl:Ur (1:2) > ChCl:EG (1:2) > NaAc:Ur (1:2) > ChCl:OA (1:2) and Ur:Ace (1:2). Meanwhile for the solvent, ChCl:Ur (1:2) showed the highest dissolution capacity for cysteine and nearly all other types of amino acids. For other amino acids, it can be seen that the relative solubilities randomly fall in the low region. This theoretically indicates that the solubility of chicken feather will be dependent on the composition of cysteine.

As seen in table 1, cysteine is only ~9 % of the total keratin fiber. Thus, other factors to improve the dissolution performance are necessary during the chicken feather treatment such as higher temperature, higher DES:chicken feather ratio, and rapid mixing process. On the other hand, although serine contributed the highest amino acid (16 %), the relative solubility of serine in DESs is rather low. This indicates another challenge in chicken feather dissolution (low solubility of the major composition), but at the same time embarks a new motivation, i.e., to search for other DESs or green solvents that can effectively dissolve serine.

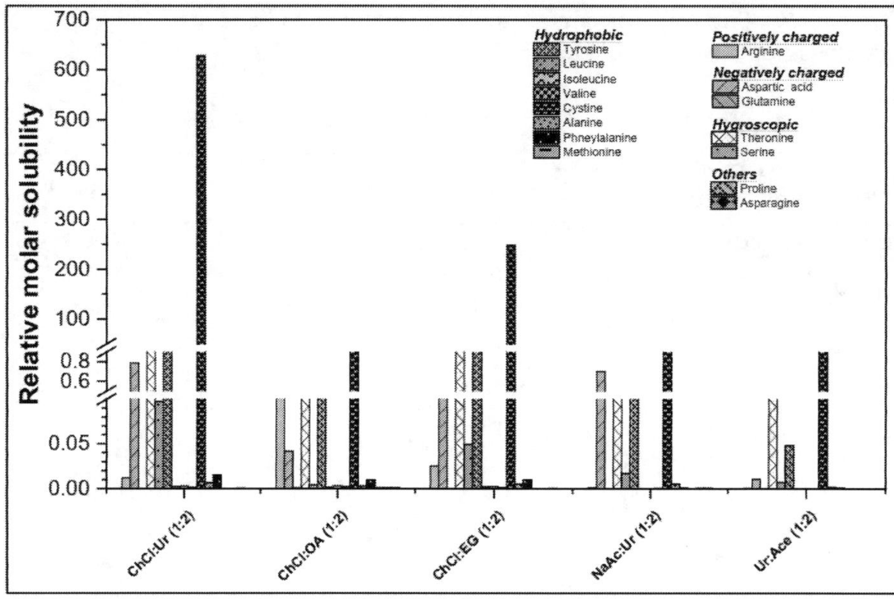

Figure 1. Relative solubilities of keratin amino acids in five DESs.

As cysteine showed the highest solubility in DES and serine was reported as the highest amino acid composition in keratin fibre, both amino acids were further compared with three types of attractive solvents: (i) water as the safest, most abundant and the cheapest solvent, (ii) ChCl:Ur (1:2) as the top ranked DES in this study, and (iii) other conventional organic solvents. The results are depicted in figure 2.

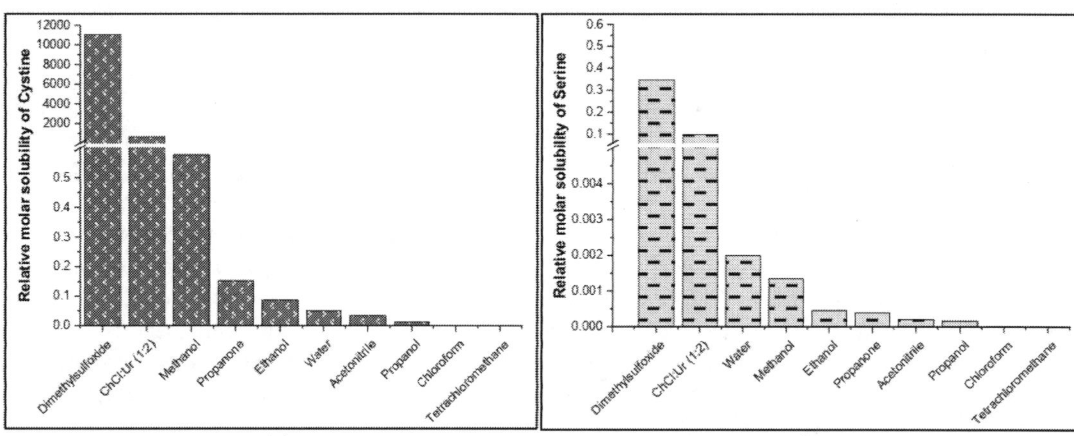

(a) (b)

Figure 2. Relative solubilities of (a) cysteine and (b) serine in ChCl:Ur (1:2) and other conventional organic solvents

Based on the solubility ranking in figure 2, dimethylsulfoxide (DMSO) relatively showed the highest dissolution capacity towards both cysteine and serine. This finding supports the well-recognised performance of DMSO as a universal and cheap solvent. However, the use of DMSO at large scale during the chicken feather valorisation is expected to be of high risk to people health and environment due to its toxic, volatile, and flammable properties.

ChCl:Ur (1:2), which is the next candidate after DMSO, would offer wiser selection of solvent due to its eco-friendly, cheap, and easy-to-prepare advantages. Despite the lower performance of ChCl:Ur (1:2) than DMSO, the treatment process using ChCl:Ur (1:2) can be potentially improved by modifying other parameters such as temperature and mixing time. It is expected that this modification will not affect the thermal stability of the DES (as it does to DMSO) because DES has negligible vapor pressure and thermally stable. The performance of other solvents to dissolve both amino acids was found ineffective as seen from its relative solubility value.

4. Conclusion

The performance of 5 DESs to dissolve 15 amino acids in keratin was computationally predicted via the COSMO-RS relative screening. For the solute, cystine showed the highest solubility compared to other amino acids, while for the solvent, ChCl:Ur (1:2) demonstrated the highest dissolution capacity than other DESs studied. ChCl:Ur (1:2) was then benchmarked with water and other volatile organic solvents to dissolve of cystine and serine, in which, the DES surpassed other screened solvents except DMSO. It is expected that the solubility of keratin amino acids can be further improved by adjusting the operating conditions. It is concluded that the use of DESs in dissolution and extraction of keratin provides more flexible dissolving capability and more environmentally friendly compared to the conventional organic solvents.

Acknowledgments

This study is supported by the GGPM-2021-010 grant. The authors also wish to acknowledge assistance from Yuveneish Siva Kumar and iLab 2021 Group KK2.

References

[1] S A Ghani, M H Jahari, and T Soo-jin 2014 Effects of sodium hydroxide treatment on the properties of low-density polyethylene composites filled with chicken feather fiber,*Journal of Vinyl and Additive Technology,* **20**, 1, 36-41

[2] S Sharma and A Gupta 2016 Sustainable management of keratin waste biomass: applications and future perspectives,*Brazilian Archives of Biology and Technology,* **59**,

[3] N Reddy 2015 Non-food industrial applications of poultry feathers,*Waste Management,* **45**, 91-107

[4] P Gerber, C Opio, and H Steinfeld 2007 Poultry production and the environment–a review,*Animal production and health division, Food and Agriculture Organization of the United Nations, Viale delle Terme di Caracalla,* **153**,

[5] K Saravanan and B Dhurai 2012 Exploration on amino acid content and morphological structure in chicken feather fiber,*Journal of Textile and Apparel, Technology and Management,* **7**,

[6] R K Donato and A Mija 2020 Keratin Associations with Synthetic, Biosynthetic and Natural Polymers: An Extensive Review,*Polymers,* **12**, 1, 32

[7] T Tesfaye, B Sithole, D Ramjugernath, and V Chunilall 2017 Valorisation of chicken feathers: Application in paper production,*Journal of Cleaner Production,* **164**, 1324-1331

[8] A S Soekoco, M Ichwan, A Hananto, and D Mustafa 2018 Application of chicken feather waste as a material of nonwoven insulator,*AIP Conference Proceedings,* **2014**, 1, 020104

[9] M Feughelman 2002 Natural protein fibers,*Journal of Applied Polymer Science,* **83**, 3, 489-507

[10] H-R Jhong, D S-H Wong, C-C Wan, Y-Y Wang, and T-C Wei 2009 A novel deep eutectic solvent-based ionic liquid used as electrolyte for dye-sensitized solar cells,*Electrochemistry Communications,* **11**, 1, 209-211

[11] F A Vicente, B Bradić, U Novak, and B Likozar 2020 α-Chitin dissolution, N-deacetylation and valorization in deep eutectic solvents,*Biopolymers,* **111**, 5, e23351

[12] M Diedenhofen and A Klamt 2010 COSMO-RS as a tool for property prediction of IL mixtures—A review,*Fluid Phase Equilibria,* **294**, 1–2, 31-38

[13] Z Salleh, M Hadj-Kali, M A Hashim, and S Mulyono 2018 Ionic liquids for the separation of benzene and cyclohexane – COSMO-RS screening and experimental validation,*Journal of Molecular Liquids,* **266**, 51-61

[14] Z Salleh, I Wazeer, S Mulyono, L El-blidi, M A Hashim, and M K Hadj-Kali 2017 Efficient removal of benzene from cyclohexane-benzene mixtures using deep eutectic solvents - COSMO-RS screening and experimental validation,*Journal of Chemical Thermodynamics,* **104**, 33-44

[15] F Eckert 2019 COSMOtherm user's manual version C2. 1, Release 01.05, COSMOlogic GmbH & Co. Kg: Leverkusen, Germany,(2005),*COSMOlogic GmbH & Co. Kg,*

Accelerated Weathering of Limestone Bicarbonate Induced $CaCO_3$ Precipitation in A Mixture of Sand, Fly Ash and CaO

S Y W Chai[1], *L H Ngu[1]

[1]Faculty of Engineering, Computing and Science, Swinburne University of Technology Sarawak Campus, Malaysia

*Corresponding author: lngu@swinburne.edu.my

Abstract. Excessive generation of CO_2 and industrial wastes has sparked environmental concerns around the globe. In this research, Accelerated Weathering of Limestone (AWL) process uses calcium carbonate ($CaCO_3$) to capture CO_2 from ambient air. It achieved an absorption efficiency of 31.88 %. The calcium bicarbonate effluent produced *via* AWL is used to precipitate calcium carbonate in lime mortar (LM) mixture containing sand and calcium oxide (CaO) with fly ash, an industrial waste. This helps develop a novel utilization pathway for the respective wastes *via* production of a novel green construction material termed bicarbonate lime mortar infused with fly ash (BLMFA). The fly ash addition percentage in bicarbonate lime mortar (BLM) mix are 12.12, 17.15 and 21.63 wt.%. This project also studies the effect of curing time (7, 28 and 88 days) on development of BLMFA strength. Experimental results found that at 7 days, the strength of BLMFA (0.44 to 0.48 MPa) is lower than the conventional LM, which has a strength of 0.52 MPa. However, on 88 days, 17.15 wt.% BLMFA achieved a strength of 0.82 MPa, which is 0.060 MPa higher than the conventional LM sample. In addition, BLMFA also possesses high $CaCO_3$ content of 36.7 to 44 wt.% out of the targeted 54.37 wt.% compared to conventional LM, which could only achieve 27.14 wt.% $CaCO_3$. This indicates that infusion of fly ash can enhance the carbonation efficiency of calcium sources to $CaCO_3$. All in all, AWL process and BLMFA precipitation possess the capability to diminish anthropogenic CO_2 while beneficially utilizing industrial waste.

Keywords: Accelerated Weathering of Limestone (AWL), Lime mortar, Fly Ash, Bicarbonate effluent.

1. Introduction

Due to the over-emission of carbon dioxide (CO_2) from power plants and cement production plant, the environment has been suffering from various environmental dangers such as ocean acidification and accumulation of greenhouse gases. In the United States [1] and Germany [2], a low-tech CO_2 capture alternative namely, accelerated weathering of limestone (AWL) process has been developed to capture anthropogenic CO_2. The end product (calcium bicarbonate effluent) of the AWL process is an amphoteric compound, thus, it is typically disposed into the ocean to remedy ocean acidification without any undesired consequence [3]. However, this is one reason that has limited the application of the AWL process near the coastline. In another study conducted by Kirchner et al. [3], it was stated that up to 50 % of the capture CO_2 from the disposed effluent re-enters the atmosphere. Thus, the aim of this project is to develop a novel utilization pathway for the effluent in the form of bicarbonate lime mortar (BLM) production. This can help permanently sequestrate the captured CO_2 in the form of carbonate products.

Other than that, lime mortar (LM) usage is also an alternative to reduce CO_2. For example, LM possesses natural carbon capturing properties as it can absorb CO_2 during its hardening process [4]. Apart from that, the switch of Ordinary Portland Cement (OPC) mortar to LM can also indirectly reduce CO_2 as the production of OPC produces a tremendous amount of CO_2 (2.53 billion tCO_2 as of 2020) [5]. However, the strength of lime mortar (1.06 to 1.68 MPa) [6] is inferior compared to OPC mortar which have a strength of approximately 24.80 MPa [7]. Although the strength of LM is relatively low, strength is not as important for LM because it is mainly used for restoration of historical buildings. For this application, lower strength is favorable as it would allow expansion and contraction of the building to occur with less resistance [8]. Nevertheless, strength is still an undeniably important property for mortar.

To improve strength of mortar, the infusion of fly ash into mortar production have also been proven experimentally by Bendapubi and Saha [9] to help increase strength of mortar. Apart from that, fly ash is also a superb cementitious substance that have been reported to offer several other beneficial advantages to mortars such as enhanced workability [10], stronger bond strength [11], reduced drying shrinkage [12] and lower water consumption [9].

Despite all these potentials to promote the usage of lime mortar, lack of comprehensive analyses on these potentials have allowed OPC mortar to be the preferred mortar in the construction industry due to its superior strength and setting time. Therefore, in this project, these potentials will be exploited through 3 main objectives which are to determine optimum fly ash content for strength improvement in fly ash infused bicarbonate lime mortar, to determine the effect of curing time on fly ash infused bicarbonate lime mortar strength improvement and to compare strength and cost improvement of fly ash infused bicarbonate lime mortar to conventional lime mortar and bicarbonate lime mortar.

2. Methodology

2.1. Operation of absorption tower for AWL process

An absorption tower (height = 1 m and diameter = 0.025 m) connected to a liquid pump (Longer, China), compressed gas inlet, air outlet and liquid outlet was operated as a continuous and co-current flow absorber. The accelerated weathering of limestone (AWL) process was operated under atmospheric condition (pressure = 1 atm and temperature = 25 °C). Before commencement of the absorption process, 143 g of limestone palette was packed into the absorption tower. Next, compressed air was pumped into the tower at a volumetric flow rate of 45 mL/s and a water pump was utilized to pump 3 L of distilled water into the tower at a rate of 3 mL/min. The collected effluent was recycled for up to 3 times for 3 subsequent runs. The CO_2 concentration of the post-absorption air was quantified using a gas analyzer (Bacharach, Australia) and the resulting calcium bicarbonate effluent was collected at the liquid outlet for further testing.

2.2. Calcium carbonate precipitation

The precipitation of mortar samples was carried out using a 3-cavity stainless steel mould which was compressed between two modified high-density polyethylene (HDPE) plates. Each HDPE plate was modified with four holes to allow calcium bicarbonate effluent to be injected into the top and out of the bottom plate. The effluent was injected into the mould *via* 4 mm diameter silicon tubes that were inserted onto the top HDPE plate. The main ingredients used in the precipitation of mortars were calcium oxide (Qrec, Malaysia), fly ash (Sarawak Energy, Malaysia) and sand (Wee Yang, Malaysia). The five mortar samples produced are divided into two control samples (conventional LM and BLM) and three BLMFA samples (with 12.12 (BLMFA 1), 17.15 (BLMFA 2) and 21.63 % (BLMFA 3) fly ash replacement). The specified amount of dry ingredients (CaO, fly ash and sand) required for each sample were weighed out with an electronic balance and thoroughly mixed. Before pouring the mixture into the mould, filter paper was inserted into each cavity to cover the holes of the bottom HDPE plate. Once the dry mixture was poured into the mould, a tamping tool was placed on top of the mixture and pounded approximately 20 times to ensure proper compression. Then, the top HDPE plate was placed on top of the mould and 500 mL of calcium bicarbonate effluent was injected into the mould. Subsequently, the entire precipitation set-up was placed on top of an effluent collection tank to collect precipitation effluent for further testing.

2.3. *Curing process*
The precipitated mortar samples were transferred into an oven to undergo drying process at a temperature of 30 °C for 24 hours. After drying, samples were cooled at 20 °C for 1 hour before the moulds were removed. The mortar samples were cured in the environmental conditions as specified by ASTM STP 1355 (i.e., 20 °C and 65 % relative humidity).

2.4. *Sample testing*
The tests conducted for sample analysis can be divided into three main parts namely, effluent test, strength test and composition test. The effluent tests are subdivided into two tests: alkalinity test and pH test. These tests are to determine the condition and concentration of the AWL process effluent. The strength test was conducted to determine the compressive strength of the mortar samples. Then, the composition tests, which consists of acid wash and XRD analysis, were conducted to determine the compound composition of the mortar samples. The detailed procedure of the respective tests is discussed in the following subsections.

2.4.1. Alkalinity test. Alkalinity test was carried out by titrating 25 mL of collected effluent with 0.05 M of HCl (Qrec, Malaysia) in an Erlenmeyer flask. Prior to the titration, a pH meter (Mettler Teledo, Switzerland) was used to determine the pH of the effluent. Based on the pH of effluent, a suitable indicator (pH > 8.3: phenolphthalein, pH < 8.3: bromocresol green) was added accordingly. The titrant volume required to turn the colour (phenolphthalein: pink to colourless, bromocresol green: blue to yellow) of the solution was recorded to determine the alkalinity of the solution.

2.4.2. Compressive strength test. For unconfined compressive strength test, an automated strength compression machine (NL Scientific Instruments, Malaysia) was used to determine the strength of cured mortar samples. Samples were placed at the centre of the machine before turning on the machine. When the machine starts, weight will be gradually added until the sample break. The highest load required to break the sample was recorded to calculate the strength of samples. The crushed samples were collected for further testing.

2.4.3. Composition test. The samples crushed during the strength test were used for the acid wash and XRD analysis. Prior to both tests, the crushed samples were pulverized into fine powder by using a mortar and pestle. For acid wash test, 10 g of powder sample was transferred onto a filter paper and weighted on an electronic balance. After weighing, the sample was transferred into a 100 mL beaker and 20 mL of 0.5 HCl was poured into the beaker. The mixture was set aside for 5 minutes to allow the acid to dissolve the calcium carbonate ($CaCO_3$) content in the sample. Next, the washed sample was then filtered and dried for 24 hours at 20 °C. The dried sample was then re-weighed and reading was recorded to determine the $CaCO_3$ content of the respective samples. For the XRD analysis, samples were sent to an external laboratory to undergo the analysis. The Smartlab XRD machine (Rigaku Corporation, Japan) utilizes a continuous scan mode with an X-Ray setting of 40 kV, scan speed of 8.2551 deg/min and a D/teX Ultra 250 detector.

3. Result and discussion

3.1. *Effluent test*

For the operation of AWL process, two methods, namely, semi-batch with counter-current flow method and continuous with co-current flow method was tested. Figure 1 shows the bicarbonate ion, HCO_3^- alkalinity of the effluent obtained by the respective methods. Based on the average bicarbonate concentration obtained, it was shown that semi-batch method can produce a higher HCO_3^- alkalinity effluent as compared to a continuous method. The reasoning for this is the difference in retention time (semi-batch process: 8 hours, continuous process: ~ 10 s) between the reactants (carbonic acid and limestone pallets). In a semi-batch configuration with no liquid outlet, the reaction time between carbonic acid and limestone pallet is much higher compared to a continuous process. This allow the limestone weathering process to be rapid, thus, producing a much higher alkalinity. However, due to the drawback of a semi-batch process (0.05 L) that can only generate a very limited amount of bicarbonate effluent, the continuous process (3 L) was selected. This is to ensure that sufficient bicarbonate effluent was generated for the precipitation process. In order to overcome the low HCO_3^- alkalinity produced from the continuous process, a recycling process was incorporated to increase the alkalinity of the effluent. This alternative involves recycling the calcium bicarbonate effluent as the water source for the subsequent AWL process. This process was repeated up to 3 times. Based on the results in Figure 1 and table 1, the recycling process increased the HCO_3^- alkalinity of the effluent from 42.68 to 111.78 mg/L while capturing CO_2 with an efficiency of 31.88 %.

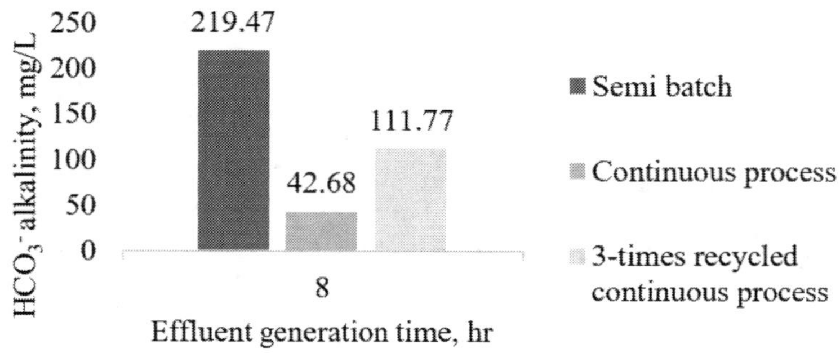

Figure 1. Bicarbonate concentration of bicarbonate effluent generated *via* semi-batch, continuous process and 3-times recycled continuous method.

Table 1. AWL process experimental data.

	Run 1	Run 2	Run 3	Run 4
Titrant volume, mL	0.35	0.52	0.68	0.92
Total alkalinity, mg/L as $CaCO_3$	35	52	68	92
HCO_3^- alkalinity, mg/L	42.68	63.40	82.91	112.17
CO_2 absorption efficiency, %			31.88	

3.2. *Strength test*

Based on figure 2, it is shown that on the 7th day, strength of mortar samples achieved is 58.12 to 67.50 % of its full strength which is comparable to the theoretical value (70 %) stated by Zemajtis [13]. Other than that, it can also be seen that at 7 days, the strength of all BLMFA is inferior compared to conventional LM and BLM. This is due to the higher sand content in LM and BLM samples compared to BLMFAs sample. It was stated by Lin [14] that mortar with a higher sand ratio shows a higher initial strength. However, it was observed that at approximately 14 days, BLMFA 2 overtook the strength of conventional LM by 0.014 MPa while reducing the strength difference with BLM from 12.97 (7 days) to 8.94 % (14 days). This phenomenon happens due to the presence of fly ash in the BLMFA samples. The replacement of fly ash with sand aggregate causes the total percentage of cementitious material in the mortar sample to increase. This would in turn aid in the increase of the long-term strength development of samples [15]. Other than that, from the strength result, it was also deduced that BLMFA 2 with a 17.15 % fly ash replacement is the optimal replacement percentage as the strength achieved by the sample is the highest among the three replacement percentages.

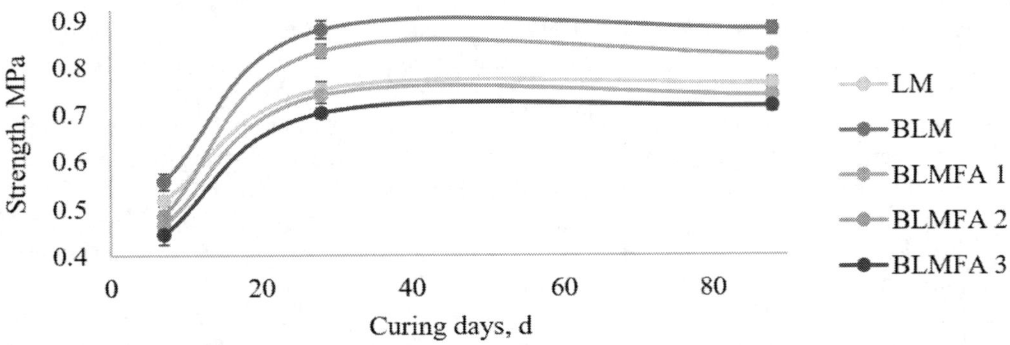

Figure 2. Strength of samples at different curing days.

3.3. Acid wash test

Based on table 2, it is shown that conventional LM sample have a much lower $CaCO_3$ content compared to BLM and BLMFA. One of the factors is due to the bicarbonate concentration in the solution used in the precipitation process. As part of the $CaCO_3$ formed is facilitated by the reaction between CaO and HCO_3^-, the solution with higher bicarbonate concentration will theoretically result in a higher $CaCO_3$ formation [16]. As bicarbonate concentration of bicarbonate effluent is much higher than distilled water, samples precipitated by bicarbonate effluent will achieve a higher $CaCO_3$ content. From the XRD results, a similar trend was also shown. BLMFA has a superior $CaCO_3$ content compared to conventional LM and BLM. This is due to the incorporation of fly ash which acts as an additional calcium source of approximately 24.31 μgCa/gFlyAsh [17].

Table 2. Experimental result of $CaCO_3$ content (wt. %) and expected result.

Sample	**$CaCO_3$ content (wt. %)**	
	Acid Wash (Experimental)	**XRD (External Lab)**
LM (control)	20.93	27.14
BLM (control)	29.31	26.17
BLMFA 1	29.79	36.7
BLMFA 2	29.31	44
BLMFA 3	29.35	42.6
Targeted $CaCO_3$ content	54.37	

4. Conclusion

This research project concluded that a continuous and co-current AWL process can generate a sufficient volume of effluent with plausible bicarbonate concentration (112.17 mg/L) for precipitation purposes while capturing CO_2 at an efficiency of 31.88 %. Moving onto one of the objectives of this project, it was determined that 17.15 % fly ash replacement percentage is the optimal replacement. The highest achievable strength of the optimum BLMFA was approximately 0.82 MPa which is higher compared to the strength of the 12.12 and 21.63 % BLMFA which are 0.74 and 0.72 MPa respectively. In addition, only 17.15 % BLMFA can exceed the strength of the conventional LM by 0.060 MPa while coming in

IOP Conf. Series: Materials Science and Engineering 1257 (2022) 012024 doi:10.1088/1757-899X/1257/1/012024

short of 0.055 MPa compared to BLM. Looking at the effect of curing time on the strength of mortar samples, it is shown that after being cured to 88 days, strength development of BLMFA samples is larger than conventional LM and BLM. This is primarily contributed to the additional cementitious material provided by the infusion of fly ash (24.31 μgCa/gFlyAsh). Lastly, experimental data obtained support the hypothesis made as LM produced *via* infusion of fly ash and precipitation with bicarbonate effluent generated from AWL process does improve the strength of lime mortar.

References

[1] Langer W H, Juan C A S, Rau G H, Caldeira K 2009 *Accelerated weathering of limestone for CO_2 mitigation opportunities for the stone and cement industries* (Austin: University of Texas Austin) pp 310-5

[2] Kirchner J S, Berry A, Ohnemüller F, Schnetger B, Erich E, Brumsack H J r, et al. 2020 Reducing CO_2 emissions of a coal-fired power plant via accelerated weathering of limestone: Carbon capture efficiency and environmental safety *Environ. Sci. Technol.* **54** 4528-35.

[3] Kirchner J S, Lettmann K A, Schnetger B, Wolff J O, Brumsack H J 2020 Carbon capture via accelerated weathering of limestone: Modeling local impacts on the carbonate chemistry of the southern North Sea *Int. J. Greenhouse Gas Control* **92** Article 102855.

[4] Luo K, Li J, Han Q, Lu Z Y, Deng X, Hou L, et al. 2020 Influence of nano-SiO_2 and carbonation on the performance of natural hydraulic lime mortars *Constr. Build. Mater.* **235** Article 117411.

[5] IEA *Cement* [Internet] 2021 [18 August 2021]. Available from: https://www.iea.org/reports/cement.

[6] Charola A E, Henriques F M A 1999 Lime mortars: some considerations on testing standardization *ASTM Spec. Tech. Publ.* **1355** 142-51.

[7] Yerramala A, Desai B 2012 Influence of fly ash replacement on strength properties of cement mortar *Int. J. Eng. Sci. Technol.* **4** 3657-65.

[8] BuildBright *Lime Mortar Vs Cement Mortar* [Internet] [23 January 2022]. Available from: https://www.buildbright.co.uk/lime-mortar/lime-vs-cement-mortar.

[9] Bendapudi S C K, Saha P 2011 Contribution of fly ash to the properties of mortar and concrete *Int. J. Earth Sci. Eng.* **4** 1017-23.

[10] Chandra L, Hardjito D 2015 The impact of using fly ash, silica fume and calcium carbonate on the workability and compressive strength of mortar *Procedia Eng.* **125** 773-9.

[11] Li G, Wu X 2005 Influence of fly ash and its mean particle size on certain engineering properties of cement composite mortars *Cem. Concr. Res.* **35** 1128-34.

[12] Mehta P K 2004 *High-performance, high-volume fly ash concrete for sustainable development* (Iowa: Iowa State University Ames) pp 3-14

[13] Zemajtis J Z *Role of Concrete Curing* [Internet] [9 January 2022]. Available from: https://www.cement.org/learn/concrete-technology/concrete-construction/curing-in-construction.

[14] Lin W T 2020 Effects of sand/aggregate ratio on strength, durability, and microstructure of self-compacting concrete *Constr. Build. Mater.* **242** Article 118046.

[15] American C A A 2003 *Fly ash facts for highway engineers* (Aurora: US Department of Transportation) pp 33-6

[16] Rackley S A 2017 *Carbon capture and storage* 2 ed (Oxford: Elsevier) pp 9 - 273

[17] Jadhav U, Hocheng H 2015 Analysis of metal bioleaching from thermal power plant fly ash by Aspergillus niger 34770 culture supernatant and reduction of phytotoxicity during the process *Appl. Biochem. Biotechnol.* **175** 870-81.

Synthesis of silver nanoparticles from microalgae for potential photocatalytic dye removal

H Y Teh[1], M K Lam[1,2*], U Suparmaniam[1,2], Y H Chai[1,2], J W Lim[2,3], P L Kiew[4], I S Tan[5], B L F Chin[5] and J S Y Lau[5]

[1] Department of Chemical Engineering, Faculty of Engineering, Universiti Teknologi PETRONAS, 32610 Seri Iskandar, Perak, Malaysia
[2] HICoE-Centre for Biofuel and Biochemical Research, Institute of Self-Sustainable Building, Universiti Teknologi PETRONAS, 32610 Seri Iskandar, Perak, Malaysia
[3] Fundamental and Applied Sciences Department, Universiti Teknologi PETRONAS, 32610 Seri Iskandar, Perak, Malaysia
[4] Department of Chemical and Environmental Engineering, Malaysia - Japan International Institute of Technology, Universiti Teknologi Malaysia, Jalan Sultan Yahya Petra, 54100 Kuala Lumpur, Malaysia
[5] Department of Chemical and Energy Engineering, Faculty of Engineering and Science, Curtin University Malaysia, CDT 250, 98009 Miri Sarawak, Malaysia

*Email: lam.mankee@utp.edu.my

Abstract. Silver nanoparticles (AgNPs) with excellent optical and electronic properties is gaining significant attention by researchers for wastewater treatment. In recent years, the production of AgNPs has shifted towards green synthesis approach by using bio-reducing agent (e.g. plant extract). In this study, live cells *Chlorella vulgaris* were used as the bio-reducing agent to produce AgNPs from precursor (silver nitrate ($AgNO_3$) solution). Two process parameters were investigated, namely incubation time and pH. The results affirmed that the optimum synthesis condition of AgNPs was at 24 hr incubation time and pH 8.4 in the mixture of live cells microalgae and 3 mM $AgNO_3$ (3:1 v/v). The synthesized AgNPs showed Localized Surface Plasmon Resonance (LSPR) peak at 412.5 nm with absorption maxima of 0.81 a.u. as indicated using UV-Vis spectrophotometer. The synthesized AgNPs had shown 80.7% of photocatalytic dye degradation in 5 mg/L MB solution and 32.1% in 25 mg/L MB solution within 9 hr of incubation time. The dye removal rate with the addition of AgNPs was 1.2 times faster in 5 mg/L MB solution and was 3.5 times faster in 25 mg/L MB solution as compared to control set.

1. Introduction

In recent decades, the application of nanoparticles in dye removal has gained attention in reviewing the nanoparticles' potential in catalysing the degradation of dyes in water sources. Inorganic nanoparticles, especially those that are purely constituted from gold, silver and copper possess excellent photocatalyst properties owing to their well-known localized surface plasmon resonance (LSPR) and broad absorption spectrum in the visible region [1]. In comparison, silver nanoparticles (AgNPs) exhibit superior optoelectronic property than that of copper and AgNPs are more cost-effective than gold. Besides, AgNPs possess antibacterial and antifungal properties justifiable with the application of AgNPs in biomedical realm [2,3] which further hinder the argument of the production of secondary heavy metal pollutant in effluent treatment.

Photocatalysis is a sustainable, environment friendly and energy-saving technique that has been proven to be an effective alternative in the degradation of organic pollutants. Silver nanoparticles (AgNPs) exhibit photocatalytic activity under the irradiation of visible light due to the localized surface plasmon resonance (LSPR) of AgNPs. LSPR occurs when plasmonic nanoparticles size are significantly smaller than the wavelength of the incident [4]. The LSPR effect excite the oscillation of electrons in the outermost band to a higher energy result in the interaction with oxygen molecules in water and promote the generation of free radicals. Dye is reduced with the transfer of electron from adsorbed dye to positive holes created under the electron excitation, followed by the oxidation of the dye by other generated free radicals [5].

Nowadays, the green synthesis of AgNPs appears to be a more promising alternative than the chemical and physical synthesis methods due to its ease of availability, simplicity, cost effective and environmental friendliness [6]. Green synthesis of AgNPs can be mediated from bacteria, fungi, yeast, algae and plants without releasing any harmful by-products. The phytochemicals of the organisms play the roles of reducing agent and stabilizer in producing AgNPs from silver salt [7]. Of these, synthesis of AgNPs by microalgae is outstand with microalgae's suitability in serving as industrial raw material. Microalgae are well known with its fast growth rate and the cultivation of microalgae does not require fertile land, large quantity of freshwater and pesticide compared to other plants, which thus giving no competing or threatening consideration to human basic resources and food chain. Moreover, microalgae are non-pathogenic organisms [8]. Thus, microalgae are recognized as a reliable and potential candidate to synthesize AgNPs in a green and sustainable way.

2. Methodology

2.1. Cultivation of Microalgae

Chicken compost was used as the main nutrient source to cultivate *Chlorella vulgaris*. The cultivation preparation and condition were following the published work in [9].

2.2. Preparation of Live Cells Microalgae as Reducing Agent

Live cells reducing agent was prepared by maintaining microalgae strains in their culturing media where macro-particulates were filtered and removed using filter paper (Filtres Fioroni Grade 601). Algae strains smaller than 13 μm passed through the filter paper. The prepared live cells reducing agent appeared in light green solution.

2.3. Synthesis of Silver Nanoparticles (AgNPs) from Microalgae

The green synthesis of silver nanoparticles (AgNPs) from microalgae was studied by manipulating two reaction parameters: incubation time (1, 5, 10, 20, 24 hr) and pH (pH 3, 6, 8.4, 10, 12) using "one factor at a time" method. To synthesize AgNPs, 24 mL of live cells microalgae extract was mixed with 8 mL of 3 mM $AgNO_3$ solution (3:1 v/v) in a 50 mL beaker and the reaction mixture was stirred with glass rod. The appearance of AgNPs in the reactant solution was examined with the colour change and spectrophotometry result using UV-Vis Spectrophotometer (UV-2600 – Shimadzu). The experiments were repeated three times to affirm the attained results.

2.4. Photocatalytic Degradation of Methylene Blue (MB)

The AgNPs synthesized under the optimum incubation time and pH were recovered through centrifugation at 13,000 rpm for 15 min, followed by thermal drying at 40 °C for 24 hr. To study the effectiveness of the produced AgNPs in photocatalytic dye removal, two different concentrations (5 and 25 mg/L) of methylene blue (MB) solutions were prepared. The initial optical density (A_0) of the MB solution was determined using UV-Vis Spectrophotometer (UV-2600 – Shimadzu). Subsequently, 1 mg of the recovered AgNPs were introduced into a test tube comprising 6 mL of MB solution with the studied concentration and the suspension was mixed using Stuart Vortex Mixer. The suspension was then incubated under the illumination of LED cool daylight (Philips LED Bulb 12 W E27 6500K, 1360 lumen) to initiate the photocatalytic dye removal activity. Every 3-hr interval, the colour changes of the MB solutions were observed and MB absorbance at 663 nm (A) were assayed with UV-Vis Spectrophotometer. The percentage of MB decolourization was calculated using equation (1). The experiments of MB photocatalytic degradation were repeated three times and average data were reported.

$$\% \, Decolourization = \left(\frac{A_o - A}{A_o}\right) \times 100 \quad (1)$$

3. Results and Discussion

3.1. The Effect of Incubation Time on the Synthesis of Silver Nanoparticles (AgNPs)

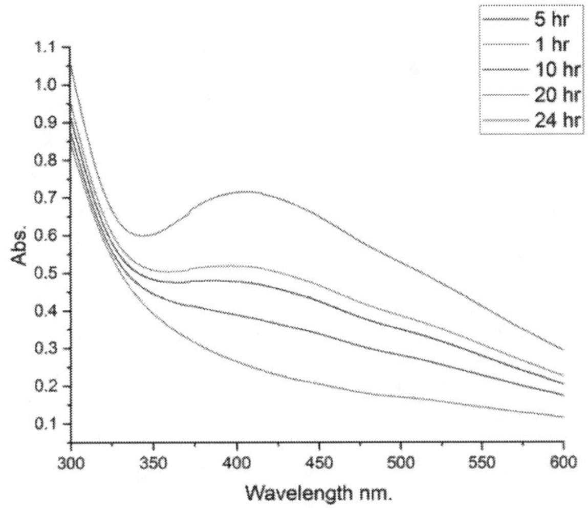

Figure 1. UV-Vis absorption spectrum of incubation time variation.

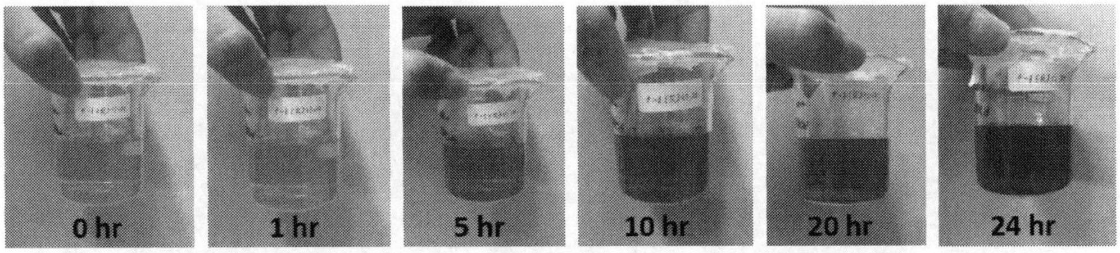

Figure 2. Colour appearance of reaction suspension of incubation time variation.

From Figure 1 and 2, the intensity of the reactant solution changed gradually from light yellowish brown to dark reddish brown within 24 hr of incubation time. The reaction was slow initially with less significant colour change from light yellowish brown to light brown in the first 5 hr, evidenced with no peak appearance in UV-Vis spectrum result. After 10 hr of incubation time, reaction solution showed a brown colour with an absorption of 0.480 a.u. at 390.3 nm. The absorption maxima wavelength (λ) increased to 398.9 nm with an absorption peak of 0.519 a.u. at 20 hr. After 24 hr, a significant absorption peak of 0.715 a.u. was observed at 403.5 nm along with an intense colour appearance, dark reddish brown. The colour of AgNPs in solution or in colloidal form is offered by its optical and electronic properties, which is known as Localized Surface Plasmon Resonance (LSPR). LSPR occurs when surface plasmon is confined to a smaller volume, that is, to nanoparticles, which are significantly smaller in size to the wavelength of the incident ray. The surface plasmon resonance results in an absorption peak of visible light in the violet-blue spectrum (380 – 480 nm) while orange-red light spectrum (600 – 740 nm) is reflected, yielding a yellow-brown colour optical appearance [10]. Therefore, the intensity of the colour change is associated with the absorption maxima, higher absorption peak reflects orange-red light spectrum in a higher degree, resulting a more intense yellow-brown colour solution. Hence, the 24 hrs incubation time yielded the highest absorption maxima accompanied with the most intense colour change, dark reddish brown was justified as the optimum incubation period.

3.2. The Effect of Incubation pH on the Synthesis of Silver Nanoparticles (AgNPs)

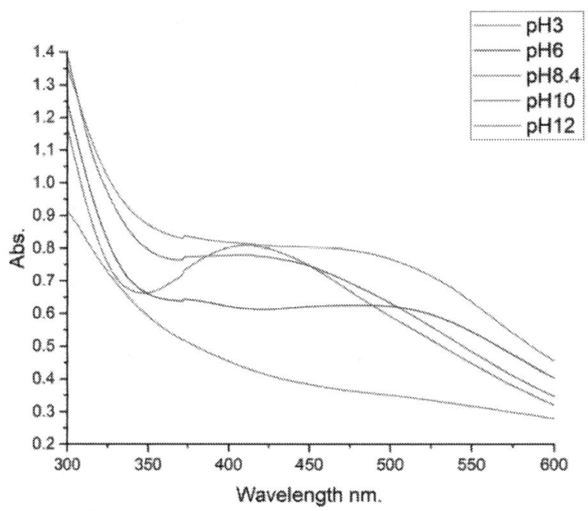

Figure 3. UV-Vis absorption spectrum for the study of incubation pH variation.

Figure 4. Colour of the reactant solutions at time = 24 hr.

Based on the spectrophotometry result as shown in Figure 3, the optimum pH condition for the fabrication of AgNPs using *C. vulgaris* was pH 8.4, justified with the significant appearance of the

LSPR peak at 412.5 nm with 0.81 a.u., accompanied with the dark reddish brown colour appearance in the synthesized colloidal suspension as illustrated in Figure 4. In the acidic condition, no AgNPs characteristic was observed in pH 3 in terms of spectrophotometry result and the optical appearance, no peak was observed in Figure 3 and the suspension colour showed a transparent and light brown colour after the incubation time of 24 hr. As way proceed in increasing the pH values to alkaline condition, coagulation was observed as depicted in Figure 4, this indicates that the synthesized silver particles were too large, nano-size was no longer applicable. The hypothesis was further strengthened with the result of UV-Vis spectrum of pH 10 and pH 12 which show a broad absorption band. The addition of negatively charged hydroxide ions (OH^-) neutralize the charges on the AgNPs surface and then eventually destabilize the nanoparticles, aggregation took place. To ensure the production of small size nanoparticles, electrostatic repulsion in the colloidal suspension must be maintained. AgNPs tend to aggregate in nature, therefore, the organic functional group of the reducing agent play the role of stabilizer or capping agent. The nucleation process in acidic condition (pH 3 and 6) were too slow causing the absence of characteristic AgNPs formation. However, further addition of excessive negative ions in the mixture beyond pH 10 caused an irreversible coagulation of AgNPs. Thus, the optimum pH incubation condition was at pH 8.4.

3.3. Photocatalytic Degradation of Methylene Blue (MB)

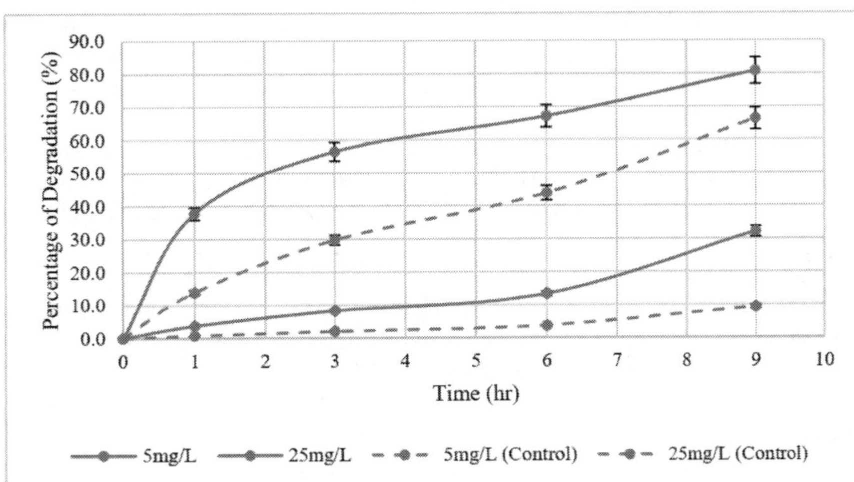

Figure 5. Comparison of MB photocatalytic degradation with and without AgNPs.

Figure 5 shows the average MB degradation rate of 5 mg/L and 25 mg/L experimental sets with the addition of AgNPs and their respective control set. Based on the results, the experimental set with the addition of 1 mg AgNPs exhibit higher and faster MB photocatalytic degradation rate than the control sets. At the incubation time of 9 hr, the percent of MB degradation reached 80.7% for 5 mg/L with AgNPs, 66.3% for 5 mg/L control set, 32.1% for 25 mg/L with AgNPs, and 9.3% for 25 mg/L control set. The dye removal rate with the addition of AgNPs was 1.2 times faster in 5 mg/L MB solution and was 3.5 times in 25 mg/L MB solution. Higher degradation rate was achieved by the LSPR properties possessed by AgNPs that induce more electron (e^-) – hole (h^+) pairs for the generation of free radicals in degrading the dye molecules.

4. Conclusion

Current study unveiled the potential of live cells *Chlorella vulgaris* in reducing silver ions (Ag^+) in silver nitrate ($AgNO_3$) solution to elemental silver particles in nano-size. The UV-Vis spectrum results had affirmed that the optimum synthesis condition of AgNPs was at 24 hr incubation time and pH 8.4 in the mixture of 24 mL live cells microalgae and 8 mL 3 mM $AgNO_3$ (3:1 v/v). The AgNPs synthesized under the optimum incubation time and pH were recovered through centrifugation and drying to study the effectiveness of the produced AgNPs in MB photocatalytic dye removal. The addition of AgNPs had displayed an up to 80.7% of photocatalytic dye degradation in 5 mg/L MB solution and 32.1% in 25 mg/L MB solution within 9 hr incubation time. The dye removal rate with the addition of AgNPs was 1.2 times faster in 5 mg/L MB and was 3.5 times faster in 25 mg/L MB as compared to the control set.

Acknowledgements

This work was funded through Higher Institutions Centres of Excellence (HICoE) from Ministry of Education Malaysia (cost centre: 015MA0-052), Yayasan Universiti Teknologi PETRONAS (cost centre: 015LC0-192) and Murata Science Foundation (cost centre: 015ME0-236).

References

[1] Kumar B 2021 Green synthesis of gold, silver, and iron nanoparticles for the degradation of organic pollutants in wastewater *J. Compos. Sci.* **5** 219

[2] Marimuthu S, Antonisamy A, Malayandi S, Rajendran K, Tsai P, Pugazhendhi A and Ponnusamy V 2020 Silver nanoparticles in dye effluent treatment: A review on synthesis, treatment methods, mechanisms, photocatalytic degradation, toxic effects and mitigation of toxicity *J. Photochem. Photobiol. B: Biol.* **205** Article 111823

[3] Xu L, Wang Y, Huang J, Chen C, Wang Z and Xie H 2020 Silver nanoparticles: Synthesis, medical applications and biosafety *Theranostics* **10** 8996-9031

[4] Kang H, Buchman J, Rodriguez R, Ring H, He J, Bantz K and Haynes C 2019 Stabilization of silver and gold nanoparticles: preservation and improvement of plasmonic functionalities *Chem. Rev.* **119** 664-99

[5] Singh J and Dhaliwal A 2020 Plasmon-induced photocatalytic degradation of methylene blue dye using biosynthesized silver nanoparticles as photocatalyst *Environ. Technol.* **41** 1520-34

[6] Rajkumar R, Ezhumalai G and Gnanadesigan M 2021 A green approach for the synthesis of silver nanoparticles by *Chlorella vulgaris* and its application in photocatalytic dye degradation activity *Environ. Technol. Innov.* **21** Article 101282

[7] Sabri M, Umer A, Awan G, Hassan M and Hasnain A 2016 Selection of suitable biological method for the synthesis of silver nanoparticles *Nanomater. Nanotechnol.* **6**

[8] Khan M, Shin J and Kim J 2018 The promising future of microalgae: Current status, challenges, and optimization of a sustainable and renewable industry for biofuels, feed, and other products *Microb. Cell Factories* **17**

[9] Tan X, Lam M, Uemura Y, Lim J, Wong C and Ramli A 2018 Semi-continuous cultivation of *Chlorella vulgaris* using chicken compost as nutrients source: Growth optimization study and fatty acid composition analysis *Energy Convers. Manag.* **164** 363-73

[10] Hammami I, Alabdallah N, Jomaa A and Kamoun M 2021 Gold nanoparticles: Synthesis properties and applications *J. King Saud Univ. Sci.* **33** Article 101560

IOP Conf. Series: Materials Science and Engineering 1257 (2022) 012026 doi:10.1088/1757-899X/1257/1/012026

Oil palm residue as precursors for low transition temperature mixture extracted lignin

J L Wong[1], D S Khaerudini[2], R K Hocking[3], J J Chew[1,*] and J Sunarso[1]

[1]Research Centre for Sustainable Technologies, Faculty of Engineering, Computing and Science, Swinburne University of Technology, Jalan Simpang Tiga, Kuching 93350, Sarawak, Malaysia

[2]Research Center for Advanced Materials, National Research and Innovation Agency (BRIN), Bld. 440 Kawasan Puspiptek Serpong, South Tangerang, 15314 Banten, Indonesia

[3]Department of Chemistry and Biotechnology and Centre for Translational Atomaterials, School of Science, Swinburne University of Technology, Hawthorn, Victoria 3122, Australia

*e-mail: jchew@swinburne.edu.my

Abstract. Lignin is a complex biopolymer in lignocellulosic biomass. It is a promising feedstock for producing various products, including chemicals and biofuels. Lignin extraction can be achieved *via* different routes such as organosolv, acid, alkaline, ionic liquids (ILs), and low transition temperature mixtures (LTTMs). LTTM is attractive for its biocompatibility, biodegradability, more straightforward synthesis procedure, lower precursor price, and toxicity than the aforementioned options. In this work, we report the use of an LTTM made from a combination of sucrose-malic acid-water (SMAW) to harvest lignin from four oil palm residues (i.e. empty fruit bunch (EFB), trunk (OPT), mesocarp fibre (MSF), and kernel shell (PKS)). There is still a knowledge gap on the applicability of SMAW to extract lignin from different oil palm residues. Limited studies have examined the properties of the SMAW-extracted lignin, particularly the lignin purity and thermal behaviour. Our results indicated SMAW as a sustainable LTTM solvent for lignin extraction for these four oil palm residues. EFB harvested the highest SMAW LTTM-extracted lignin yield of 27.23% with a lignin purity of 74.93%. TGA curves and DTG curves implied that the extracted lignin was thermally more reactive than alkaline lignin and served as an attractive precursor for thermochemical conversion processes.

Keywords: Extraction; Lignin; Low transition temperature mixture; Oil palm residue

1. Introduction

Oil palm residues are lignocellulosic biomass components, mainly cellulose, hemicellulose, and lignin. Cellulose and hemicellulose have attracted more interest compared to lignin **[1]**. Cellulose and hemicellulose are promising feedstock for bioethanol production, a potential green alternative to fossil-derived fuels **[2]**. Compared to cellulose and hemicellulose, the potential of lignin has been overlooked given its heterogeneous structure, which hinders its conversion to value-added products **[3]**. The application of lignin is generally limited to traditional biomass applications (i.e. direct combustion). For instance, solubilised lignin in black liquor, which is generated during Kraft pulping, is typically combusted for power generation in paper mills **[3]**. Traditional biomass application prevails despite the possibility of converting lignin to biofuels and chemicals such as biohydrogen and aromatic chemicals for modern bioenergy use **[4]**.

Lignin is one of the most widely available aromatic biopolymers with complex three-dimensional structures. Various lignin extraction technologies have been developed such as organosolv (OS), acid, alkaline, and ionic liquid (IL) **[5]**. Recently, low transition temperature (LTTM) has emerged as a potential alternative lignin extraction method, which uses environmentally friendly and inexpensive materials as solvent *via* non-complicated procedures **[6]**. Yiin *et al.* **[7]** studied the pretreatment of oil palm empty fruit bunch (EFB) using sucrose-malic acid-water (SMAW) LTTM at a ratio of 4:2:2 wt.%. The LTTM pretreatment performed at 80 °C for 24 hours, can remove 50.14% of lignin from EFB. However, there is no reported application of this LTTM to the other types of oil palm residues, such as oil palm trunks (OPT), oil palm mesocarp fibres (MSF), and oil palm kernel shells (PKS). It is desirable to evaluate the lignin extraction ability of SMAW LTTM to these residues and to characterise the properties of extracted lignin.

This paper aims to study the lignin extraction efficiency of SMAW LTTM for the pretreatment of OPT, MSF, PKS, and EFB. There is a knowledge gap in the SMAW LTTM extracted lignin characteristics. The latter is an essential information for lignin's downstream conversion into pyrolysis oil and biohydrogen *via* thermochemical processes such as pyrolysis and gasification. This paper will provide a picture to what extent SMAW LTTM can extract lignin from the selected oil palm biomass and the possible application for the extracted lignin.

2. Methodology

2.1. Raw material preparation and biomass characterisation

Commercial alkaline lignin was purchased from Sigma-Aldrich (CAS: 8068-05-1). Inland Classic Estate, Sibu, Sarawak provided the oil palm residues used in this project. The raw samples were dried in an oven at 105 °C for 24 hours. Then, the dried samples were ground using a rotor mill and sieved using a vibration sieve shaker to obtain particle sizes of 0.5 to 1.0 mm. The sieved biomass samples were kept in sealed plastic containers to minimise pre-hydration until usage. The lignin content is determined using Tappi-222 om-02 **[8]** and ultraviolet visible (UV-vis) spectrometry method **[7]**. Holocellulose (combination of cellulose and hemicellulose) content was defined as the difference between the initial weight of the sample with the weight of hemicellulose and lignin.

Table 1. Lignocellulosic content of oil palm residues.

Biomass	Holocellulose (%)	Lignin (%)
EFB	83.60	16.40
OPT	78.75	21.25
PKS	51.24	48.76
MSF	68.57	31.43

2.2. LTTM preparation

DL-Malic acid 99% was purchased from Sigma-Aldrich (CAS: 6915-15-7), and sucrose 99% was purchased from R&M Chemicals (CAS: 57-50-1). Sucrose, malic acid, and distilled water were mixed with a ratio of 4:2:2 wt.% in a beaker. The mixture was heated on a heating plate with magnetic stirring until a clear, homogeneous solution was obtained.

2.3. LTTM pretreatment

1 g of dried oil palm residue was added to the LTTM with a mixing ratio of 1:20 wt.%. The mixture was placed in an oil bath heated at 100 °C with continuous magnetic stirring for 6 hours. After the pretreatment, the distilled water was added to the pretreated biomass mixture and left overnight for

lignin precipitation. Using Whatman Grade 1 filter paper, filtration was applied to separate lignin from the pretreated biomass. Distilled water was used to wash the lignin until the filtrate became clear and non-turbid. Lignin was washed off from the filter paper into a pre-weighted Petri dish. The sum of the residue weight on filter paper and Petri dish represented the weight of the extracted lignin. The pretreated biomass was washed using distilled water under vacuum filtration. The residue biomass and lignin on filter paper and Petri dish were dried in a drying oven at 105 °C for 24 hours. The lignin yield of the pretreatment was calculated using **equation (1)**:

$$Lignin\ yield\ (\%) = \frac{Lignin\ \ on\ filter\ paper + Lignin\ \ in\ petri\ dish}{Initial\ lignin\ content\ in\ biomass} \times 100 \quad (1)$$

2.4. Lignin purity analysis

The purity of the extracted lignin was determined according to the procedure described in Klason method (Tappi-222 om-2) **[8]** and UV-vis spectrometry method **[9]**. In brief, LTTM extracted lignin was mixed with 72 wt.% sulphuric acid, and the mixture was kept at 20 °C for 2 hours. The solution was then diluted to 3 wt.% and boiled for 4 hours. Water was added continuously to maintain the constant water level. After the boiling process, the mixture was allowed to stand and the solid residue to settle. The solid residue was denoted as acid-insoluble lignin (AIL). The supernatant was decanted and stored, whereas the AIL was filtered with a glass filter crucible using vacuum filtration. The filtered AIL was dried in an oven at 105 °C for 24 hours and calculated using **equation (2)**. The UV-vis spectroscopic method was applied to determine lignin that remains soluble in the decanted supernatant (i.e. acid-soluble lignin (ASL)). The absorbance, A of the solution at a wavelength of 205 nm was identified and ASL was calculated using **equation (3)**. The lignin purity was determined as the sum of AIL and ASL, as shown in **equation (4)**.

$$AIL\ (\%) = \frac{GC_{after\ filtration} - GC_{before\ filtration}}{Initial\ lignin\ weight} \times 100 \quad (2)$$

where GC refers to the weight of the glass crucible.

$$ASL\ (\%) = \frac{Absorbance \times Volume\ of\ filtrate \times Dilution\ factor}{Lignin\ Absorptivity \times Sample\ weight \times Pathlength} \times 100 \quad (3)$$

where lignin absorptivity at 205 nm is 110 L g^{-1} cm^{-1} **[10]**, and UV-vis cell pathlength is 1.0 cm.

$$Lignin\ purity\ (\%) = AIL + ASL \quad (4)$$

2.5. Thermogravimetric analysis (TGA)

TGA was performed using Simultaneous Thermal Analyser (STA) 8000 (PerkinElmer). In brief, 5 mg of lignin sample was heated from ambient temperature to 900 °C under a nitrogen flow rate of 10 mL min^{-1} with a heating rate of 10 °C min^{-1}.

3. Result and discussion

3.1. Lignin extraction efficiencies of different oil palm residue using SMAW LTTM

Figure 1(a) shows the lignin extraction yield of EFB, OPT, MSF, and PKS extracted using SMAW LTTM. The highest lignin yield is obtained using EFB (27.23%), followed by MSF (15.98%), OPT (12.03%), and PKS (7.44%). Although **Table 1** shows that PKS had the highest initial lignin content among the oil palm residues, it gave the lowest lignin yield. The rigid and robust structure of PKS can hinder the penetration of SMAW LTTM into the lignocellulosic structure and, in turn, affect the extraction of lignin. Lignin purity was determined according to the procedure described in Tappi-222 om-02 coupled with UV-vis spectrometry. Among the extracted oil palm residue lignin, EFB lignin attains a purity of 74.93%, whereas MSF lignin shows purity of 71.98%. The impurities can be due to residual cellulose and hemicellulose, chemically bonded to lignin, as indicated by the degradation occurs within 220 °C and 440 °C in DTG curve from TGA in Section 3.2. **Figure 1(b)** illustrates the lignin extraction performances of three different pretreatment technologies on oil palm residues. Compared with these methods, the yield of LTTM extracted lignin is much lower, albeit comparable purity against the organosolv-derived lignin. The EFB lignin yield obtained by IL (1-butyl-3-methylimidazolium chloride, [Bmim][Cl]) **[11]**, sequential acid/alkaline (AA) (72 vol.% sulphuric acid/5.0 M sodium hydroxide) **[12]**, OS (ethanol) pretreatment **[13]** is 1.1 to 1.5 times higher than the LTTM-derived EFB lignin in present study. In addition, the yield of IL-derived OPT lignin more than 3.7% times higher than the LTTM-derived OPT lignin obtained in the present study **[11]**. Nevertheless, the lignin purity attained by LTTM-derived EFB lignin is 6.14% higher than OS-derived EFB lignin (70.6%) **[13]**. Despite the lower lignin extraction yield, LTTM pretreatment method is attractive in terms of lower temperature, less aggressive condition, and the absence of a catalyst. Extraction using LTTM avoids the flammability, corrosion, and toxicity issues related to OS, AA, and IL pretreatment.

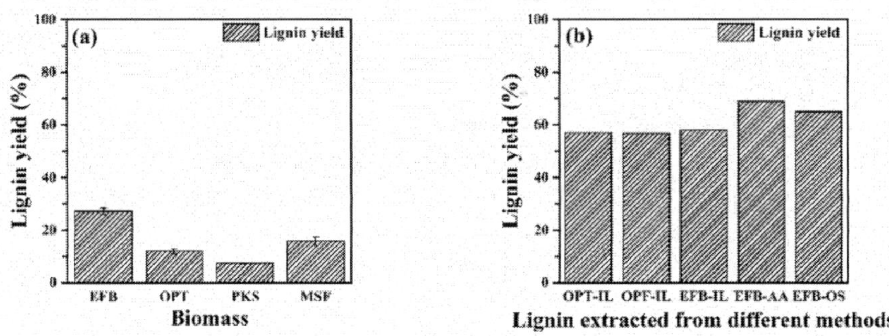

Figure 1. Lignin yields from oil palm residues using **(a)** the LTTM pretreatment method, and **(b)** other conventional methods.

3.2. Thermal degradation behaviour of LTTM extracted lignin in an inert atmosphere

Figure 2(a) shows the thermogravimetric analysis (TGA) profiles of alkaline lignin and lignin extracted derived from four different oil palm residues *via* SMAW LTTM extraction obtained at a heating rate of 10 °C min^{-1}. All extracted lignins show an almost identical weight loss trend but with a different extent of weight loss. The weight loss below 100 °C can be attributed to the removal of moisture from the lignin. Three phenomena mainly cause the weight loss after the dehydration stage: (1) the degradation of impurities (cellulose and hemicellulose) that are chemically attached to the lignin **[14]**; (2) the removal of volatiles with char formation **[15]**; and (3) the cleavage of lignin linkages such as aryl ether β-O-4 linkages, C-C bonds, and β-β bonds **[16, 17]**. On the other hand, alkaline lignin shows a char yield of 57.21%, higher than the other four LTTM extracted lignins. This indicates that the four LTTM-derived lignins exhibited a relatively higher thermal reactivity than alkaline lignin (**Figure 2(a)**). PKS lignin had the highest thermal stability among the extracted lignin with the highest char yield of 50.09%, followed by MSF lignin (43.30%), EFB lignin (39.34%), and OPT lignin (38.46%).

Figure 2(b) displays the derivative thermogravimetric (DTG) analysis profiles for the four LTTM extracted lignins compared to the alkaline lignin. Different DTG peaks and shoulders are observed, showing that the thermal degradation behaviour of the extracted lignin is affected by its biomass precursor origin. Carrier *et al.* **[18]** reported that cellulose, hemicellulose, and lignin decomposition occurs within temperature ranges of 250-350 °C, 200-300 °C, and 200-500 °C, respectively. The degradation of LTTM-derived lignin mainly occurred between 220 °C and 440 °C, with the peak temperatures at 330 °C for EFB lignin and OPT lignin and 345 °C for MSF lignin and PKS lignin. Cellulose, hemicellulose, and lignin degradations coincided within these temperature ranges. Comparatively, the degradations before and after peak temperature were less significant. Before reaching the peak temperature, degradations of cellulose and hemicellulose bonded to the extracted lignin contribute to the major weight loss. A shoulder near 290 °C is observed from the DTG curve of

MSF lignin, possibly caused by the cleavage of β-O-4 linkages that occurred within the temperature range of 200-350 °C **[16]**. After the peak temperature, the leading cause of weight loss can be attributed to lignin degradation, which took place over a broader temperature range (200-500 °C) compared to hemicellulose (200-300 °C) and cellulose (250-350 °C). In contrast, the decomposition of alkaline lignin occurred throughout a wide temperature range, i.e. from ambient to 900 °C. Thus, no significant degradation peak or shoulder was observed on its DTG profile. In short, the TGA and DTG results show that the LTTM-extracted lignin has higher thermal reactivity than the commercial alkaline lignin. Higher thermal is a desirable feed characteristic for thermochemical conversion processes (e.g. pyrolysis and gasification).

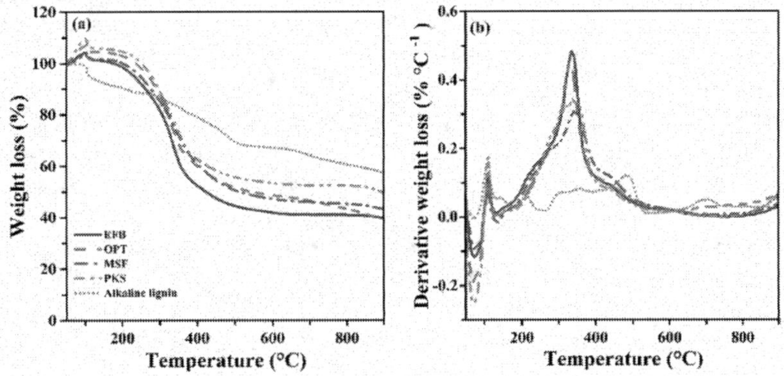

Figure 2. Thermograms for alkaline lignin and LTTM extracted lignin **(a)** TGA and **(b)** DTG curves.

4. Conclusion

Lignin is extracted successfully from four different oil palm biomass, i.e. EFB, OPT, PKS, and MSF using SMAW LTTM as an extraction solvent. The result shows that EFB yielded the most lignin with a purity of 74.93%. All extracted lignins showed different thermal behaviours, with OPT lignin experiencing the most significant thermal degradation with the lowest char yield of 38.46%. The final char yields of all extracted lignins were lower than that of alkaline lignin, implying that the extracted lignins are thermally more reactive than alkaline lignin. The major degradation of the LTTM-extracted lignin occurred within the temperature range of 220-440 °C, while no significant degradation was observed for alkaline lignin in the same temperature range. This study shows that apart from EFB, lignin extraction from other oil palm residues using SMAW LTTM was also feasible. The performance of SMAW LTTM lignin extraction and thermal degradation behaviour of lignin vary with different biomass types. More studies should be conducted to cover a broader range of feedstock biomass spectrum to explore the potential of SMAW LTTM. Also, more characterisation of LTTM-derived lignin is required to evaluate their potential as feedstock for pyrolysis and gasification.

Acknowledgement

The authors would like to acknowledge the financial support from Swinburne University of Technology Sarawak Campus Research Micro Fund (RMF) (SUTS/SoR/RMC/RMFST/2021) and the Postgraduate Research Scholarship Award for Jung Lin Wong. In addition, the authors would like to acknowledge the facilities, scientific, and technical support from Advanced Characterisation Laboratories Serpong, National Research and Innovation Agency (BRIN) through E-Layanan Sains, Badan Riset dan Inovasi Nasional.

References

[1] Alherech M, Omolabake S, Holland C M, Klinger G E, Hegg E L and Stahl S S 2021 From lignin to valuable aromatic chemicals: Lignin depolymerisation and monomer separation via centrifugal partition chromatography *ACS Cent. Sci.* **7** 1831-7.

[2] Ismail K S K, Matano Y, Sakihama Y, Inokuma K, Nambu Y, Hasunuma T and Kondo A 2022 Pretreatment of extruded Napier grass by hydrothermal process with dilute sulfuric acid and fermentation using a cellulose-hydrolysing and xylose-assimilating yeast for ethanol production *Bioresour. Technol.* **343** Article 126071.

[3] Ahmad E and Pant K K Chapter 14 - Lignin conversion: A key to the concept of lignocellulosic biomass-based integrated biorefinery *Waste Biorefinery* ed Bhaskar T, Pandey A *et al* (Elsevier) pp 409-44.

[4] Azadi P, Inderwildi O R, Farnood R and King D A 2013 Liquid fuels, hydrogen and chemicals from lignin: A critical review *Renew. Sust. Energ. Rev.* **21** 506-23.

[5] Haldar D and Purkait M K 2021 A review on the environment-friendly emerging techniques for pretreatment of lignocellulosic biomass: Mechanistic insight and advancements *Chemosphere* **264** Article 128523.

[6] Francisco M, van den Bruinhorst A and Kroon M C 2013 Low-transition-temperature mixtures (LTTMs): A new generation of designer solvents *Angew. Chem. Int. Ed.* **52** 3074-85.

[7] Yiin C L, Quitain A T, Yusup S, Uemura Y, Sasaki M and Kida T 2018 Sustainable green pretreatment approach to biomass-to-energy conversion using natural hydro-low-transition-temperature mixtures *Bioresour. Technol.* **261** 361-9.

[8] TAPPI. Standard T 222 om-02 : Acid-insoluble lignin in wood and pulp. Atlanta, GA, USA: TAPPI; 2006.

[9] Chen Y, Zhang L, Yu J, Lu Y, Jiang B, Fan Y and Wang Z 2019 High-purity lignin isolated from poplar wood meal through dissolving treatment with deep eutectic solvents *R. Soc. Open Sci.* **6** Article 181757.

[10] Yiin C L, Quitain A T, Yusup S, Uemura Y, Sasaki M and Kida T 2017 Choline chloride (ChCl) and monosodium glutamate (MSG)-based green solvents from optimised cactus malic acid for biomass delignification *Bioresour. Technol.* **244** 941-8.

[11] Mohtar S S, Tengku Malim Busu T N Z, Md Noor A M, Shaari N, Yusoff N A, Bustam@Khalil M A and Mat H B 2015 Extraction and characterisation of lignin from oil palm biomass via ionic liquid dissolution and non-toxic aluminium potassium sulfate dodecahydrate precipitation processes *Bioresour. Technol.* **192** 212-8.

[12] Coral Medina J D, Woiciechowski A, Zandona Filho A, Noseda M D, Kaur B S and Soccol C R 2015 Lignin preparation from oil palm empty fruit bunches by sequential acid/alkaline treatment – A biorefinery approach *Bioresour. Technol.* **194** 172-8.

[13] Mondylaksita K, Ferreira J A, Millati R, Budhijanto W, Niklasson C and Taherzadeh M J 2020 Recovery of high purity lignin and digestible cellulose from oil palm empty fruit bunch using low acid-catalysed organosolv pretreatment *Agronomy* **10** 674.

[14] Moustaqim M E, Kaihal A E, Marouani M E, Men-La-Yakhaf S, Taibi M, Sebbahi S, Hajjaji S E and Kifani-Sahban F 2018 Thermal and thermomechanical analyses of lignin *Sustain. Chem. Pharm.* **9** 63-8.

[15]Ház A, Jablonský M, Šurina I, Kačík F, Bubeníková T and Ďurkovič J 2019 Chemical composition and thermal behavior of kraft lignins *Forests* **10** 483.

[16] Chu S, Subrahmanyam A V and Huber G W 2013 The pyrolysis chemistry of a β-O-4 type oligomeric lignin model compound *Green Chem.* **15** 125-36.

[17] Kim J Y, Oh S, Hwang H, Kim U J and Choi J W 2013 Structural features and thermal degradation properties of various lignin macromolecules obtained from poplar wood (Populus albaglandulosa) *Polym. Degrad.Stab.* **98** 1671-8.

[18] Carrier M, Loppinet Serani A, Denux D, Lasnier J M, Ham Pichavant F, Cansell F and Aymonier C 2011 Thermogravimetric analysis as a new method to determine the lignocellulosic composition of biomass *Biomass Bioenergy* **35** 298-307.

Design of control system for water quality monitoring system for hydroponics application

M R Safira[1], M W Lim[1] and W S Chua[2]

[1]Center of Water Research, Faculty of Engineering, Built Environment, and Information Technology, SEGi University, 47810 Kota Damansara, Selangor Darul Ehsan, Malaysia
[2]Malaysian Smart Factory 4.0, Selangor Human Resource Development Centre (SHRDC), 40100 Shah Alam, Selangor Darul Ehsan, Malaysia

E-mail: limmeewei@segi.edu.my

Abstract. In hydroponic farming, optimal pH value is important to regulate nutrient availability for efficient plant growth. This study aims to design an autonomous pH monitoring and control system for maintaining an optimal pH range. The prototype was developed for data acquisition, data processing and data visualization to ensure that the measured nutritional solution parameters are within the specified range: temperature (18–35°C), pH (5.5–8.0), and dissolved oxygen level (>2 ppm). Data collected at sampling rate of 10 seconds utilising temperature sensor, pH sensor, and dissolved oxygen sensor were processed using a microcontroller on a Node-RED interface in a Raspberry Pi. The processed data were stored in InfluxDB before being displayed in Grafana. If the value exceeded the threshold, a Telegram alert was delivered to the end-user. The pH data were used to build a framework to control the pH levels within range. Two peristaltic pumps (DFRobot, DFR0523) were utilized to pump potassium hydroxide or phosphoric acid solution if the pH was out of range. The developed prototype was able to automatically control the pH within the optimum range in the nutrient solution, which will positively impact the nutrient adsorption and subsequent plant growth in a hydroponics system.

1. Introduction

Hydroponics is a well-established method of growing plants utilising a soilless growth medium and a nutrient solution (water) tailored to the crop's needs. Due to the soil-free agriculture method, the plant growth depends highly on large water volumes. Hence, maintaining constant water conditions such as pH, water temperature, dissolved oxygen concentration (DOC), electrical conductivity, and nutrient concentration in the nutrient solution is a critical feature of a hydroponic system to ensure the plant's ability to adsorb the nutrients [1]. pH level is an important parameter that impacts the plant development.

The pH of the nutrient solution must be maintained at optimal level. This is because a solution with a low pH level may contribute to plant toxicity. In contrast, a solution with a high pH level can lead to macronutrient and micronutrient deficiencies in plants. Toxicity and nutrient deficiencies in plants are physically visible through browning of the leaves, brittle leaves or loss of plant weight [2]. For example, research conducted by Krumrei (2019) had suggested that the growth of cucumber in an alkaline hydroponic solution will reduce the solubility of copper, zinc and iron in the nutrient solution, causing the cucumber to undergo chlorosis and reduce the plant yield [3]. Therefore, it can be concluded that a suboptimal pH level does not only affect a single plant but could affect the whole crop cycle.

As the nutrient solution is the main medium growth for the plants, the requirement of water quality index is stringent which indicates the need for data monitoring and control. This is in-line with

Agriculture 4.0, which applies the principles of Industrial Revolution 4.0 (IR 4.0) which applies technology such as sensors, smart tools, the Internet of Things (IoT) and remote sensing to agriculture practices to maximize productivity and quality. Also known as precision agriculture, IR 4.0 elements such as big data, cloud computing as well as artificial intelligence can be applied in precision agriculture management to for complete automation of the agricultural processes [4].

Therefore, a precise water quality monitoring and automated system, focusing on pH control had garnered attention in the management of water quality in the agricultural industry. Recent works in developing control system utilizes fuzzy logic controller for the automating the pH, electrical conductivity and humidity values [5,6]. Alternatively, the control system could also be developed using a microcontroller using the logic 'if else' [7]. However, the technologies currently available for an automated hydroponic system have complex control logic, difficult to program, and is unable to regulate to the desired pH level without human intervention. In addition, an improper designed pH control system could cause accidental overdosing or underdosing of pH buffer solution, which could cause non-linear pH changes that causes a chemical shock to the plants leading to detrimental plant growth [8].

This research therefore, aims to develop and test a completely automated pH level monitoring and control system with a visualization system, utilizing a micropump control system, microcontroller and sensors in order to create an ideal pH environment for plant growth in a hydroponics system.

2. Materials and methods

2.1. Materials

The nutrient solutions were prepared with two sets of commercially available mineral fertilizers with varying macro and micronutrients (Set I: N: 15.5%, CaO: 26%, Fe-EDTA: 13.2% (1g); Set II: NO_3: 7%, P2O5: 9%, K_2O: 37%, MgO: 3%, SO_3: 13%, B: 0.02%, Cu-EDTA: 0.004%, Fe-DTPA: 0.14%, Fe-EDDHA: 0.01%, Mn-EDTA: 0.033%, Mo: 0.003%, Zn-EDTA: 0.021%). The nutrient solution was prepared by dissolving both the set I and set II mineral fertilizers with 1L of water respectively to produce two different stock solutions. The ratio of the mineral fertilizers to the water is 1:100. Then, 200 mL from each stock solution were added into another 40L of water to obtain a 1.25 mS/cm of electrical conductivity (EC) reading. The pH buffer solutions were prepared for the control system by diluting pH up solution containing 25% potassium hydroxide and pH down solution containing 50% food grade phosphoric acid (City Farm, Malaysia) at a dilution factor of 1:8 and 1:6 respectively.

2.2. Hardware

The temperature, dissolved oxygen concentration and pH values in the hydroponics system were monitored using various sensors for data collection (Table 1). All sensors were calibrated in accordance with the manufacturer's instructions. A peristaltic pump (DFR0523, DFRobot) was subsequently used to control the addition of buffer solution to adjust the pH in the hydroponics chamber. The selected peristaltic pump includes a pulse position modulation (PPM) motor driver, whereby the control signal is compatible with a standard servo motor which was controlled using microcontrollers. The peristaltic pump which is a positive displacement pump was able to deliver up to 100 mL/min of fluid.

Table 1. Sensors Information for Data Collection.

Sensors	Temperature Sensor	DOC Sensor	pH Sensor
Brand	GI Electronic	DFRobot	Inesa Scientific Instrument
Model Number	DS18B20	SEN0237	E-201-C
Operating Temperature (°C)	-55 to +125	0 to 40	5 to 60
Pressure	-	0 to 50 PSI	-
DOC	-	0 to 20 mg/L	-
Power (V)	3 to 5.5	3.3 to 5	3.3 to 5
Accuracy	±0.5%	±0.1%	±0.1%

IOP Conf. Series: Materials Science and Engineering 1257 (2022) 012027 doi:10.1088/1757-899X/1257/1/012027

2.3. Testing and Validation

2.3.1 Peristaltic Pump Volumetric Flowrate. The relationship between the volumetric flow rate and pump rotation speed, as an improperly selected volumetric flow rate may affect the pump to pump insufficient or excess pH up or pH down buffer solution into the nutrient solution. The volumetric flow rate of water was measured at various rotational speeds ranging from 0- 180.

2.3.2 Transient Response of the Pump. A transient response test was designed using the peristaltic pump and pH buffer solutions in order to determine the response of the control system to changes in the pH [9]. A tank containing 50 L of nutrient solution was prepared according to Section 2.1, and the pH sensor was submerged on the opposite side of the peristaltic pump in the tank. The pH of the nutrient solution was manually varied to acidic conditions (pH 4,5) and alkaline condition (pH 9) which was out of the set optimal range (pH 5.5-8.0) using the buffer solutions to artificially simulate a condition when the pH value is out of the optimal range. The peristaltic pumps were then switched on at a rotational speed of 40. A total of two timings were recorded: (1) time taken to dispense out pH up or pH down solution until it reaches the optimal range and (2) time taken to stabilize the pH after the pump is deactivated. The experiments were repeated for rotation speed of 50 and 60 respectively.

3. Design and Development of Control System

Figure 1 details the conceptualization visualization of the design and development of the water quality monitoring and control system design. Data acquisition was conducted by connecting the temperature, pH, dissolved oxygen sensors and peristaltic pumps using hardware wiring to the Arduino UNO, which acts as the microcontroller for the interfacing of the sensors [10]. The command program for this board is written through the Arduino UNO in the C++ language, and then compiled and uploaded for the various components to function in the board. The microcontroller is connected to the Raspberry Pi that contains the Node-RED program (Javascript web hosted), which is a platform for programming hardware, Application Programming Interface (API), and web resources and InfluxDB for data processing and storage. Subsequently, the data is graphically visualized using Grafana as a time-based observability dashboard at a sampling time of 10 seconds.

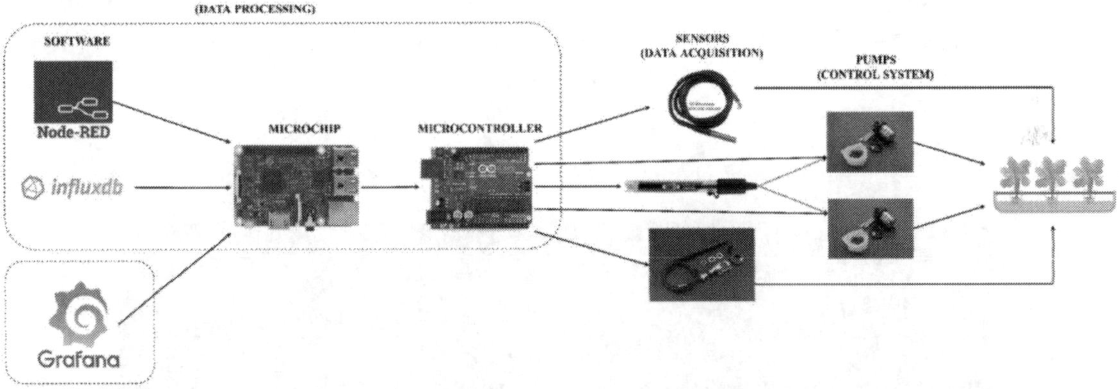

Figure 1. Overall Process Flow Diagram of Data System.

Figure 2 shows the design of an open-loop control system for the pH control system, designed to control the pH level to an optimum range of pH 5.5-8.0 for hydroponics application. The pH sensor is therefore an important regulated input data. These data are stored and processed in the Arduino UNO microcontroller device, where the range of the pH data were evaluated if it falls within the optimal range. If the pH value exceeds out of the optimal range, the microcontroller sends a trigger signal to activate the peristaltic pumps which contains the pH up and pH down buffer solution respectively until the pH sensor detects that the solution reaches the desired range. In addition, a warning message system would also be sent to the end-user via Node-RED through the Telegram program when the following parameters: temperature (18°C - 35°C), dissolved oxygen concentration (DOC) (above 2 ppm) and pH (pH 5.5 -8.0) level are not within the specific range. Figure 3 demonstrates the Grafana dashboard which consists of real-time visualization of the pH, temperature and DOC data.

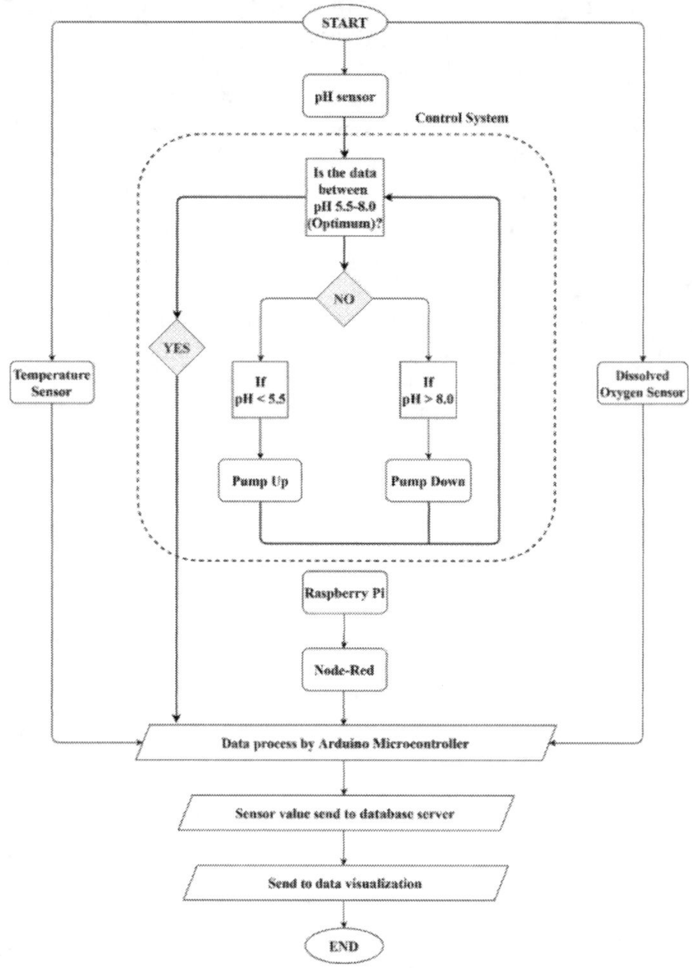

Figure 2. Block Diagram of the Overall Hydroponics System.

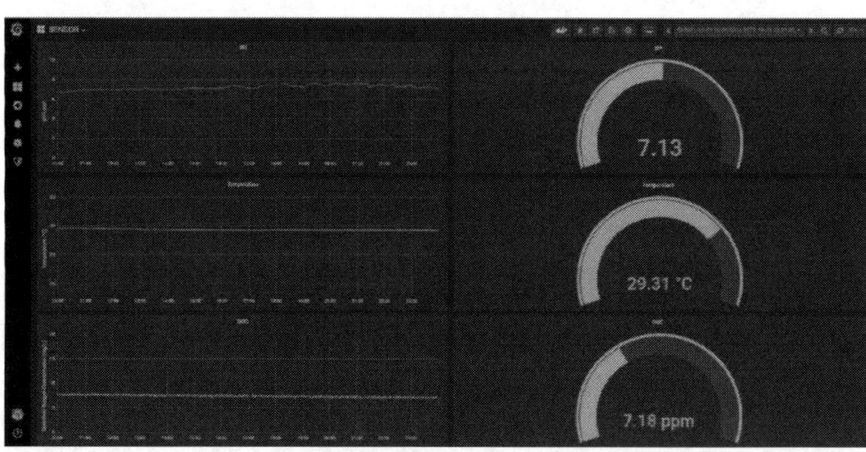

Figure 3. Grafana Dashboard.

4. Results and Discussion

4.1 Peristaltic Pump Volumetric Flowrate

Figure 4 demonstrates the volumetric flow rate at various angles of rotation of the pump. The maximum flow rate recorded was 102 mL/min, at rotational angle of 0 and 180. The rotation direction of the peristaltic pump is clockwise for rotation angle of 0-89, and anticlockwise for rotation angle of 91-180. At rotation speed of 90, the pump does not rotate. The volumetric flow rate data provides an indication on the transient response of the pump and the volume dispersed to control the pH to an optimum level.

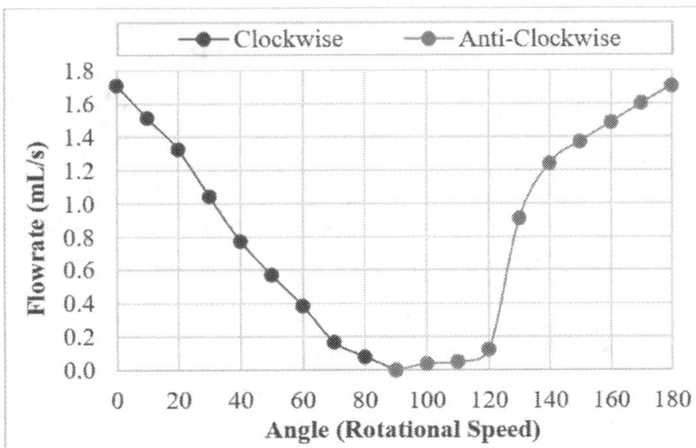

Figure 4. Volumetric Flow Rate of Pump at Different Rotational Angle.

4.2. Transient Response of the Pump

Table 2 summarizes the transient response of the peristaltic pump at varying rotational speed angle of 40,50 and 60. These speeds was able to dispense the buffer solution at controllable volumetric flowrates. At varying initial nutrient solution pH of 4, 5 and 9, the pH control system successfully control the pH values until it reaches the optimal range. For all rotational speed angles and initial pH solution, the final pH after the addition of the pH buffer lies in the mildly acidic region, averaging between pH 6.0 – 6.5.

Table 2. Transient Response at Different Speed.

Rotational Speed Angle	Initial pH	Final pH	Time Taken to Dispense pH buffer (seconds)	Time Taken to Stabilize (seconds)
	4	6.5	190	150
40	5	6.5	130	70
	9	6.3	180	240
	4	6.2	310	80
50	5	6.3	140	250
	9	6.2	80	310
	4	6.5	450	180
60	5	6.0	210	330
	9	6.5	190	250

Koehorst, Laubscher, and Ndakidemi (2010) reported that the highest chlorophyll was produced at pH 6.5 by *Artemisia Afra Jacq.* [11]. Similarly, Anderson et al. (2017) had showed butterhead lettuce grown in hydroponics system at pH 5.8 gave a higher fresh and dry weight by 26% and 18% respectively as opposed to plants grown at pH 7.0 [12]. Seyfferth et al. (2008) also indicated that butterhead lettuces grown under pH 5.5 – 6.5 demonstrated a higher nutrient uptake [13]. Hence, the final pH value after pH control was in-line with the optimal pH of pH 5.0-7.0 for hydroponics plant growth.

Based on the results, the optimal rotational speed angle for the pH control was 40 as this setting resulted in the lowest average time taken to dispense the pH up/down solution into the tank, at approximately 167±32 seconds, as opposed to 177±119 seconds and 283±144 seconds for rotational speed angle of 50 and 60 respectively. In addition, the average time taken to stabilize the pH data after the pump reaches optimal range for rotational speed angle of 40 is the shortest at 153 seconds, followed by 213 seconds and 253 seconds at rotational speed angle of 50 and 60. The results suggested that at the rotational angle of 40 was able to dispense the pH buffer solution swiftly to meet the desired pH range.

5. Conclusion

This study utilizes an Internet of Things (IoT) based control system for monitoring water quality in a hydroponic system. The Arduino UNO microcontroller effectively acquires pH, temperature, and dissolved oxygen sensor data, which is displayed in Grafana through the InfluxDB database. In addition, an alert message was successfully implemented whereby an alert message was sent to Telegram when the temperature, dissolved oxygen concentration (DOC) and pH level were not within the range. The control system was developed by analyzing the data obtained from the pH sensor. If the pH data was not within the desired range, pH up or pH down buffer solution will dispense out into the nutrient solution until the pH sensor detects that the pH value is within the optimal range. The transient response test of the peristaltic pump suggested for an initial nutrient solution pH value of 4, 5 and 9, the final pH after the addition of the pH buffer solution averages between pH 6.0 – 6.5 which is advantageous for hydroponics plant growth. It is suggested that the optimal rotational speed angle for the pH control is 40 as this setting resulted in the lowest average dispense and stabilization time. Results indicated that the control system was able to successfully control the pH value of the nutrient solution but further work is necessary to further optimize the physical setup for hydroponics application.

Acknowledgement

The authors gratefully acknowledge the financial support of SEGi University, Malaysia (SEGiIRF/2021-03/FOEBEIT-04/109) and Malaysian Smart Factory 4.0 at Selangor Human Resource Development Centre for providing the research facilities.

References

[1] Gagnon V, Maltais-Landry G, Puigagut J, Chazarenc F and Brisson J 2010 *Water Air Soil Pollut.* **212** 483-90
[2] Sardare M 2013 *Int. J. Eng. Res. Technol.* **2** 299-304.
[3] Krumrei M 2019 *Hydroponics: Effect of pH on Different Cucumber Varieties* (Nebraska: University of Nebraska – Lincoln) p 12
[4] Zambon I, Cecchini M, Egidi G, Saporito M and Colantoni A 2019 *Processes.* **7** 36-52
[5] Chen C, Jeng S and Lin C 2022 *Appl. Sci.* **12** 405-17
[6] Fakhrurroja H, Mardhotillah S, Mahendra O, Munandar A, Rizqyawan M and Pratama R 2019 *Int. Conf. Comput. Control Inform. Appl.: Emerg. Trends Big Data Artif. Intell (Indonesia)* vol 1 (IEEE Explore) p 156-61.
[7] Sihombing P, Karina N, Tarigan J and Syarif M 2018 *J. Phys.: Conf. Ser. (Indonesia)* vol 978 (IOP Publishing) 012014
[8] Ighalo J, Adeniyi A and Marques G 2021 *Internet of Things for Water Quality Monitoring and Assessment: A Comprehensive Review* vol 912, ed Hassanien A, Bhatnagar R, and Darwish A (New York: Springer)
[9] Chowdhury M and Khandakar A 2020 *Sensors.* **20** 1-24
[10] Azimi F, Aw S, Adiimah S, Aris R, and Safiyah L 2020 *Int. J. Synergy Eng. Tech.* **1** 1-10
[11] Koehorst R, Laubscher C and Ndakidemi P 2010 *J. Med. Plant Res.* **4** 1617-23
[12] Anderson T, Villiers D and Timmons M 2017 *Horticulturae* **3** 1-19
[13] Seyfferth A, Henderson M and Parker D 2007 *Plant. Soil.* **32** 139-48

IOP Conf. Series: Materials Science and Engineering 1257 (2022) 012028 doi:10.1088/1757-899X/1257/1/012028

Synthesis of Integrated Flower Waste Biorefinery: Optimisation with Economic Consideration

EHY Chong[1], DKS Ng[1], V Andiappan[1,2], LY Ng[1,3] and P Shivaprasad[4].

[1]School of Engineering and Physical Sciences, Heriot-Watt University, No.1, Jalan Venna P5/2, Precint 5, 62200 Putrajaya, Malaysia.
[2]Research Centre of Sustainable Technologies, Faculty of Engineering, Computing and Science, Swinburne University of Technology, 93350 Kuching, Sarawak, Malaysia.
[3]Faculty of Engineering, Technology & Built Environment, UCSI University, Puncak Menara Gading, Taman Connaught, 56000 Cheras, Malaysia
[4]Faculty of Engineering, University of Nottingham, Nottingham, NG7 2RD, United Kingdom

Abstract. Disposal of flower waste is a major environmental issue especially in India. Improper disposal of flower waste from heritage and cultural activities has resulted in an imbalance natural ecosystem causing water, soil, and air pollution. Hence, it is critical to minimise the amount of flower waste produced on daily basis by converting the flower waste into value-added products. These value-added products can be formed in an integrated flower waste biorefinery. Hence in this work, a systematic optimisation model for the synthesise of an integrated flower waste biorefinery with maximum economic performance is presented.

1. Introduction

In India, flowers are commonly utilised as holy entities to be offered to Gods and idols to represent the devotion and loyalty that a devotee has [1]. Thus, large amount of flowers are left as offerings and accumulated at religious sites. According to Statista [2], flower production in India is expected to reach about 3 million metric tonnes in year 2021. Most wilted and unused flowers in India are disposed to landfill or released to water bodies [3]. Such practice causes water contamination from toxic pesticide and insecticides found in the flower waste.

Waste valorisation is a promising way to reduce the environmental impact caused by the flower waste [4]. Apart from using flowers for decorative and religious purposes, flowers can be converted into valuable products for different application: agricultural, bioenergy, pharmaceutical, food and beverage due to the rich source of biologically active substances.

Research works have been conducted in the past to recover flower waste into value added products. According to Prasad el al. [5] and Dilta el al. [6],flowers can be utilised as edible flower which conserves the medical constituents. In order to preserve the flower, the water content in flower can be removed via drying process. Purohit et al. [7] presented various extraction methods to secure the biological active compound found in flower (i.e., hydrodistillation, hydrolysis, supercritical carbon dioxide, membrane extraction etc.). Meanwhile, Slavov et al. [8] highlighted the possible approaches to valorise rose flower waste into bioenergy application through fermentation, gasification, biosorption and composting. All the possible pathways are compiled and presented in flower waste biorefinery.

Another group of researchers Alig et al. [9] applied life cycle assessment (LCA) to determine the environmental impacts of production of different type of cut roses. Meanwhile, Russo et al. [10] has also conducted similar LCA towards Roses and Cyclamens in greenhouse cultivation. Note that both works [9] and [10] evaluated the environment impact; however, economic evaluation of flower waste biorefinery is not considered. Economic evaluation is important since it allows the comparison of cost and health outcomes of different technologies. In order to address the gaps, an optimization model is presented in this work to synthesize an integrated flower waste biorefinery with maximum economic performance.

2. Mathematical Optimisation Model

In this work, a set of flower wastes is converted into value-added products via different conversion technologies. The economic performance of the synthesized flower waste biorefinery is determined based on revenue of products as well as intermediates.

Given a set of flower wastes i is converted into intermediates k via biomass conversion technologies j. The intermediates k can be further converted into products k' via technologies j'.

2.1. Model Formulation

The total amount of flower waste i, $F_i^{Biomass}$ is sent to different biomass conversion technologies j to form intermediates k with the mass flow rate of F_k and then further converts to final products, k' with the mass flow rate $F_{k'}$. Equation (1) represents the flow distributions within the synthesized integrated biorefinery:

$$F_i^{Biomass} = \sum_j f_{ij} \qquad \forall I \qquad (1)$$

Where f_{ij} is the mass flowrate of flower waste to technologies j.

$$F_k = \sum_j \sum_i f_{ij} X_{ijk} \qquad \forall k \qquad (2)$$

Where X_{ijk} is the conversion factor of flower waste i to intermediate k.

For the intermediate k which required to be further processing in technologies j', the mass flowrate of final product k' is given as below:

$$F_k = \sum_{j'} f_{kj'} \qquad \forall k \qquad (3)$$

Where $f_{kj'}$ is the mass flowrate of flower waste intermediate to technologies j'.

$$F_{k'} = \sum_{j'} \sum_k f_{kj'} X_{kj'k'} \qquad \forall k' \qquad (4)$$

Where $X_{kj'k'}$ is the conversion factor of intermediate k to final product k'.

Note that the above mathematical model can be modified to suit for different technologies j and j'. For example, in drying process water within the biomass is performed to determine the water loss during the process and total mass weight changed. For the technologies which convert the materials into energy, the conversion factor, X can be replaced by energy conversion factor.

Other than tracking the final products, waste generated throughout the biorefinery is also determined. In this work, it is assumed that all the waste, i which cannot be converted into value added products will be sent to landfill.

Equation below describes the mass flow of the collected waste to landfill, $F^{\text{Total Landfill}}$ from intermediates k and final products k' in technologies j and j'.

$$F^{\text{Landfill}} = \sum_j \sum_i f_{ij}(1\text{-}X_{ijk}) + \sum_{j'} \sum_k f_{kj'}(1\text{-}X_{kj'k'}) \tag{5}$$

To determine the economic performance of the synthesized integrated biorefinery (EP), the revenue, REV and landfill cost, $COST^{\text{Landfill}}$ are determined as below:

$$REV = \sum_k F_k \times \text{P}_k + \sum_{k'} F_{k'} \times \text{P}_{k'} \tag{6}$$

$$COST^{\text{Landfill}} = F^{\text{Landfill}} \times \text{TR} \tag{7}$$

$$EP = REV - COST^{\text{Landfill}} \tag{8}$$

Where P_k and $\text{P}_{k'}$, TR are selling price of intermediate k, final product k' and disposure cost of landfill.

To determine the maximum economic performance of the synthesized integrated biorefinery, the optimization objective is set as:

$$\text{Maximise } EP \tag{9}$$

To illustrate the proposed approach, a case study is solved in the following section.

3. Case Study

This hypothetical case study is developed based on the statistic of flower waste produced in India. According to Statista [2], flower production can reach 3 million metric tonnes per year, which amounts to 8,220 metric tonnes of flower produced daily. According to Waghmode's [11], 40% of India's entire flower production is sold, amounting to 3,300 metric tonnes of flower is sold each day. Assuming that half of the flower waste sold is utilised for decorative purposes and for heritage activities, this will result in 1,640,000 kg of waste produced per day, and roughly 68,500 kg of flower waste is produced every hour.

Based on detailed literature review, all alternatives' technologies which convert flower waste into value added product are summarised in the superstructure shown figure 1. Note that the entire superstructure can be subdivided into several parts: pre-treatment of flower waste, biomass conversion technologies (which also known as main process) and upgrading process to form final product.

4. Result and Discussion

In this work, the mathematical optimisation model is solved via a commercial optimisation software LINGO version 18. Solving equation (9) subject to equation (1) to (8), the optimised results with the maximum economic potential (EP) can be determined.

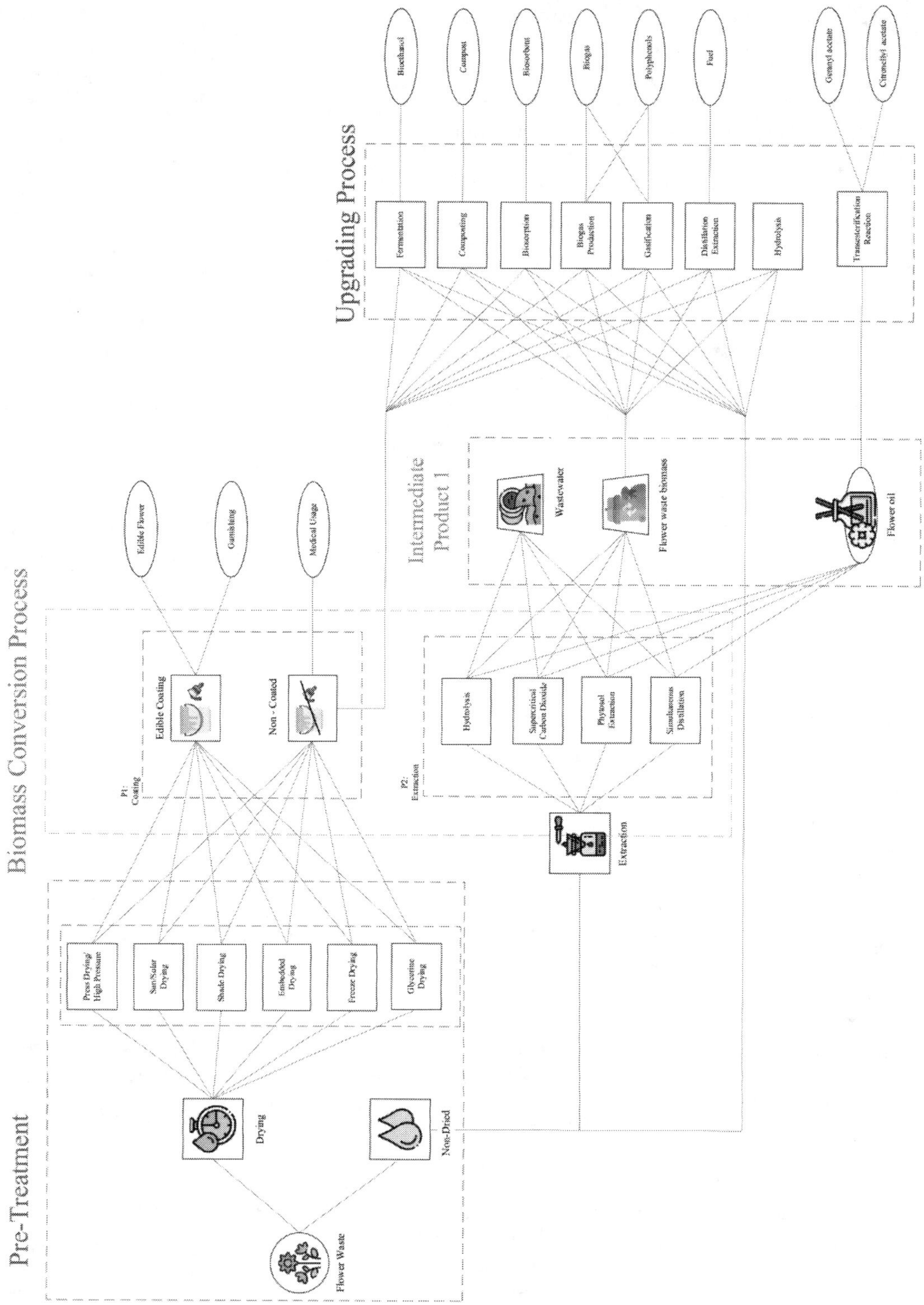

Figure 1: Overall superstructure for synthesis of integrated flower waste biorefinery.

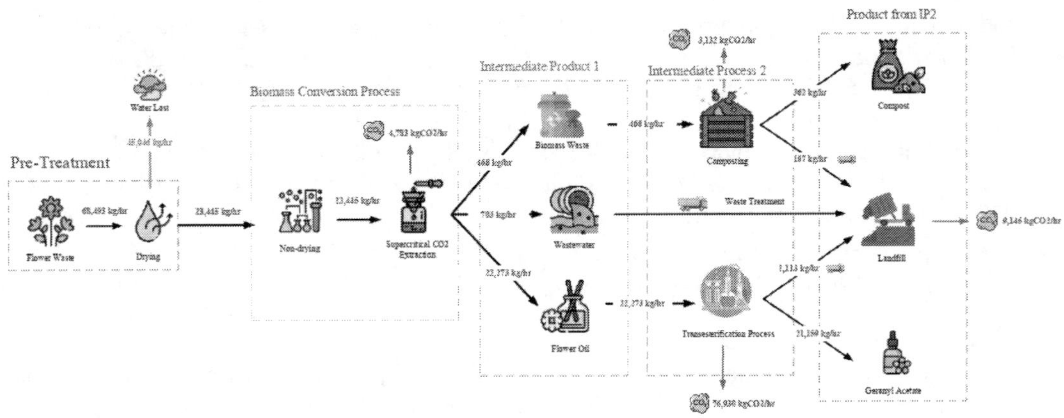

Figure 2: Optimisation based on maximum economic performance.

Figure 2 shows the optimum pathway which convert flower waste into value-added products. Based on the optimised result, 23,445 kg/hr of pre-dried flower waste is sent to supercritical carbon dioxide extraction process after the first stage drying. Once the flower waste is extracted, around 22,273 kg/hr of flower essential oil is formed, follow by 703 kg/hr of wastewater and 468 kg/hr of biomass waste. Then, the flower waste is converted into different product via upgrading process composting and transesterification processes. Such processes produce compost and geranyl acetate. Via the optimum integrated biorefinery, the total economic performance (*EP*) is determined as $7,874,455.

5. Conclusion

This paper presented a systematic synthesis approach for an integrated flower waste biorefinery. Based on the proposed approach, all process alternatives are first summarised into a superstructure. Then, mathematical optimisation model is formulated in optimisation software to determine the optimum process pathway based on optimisation objective. In this work, a hypothetical case study is solved to illustrate the proposed approach. To enhance the strength of the synthesized integrated biorefinery, environment emission can be included, and multi-objective optimisation can be applied in the future work. In addition, as this work only consider the integrated flower waste biorefinery in India, the model can further revise to suit different location requirement.

References

[1] Menon S 2018 *Why Do We Offer Flowers to Gods?* [Online] Available on: https://www.boldsky.com/yoga-spirituality/faith-mysticism/2018/why-do-we-offer-flowers-to-god-121808.html [Accessed on 30th November 2021]

[2] Statista Research Department 2021 *Production Volume of Flowers in India Between Financial Year 2016 and 2020, With an Estimate for 2021.* [Online] Available on: https://www.statista.com/statistics/1036938/india-production-of-flowers/ [Accessed on 30th November 2021]

[3] Frost R 2020 *Waste from Worship: Solving India's Unique River Pollution Problem.* [Online] Available on: https://www.euronews.com/green/2020/02/26/waste-from-worship-solving-india-s-unique-river-pollution-problem [Accessed on 30th November 2021]

[4] Kabongo J D 2013 *Waste Valorization. Encyclopedia of Corporate Social Responsibility*, 2701–

2706. doi:10.1007/978-3-642-28036-8_680

[5] Prasad V J K, Pal R K and Voleti S R 1997 *Drying of Flowers: An Upcoming Industry, Dry Flower Industry*

[6] Dilta B S, Sharma B P, Baweja H S and Kashyap B 2011 *Flower Drying techniques – A review, International Journal of arm Sciences* 1(2): 1-16, 201

[7] Purohit S R, Rana S S, Idrishi R, Sharma V and Ghosh P 2021 *A review on Nutritional, Bioactive, Toxicological Properties and Preservation of edible flowers, Future Foods*, Volume 4, 2021 doi.org/10.1016/j.fufo.2021.100078.

[8] Slavov A, Vasileva I, Stefanov L and Stoyanova A 2017 *Valorization of Wastes from The Rose Oil Industry. Rev Environ Sci Biotechnol (2017)* 16:309-325 doi:10.1007/s11157-017-9430-5

[9] Alig M, Frischknecht 2018 *Life Cycle Assessment Cut Roses. Treeze Ltd., Kanzleistrasse 4, CH-9810 Uster,* www.treeze.ch

[10] Russo G, Buttol P and Tarantini M 2008 *LCA (Life Cycle Assessment) of Roses and Cyclamens in Greehouse Cultivation. Acta Hortic.* 801, 359-366 doi:10.17660/ActaHortic.2008.801.37

[11] Waghmode M S, Gunjal A B, Nawani N N, and Patil N N 2016 *Management of Floral Waste by Conversion to Value-Added Products and Their Other Applications. Waste and Biomass Valorization*, 9(1), 33–43. doi:10.1007/s12649-016-9763-2

Design of lubricants in electric and hybrid vehicles via computer-aided molecular design (CAMD) technique

K W Wong[1], L Y Ng[2*], C H Lim[1], D K S Ng[1] and V Andiappan[3]

[1] School of Engineering and Physical Sciences, Heriot-Watt University (Malaysia Campus), 62200, Putrajaya, Wilayah Persekutuan Putrajaya, Malaysia.

[2] Department of Chemical & Petroleum Engineering, Faculty of Engineering, Technology and Built Environment, UCSI University (Kuala Lumpur Campus), 56000 Cheras, Kuala Lumpur, Malaysia.

[3] Faculty of Engineering, Computing and Science, Swinburne University of Technology, Jalan Simpang Tiga, 93350, Kuching, Sarawak, Malaysia.

*NgLY@ucsiuniversity.edu.my

Abstract. The market of electric and hybrid vehicles (EVs and HEVs) is expanding exponentially in recent years. It serves as a greener solution in the transportation sector as compared to internal combustion engine vehicles' (ICEVs). Conventional lubricants used for ICEVs are not suitable for the EVs and HEVs as the latter operates on electric motors instead of combustion engines. Performance characteristics such as the electrical conductivity, thermal conductivity and copper compatibility are not considered by the ICEVs lubricant but are important for the EVs and HEVs. In this research, computer-aided molecular design technique is used to design the lubricant that is compatible with the EVs and HEVs. The lubricant properties considered are thermal conductivity, specific heat capacity, viscosity, density, and flash point. Group contribution (GC) property prediction model is utilized to ensure that the desired properties are possessed by the generated lubricant. There are six possible solutions of lubricant being generated and discussed. Other properties such as the electrical conductivity, copper corrosivity etc. are reviewed for future development. Additionally, the advantages and challenges of utilizing GC method in designing EVs and HEVs lubricant are discussed. This research serves as guidance for the comprehensive design of advanced lubricant of EVs and HEVs.

1. Introduction

According to Paris Agreement, global temperature increment in this century is limited to 2 °C by consistently reducing the emission of greenhouse gas globally [1]. Transportation sector is the primary contributor to greenhouse gases emission, while traditional internal combustion engine vehicles (ICEVs) contributed to 75 % of carbon monoxide emission [2]. Therefore, the invention of electric vehicles (EVs) and hybrid electric vehicles (HEVs) are encouraged to be invented as it will serve as a greener solution. The market of EVs is favorable due to advantages such as environmental friendliness, absence of greenhouse gas emissions etc. [3]. Depending on assumptions made on fuel economy, the most aggressive scenario predicted that 95 % of all vehicles would be EVs by year 2042

[4]. The driving force of inventing new EV lubricants is to overcome the difference of requirements between ICEVs and EVs in heat transfer properties, electrical conductivity, copper, and elastomer compatibility [5]. The most suitable EV lubricant must be designed to possess higher thermal transfer properties to release high heat generated by electric motor. At the same time, it's viscosity should be low to handle the high rotational speed of electric motor. Electrical conductivity must be lowered to avoid short circuit, but not too low as accumulation of bearing current could occur. In addition, reaction between copper components, plastic polymers and the lubricant must be prevented to avoid corrosion, while the flash point of EV lubricant must be high to avoid ignition in the vehicles.

According to previous studies, several possible lubricants to be used in EVs have been proposed. Firstly, a hybrid electric transmission fluid (HEVTF) using mineral oils as base oil, and calcium detergents and dispersants addictive have been proposed [6]. Electrical properties, viscosity and anti-wear properties of the HEVTF are improved to be compatible with EVs. Then, bio-lubricants such as plant oil and animal fats are proposed. These lubricants provided adequate anti-wear, anti-friction and thermal stability [7]. Bio-based ionic liquids derived from fatty acids are also trend in recent research and are promising advanced lubricants due to the high thermal stability, low melting point etc. [8] [9]. Besides, synthetic-based oils such as polyalphaolefins (PAOs) and polyalkylene glycol (PAG) are incorporated with boron compounds to improve oxidative stability [10]. These possible solutions possess exceptional performance in terms of anti-wear and stability properties. However, the electrical properties are not well studied except for the HEVTF. Important properties such as thermal transfer properties, anti-corrosion of copper and plastic polymer are not addressed.

In this paper, lubricant that is compatible to be used by EVs and HEVs will be designed using computer-aided molecular design techniques (CAMD). It will be mentioned as EV lubricants in this research as HEV would share the same lubricant with EVs. Group contribution (GC) property prediction models will be utilized to obtain desired base oil properties, specifically thermal conductivity, viscosity, density, specific heat capacity, flash point and acute toxicity (LC_{50}). Then, advantages and challenges of utilizing GC method in designing EV lubricant will be analysed in detail, and the possible suggestions to improve EV lubricant performance will be critically discussed.

2. Methodology

2.1. Design Procedure

To obtain a suitable EV lubricant, the targeted properties are chosen while considering several different aspects such as functionality, environmental and safety aspects. These aspects must be well balanced to obtain reliable EV lubricants. As the product being designed in this research is a single chemical component instead of a mixture, the procedure of molecular design will be followed as referred to Chapter 4 in the Thesis by Ng [11].

2.2. Target Properties and Ranges Identification

The chosen properties in this research are the thermal conductivity, viscosity, density, specific heat capacity, flash point and the acute toxicity to fathead minnow. The purposes of the selected properties are explained in table 1. To determine the target property ranges, the typical specifications of lubricant used by ICEVs are analysed and shown in table 2. The properties data are acquired by the top ICEVs lubricant companies such as Castrol, Mobil, Valvoline and ACDelco. The thermal conductivity of lubricants is to be maximised in this design. The other target property ranges are determined and described in table 3.

Table 1. Targeted properties for the EV lubricants design.

Class	Properties	Purpose	Reference
Functionality	Viscosity	To increase cooling performance and support the high rotational speed of motor.	[12]
	Specific heat capacity	To absorb the higher amount of heat generated by motor, while not increasing the temperature of lubricant.	[13]
	Thermal conductivity	To release the high amount of heat generated by motor.	[14]
	Density	To ensure the feasibility, and reliability of product.	[15]
Safety	Flash point	To avoid ignition at high operating temperature of motor.	[16]
Environment	Fathead Minnow 96-h LC_{50}	To minimize the environmental impact to water source in the event of leakage.	[17]

Table 2. Average property values of ICEV and EV lubricants.

Properties	ICEV Lubricants	EV Lubricants	Unit
Density	852	848	kg/m^3
Flash Point	492	494	K
Dynamic viscosity @ 40 °C	54.5	26.9	cPs

Table 3. Target property ranges for the EV lubricants design.

Property	Target property range		Description
	Lower	Upper	
Thermal conductivity (W/ m K)	To be maximised		To obtain high thermal properties.
Dynamic viscosity @ 40 °C (cPs)	21.5	26.9	Lower boundary is set 20 % lower.
Density (kg/m^3)	680	848	
Flash point (K)	494	592	Upper boundary is set 20 % higher.
Specific heat capacity (kJ/ kg K)	2.00	3.00	
Acute toxicity LC_{50} (mg/L)	1.00	n/a	Lower than 1 mg/L is highly toxic.

2.3. Property Prediction Model Identification
To have a reliable designed product, accurate property estimation method is crucial. The GC method is chosen as it provides quick estimation of properties without needing intensive computational resources. The property prediction models utilised for this work are available from the references shown in table 1.

2.4. Molecular Building Blocks Selection
Suitable molecular building blocks are to be selected for the design problem. As the base oil to be designed is mineral oil, formed by paraffins, alkanes are selected as the target molecule. Only molecular groups with single bond are considered in the design which are C-, CH-, CH_2- and CH_3-.

2.5. Property Operators Normalisation
Transformation of the property prediction models into normalised property operators, and target property ranges is crucial as the values must be obtained to form feasible formulas in the LINGO

mathematical model. By normalisation, linear equations can be obtained, and the property constraint equation be used to ensure the specifications of product fall within the desired range.

2.6. Structural Constraint Formation

Structural constraint must be formulated to ensure that the molecule can be completed without the presence of free bonds in the structure. The handshaking lemma equation must be obeyed [18].

3. Results and Discussion

3.1. Product Analysis

The generated best six solutions are shown in table 4. Integer cut is utilized to obtain additional solutions. Only the simplest isomer of the products will be shown. Table 5 displays the product specifications for the best six solutions. According to table 5, all generated lubricants possessed properties within the target property range except for solution C which possessed a slightly lower flash point. It indicates that EVs lubricants possessing desired properties, can be successfully designed. The best solution can be selected based on different needs, either functionality, safety, or environmentally. By comparing with table 2, the viscosity of the designed EV lubricants is 1.3 % to 18.36 % lower than the EVs lubricants available in the market and is also 105.6 % to 140.2 % lower than the ICEVs lubricants. In addition, the density and flash point of the designed EV lubricants are 5.8 % and 1.2 % more desirable for EV vehicles, respectively.

Table 4. Molecular structures of generated solutions.

Solutions	Molecular Structure	Name (Simplest isomer)
A	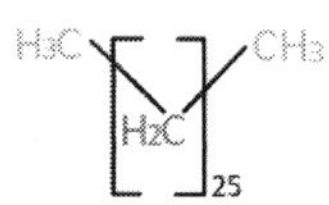	Heptacosane
B	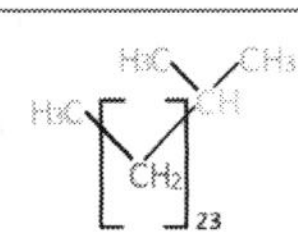	2-methylhexacosane
C	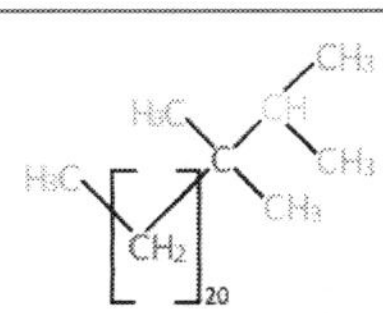	2,3,3 – trimethyltetracosane
D	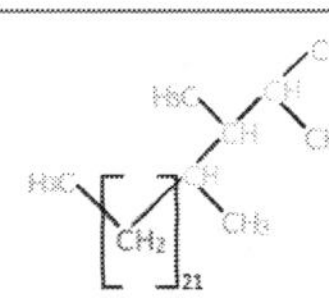	2,3,4 - trimethylhexacosane
E	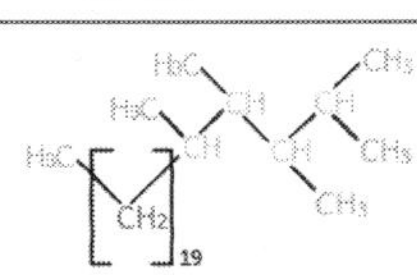	2,3,4,5 - tetramethylpentacosane
F	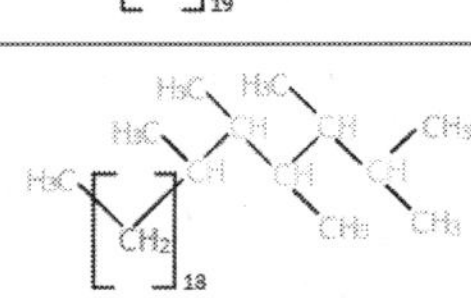	2,3,4,5,6- pentamethylpentacosane

Table 5. Product specifications of best six solutions.

	Property					
Solution	**k (W/mK)**	**η (cPs)**	**ρ (kg/m³)**	**F_p (K)**	**C_P (kJ/kgK)**	**LC_{50} (mg/L)**
A	0.13	25.97	800.76	502.18	2.19	2.33
B	0.12	22.69	801.51	495.77	2.19	4.01
C	0.10	25.63	802.40	484.95	2.16	45.37
D	0.09	26.51	806.37	505.98	2.17	3.38
E	0.07	23.17	807.08	499.57	2.16	5.81
F	0.06	25.31	809.27	504.67	2.16	5.33

3.2. Relationship of Presence of C- Molecular Group to Product Properties
For research purpose, there are three additional solutions that fulfil the target property ranges of EVs lubricants being generated. It is observed that the flash point of molecule decreased from 484.9 K to 424.3 K as the occurrence of C- molecular group increased from 1 to 4. It is suggested that the presence of C- molecular group will lead to the presence of alkane branches as the C- molecular group cannot be linked with Hydrogen atom. As the branching increased, surface area of molecule will be decreased, which smaller area of contact between molecules will be resulted, leading to weaker Van der Waals forces and lower boiling point [19]. While the flash point is in linear relationship with the boiling point [20]. Then, it is noticed that the thermal conductivity and specific heat capacity of molecule decreased with the increasing occurrence of C- molecular groups. As occurrence of branching increases, the number of vibrational and rotational quantum states will decrease, leading to the decrease in thermal properties of molecule. From this hypothesis, it is justified that the occurrence of C- molecular groups must be low when designing EVs lubricants.

3.3. Group Contribution Methods for EV Lubricants Design
Throughout the design of EV lubricants using GC property prediction model, several advantages and challenges have been observed and listed in table 6.

Table 6. Advantages and disadvantages of lubricants design using GC methods.

Advantages	**Disadvantages**
Simpler design procedure due to lesser data needed to obtain relatively accurate results	Properties such as electrical conductivity, copper corrosion resistivity etc. cannot be predicted as property prediction model has not been developed yet
Cost of experimental product design and measurement can be reduced	The combination of generated molecular groups for the EV lubricants can form various type isomers, which possess different chemical and physical properties

By using the predicted results provided by GC methods, number of feasible solutions required to be produced through the traditional generate-and-test product synthesis can be greatly lessened, which directly reduces the costs of experimental product design. Therefore, GC methods can be used as the first step of EV lubricants design, to aid the whole design process by narrowing down the possibilities and options. Then, up to date test methods and equipment can be developed to examine these critical fields of interest for EV lubricants to overcome the disadvantages of GC model. Besides, the generated products are in terms of occurrence of molecular groups, and the arrangement are not known. To overcome this weakness, 2nd order groups, and 3rd order groups are used to improve the accuracy and reliability of the GC models

4. Conclusion and Future Work
The lubricant properties that are of concern by EVs are the thermal conductivity, viscosity, density, flash point, specific heat capacity, toxicity, electrical conductivity, and copper compatibility. 6

solutions are generated using the GC methods to obtain property values within the desired target ranges, with the Heptacosane being the lubricant base oil with highest thermal conductivity value of 0.134 W/ mK. All specifications of the generated solutions fulfilled the desired property ranges, it can be selected based on different needs. In addition, the trend of product properties with respect to the occurrence of C- molecule is justified. It is suggested the decrement of the mentioned properties is due to the increase in the number of branching of the lubricant structure. Recommendations to improve properties that cannot be predicted by GC methods are proposed. Next, it is concluded that GC methods can be used to reduce the cost of product design by identifying trends of molecule groups in affecting desired properties. However, there are some properties that are cannot be predicted by the model and the corresponding test methods have been suggested to overcome this issue. Additionally, the implementation of higher order groups in the GC model can effectively distinguishing between isomers, and more accurate structural information of the EV lubricants can be obtained. As the EVs and HEVs market continues to evolve, the requirements of the lubricants could also differ. For future work, more properties, and contributions of higher order groups can be included in the GC model to obtain a more compatible and reliable EV lubricant. For instance, additional safety properties can be incorporate into the model for better environmental quality of the EV lubricants.

References

[1] Paris Agreement 2015 Treaty no. 54113. United Nations Treaty Series. Available from: https://treaties.un.org/Pages/ViewDetails.aspx?src=IND&mtdsg_no=XXVII-7-d&chapter=27&clang=_en

[2] Shah R, Gashi B, González-Poggini S, Colet-Lagrille M and Rosenkranz A 2021 *Advances in Mechanical Engineering* **13**

[3] Gelmanova Z, Zhabalova G, Sivyakova G, Lelikova O, Onishchenko O, Smailova A and Kamarova S 2018 *Journal of Physics: Conference Series* **1015** 1-5

[4] Becker E P 2019 *Tribology & Lubrication Technology* **75** 60

[5] Van Rensselar J 2019 *Tribology & Lubrication Technology* **75** 34-43

[6] Tang T, Devlin M, Mathur N, Henly T and Saathoff L 2013 *SAE International Journal of Fuels and Lubricants* 6289-94

[7] Reeves C, Menezes P, Jen T and Lovell M 2015 *Tribology International* **90**

[8] Gusain R and Khatri O 2016 *RSC Advances* **6** 3462-9

[9] Song Z, Liang Y, Fan M, Zhou F and Liu W 2014 *RSC Advances* **4**

[10] Farfan-Cabrera L 2019 *Tribology International* **138** 473-86

[11] Ng LY 2015 *Novel Integrated Design Techniques For Biorefineries.* PhD Thesis. University of Nottingham.

[12] Conte E, Martinho A, Matos H and Gani R 2008 *Industrial & Engineering Chemistry Research* **47** 7940-54

[13] Štěpina V and Vesely V 1992 *Lubricants and Special Fluids - Chapter Two The General Properties of Lubricants* (Amsterdam: Elsevier) p 9-124

[14] Nagvekar M and Daubert T 1987 *Industrial & Engineering Chemistry Research,* **26** 1362-5

[15] Elbro H, Fredenslund A and Rasmussen P 1991 *Industrial & Engineering Chemistry Research,* **30** 2576-82

[16] Hukkerikar A, Sarup B, Ten Kate A, Abildskov J, Sin G and Gani R 2012b *Fluid Phase Equilibria* **321** 25-43

[17] Hukkerikar A, Kalakul S, Sarup B, Young D, Sin G and Gani R 2012a *Journal of Chemical Information and Modeling* **52** 2823-39

[18] Chemmangattuvalappil N, Solvason C, Bommareddy S and Eden M 2010 *Computers & Chemical Engineering* **34** 2062-71

[19] Santak P and Conduit G 2019 *Fluid Phase Equilibria* **501**

[20] Ding C, Zhou D,Zhang X, Ma X and Wang J 2014 *Applied Mechanics and Materials* **664** 210-4

IOP Conf. Series: Materials Science and Engineering 1257 (2022) 012030 doi:10.1088/1757-899X/1257/1/012030

Formulation of Chitosan-Saponin Microbeads using Microfluidic Technology

J Yong[1], *K S Koh[1], V L Wong[1] and S S Lim[2]

[1] School of Engineering and Physical Science, Heriot-Watt University Malaysia, 1, Jalan Venna P5/2, Precinct 5, 62200, Putrajaya, Malaysia
[2] Department of Chemical and Environmental Engineering, Faculty of Science and Engineering, Jalan Broga, 43500, Semenyih, Selangor, Malaysia

*Corresponding author's e-mail: k.koh@hw.ac.uk

Abstract. Type II diabetes mellitus caused adverse impact on almost 3.5 million patients locally. Conventional treatment for such disorder includes oral administration of anti-diabetic drugs or herbs supplement. However, the performance of this drugs in regulating blood glucose level is not too reliable due to the occurrence of weight gain, hypoglycaemia and low tolerability as well as structural breakdown of herbal supplements during consumption. In this project, an alternative approach was attempted by immobilising saponin, a known insulin stimulant extracted from plants (e.g. bitter gourd, eggplant or soapbark) onto chitosan beads via microfluidic technology. This technology offers advantages of producing micro-sized droplet with high surface area to volume ratio and minimal coefficient of variation for more accurate dosage. The droplets undergo solidification via cross-linking with NaOH as solvent. These beads are approximately 550 microns with 5.12 % polydispersity in size. The solidified beads were used as carrier for the immobilisation of saponin. This was achieved through altering the surface charge of the microbeads with Phosphate Buffer Solution (PBS), a commonly used electrolyte for the surface modification of chitosan before coating saponin onto its surface. Successful adsorption is rectified through FTIR, FESEM and EDX analysis.

1. Introduction

Diabetes mellitus (DM) has affected approximately 3.6 million Malaysians of age 18 and above [1]. DM are segregated into 2 types which relates to their respective severity. Type I DM (T1DM) is caused by defective pancreas that are unable to produce insulin. This type is often hereditary, and there is by far no preventive actions to be done to avoid such disorder. As a contrary, Type II DM (T2DM) patients have high blood sugar level either due to insulin resistance or lack of insulin production by the pancreas. 99% of diabetic cases in Malaysia are the second type [2]. The focus of this research is a to establish a preventive measure for T2DM. Insulin plays an important role to regulate blood glucose level by allowing glucose to be taken up by cells and be used for energy. One of the new trend of the development in the field includes an introduction of herbs supplements in the form of tablets, capsules or liquid can be used to replace these drugs. They are the preferred alternatives for diabetic patients at mild stage [3]. However, the antidiabetic compounds from herbs experiences structural breakdown and low absorption in digestion system during consumption [4]. This drawback causes fluctuation in serum glucose levels. An improvement on the compound release has been proposed by immobilization of herbal extract onto solid carriers as a structural support. One of the most effective antidiabetic herbal extracts discovered is from bitter gourd (*Momordica Charantia sp.*) [5]. Through some vivo tests, saponin, the extract helps in the release of pancreatic insulin via increase in plasma insulin level [6] and hinders glucose formation in the bloodstream [7]. Other health benefits of the saponin and clinical studies are well documented in *Alam et al.* [8]. Chitosan, a derivative from chitin, is a natural bio-membrane extracted from exoskeleton of crabs, shrimps, or other crustaceans. It has been broadly applied as therapeutics material as it is biodegradable, non-toxic, antacid, and low antigenic characteristic [9]. Usage on chitosan in oral sustained release preparation was also established [10] using microfluidic approach. Microfluidics is the study of the behavior of fluids interactions at microscale with application including imaging, drug delivery and diagnostics [11].

However, the immobilization of saponin onto chitosan-based carrier has yet to be reported. This project therefore attempts to produce a controllable size carrier for the immobilization of saponin. To do so, chitosan droplets were first produced with a co-axial flow geometry device in dripping regime using microfluidic technologies. The microfluidic device was configured via off-the-shelf technology with dispensing needles and a mini crosslink. In the second stage, droplets generated were solidified with NaOH solution to produce chitosan microbeads. In the third stage, chitosan microbeads were used as a carrier for the immobilization of saponin via PBS soaking.

2. Experiment Setup

2.1. Droplet Generation

The needle-based microfluidic devices fabricated based on co-flow geometry. The plastic head of a 19G needle was removed from the stainless-steel needle as shown in figure 1 (a) below. The dissembled needle and a 23G dispensing needle were inserted into two opposite ends of a cross-link and affixed with ultraviolet-curing adhesive. Needles are aligned to the center of the junction before exposed to an UV torch light for 3 minutes for cross-link purpose, as seen in figure 1 (b). From Figure 1 (c), 3 device size combinations were fabricated (i.e. 23G-19G, 25G-20G and 27G-21G.

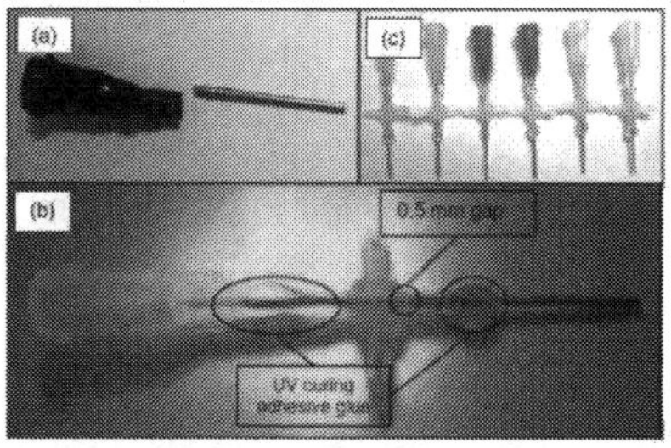

Figure 1. (a) Dissembled 19G dispensing needle; (b) Illustration of fabricated and assembled 23G-19G device; (c) 3 pairs of devices configuration assembled (23G-19G, 25G-20G and 27G-21G).

2.2. Chitosan Microdroplet and Solidification.

0.2M acetic acid was first diluted to prepare 2 wt% chitosan solution. For chitosan solution, 2 g of chitosan powder was added into 98 g of 0.2 M acetic acid. The mixture was stirred for 3 hours using magnetic stirrer (IKA C-MAG HS 7) until fully dissolved to form chitosan solution. Meanwhile, in a separate beaker, 20 mL of 10 wt% NaOH solution was topped with 2.5 mL of sunflower oil. The experimental set up was done with chitosan solution as disperse phase. Flowrate of chitosan solution and sunflower oil was set as 1 mL/hr and 10 mL/hr respectively. After 30 minutes, the beaker was swirled gently to allow settling of the chitosan microbeads. The beads were set aside for another 2 hours to allow complete crosslinking. Next, the beads will undergo washing step with 5 mL distilled water repeatedly to remove residual alkali, oil and soap. They were stored in 5 mL of distilled water in the centrifuge tube for later use.

2.3. Immobilization of Saponin onto Chitosan Microbeads

10x Phosphate Buffer Solution (PBS) was prepared by adding 20 g KCl, 800 g NaCl, 24 g of KH_2PO_4 and 144 g $Na_2HPO_4 \cdot 2H_2O$ into 8 liters of distilled water. The solution was topped up to 10 liters as the final PBS solution after complete mixing. Microbeads from stage 2 were removed from distilled water medium and soaked in 15 mL of PBS solution for 24 hours. The beads were rinsed 3 times and dried on a petri dish in a sealed box with silica gel (figure 2 (a)) for 3 days. Saponin solution was prepared by mixing 0.5 g of saponin powder with 100 mL of liquid ethanol. The solution was stirred with magnetic stirrer for 3 hours. A colloidal saponin solution is then formed. Undissolved saponin powder were left for a day to settle (figure 2 (b) & (c)). The top layer of the settled saponin solution were used to carry out saponin adsorption. Dried microbeads were transferred to a beaker together with 40 mL of top layer saponin solution. To prevent spillage, the beaker was covered with parafilm and inserted into a larger sized plastic beaker with Styrofoam and Blu-Tack as shown in figure 2 (d). The beaker was secured on a shaker (IKA KS 4000) for 6 hours at 200 RPM.

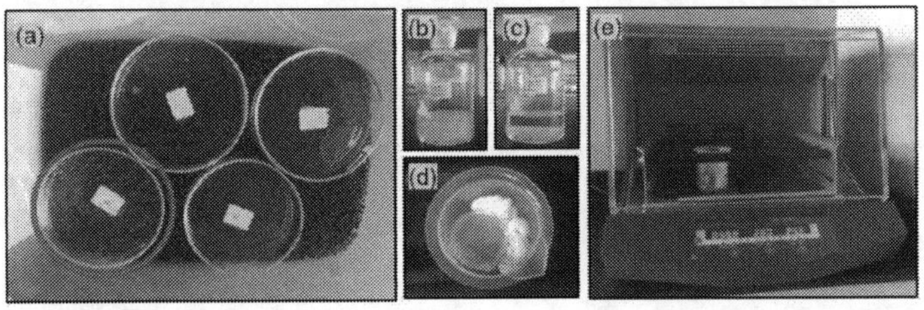

Figure 2. (a) Beads dried on petri dish in a sealed box with silica gel; (b) Colloidal saponin solution formed from saponin powder and ethanol solution; (c) Settled saponin solution forming 2 layers; (d) Top view of beaker secured on the shaker top view; (e) Side view of the shaker.

3. Result and Discussion

3.1. Droplet Generation

Several co-flow geometry devices were fabricated to generate droplets through dripping regime. Study was carried out to determine the relationship between Q_d/Q_c (above 0.1) and the frequency of droplet generated by different size configuration of devices. The introduced continuous phase was oil whereas the disperse phase was distilled water with 0.5 wt% of surfactant sodium dodecyl sulphate (SDS) to avoid coalescence of the droplets generated.

From figure 3, the device with larger diameter (23G-19G) produces the greatest number of droplets at constant flow ratio. This is because device with larger needles restricts the flow of the outer continuous phase. Thus, at constant Q_d/Q_c, smaller continuous phase flow area induces a higher continuous phase flow velocity, U_c. Since U_c is directly proportional to Ca_c, the droplet size generated will decrease non-linearly with the increase of Ca_c. Smaller droplet size represents a higher breakup frequency and ultimately higher number of droplets generated, which is in line with the reported literature [12]. On the other hand, by using the same device size configuration, an increase in Q_d/Q_c causes U_c to be relatively lower. By using the relationship as explained above, the frequency of droplets generated should reduce. Hence the results shown in Figure 4 is further clarified.

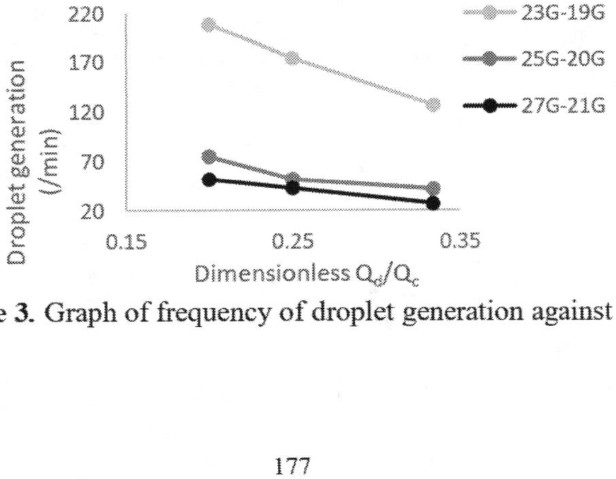

Figure 3. Graph of frequency of droplet generation against Q_d/Q_c.

3.2. Chitosan Solidification

The solidified microbeads were characterized with an optical microscope (Nikon, H600L). The mean ± standard deviation was calculated to be 552 ± 28 microns. The size range distribution of bead diameters is 499 to 613 microns with a polydispersity index of 5.12%. In the solidification step, uneven shrinking of chitosan beads between 11 to 13% were reported due to crosslinking of microbeads [13]. The differences of the mean diameter as compared to the literatures was due to the type of disperse and continuous phase used and their respective flowrates. From figure 4 (a), majority of beads produced are either droplet like shaped or spherical with a small tail whereas some are oval shaped. This is caused by the suspension of beads between oil and NaOH layer. The interfacial tension as mentioned and low-density difference between chitosan droplet and NaOH solution. Thus, cross-linking took place at the bottom of the drops to form a perfect spherical bottom whereas the top surfaces of the beads were immersed in oil layer are yet to exposed to NaOH solution as illustrated in figure 4 (b). Solidification of the top surface only initiated once the bead penetrated the oil – NaOH interface. For droplet shaped spherical beads, the tail was ignored while taking 2 diameter readings (d_1 and d_2). Both diameter readings are taken perpendicularly and pass through the centre of the circle. An average diameter (d_{avg}) between d_1 and d_2 was taken for a single bead. For oval or ellipsoidal beads, Feret diameter approached was used by measuring the length of both major and minor axis of the bead, average values were taken as the d_{avg} reading as seen in figure 4 (c).

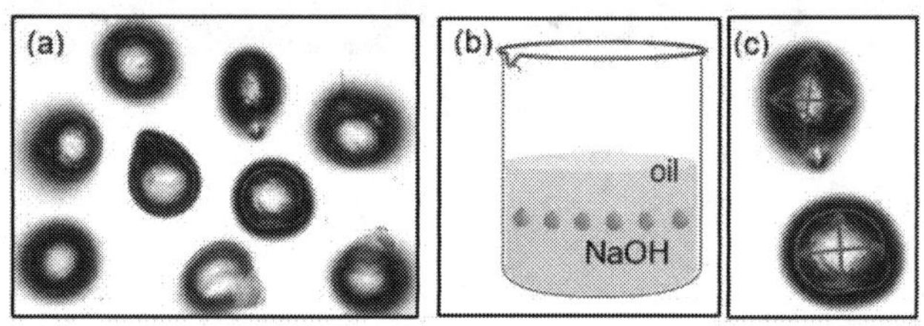

Figure 4. (a) Microscope image of the microbeads; (b) Illustration of droplet shaped microbeads stuck at interface; (c) Method of reading d_1 and d_2.

3.3. Saponin Immobilization

Chitosan microbeads acts as a bio-membrane for the adsorption of saponin. The blank chitosan beads are soaked into 10x PBS solution for 24 hours. Surface modification causes the chitosan microbeads to have positively charged surface. This allows electrostatic interaction to occur between PBS and microbeads surface. Adsorption of anionic saponin molecules onto the cationic microbead surface is then possible as illustrated in figure 5 (a).

The blank beads were modified with PBS and saponin solution. The saponin coated chitosan beads were analyzed with FTIR (PerkinElmer) spectroscopy with pure saponin powder as a control. From figure 5 (b), saponin coated chitosan beads has similar FTIR trend as the pure saponin powder (control sample). The peak observed close to 1000 cm^{-1} indicated the presence of -C-O-C- bond from saponin molecules [14].

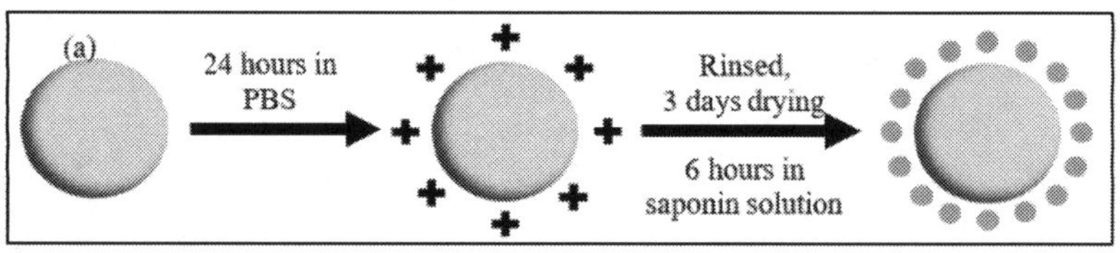

(b)

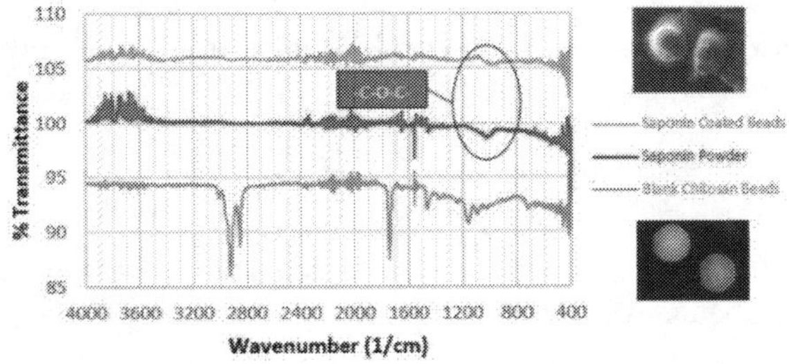

Figure 5. (a) Immobilization of saponin onto chitosan microbeads mechanism; (b) FTIR tests on blank chitosan beads (without saponin adsorption), saponin powder and saponin coated beads with FESEM images of blank chitosan beads and saponin coated beads.

3.4. FESEM and EDX Analysis

The surface morphology of the blank and modified microbeads was observed under FESEM (Quanta 400F, FEI) and analyzed with EDX (Quanta 400F, FEI) to determine the sphericity and elements present on the surface of the beads as shown in Table 1. For blank chitosan beads, the amount of C atoms and O atoms are approximately 85% and 15%. This is because chitosan molecules are mainly made up of those 2 atoms as seen in literature [15]. Oxygen atom present in both modified beads was comparatively more than those in the blank microbeads. This can be explained with the molecular structure of saponin itself as seen in literature [16]. The O element present in a single saponin molecule are more than that in chitosan molecule. This further validate the successful surface adsorption done on chitosan microbeads.

Table 1. EDX analysis of microbeads.

FESEM image	Beads type	No.	Normalised atomic %		Remark
			C	O	
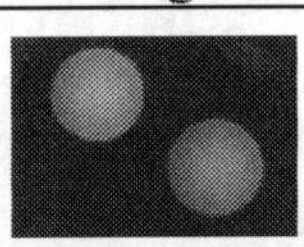	Blank Microbeads	1	86.95	13.05	-
		2	86.84	13.16	-
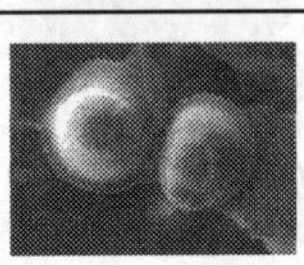	Saponin Modified Microbeads	1	65.69	34.31	With traces of Cl element
		2	74.22	25.79	With traces of Si element

4. Conclusion

Microfluidics technology has been well established but the report of using dispensing needles as microfluidic devices is very limited. This work has proven the reliability of this device to generated highly monodisperse droplet. Chitosan droplet has been generated by using a 23G-19G microfluidic device, cross-linked with NaOH solution and immobilized with saponin extracted from Momordica Charantia species. Inevitable saponification during the solidification process has hindered the crosslinking of microbeads, which was resolved by gently swirling of the NaOH solution. Microbeads produced are in an acceptable size range with polydispersity index of 5.12% despite the uneven shrinking from chitosan cross-linkage. Successful immobilization has been proven through FTIR and FESEM analysis. The ongoing research in process optimization can be done on several aspects to produce drugs with the best effect. Of the current studied parameters of the drug production process, the type of buffer electrolyte used, its concentration and soaking time can be varied to allow uniform and maximum saponin adsorption. Other parameters such as saponin concentration, its adsorption time and chitosan carrier's concentration can also be manipulated for better saponin adsorption. In addition, there is much analytical work to be carried out such as the effectiveness of saponin from the synthesis route to ensure the function of saponin is not compromised from this synthesis route.

References

[1] Akhtar S, Nasir J A, Ali A, Asghar M, Majeed R and Sarwar A 2022 Prevalence of type-2 diabetes and prediabetes in Malaysia: a systematic review and meta-analysis. *PLoS ONE*, **17(1)**, Article e0263139

[2] Bernama. *Diagnosing diabetes*. New Straits Times [Internet]. 2016 [cited 2016 May 9] Available from: https://www.nst.com.my/news/2016/05/144687/diagnosing-diabetes.

[3] Gupta P, Bala M, Gupta S, Dua A, Dabur R, Injeti E and Mittal A 2016 Efficacy and risk profile of anti-diabetic therapies: Conventional vs traditional drugs—A mechanistic revisit to understand their mode of action. *Pharmacological Research*, **113**, 636-74

[4] Home P 2003 The challenge of poorly controlled diabetes mellitus. *Diabetes Metabolism*, **29**, 101-9

[5] Zhang F, Lin L and Xie J 2016 A mini-review of chemical and biological properties of polyssaccharides from Momordica charantia. *International Journal of Biological Macromolecules*, **92**, 246-53

[6] Fu X, Wen M, Han X, Yanagita T, Xue Y, Wang J, Xue C and Wang Y 2016 Effect and potential mechanism of action of sea cucumber saponins on postprandial blood glucose in mice. *Bioscience, Biotechnology, and Biochemistry*, **80(6)**, 1081-7

[7] Rajalakshmi K, Christian G, Priya P S and Gladys R J 2015 Validation of anti-diabetic potential of avirai kudineer a siddha herbal formulation-a review. *IOSR Journal of Dental and Medical Sciences (IOSR-JDMS)*, **14(7)**,7-15

[8] Alam M A, Uddin R, Subhan N, Rahman M M, Jain P and Reze H M 2015 Beneficial role of bitter melon supplementation in obesity and related complications in metabolic syndrome. *Journal of Lipids*, **2015**, Article 496169

[9] Liu Z, Zhang T, Tang C and Yin C 2016 Amphiphilic nanoparticles based on poly(vinyl pyrrolidone) and stearoyl modified chitosan as drug vehicles for paclitaxel delivery. *Materials Letters*, **185**, 226-9

[10] Li W, Peng H, Ning F, Yao L, Luo M, Zhao Q, Zhu X and Xiong H 2014 Amphiphilic chitosan derivative-based core–shell micelles: Synthesis, characterization, and properties for sustained release of vitamin D3. *Food Chemistry*, **152**, 307-15

[11] Chou W L, Lee P Y, Yang C L, Huang W Y and Y.-S. Lin 2015 Recent advances in applications of droplet microfluidics. *Micromachines*, **6**, 1249-71

[12] Hong Y and Wang F 2007 Flow rate effect on droplet control in a co-flowing microfluidic device. *Microfluid Nanofluid,* **3**, 341-6

[13] Yang C H, Wang L S, Chen S Y, Huang M C, Li Y H, Lin Y C, Chen P F, Shaw J F and Huang K S 2016 Microfluidic assisted synthesis of silver nanoparticle-chitosan composite microparticles for antibacterial application. *International Journal of Pharmaceutics,* **510**, 493-500.

[14] Munichandran M, Gangadhar B and Naidu G 2016 Bioremoval of nickel and lead using bitter gourd (momordica charantia) seeds. *International Journal of Advance Research in Science, Engineering and Technology,* **3(8)**, 2475-84.

[15] Akinyeye O J, Ibigbami T B and Odeja O 2016 Effect of Chitosan powder prepared from snail shells to remove lead (II) ion and nickel (II) ion from aqueous solution and its adsorption isotherm model. *American Journal of Applied Chemistry*, **4(4)**, 146-56.

IOP Conf. Series: Materials Science and Engineering 1257 (2022) 012030 doi:10.1088/1757-899X/1257/1/012030

[16] Chukwuebuka E and Chinenye I J 2015 Biological functions and anti-nutritional effects of phytochemicals in living system. *IOSR Journal of Pharmacy and Biological Sciences*, **10(2)**, 10-9.

The Effect of Printing Temperature on Bonding Quality and Tensile Properties of Fused Deposition Modelling 3d-Printed Parts

R Pang, M K Lai*, K I Ismail, T C Yap

School of Engineering and Physical Sciences, Heriot-Watt University Malaysia, Jalan Venna P5/2, Precinct 5, 62200, Putrajaya, Malaysia.

*e-mail: m.lai@hw.ac.uk

Abstract. FDM 3D-printing, is an additive manufacturing technology that is being increasingly adapted in the engineering industry due to its ability to produce complex design at lower costs as the materials are thermoplastic based. Due to the nature of its sequential layer deposition, the bonding quality is highly dependent on the temperature development during printing. In this paper, an experimental investigation was conducted to study the effect of printing temperature on bonding quality with regards to dimensional accuracy and tensile behaviour of the fabricated parts. The test specimens were fabricated using PLA material at different printing temperature ranging from 180°C-240°C at intervals of 10°C. Uniaxial tensile test according to ASTM D638-14 standard was conducted at a strain rate of 1 mm/min and was repeated five times for each variable. Results show that specimens fabricated at higher printing temperature have better tensile properties. The ultimate tensile strength recorded for specimens fabricated at T=240°C and T=180°C were 36.97 MPa and 17.47 MPa respectively. Fracture surfaces were analysed and related with the tensile results to deduce the failure mechanism. The dimensions of the sample at printed different printing temperatures were measured using digital callipers and compared with the designed specification. Specimens printed at T=240°C had the worst dimensional accuracy, with a percentage difference of 10% although it yielded the highest tensile strength. The optimum printing temperature that yielded the best tensile behaviour while maintaining its dimensional accuracy was when T=220°C.

Keywords: Additive manufacturing, fused filament fabrication, PLA, tensile strength, printing temperature, dimensional accuracy, FDM.

1. Introduction

In its pioneering stage, the development of additive manufacturing technologies was driven by the need to produce proof-of-concept prototypes and complex shaped parts which are difficult to be produced using traditional manufacturing methods[1]. It has since developed into an alternative manufacturing process and is especially useful for producing parts with complex geometries and small to medium batch production [2]. This technology is increasingly adapted in the engineering industry due to its ability to produce complex designs at lower costs and lower material wastage. Therefore, it is especially cost effective for low-batch productions and is commonly used to produce conceptual prototypes. Fused deposition modelling (FDM) has proven itself as a manufacturing method to produce complex parts, particularly using thermoplastic materials. However, there are limitations to this technology as the core process relies on the heating and cooling of materials through a sequential layer-by-layer deposition process. A major drawback of this technology is the poor mechanical properties resulting in the lack of functionality of the printed parts [3]. The poor mechanical properties are due to the weak bonding between layers which are heat dependent and the presence of void spaces formed between extruded filaments due to the sequential layer process. The mechanical properties of FDM fabricated parts depends on the combination of various process parameters such as nozzle temperature, build platform temperature, printing speed, layer thickness, build orientation, raster angle, road width and gap etc [4]. The quality and strength of the printed parts can be further enhanced by optimizing these parameters. Previous studies investigated the possibility of increasing the printing temperature to reheat the previous layer and concluded that inter-filament bonding was stronger when the printing temperature was higher. However, an excessively high printing temperature will result in a reduced dimensional accuracy of the printed parts [4–7]. Therefore, the research will focus on determining the range of working temperature which will enable a good balance between mechanical properties and dimensional accuracy. In the present work, the effect of printing temperature on the tensile properties and dimensional accuracy has been investigated. The information from the internal geometry and the fracture surface was used to analyze the relationship between printing temperature and tensile properties as a result of bonding quality between filaments.

2. Materials and Method

2.1 Printer & Filament Material

The 3D-printer utilized for this study is the Ultimaker S3. The study mainly investigated the role of printing temperature with respect to bonding quality. In order to reduce the complexity of the investigation, other processing parameters were kept the same throughout the study. The material used for the investigation was polylactic acid (PLA) by PolyMaker which is a thermoplastic material. The filament and nozzle diameter used was 2.85 mm and 0.4 mm respectively. The material was stored in a humidity-controlled box to keep them dry prior to the printing process.

2.2 Printing Procedure and Parameters

The 3D-model of the test specimen was created using Autodesk Inventor according to ASTM D638 Type 1 specification for tensile testing. The file was then sliced using Ultimaker Cura software and printed using the 3D-printer. Only the printing temperature was varied to investigate its role on the bonding quality, while other processing parameters were kept the same. The fixed parameters are printing speed, layer height, build orientation, infill density & pattern, raster angle and platform temperature as shown in table 1. The test specimens were printed each printing temperature ranging from 180 °C to 240 °C at intervals of 10°C.

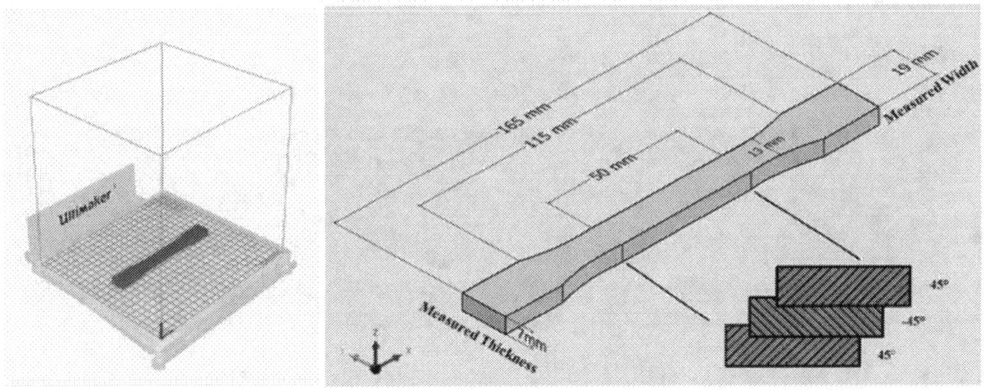

Figure 1. Print orientation & dimensions of specimen according to ASTM D638-14 Type 1 Standard.

Table 1. Process parameters & settings used for investigation.

Process Parameter	Setting
Nozzle Printing Temperature	180 °C to 240 °C (at intervals of 10°C)
Print Speed	50 mm/s
Layer Height	0.2 mm
Build Orientation	Flat
Infill Density	100 %
Infill Pattern	Lines
Platform Temperature	50 °C
Raster Angle	45 °/ - 45 °

2.3 Experimental Testing

The methods used to assess the FDM fabricated specimens in this investigation are: (i) dimensional accuracy, (ii) tensile properties and behaviour. The width and thickness of the fabricated specimens as shown in figure 1 were measured using digital callipers with an accuracy of ±0.01 mm and compared with the designed specification. Tensile test was conducted using a Quasar 25 Universal testing machine using a strain rate of 1 mm/min and repeated 5 times for each printing temperature. The corresponding results containing the stress and strain values were exported and analysed. Lastly, the fracture surface of the specimens was analysed to determine the fracture behaviour, and the bonding between filament was observed under scanning electron microscope LEO 1455VP.

3. Results and Discussion

3.1 Part Quality

The dimensional accuracy of the specimens was examined by measuring their deviation from the designed specification. A poor part quality on the side walls was observed on specimens fabricated at T=180°C & 190°C shown in figure 2. The outer wall showed regions of irregularities which is due to the inconsistent deposition of material and the quality of the parts improved with increasing printing temperature. The same phenomenon can also be observed in the cross-section of the specimens. The measured width and thickness of the specimens was closest to the designed specification when the specimens were printed at printing temperature of T=210°C, 220°C & 230°C, where the dotted line in figure 3 represents the designed width and thickness. All the specimens printed had lower measured thickness when compared with the specified dimensions except for T=180°C. Specimens printed at T=180°C and 240°C exhibited higher deviation in values from the designed dimensions. At low printing temperatures, polymer melts exhibit low fluidity and high viscosity which leads to under extrusion of the polymer melt [5]. This causes the formation of void spaces which leads to poor surface finish. At high temperatures, the fluidity of the material increases and expands freely in the XY direction decreasing the measurement on the z-axis [8].

Figure 2. Side wall of specimens at different printing temperatures. (a) 180°C, (b) 190°C, (c) 220°C.

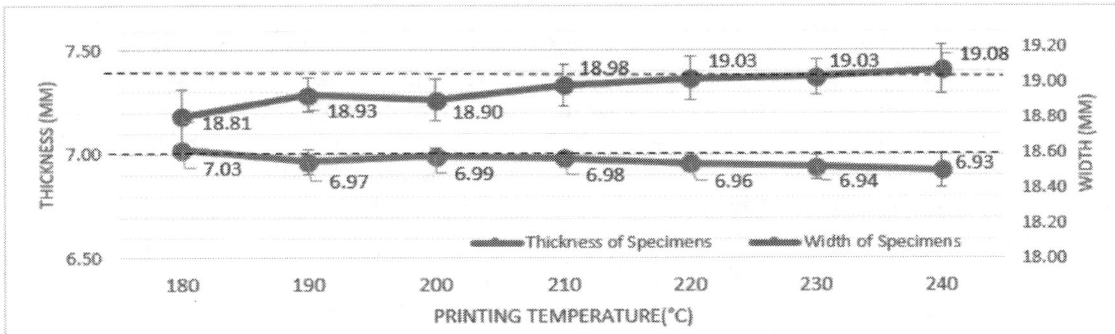

Figure 3. Average measured width and thickness of specimen.

3.2 Tensile Properties

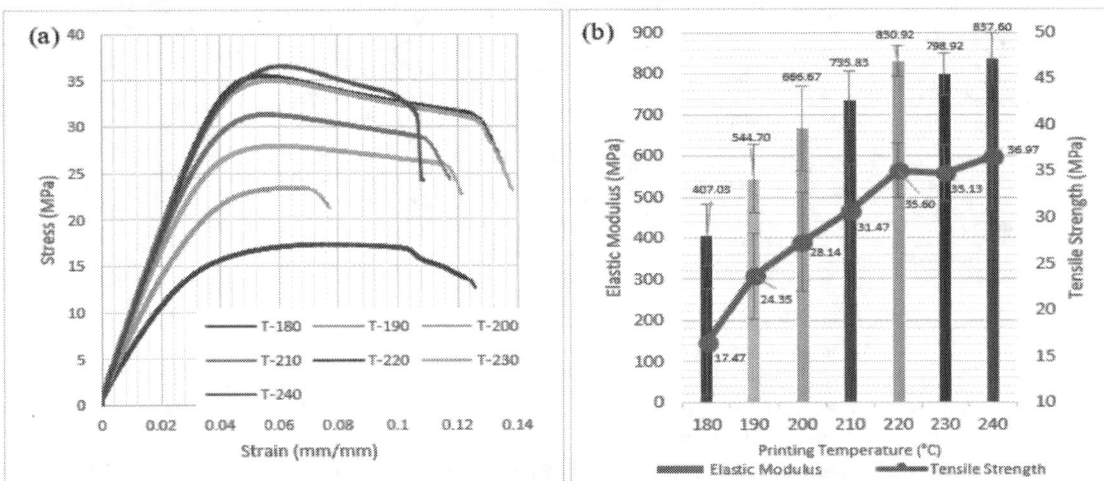

Figure 4. (a) Stress-strain curve (b) Elastic modulus and Ultimate Tensile Strength at different temperatures.

Figure 4 shows the tensile properties (elastic modulus and tensile strength) of PLA at different printing temperatures. The tensile strength, elastic modulus and strain at break increases with printing temperature shown in figure 4 (a) and the stress-strain curves shows that PLA exhibit an elastic-plastic response. Specimens printed at T=190°C had the lowest strain at break. It can be observed that the tensile property of PLA increases up till T=220°C and further increase in printing temperature does not improve the tensile properties significantly. Although specimens printed at T=240°C showed the highest tensile strength, the strain at break was slightly lower. Based on the results obtained, the specimens exhibit poor bonding between adjacent and preceding deposited filament at low printing temperatures [6]. As printing temperature increases, the viscosity of the polymer decreases which allows better fusion between the deposited material. The inhibiting factor is the degradation temperature where a reduction in molecular weight occurs. Above the degradation temperature, the material are less ductile [9].

3.3 Fracture Analysis

Figure 5 shows the fractured PLA specimens at different printing temperatures. Figure 5 (a) and (b) shows that failure occurs in between the layers (inter-layer) first, in the form of delamination as the bonding between layers are weaker at lower printing temperatures. As the printing temperature increases, the bonding between filament is more homogenous leading to smoother fracture surface as shown in figure 5 (c) and (d), therefore the failure occurs within the layers (in-layer) [10]. When a new layer is deposited at temperatures greater or equal to the polymer crystallization temperature, the inter-filament bonding strength is higher [7]. The filament spreads into an oblong shape, forming small voids in the part. Spreading rate is dependent on the viscosity of the melt which is affected by the extrusion temperature [11]. As FDM 3D-printing is a layer-by-layer process, the main bonding between the filament material can be categorized into in-layer and inter-layer bonding. In-layer bonding is defined as the bonding between adjacent filaments within the same layer while inter-layer bonding is defined as the bonding between two successive layers [12]. When the melt is extruded form the nozzle, it is subjected to convective cooling with the environment and conductive cooling with the previous layer [13]. Heat from the deposited material is transmitted to the previous layer to enable bonding through sintering [14]. At low printing temperatures, the previously deposited layer cools down faster before a new layer is deposited which reduces adhesion between the filaments [15].

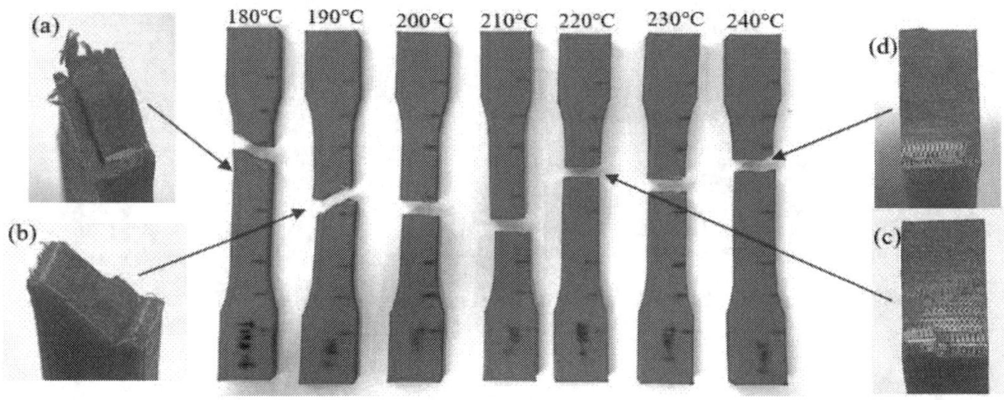

Figure 5. Fractured PLA specimens at different printing temperature [middle] and & Fracture surface of samples at: (a) 180°C, (b) 190°C, (c) 220°C, (d) 240°C.

Figure 6 shows the cross-section fracture surfaces at a magnification of 50x. It can be observed from that the contact area between filaments at T=180°C is significantly lesser compared to when T=220°C and 240°C. Circular filaments can be seen clearly at printing temperature T=180°C while the filaments were fused more closely together at T=220°C and 240°C. It can be seen that the fracture surface becomes smoother with increasing printing temperature. This is because the force subjected on the specimens during tensile testing are distributed more equally as bonding between layers are stronger at higher printing temperatures resulting in an in-layer failure. The mechanical properties of the part depend on the bonding between neighbouring filaments. This is influenced by road spreading, where a higher contact area or lesser voids result in stronger bonds [16].

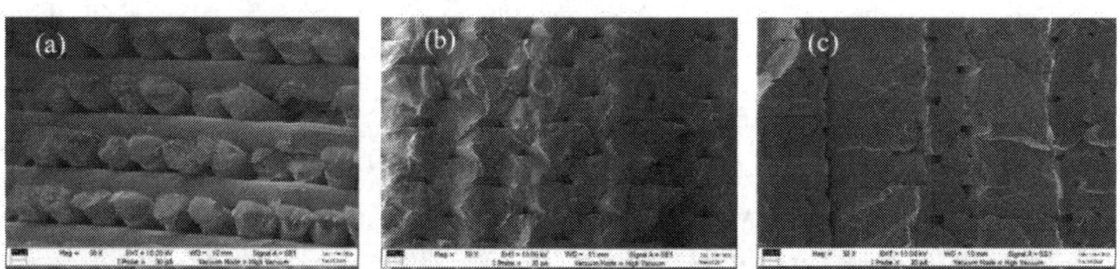

Figure 6. FE-SEM images of fracture surface at (a) 180°C, (b) 220°C and (c) 240°C.

4. Conclusion

In the current work, the contribution of printing temperature towards the bonding between filaments has been identified. The bonding quality at different printing temperature can be attributed to the presence of void spaces. At low printing temperatures, inter-layer failure is observed as the bonding between layers are weak. Tensile properties of the specimens increase with printing temperature as the fusion between filaments are stronger with less void spaces present. Although specimens printed at T=240°C showed the highest tensile strength, the resulting dimensional accuracy of the printed parts were the worst. The optimum printing temperature that yielded the best tensile behavior while maintaining its dimensional accuracy was when T=220°C. This research is a preliminary study to understand and improve mechanical properties of FDM fabricated parts by optimizing temperature related parameters. The methodology is being improved to include quantitative analyses to determine the percentage of porosity as well as the bond length between roads at different printing temperatures. The relationship between crystallinity of material and mechanical properties at different printing temperatures will also be studied.

5. Acknowledgement

This work was supported by the Ministry of Higher Education Malaysia under the Fundamental Research Grant Scheme (FRGS/1/2019/TK05/HWUM/02/2).

References

[1] Wong K V and Hernandez A 2012 A Review of Additive Manufacturing *ISRN Mech. Eng.* **4**

[2] Gardan J 2016 Additive manufacturing technologies: State of the art and trends *Int. J. Prod. Res.* **54** 3118–32

[3] Calignano F, Manfredi D, Ambrosio E P, Biamino S, Lombardi M, Atzeni E, Salmi A, Minetola P, Iuliano L and Fino P 2017 Overview on additive manufacturing technologies *Proc. IEEE* **105** 593–612

[4] Gao X, Qi S, Kuang X, Su Y, Li J and Wang D 2021 Fused filament fabrication of polymer materials: A review of interlayer bond *Addit. Manuf.* **37** 101658

[5] Hsueh M H, Lai C J, Wang S H, Zeng Y S, Hsieh C H, Pan C Y and Huang W C 2021 Effect of printing parameters on the thermal and mechanical properties of 3d-printed pla and petg, using fused deposition modeling *Polymers (Basel).* **13**

[6] Zekavat A R, Jansson A, Larsson J and Pejryd L 2018 Investigating the effect of fabrication temperature on mechanical properties of fused deposition modeling parts using X-ray computed tomography *Int. J. Adv. Manuf. Technol. 2018* **100** 287–96

[7] Vanaei H R, Raissi K, Deligant M, Shirinbayan M, Fitoussi J, Khelladi S and Tcharkhtchi A 2020 Toward the understanding of temperature effect on bonding strength, dimensions and geometry of 3D-printed parts *J. Mater. Sci.* **55** 14677–89

[8] Valerga A P, Batista M, Salguero J and Girot F 2018 Influence of PLA filament conditions on characteristics of FDM parts *Materials (Basel).* **11**

[9] Pantani R, De Santis F, Sorrentino A, De Maio F and Titomanlio G 2010 Crystallization kinetics of virgin and processed poly(lactic acid) *Polym. Degrad. Stab.*

[10] Yao T, Ye J, Deng Z, Zhang K, Ma Y and Ouyang H 2020 Tensile failure strength and separation angle of FDM 3D printing PLA material: Experimental and theoretical analyses *Compos. Part B Eng.* **188** 107894

[11] Crockett R S and Calvert P 1997 The liquid-to-solid transition in stereodeposition techniques *Dep. Mater. Sci. Eng.* 221

[12] Gurrala P K and Regalla S P 2014 Part strength evolution with bonding between filaments in fused deposition modelling: This paper studies how coalescence of filaments contributes to the strength of final FDM part *Virtual Phys. Prototyp.* **9** 141–9

[13] Lepoivre A, Boyard N, Levy A and Sobotka V 2020 Heat transfer and adhesion study for the FFF additive manufacturing process *Procedia Manuf.* **47** 948–55

[14] Rosenzweig N and Narkis M 1981 Sintering rheology of amorphous polymers *Polym. Eng. Sci.* **21** 1167–70

[15] Morales N G, Fleck T J and Rhoads J F 2018 The effect of interlayer cooling on the mechanical properties of components printed via fused deposition *Addit. Manuf.* **24** 243–8

[16] Turner B N and Gold S A 2015 A review of melt extrusion additive manufacturing processes: II. Materials, dimensional accuracy, and surface roughness *Rapid Prototyp. J.* **21** 250–61

IOP Conf. Series: Materials Science and Engineering 1257 (2022) 012032 doi:10.1088/1757-899X/1257/1/012032

Surface modification of PVDF membrane via graft polymerization of acetic and acrylic acid

***S M Saleh[1], P C Oh[2], A S Zulkifli[1]**

[1] PETRONAS Research Sdn Bhd, Lot 3288 & 3289, Off Jalan Ayer Itam, Kawasan Institusi Bangi, 43000 Kajang, Selangor, Malaysia

[2] CO_2 Research Centre (CO2RES), Chemical Engineering Department, Universiti Teknologi PETRONAS, 32610 Tronoh, Perak, Malaysia

*Email: syafiqa_saleh@petronas.com.my

Abstract. Surface modification of polyvinylidene fluoride (PVDF) hollow fibre membranes through simple chemical treatment is carried out to graft the surface with the presence of carboxylic acid groups. This is to provide means for TiO_2 immobilisation in subsequent stages for photocatalysis application. The focus of this study was to identify the highest degree of grafting in order to provide abundant potential sites for subsequent TiO_2 assembly. Both acetic and acrylic acid, at varying concentrations, were used and compared as the grafting monomer. It was observed that acrylic acid provides significantly higher grafting degree compared to acetic acid. The highest degree 238 μg/cm^2 was obtained at 70% acrylic acid in water, where the surface contact angle significantly reduced to 37°. The highest grafting degree obtained using acetic acid was at 60% in water, which only achieved 31 μg/cm^2. Minor improvement was also observed when the solvent for grafting solution was changed from water to toluene; the grafting degree slightly improved by 2% from 238 μg/cm^2 to 243 μg/cm^2. Higher grafting degree shall allow for more of TiO_2 to be immobilized onto the membrane in subsequent work, hence potentially leading to better photocatalysis performance.

1. Introduction

Polyvinylidene fluoride (PVDF) is a thermo-stable and durable membrane, with resistance to ultraviolet rays. Thus, it has the potential to be used in photocatalytic membrane reactor [1,2]. However, its polymer backbone is not conducive for TiO_2 photocatalyst self-assembly and require surface modification in order to provide the necessary binding sites [3]. Surface modification allows the membrane surface properties to be improved, whilst keeping the membrane bulk not significantly affected [4].

Grafting polymerization can be described as a covalent attachment process where the technique can alter the surface of polymeric materials by introducing new functional groups on the backbone of the carbon chain [5]. Surface of polymers with –COOH, –SO_2OH, and ether bond provides self-assembly behaviors for TiO_2 [6]. Chemical reagent-induced graft polymerization can introduce functional groups such as –C–O and –C=O only on the surface [7]. The grafting conditions that yield the highest grafting efficiency has to be selected to maximize the number of active binding sites [3].

Prior to grafting, PVDF first needs to go through dehydrofluoronation, where the polymer is degraded by loss of hydrogen fluoride (HF) that leads to several chemical reactions, including the formation of carbon–carbon double bonds (–C=C–) or cross-linking of the polymer [8,9]. A variety of methods, including the wet-chemical method, plasma treatment, and irradiation, are found in literature for the modification of PVDF membrane [10].

Many studies on grafting have been conducted using plasma [3,11,12] and a variety of monomers have been used for grafting, such as styrene [13], polyethylene glycol diacrylate [14], acetic acid [15] and acrylic acid [3]. This study however will use a simple chemical treatment for PVDF grafting process as it is more economical; and shall compare the use of different concentrations of acetic and acrylic acid as grafting monomer to achieve high grafting degree. Acrylic and acetic acid solution have hydrophilic properties, can initiate in-situ polymerization that is strongly bonded onto membrane and provides the necessary functional groups for TiO_2 immobilization in subsequent work [5][15].

2. Experimental Methodology

2.1. Materials

The PVDF hollow fibre membranes were purchased off shelves from Hygienix Appliances, with properties of the membrane as shown in Table 1. Strong alkali potassium hydroxide (KOH) was purchased from ACROS Organics and used as received. Acetic acid (95% A.R) and acrylic acid (99% A.R.) for the grafting process were purchased from R&M Chemicals and used as received.

Table 1. Properties of PVDF membrane hollow fibre membrane

Inner diameter (mm)	0.57
Outer diameter (mm)	0.93
Porosity (%)	73%

2.2. Membrane surface dehydrofluorination

56 g of potassium hydroxide (KOH) was added into 944 mL of distilled water to produce 1M KOH solution. PVDF membranes were immersed into beakers containing the solution and submerged in water bath at 60°C for 2 hours [16]. After the alkali treatment, the hollow fibers were rinsed with distilled water three times and dried at 60°C in an oven for 2 hours.

2.3. Membrane surface graft polymerization

Acetic acid solutions and acrylic acid solutions with various concentration in distilled water (from 10 to 70 wt.%) were prepared. On top of the 14 samples tests, one additional run was done by changing the solvent from distilled water to toluene to observe for significant difference [17]. The pre-treated PVDF hollow fibers were immersed in the solution at 60°C for 2 hours. Lei and Liao [18] reported that grafting degree levels off at 2 hours; whilst You et al. [3] reported 60°C is most suitable for PVDF graft polymerisation. The immersed membranes were then rinsed with distilled water three times and dried at 60°C in an oven for 2 hours.

2.4. Characterisation

Characterisation is first foremost is to evaluate the grafting degree. Modified membrane with highest grafting degree shall be further characterised through its morphology, functional group analysis and contact angle measurement to observe the effect of the graft polymerization on the membrane properties.

2.4.1. Determination of grafting degree. The grafted PVDF membranes (4 strands with total area of 13cm^2) were soaked in 20ml of 0.01 M NaOH for 2 h. Then 5 ml of the supernatant were back titrated with 0.01 M HCl [12] and replicated 3 times. Grafting degree was calculated based on Equation (1) [18]:

$$Grafting\ degree\ (\mu g/cm^2) = \frac{[(V_{NaOH}C_{NaOH} - V_{HCl}C_{HCl}) \times 10^3 M]}{S} \quad (1)$$

where S is the area of PVDF membrane in cm^2, M is the molecular weight of graft chemical (i.e., 60g/mol for acetic acid or 72g/mol for acrylic acid), V_{NaOH} and V_{HCl} are the volume of NaOH (i.e., 5 ml) and HCl solutions, respectively, C_{NaOH} and C_{HCl} are concentrations of NaOH and HCl solution, respectively.

2.4.2. Contact angle measurement. Tensiometer (Theta Lite Optical Tensiometer TL100) was used to perform contact angle measurements. Sessile drop method was done with distilled water on 10 cm of grafted membrane samples, using Young-Laplace setting to interpret the image. Minimum of seven readings were taken and averaged, similar to other researchers in membrane study [19 - 21].

2.4.3. Porosity. The overall porosity (ε) of the membranes was determined by the gravimetric method [22][23] where the membranes were immersed in distilled water overnight to ensure liquid penetration into membrane pores. Porosity was calculated based on initial dry weight and the wet weight of the membrane using Equation (2) [22]:

$$\varepsilon(\%) = \frac{(m_w - m_d)/\rho_{H_2O}}{(m_w - m_d)/\rho_{H_2O} + m_d/\rho_P} \times 100 \quad (2)$$

where m_w and m_d are the weights of the wet and dry membranes, respectively, and ρ_{H2O} and ρ_p are the water (1.0 g/cm^3) and PVDF polymer (1.78 g/cm^3) densities, respectively.

2.4.4. Attenuated Total Reflection Fourier Transform Infrared (ATR-FTIR) Spectroscopy. ATR-FTIR (Perkin Elmer, Spectrum 100 FTIR) was used to determine presence of functional groups in the sample. Samples were pressed against ATR crystal in transmittance mode 450-4000 cm^{-1} at ambient temperature.

2.4.5. Field Emission Scanning Electron Microscope (FESEM). Field Emission Scanning Electron Microscope (FESEM) (Hitachi, Model SU7000) was used to observe the morphology before and after the surface modification. The membrane samples were fractured using liquid nitrogen to obtain the cross-sectional image.

3. Results and Discussion

3.1. Graft polymerisation

Figure 1 shows the grafting degree of PVDF membrane with acetic or acrylic acid solution from 10% to 70% concentration. It was observed that acetic acid provides significantly lower grafting degree compared to acrylic acid. This is likely due to acetic acid possessing only two carbons that could covalently attach to PVDF, whereas acrylic acid has three carbons, which resulted in higher grafting yield from this monomer, as observed by Ismail et al. [5].

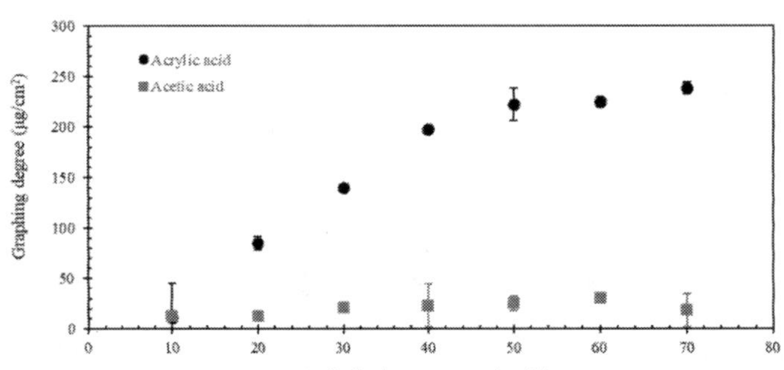

Figure 1. Grafting degree of PVDF with acetic and acrylic acid at varied concentrations, i.e., 10 – 70%

For acrylic acid, the highest degree was obtained at 70% concentration (i.e., 238 µg/cm^2); whilst for acetic acid at 60% concentration (i.e., 31 µg/cm^2). It was postulated that homopolymerization process competes with graft polymerization, where under certain grafting conditions, homopolymerisation is preferred, leading to large amounts of ungrafted carboxylic acid [3,17]. At low acetic acid and acrylic acid concentration, there is higher chance of surface graft polymerization to occur than homopolymerization. However, at high concentration, homopolymerization becomes more dominant [12]. Thus, past certain concentration, grafting degree decreases, as observed for acetic acid after 60%.

3.2. Effect of surface grafting to membrane contact angle

The contact angle of the modified PVDF was observed to significantly reduced with acrylic acid concentration. The initial contact angle after dehydrofluorination step was 56.2°; which had reduced to below 40° for all acrylic acid grafted PVDF membrane. The modified membrane was observed to become more hydrophilic when treated with acrylic acid at higher concentration, i.e., 36.7° at 70% acrylic acid immersion, as shown in Figure 2. This trend observed is similar to observation by other researchers since homopolymers of acrylic acid are known to be hydrophilic [18,24].

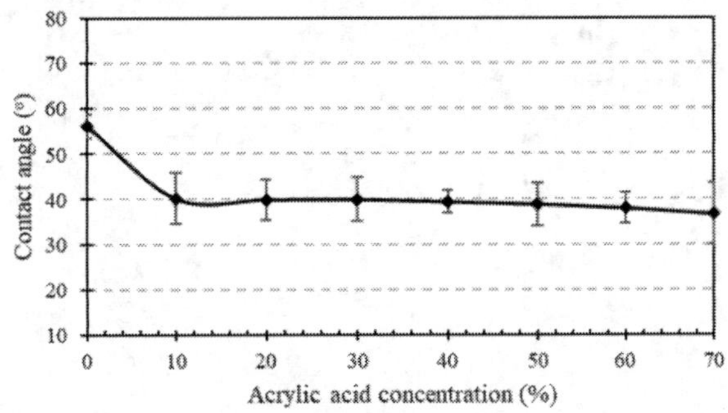

Figure 2. Contact angle of grafted PVDF at varied acrylic acid concentration

3.3. Compositional and functional group analysis

Figure 3 shows the FTIR spectra of the neat PVDF sample, and for the modified membrane after grafting with 70% acrylic acid. Significant difference was observed between the two spectra. Peak at 1106cm^{-1} indicates C – F in the neat PVDF membrane; this peak was not observed in the modified membrane, wherein due to the alkali treatment and dehydrofluoronation step, loss of hydrogen fluoride had occurred. In addition, peak at 2885 cm^{-1} have also disappeared from the neat to modified membrane denoting reduction in $C - H_2$ bonds.

New peaks were observed in the spectra of modified membrane at 1725 cm^{-1} which indicates the presence of C = O bond, and at 1181 cm^{-1} which indicates the presence of C – O bond from grafting with acrylic acid. Intensity of the peak at 1546 cm^{-1} has increased when compared the neat against modified membrane, indicating formation of more C=C bonds on the membrane.

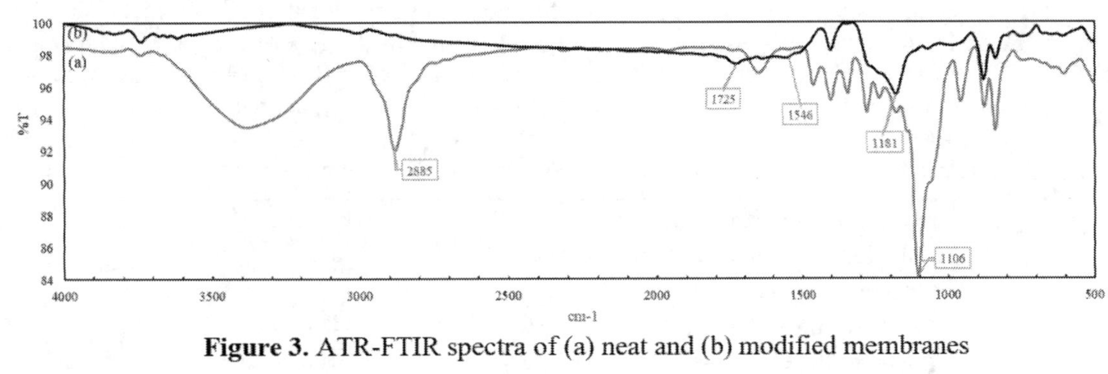

Figure 3. ATR-FTIR spectra of (a) neat and (b) modified membranes

3.4. Morphology and porosity of grafted membrane

Based on measurement of the grafted membrane (with 70% acrylic acid), the cross-sectional outer diameter of the hollow fibre membrane was observed to reduce from the initial 0.93mm to 0.92mm; and the inner diameter reduced from 0.57mm to 0.55mm. Besides that, using the gravimetric method, it was found that the membrane porosity has increased from 73% to 79%.

In the FESEM images in Figure 4, the cross section of the grafted membrane was seen to be denser. This corresponded to the smaller diameter of the hollow fibre membrane. Additionally, based on the membrane outer surface, more visible pores were observed for the grafted membrane compared to neat membrane. This corresponded to the increase in porosity measured.

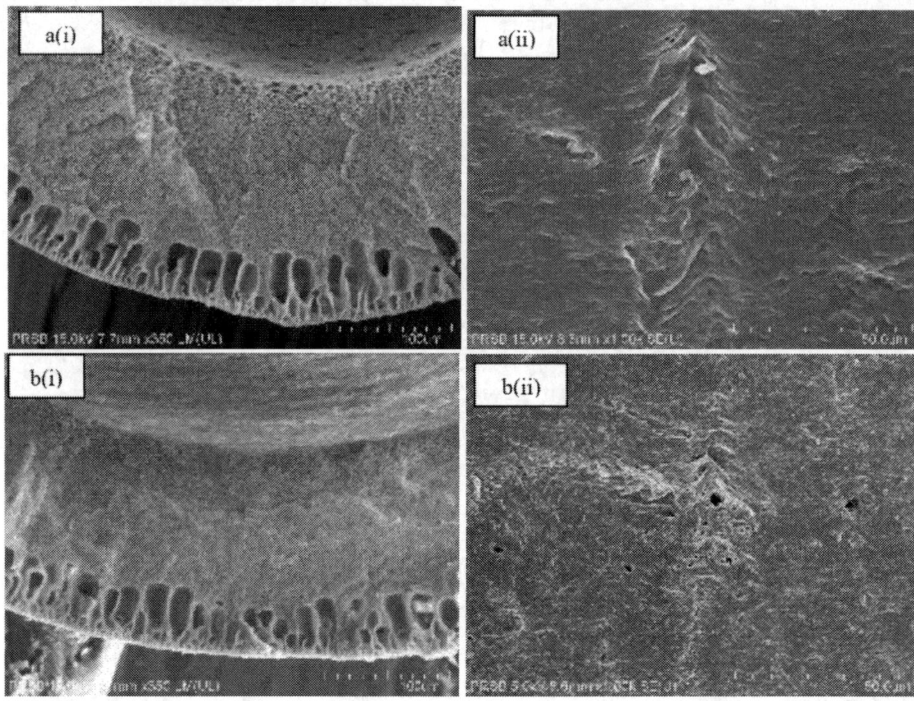

Figure 4. FESEM images of (a) neat and (b) modified membranes (with 70% acrylic acid); (i) denotes the cross sectional and (ii) the outer membrane surface at magnification 350x and 1000x, respectively

3.5. Effect of different solvent for the grafting solution

Changing the medium for grafting solution from water to toluene has slightly improved the grafting degree by 2%, i.e., from 238 μg/cm^2 to 243 μg/cm^2, as shown in Table 2. Toluene was chosen in this study since Xu et al. reported it is the best medium for the graft polymerization of acrylic acid in their study, compared to dichloroethane, and ethanol [17]. Toluene has a lower viscosity compared to water, i.e., 0.55 cP against 0.89 cP, respectively. Therefore, it accelerated the diffusion of monomer from bulk solution to the interface of membrane [17], resulting in the higher grafting degree.

Table 2. Comparison of water and toluene as solvent for membrane grafting with 70% acrylic acid

Solvent for grafting solution	Water	Toluene
Acrylic acid concentration	70%	70%
Grafting degree (μg/cm^2)	238	243
Contact angle	37°	34°
Porosity (%)	79%	80%

In terms of hydrophobicity, the grafted membrane using toluene as solvent has a slightly lower contact angle (i.e., 34° compared to 37°), which relates to the grafting degree of the membrane. Porosity of both grafted membranes were rather similar. Since the improvement from changing solvent base was too small, it is therefore recommended to use water as solvent medium due to economic sense and in view of managing chemical waste.

4. Conclusion

In this study, PVDF was grafted with acrylic and acetic acid at varying concentrations. It was found that the highest grafting degree 238 μg/cm^2 was obtained at 70% for acrylic acid concentration. Membrane grafted with 70% acrylic acid was observed to be more hydrophilic with contact angle of 37° and has increased porosity with more visible pores. Slight improvement by 2% was observed when the grafting solution medium was changed from water to toluene, from grafting degree of 238 μg/cm^2 to 243 μg/cm^2. Higher grafting degree shall allow for more of TiO_2 to be immobilized onto the membrane in subsequent work, hence potentially leading to better photocatalysis performance.

Acknowledgement

This work was jointly produced by PETRONAS Research Sdn Bhd and Universiti Teknologi PETRONAS and funded by Petroliam Research Fund. We would like to express our gratitude for the financial support from Gastech PETRONAS (MRA Grant No. 015MD0-065).

References

[1] Lien M, Fu C, Wei T, Hsieh C, and Jung R 2020 *J. Water Process Eng* **37** Article 101488
[2] Dekkouche S, Morales-Torres S, Ribeiro A R, Faria J L, Fontàs C, Kebiche-Senhadji O, and Silva A M T 2022 *Chem. Eng. J* **427** Article 131476
[3] You S, Uy G, Lu S, R. Damodar R A, and We T 2012 *J. Hazard. Mater.* **237–238** 10–19
[4] Kang G and Y. Cao Y 2014 *J. Memb. Sci.* **463** 145–165
[5] Ismail Z, Hidzir N M, Radzali N A M, and Rahman I A 2017 *Malaysian J. Anal. Sci.* **21** 527–534
[6] Shi Y, Huang J, Zeng G, Cheng W, and Hu J 2019 *J. Memb. Sci.* **584** 364–392
[7] Wang S, Li J, Suo J, and Luo T 2010 *Appl. Surf. Sci.* **256** 2293–8
[8] Marshall J E, Zhenova A, Roberts S, Petchey T, Zhu P, Dancer C E J, Mcelroy C R, Kendrick E, and Goodship V 2021 *Polymers* **13** 1–31
[9] Liu F, Hashim N A, Liu Y, Abed M R M, and Li K 2011 *J. Memb. Sci.* **375** 1–27
[10] Feng S, Zhong Z, Wang Y, Xing W, and Drioli E 2017 *J. Memb. Sci.* **549** 332–349
[11] Hidzir N M, Hill D J T, Taran E, Martin D, and Grøndahl L 2013 *Polymer* **54** 6536–6546
[12] Wang C and Chen J 2007 *Appl. Surf. Sci.* **253** 4599–4606
[13] Chen J and Seko N 2019 *Polymers* **11** 1–17
[14] Ma N, Cao J, Li H, Zhang Y, Wang H, and Meng J 2019 *Polymer* **167** 1–12
[15] Chi L, Qian Y, Guo J, Wang X, Arandiyan H, and Jiang Z 2018 *Catal. Today* **335** 527–537
[16] Xu Y, Lin Y, Lee M, Malde C, and Wang R 2017 *J. Memb. Sci.* **552** 253–264
[17] Xu Z, Wang J, Shen L, Men D, and Xu Y 2002 *J. Memb. Sci.* **196** 221–9
[18] Lei J and Liao X 2001 *Eur. Polym. J.* **37** 771–9
[19] Zhu H, Wang H, Wang F, Guo Y, Zhang H, and Chen J 2013 *J. Memb. Sci.* **446** 145–153
[20] Wang H, Ding S, Zhu H, Wang F, Guo Y, Zhang H, and Chen J 2014 *Sep. Purif. Technol.* **126** 82–94
[21] Naim R, Ismail A F, Cheer N B, and Abdullah M S 2014 *Chem. Eng. Res. Des.* **92** 1391–8
[22] Silva T L S, Morales-torres S, Figueiredo J L, and Silva A M T 2015 *Desalination* **357** 233–245
[23] Salim N E, Jaafar J, Ismail A F, Othman M H D, Rahman M A and Yusof N 2018 *Chem. Eng. J.* **335** 236–247
[24] Beaudoin G, Lasri A, Zhao C, Liberelle B, Crescenzo G D and Zhu X X 2021 *Macromolecules* **54** 7963–9

Simulation of acid gas removal unit using DIPA+TEA amine solvent

A S Farooqi[1,2*], R M Ramli[1,2*], S S M Lock[1,3], N Hussein[4], A S Farooqi[1,3] and S M Wajahat[1,3]

[1]Department of Chemical Engineering, Universiti Teknologi PETRONAS, Bandar Seri Iskandar, 32610 Seri Iskandar, Perak, Malaysia.

[2]Centre of Innovative Nanostructures & Nanodevices, Universiti Teknologi PETRONAS, Bandar Seri Iskandar, 32610 Seri Iskandar, Perak, Malaysia.

[3]CO_2 Research Center (CO2RES), Universiti Teknologi PETRONAS, 32610 Seri Iskandar, Malaysia.

[4]Group Technical Solutions, Project Delivery & Technology, Tower 3, PETRONAS Twin Towers, KLCC, 50088 Kuala Lumpur, Malaysia.

*e-mail: abid_20001276@utp.edu.my; raihan.ramli@utp.edu.my

Abstract. Due to the inexpensive cost of amine solvent, more than 95 % of natural gas (NG) processing plants use an acid gas removal unit that utilizes an aqueous amine solvent to remove sour gas components such as carbon dioxide (CO_2) and hydrogen sulfide (H_2S). Different technologies are available to capture CO_2 from NG. However, chemical absorption is the most reliable and used technology all over the world. However, it is challenging to determine the amine blend's optimal composition for the effective removal of CO_2 and H_2S and solvent regeneration. This is mainly due to the difference in reservoir gas compositions, affecting gas removal efficiency and solvent regeneration energies. The present investigation addresses the performance of using a novel solvent blend of diisopropanolamine (DIPA) and Triethanolamine (TEA) to determine the absorption capacity of CO_2 & H_2S using Aspen HYSYS software. A study on the effects of solubility on CO_2 absorption was performed at varying pressure (10-80 bar) and temperature (25°C to 50°C). The percentage of CO_2 removal increased from 80% to 98% as the temperature increased from 25°C to 50°C. The results revealed that the concentration of CO_2 and H_2S in sweet gas decrease with the increase in pressure while the concentration of CO_2 and H_2S increases with the increase in temperature.

1. Introduction

Natural gas (NG) is known as the clean energy source producing 56% lesser emissions as compared to coal and oil [1]. To ensure that the natural gas is environmentally safe and suitable to be used, it is essential to remove most of the contaminants to comply with the sales gas specifications. This is because the presence of contaminants can affect its utilization and cause asset integrity issues such as corrosion, erosion, plugging as well as health and environmental hazards [2]. NG is traditionally extracted from subsurface sources along with crude oil. Associated gas is NG that has been produced along with oil, whereas non-associated gas has been generated without much oil. Associated gas typically contains less methane and more high molecular weight hydrocarbons than non-associated gas [3].

Depending on the feed composition, operating conditions, and environment, acid gases are removed through different methods. Based on the properties and applications, there are different kinds of processes used for CO_2 separation, like permeation through porous membranes, cryogenic separation,

IOP Conf. Series: Materials Science and Engineering 1257 (2022) 012033 doi:10.1088/1757-899X/1257/1/012033

adsorption on solids, absorption with solvents, and chemical conversion. Each process has its own merits and demerits. Chemical amines, physical amines, and a combination of chemical and physical amines are the most popular methods for removing acid gases. Generally, feed gas at high pressure is treated with physical solvents and low-pressure feed gas can be treated with chemical solvents [4].

On an industrial scale, chemical solvents are used for the separation of CO_2 which is based on the amine scrubbing technique. The aqueous solution of the amine-based scrubbing technique is a highly mature and mostly practiced technique in the CO_2 capture process on an industrial scale [5]. Amines are further divided into three different categories; primary, secondary, and tertiary amines, depending upon their one, two, or three hydrogen atoms which are replaced by the alkyl functional group. Monoethanolamine (MEA), diethanolamine (DEA), N-methyl diethanolamine (MDEA), diglycolamine (DGA), diisopropanolamine (DIPA), 2-amino-2-methyl-1-propanol (AMP), and piperazine (PZ) are some of the most commonly utilized aqueous alkanolamines in the absorption process [6].

Even though alkanolamines have numerous applications industrially, and are frequently used solvent, still have many demerits, which includes low loading of CO_2, high loss of vapors, corrosion of equipment, and consumption of high energy, thermal and oxidative degradation [7]. According to the stoichiometric analysis, the loading capacity of primary and secondary amines (MEA and DEA) is restricted to 0.5 moles of CO_2 per mole of amine. The tertiary amines MDEA and TEA have recently become very important amines because it is selective and more ideal for H_2S but having the high capacity to remove CO_2, loss of low vapors, enthalpy of reaction with CO_2 is very low, high thermal stability, low vapors loss, and low enthalpy of reaction with CO_2, the most vital property in the regeneration of amines [8].

In this case, the existence of H_2S and CO_2 together in the NG, tertiary amines are preferred because they can remove both H_2S and CO_2 simultaneously. Many studies have been done on increasing the rate of CO_2 absorption into MDEA solutions, however, the usage of TEA in the literature is quite limited. The majority of them are focused on adding activators to the solution. Primary and secondary amines are the most often utilized activators. Activator DIPA is added to increase the reaction rate, and also the loading capacity of TEA. CO_2 reacts rapidly to form carbamate when reacting with primary and secondary amine [9].

In the past, different studies are done to investigate the performance of TEA solvent to clean acid gas. Rinprasertmeechai et al. evaluated the CO_2 absorption capacities of several amines and found that the primary amine (MEA) had the highest CO_2 absorption capacity, but the regeneration energy required is high [10]. Erfani et al. compared different amines i.e., DIPA, DEA, MDEA, DGA, and MEA. The comparison revealed that the blend of DIPA-MEA lessens the regeneration energy as compared to the standard 30% MEA solution [11]. In the same way, Sarker et al. [12] revealed that using a mixture of amines can give the best results as compared to using a single amine for the sweetening of natural gas. However, these studies only looked at CO_2 in the NG but did not look at the influence of H_2S in NG.

This study aims to investigate the capture of both CO_2 and H_2S using a novel amine blend to find out the solubility of CO_2 and H_2S. TEA is selected as the main amine in this study with a combination of secondary amine DIPA as an activator. The concentration of H_2S and CO_2 in sweet gas using aqueous alkanolamine mixtures of TEA+ DIPA at 25°C to 50°C and a pressure range of 10-80 bar was investigated by using the Aspen HYSYS simulator.

2. Methodology

2.1. *Software*
The present work was completed by using Aspen HYSYS V12.1, a commonly used software for acid gas removal unit (AGRU) in different oil and gas fields. The Acid Gas (chemical solvent) built-in thermodynamic package was used for this simulation because both solvents are chemical solvents and the Acid Gas package produces output with lesser deviation.

2.2. *Process Description*
In the absorption tower (T-100), the sour gas feed containing acid gases will pass through the solvent and the gases will be absorbed selectively by the solvent depending on its properties. The gases in the solvent later on will be stripped in the stripper unit [13]. figure 1 illustrates the general process flow diagram of an amine-based AGRU.

After the rich amine solution leaves the absorber unit, the solution will be fed into the stripper (T-101) to strip the CO_2 and H_2S from the solvent and return the lean solvent to the absorber unit. In the stripper, the rich amine solution is heated adequately to release CO_2 and H_2S from the amine solvent. The stripping vapors, which also contain minor amounts of solvents, CO_2, H_2S as well as water vapors, are regenerated in the reboiler and leave the column from the top of the stripper. Through countercurrent flow, the stripping vapors and the rich amine solution come into contact, where the solution absorbs energy from the steam for CO_2 desorption. The CO_2 released from the solvent is compressed for transport and sequestration while the leftover vapors are condensed at the top of the column in the overhead condenser. [14]. Lastly, the regenerated lean solvent is then cooled to around 40°C to 50°C and reused in the absorber column for the continuous acid gas removal process.

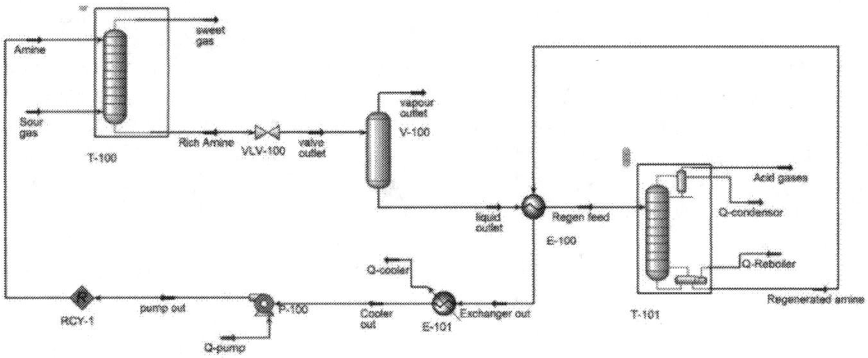

Figure 1: Aspen HYSYS flow sheet of AGRU.

2.3. *Simulation Basis*

The two most common approaches for simulating columns are the equilibrium-stage model and the rate-based model. The liquid and vapor leaving each level of the column are thought to be in balance in an equilibrium-based model. For accurate results, equilibrium-based models often require empirical parameter tunings. In contrast, rate-based models analyze heat and mass transfer phenomena using tray geometry, transport properties, and concentration gradients between contacting phases. This allows them to predict processing parameters under different operational circumstances and brings the rate-based model outputs closer to the original data. The acid gas removal using chemical absorption of amines is based on the theory of acid-base reaction. CO_2 and H_2S are known as acid gases as they dissociate to form a weak acidic solution in an aqueous solution given as (1), (2), and (3). In the chemical absorption process, the amines act as the weak organic bases to react with the acid gas components via exothermic reactions and form soluble acid-base complexes in the treating solvent. The acid-base complexes are then reversed to strip off the acid gas components from the amine solvent in the stripper column through a thermal regeneration process [15]. Following reaction between amines with CO_2 and H_2S taking place.

Ionization of water:

$$H_2O \leftrightarrow H^+ + OH^- \qquad (1)$$

Ionization of dissolved H_2S

$$H_2S \leftrightarrow H^+ + HS^- \qquad (2)$$

Ionization and Hydrolysis of dissolved CO_2

$$CO_2 + H_2O \leftrightarrow HCO_3{}^+ + H^+ \qquad (3)$$

Protonation of amine

$$RNH_2 \leftrightarrow H^+ + RNH_3{}^+ \qquad (4)$$

Carbamate formation

$$RNH_2 + CO_2 \leftrightarrow RNHCOO^- + H^+ \qquad (5)$$

According to the degree of substitution on the central nitrogen by organic groups, amines or alkanolamines are classified as primary, secondary, or tertiary. Primary amine has one alkyl group on the nitrogen atom, followed by two for secondary amine and three for tertiary amine. While alkanolamines consist of at least one amino group and one hydroxyl group where the amino group provides alkalinity in the aqueous solution for the acid-base reaction and the hydroxyl group allows for solubility in water [16].

2.4. Process simulation

To optimize the process, the simulation models are checked to see whether the use of an amine blend is more useful than compared to a single amine. The base amine and additive amine blends are mixed by keeping the overall amine strength constant at 50 wt.%. The sour gas enters the absorber at 25°C with a flow rate of 1245 kmol/h. specification of simulation of AGRU is given in table 1. The base amine in this case is TEA, while the additive amine is DIPA. In the TEA+DIPA mixture, the concentration of TEA is altered from 35 to 50 wt.%, while the concentration of DIPA is modified from 0-15 wt.%. The condenser and reboiler temperatures in the regenerator are initially specified to get convergence of the base case models. The influence of temperature and pressure were then investigated to better understand their impact on process performance.

Table 1: Input specification for Aspen HYSYS simulation.

Parameters	
Gas flowrate (kmol/hr)	1245
Gas temperature (°C)	25
Gas pressure (bar)	69
Composition of CO_2 in feed gas (mole fraction)	0.46
Composition of H_2S in feed gas (mole fraction)	0.06
Solvent flowrate (m^3/h)	251
Solvent temperature °C	35
solvent pressure (bar)	68.5
Absorber trays	22
Regenerator trays	18
Condenser temperature °C	34
Reboiler temperature °C	97

3. Result and discussion

3.1. Effect of Temperature on the concentration of CO_2 and H_2S

The solubility data of the amine solvent is investigated at six different temperatures from (25°C to 50°C). The effect of the temperature on the concentration of CO_2 and H_2S in sweet gas is shown in figure 2. A general trend was observed in figure 2, with the increase in temperature of the system the loading capacity of the amine solvents decreases to a significant value means the concentration of CO_2 and H_2S increases with an increase in temperature. It is well-established phenomenon that the lower temperatures always favor the absorption process, while the increased temperature showed a negative effect on the absorption process. At higher temperatures less diffusion of the gas molecules take place in chemical solvents. The other reason might be the high vapor pressure of the solvents increased, which reduces the loading capacity of the amine solvents.

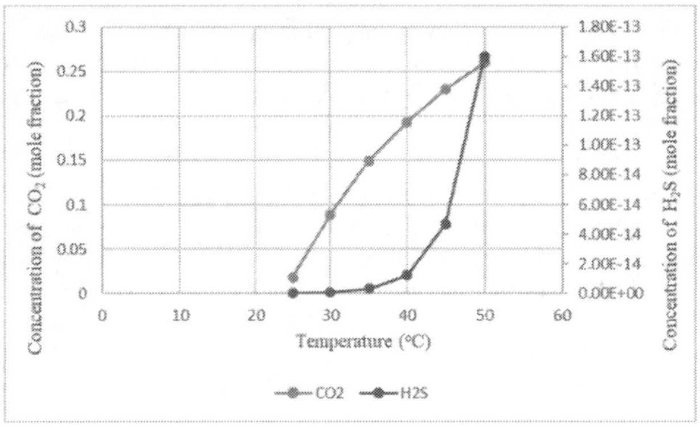

Figure 2: Effect of temperature on the concentration of CO_2 & H_2S in sweet gas.

3.2. *Effect of Pressure on the concentration of CO_2 and H_2S*

Over a broad pressure range, the impact of operating pressure on the absorption capabilities of the amine solvent was examined (10-80 bar). To examine the performance of the amine solvents to effectively absorb CO_2 gas, the absorption performance of the hybrid solvents was assessed in terms of loading capacity, a crucial attribute in the absorption process. According to the simulation analysis, the system higher pressure encouraged the absorption process. figure 3 shows that the concentration of CO_2 and H_2S in sweet gas is significantly impacted by an increase in pressure. Additionally, it was noted that the amount of CO_2 and H_2S in the sweet gas reduces as pressure rises. The literature supports that the absorption process was facilitated by increased pressure when CO_2 partial pressure rises.

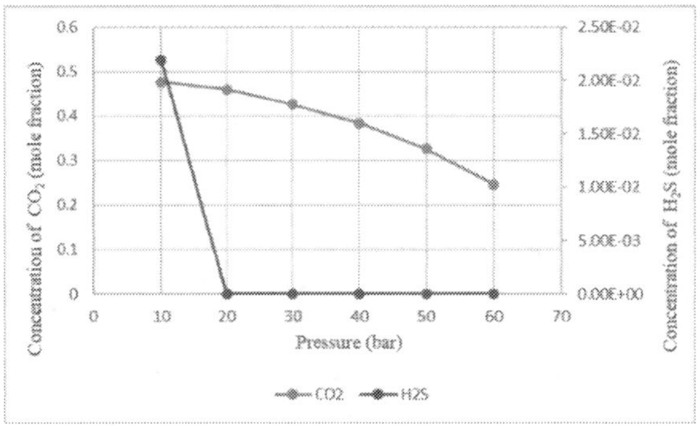

Figure 3: Effect of pressure on the concentration of CO_2 & H_2S in sweet gas.

4. Conclusion

The overall CO_2 and H_2S absorption in a binary mixture of TEA+DIPA were investigated using Aspen HYSYS simulation software. Operating parameters i.e., Pressure and temperature effect on the concentration of CO_2 and H_2S in sweet gas were studied. It is investigated that the pressure has a positive impact on the CO_2 and H_2S concentration in sweet gas while temperature shows a negative effect as the temperature increases the concentration of CO_2 and H_2S increases in sweet gas. This solubility data will be further used to study the effect of solvent circulation rate and concentration of amines for solvent regeneration in the stripper column to lessen the reboiler duty.

Acknowledgment
The authors gratefully acknowledge financial support from Universiti Teknologi PETRONAS, Malaysia, under the Yayasan Universiti Teknologi PETRONAS (YUTP) grant number 015LC0-293.

REFERENCES

1. Lim W, Choi K, Moon I. 2013 Current Status and Perspectives of Liquefied Natural Gas Plant Design. *Ind Eng Chem Res* **52** 3065-3088.
2. Al-Lagtah NMA, Al-Habsi S, Onaizi SA. 2015 Optimization and performance of Lekhwair natural gas sweetening plant using Aspen HYSYS. *J Nat Gas Sci Eng* **26** 367-381.
3. Kidnay AJ, Kidnay AJ, Parrish WR. 2006 Fundamentals of Natural Gas Processing. *Fundam Nat Gas Process* **27** 65-71.
4. Park J, Yoon S, Oh SY, Kim Y, Kim JK. 2021 Improving energy efficiency for a low-temperature CO_2 separation process in natural gas processing *Energy 29 45-51.*
5. Abkhiz V, Heydari I. 2014 Comparison of amine solutions performance for gas sweetening. *Asia-Pacific J Chem Eng* **9** 656-662.
6. Zahid U, Al Rowaili FN, Ahmed U. 2017 Simulation and parametric analysis of CO_2 capture from natural gas using diglycolamine. *Int J Green house Gas Control* **57** 42-51.
7. Zahid U. 2020 Simulation of an acid gas removal unit using a DGA and MDEA blend instead of a single amine. *Chem Prod Process Model* **15** 13-18.
8. Zoghi AT, Feyzi F, Dehghani M. 2012 Modeling CO_2 solubility in aqueous N-methyldiethanolamine solution by electrolyte modified peng-robinson plus association equation of state. *Ind Eng Chem Res* **51** 9875-9885.
9. Rinprasertmeechai S, Chavadej S, Kulprathipanja S. 2019 Carbon Dioxide Removal from Flue Gas Using Amine-Based Hybrid Solvent Absorption. *J Nat Gas Sci Eng* **18** 18-22.
10. Erfani A, Boroojerdi S, Dehghani A. 2016 simulation of an operational amine-based CO_2 removal plant of CO_2 capture at coal fired power plants. *J Chem Eng Data* **49** 1331-1264.
11. Sarker NK. 2016 Theoretical effect of concentration, circulation rate, stages, pressure and temperature of single amine and amine mixture solvents on gas sweetening performance. *Egypt J Pet* **25** 343-354.
12. Torabi Angaji M, Ghanbarabadi H, Karimi Zad Gohari F. 2013 Optimizations of Sulfolane concentration in propose Sulfinol-M solvent instead of MDEA solvent in the refineries of Sarakhs. *J Nat Gas Sci Eng* **15** 22-26.
13. Lunsford KM, Bullin JA. 2006 *Optimization of Amine Sweetening Units Ind Eng Chem Res* **43** 987-991.
14. Abkhiz V, Heydari I. 2014 Comparison of amine solutions performance for gas sweetening. *Asia-Pacific J Chem Eng* **9** 656-662.
15. Rho SW, Yoo KP, Lee JS, Nam SC, Son JE, Min BM. 1997 Solubility of CO_2 in aqueous methyldiethanolamine solutions. *J Chem Eng Data* **42** 1161-1164.
16. Rayer A V, Henni A, 2012 High pressure physical solubility of carbon dioxide (CO_2) in mixed polyethylene glycol dimethyl ethers. *Can J Chem Eng* **90** 576-583.

IOP Conf. Series: Materials Science and Engineering 1257 (2022) 012034 doi:10.1088/1757-899X/1257/1/012034

Quality prediction for polypropylene extrusion based on neural networks

C H Tan[1]**, K M Yusof**[1] **and S R W Alwi**[1]

[1] School of Chemical and Energy Engineering, Faculty of Engineering, Universiti Teknologi Malaysia, 81310 UTM Johor Bahru, Johor, Malaysia

Corresponding author's e-mail: khairiyah@cheme.utm.my

Abstract. In the polypropylene (PP) industry, melt index (MI) is considered the most important quality variable. Different grades of PP have their specific range of MI. Accurate prediction of MI is essential for efficient monitoring and off-grade reduction. Neural Networks (NN) modelling is proposed as the technique for MI estimation. It has powerful adaptive capabilities in response to nonlinear behaviours. By training the NN, it can discover the relationship between inputs and outputs and makes it capable of function approximation. The goal of this research is to develop NN model to predict the MI based on PP extrusion parameters. Different types of NN such as artificial neural networks (ANN), stacked neural networks (SNN) and deep neural networks were trained and compared to understand their efficiency in solving the problem. The simulation results show that deep neural networks can perform the highest accuracy prediction with the lowest root mean square error (RMSE), followed by SNN and ANN. All three modelling proved that NN can perform non-linear function approximation for polymer extrusion.

1. Introduction

Polypropylene (PP) is a versatile thermoplastic material with various applications, including food packaging, appliances, injection moulded parts for automobiles, textiles, labelling and toy making. It has intrinsic properties of good tensile strength, inert towards acid, high stiffness, and good processability. The polypropylene market is expected to have a compound annual growth rate (CAGR) of more than 6% expected over the forecast period (2021-2026).

Different grades of PP have their own specific range of melt index (MI) that will be changed during product grade transition. MI is the weight of the polymer in grams extruded in 10 min under the application of a dead weight through a melt flow apparatus. It is the most widely used specification in the plastics industry to differentiate between different polymer grades. It has long been used by PP manufacturers to define the best end application for a particular grade of polymer. Material exceeding or lower than the intended MI can have a negative impact on the quality of the end product and receive customers' complaints and rejects. As a result, it is critical to guarantee that MI is within the specification. Due to the high demand for different PP grades, the frequency of grade transition reaches almost four transitions per week. Therefore, fast and efficient quality analysis is important in order to reduce the off-grade quantity. Yet, the problem is that MI can only be evaluated offline with the analytical procedure in the laboratory. Not only the procedure is costly, but it is also time-consuming, requiring between 2 to 4 hours [1]. Therefore, there is a need to come out with methods to predict product quality in real-time.

For past research, in order to predict the MI for PP, soft sensors to measure MI in industrial polypropylene loop reactors using artificial neural network (ANN) model, stacked neural network (SNN) model and serial hybrid neural network (HNN) model were presented by Jumari and Mohd-Yusof [2]. On the other hand, Li and Liu [3] presented an optimized radial basis function (RBF) neural network to predict the MI of polypropylene produced by Hypol process. Based on the past research, modelling studies were focused on the MI estimation of the PP virgin flakes using only the reactor polymerization variables as the input variables and missed out on the extruder part in their research.

In the PP production plant, after producing virgin polymer from polymerization reaction, extrusion is a process to melt and mix the PP with the additives, such as anti-oxidant, nucleating agent, lubricant, and slip agent to suit the commercial purpose, and finally, convert the polymer to their standard commercial resin size. Shearing and additive mixing in the extruder can cause the MI to shift and slightly deviated from its original PP flakes. Prediction of MI by using reactor and extruder variables are equally important. Therefore, this study proposes to look into extrusion parameters for MI prediction.

There are several types of extruders available, such as single/multi screw extruders, disk/drum extruders, etc [4]. The extruder focused in this research is a twin screw extruder. Extruders are highly complex. To date, there have been no fully-resolved first-principles simulations of entire twin-screw extruders reported [5]. Due to the high complexity and non-linearity of the PP extrusion process, first principle modelling is not encouraged to be used as PP MI prediction. Therefore, instead of creating a model-driven predictor, a data-driven predictor would be a better choice. In this work, the NN model is the data-driven model chosen for the MI predictor. NN learns the past relationship between the extruder process data and MI results and predicts the future MI. NN approach has become more popular recently due to the ability to simulate a wide variety of complex engineering and scientific problems. NN focused in this research are artificial neural networks (ANN), stacked neural networks (SNN) and deep learning.

2. Process Description

The type of extruder focused in this research is a co-rotating twin screw extruder with a gear pump to assist the polymer flow. Figure 1 shows the diagram of an extruder.

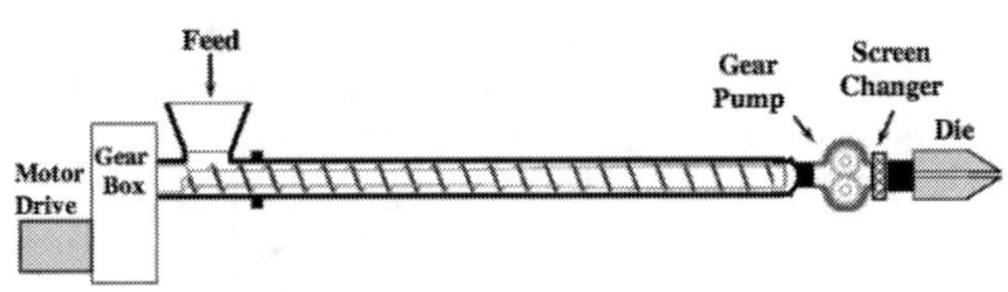

Figure 1. Extruder Diagram.

The most significant variable that will respond to MI is gear pump discharge pressure. By observing the process history, the gear pump discharge pressure increases when the MI is low, and vice versa. This can be explained by the viscosity of the polymer. Lower MI polymer is higher in viscosity and subsequently creates a stronger back pressure when the polymer is squeezed through the die hole. In order to have a better prediction, polymer temperature and pressure before the die plate are also taken as the input variables because lower MI polymer with higher backpressure will cause higher polymer temperature and pressure when the polymer is squeezed through the die plate.

The next variable that will be taken is motor drive kW and extruder feed rate. Higher viscosity polymer (lower MI) and higher feed rate can lead to extra energy (kW) exerted by the motor, and vice versa. Next, slot control will also be one of the variables because slot opening will affect the shearing of the polymer. Lower slot opening causes high shearing, which slightly affects the MI as shear would possibly break some of the polymer chains, causing the polymer becomes less viscous.

Lastly is the organic peroxide (OP) flow rate. OP would cause chain scission, which narrows the molecular weight distribution of the PP. Simultaneously, the average molecular weight would reduce and resulting in the melt viscosity of the PP to decrease and the MI to increase. This process is called 'visbreaking'. Table 1 shows the summary of the inputs that will be used in developing the predictor.

Table 1. Extruder variables required in developing neural networks.

Type	Variable	Function
Input 1	Extruder feed rate	The rate of PP production through the extruder
Input 2	Organic peroxide flow rate	Organic peroxide is used for controlled polymer degradation during special PP grade
Input 3	Mixer motor ampere	Ampere needed to extrude polymer in different MI
Input 4	Mixer motor kW	Energy needed to extrude polymer in different MI
Input 5	Slot	Control of the polymer mixing during extrusion
Input 6	Specific Energy for Extruder	Energy needed for 1MT of PP produced
Input 7	Gear pump discharge pressure	Back pressure of the polymer generated during extrusion
Input 8	Gear pump suction pressure	The suction pressure of extruder gear pump
Input 9	Polymer die pressure	The polymer pressure before the extruder die
Input 10	Polymer die temperature	The polymer temperature before the extruder die
Output	Melt Index (MI)	Polymer specification of desired polymer grade

3. Model Development

Data collected for the NN must be big and significant enough to cover the possible known variation in the problem domain [6]. The data collected will be in the MI range of 1 to 55 g/10mins, with the input parameters (refer Table 1). Collected data will be divided into two distinct sets for training and testing.

3.1. *Development, Simulation and Validation of ANN Models*

ANN models were developed to predict MI using extruder parameters collected. For the ANN topology, the hidden layer is log sigmoid transfer function and the output layer was pure linear transfer function. This type of topology can be trained to estimate most of the functions well [7]. The ANN was trained using "nntools" function in MATLAB. Feed-Forward Back Propagation (FFBP) was used as the network architecture and Levenberg-Marquardt as the training method because of the fast back-propagation algorithm and is highly suggested as the first choice algorithm [2]. Data normalization is used before the training of the NN. The normalization formula used is as follows:

$$X_{normalized} = \frac{X - X_{minimum}}{X_{maximum} - X_{minimum}} \quad (1)$$

where $X_{normalized}$ is the normalized data. X is input data. $X_{minimum}$ is the lowest value within the input data range. $X_{maximum}$ is the highest value within the input data range.

Figure 2 shows the input variables and the output variables of the ANN modelling. In the MATLAB "nntool", "input data" and "target data" were specified with input and output variables respectively prior to the NN being created. Then the NN model is configured and trained. After the training, the number of hidden nodes is systematically altered between 10 ~ 30 nodes and re-trained. Prediction from each ANN with different numbers of nodes is cross-validated with the unseen data outside the training inputs. The number of nodes with the best prediction will be presented.

3.2. *Development, Simulation and Validation of SNN Models*

SNN model was presented to increase the accuracy of the model. Model accuracy was improved by combining several NN together. Figure 3 shows the model for SNN.

In order to produce a SNN, different ANN model (ANN1, ANN2, ANN3, ANN4, ANN5 and ANN6) are trained with a different number of hidden layer nodes. ANN1 and ANN2 were trained by using the

MI range of 1 – 5, ANN3 and ANN4 were trained using the MI range between 11 – 17, ANN5 and ANN6 were trained using MI 45 - 55. Subsequently, all the MI results predicted from each ANN will serve as the training input data for the last ANN (ANN Level 1), along with the original input variables. The reason for choosing MI ranges 1 – 5, 11 – 17 and 45 - 55 is due to there are commercial products within this range and more input data are available for training and testing within these ranges.

For ANN1 and ANN2, the number of hidden nodes is systematically altered between 10 ~ 30 nodes and re-trained. The two best NN models which have the best prediction will be chosen as ANN1 and ANN2. The method is then repeated for ANN3 – 4 & 5 – 6 for MI range 11 – 17 & 45 – 55 respectively.

Lastly, the number of hidden nodes for ANN Level 1 is systematically altered between 16 ~ 32 nodes and re-trained. Prediction from each ANN with different numbers of nodes is cross-validated with the unseen data outside the training inputs. The number of nodes with the best prediction will be presented.

3.3. Development, Simulation and Validation of Deep Learning Models

Deep learning was proposed to solve the problem of vanishing gradients and overfitting. Different from ANN, Rectified Linear Unit (ReLU) transfer function was used instead of log sigmoid transfer function. ReLU transfer function is used in order to solve the problem of vanishing gradients, which arises when applying the chain rule in layered networks of sigmoidal units. The deep learning topology that is focused on is deep neural networks. Figure 4 shows the deep neural networks topology.

For deep neural networks training in MATLAB, instead of using "nntool" it trains using "trianNetwork" function. It has the full connection between all the input, hidden and output layers, and no connections between the same node in the same layer. The input and output variables will be the same as those in ANN modelling. After the training, the number of 1st and 2nd hidden nodes are systematically altered between 10 ~ 30 nodes and re-trained. Both 1st hidden and 2nd hidden layers will be the same number of nodes to prevent the combination between the 1st and 2nd hidden nodes to be too complicated. Prediction from each ANN with different numbers of nodes is cross-validated with the unseen data outside the training inputs. The number of nodes with the best prediction will be presented.

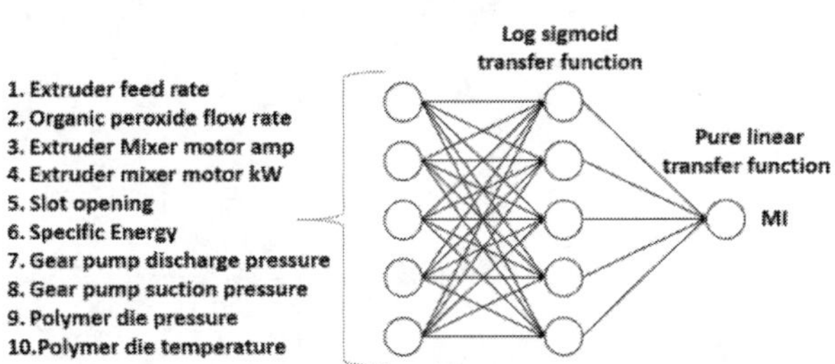

Figure 2. ANN Topology.

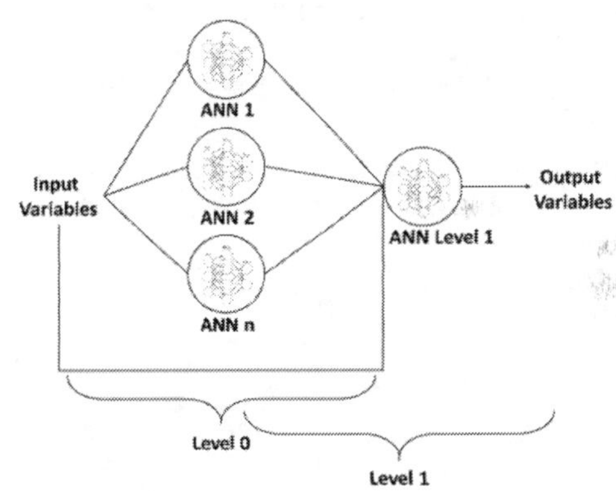

Figure 3. SNN Topology.

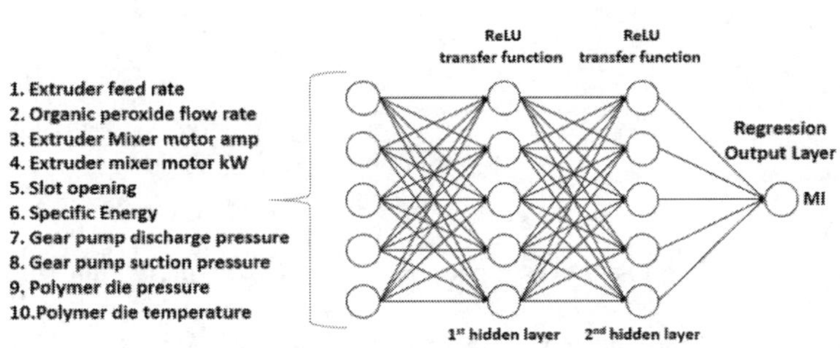

Figure 4. Deep Learning Topology.

4. Results and Discussion

The NN models were developed by using extruder data set collected from PP Plant. 383 data sets for training and 124 data sets for testing with the MI ranges from 1 - 55 g/10mins.

Table 2 shows the performance comparison of different models on the extruder dataset. The model with the lowest root mean square error (RMSE) is the best model developed to estimate the MI. The RMSE result for ANN, SNN and deep neural networks are 0.0240, 0.0204 and 0.0189 respectively, while the R^2 results are 0.9982, 0.9979 and 0.9978. Figures 5 to 7 show the graph for the actual and predicted MI comparison. Actual MI and the predicted MI are plotted in the graphs for comparison.

Table 2. Performance comparison of different models on extruder dataset

Model	RMSE	R^2
ANN	0.0240	0.9982
SNN	0.0204	0.9979
Deep neural networks	0.0189	0.9978

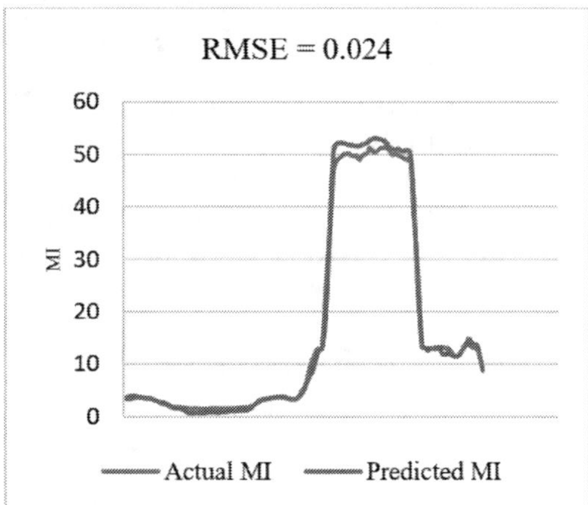

Figure 5. Predicted and Actual MI (ANN with 10 hidden nodes).

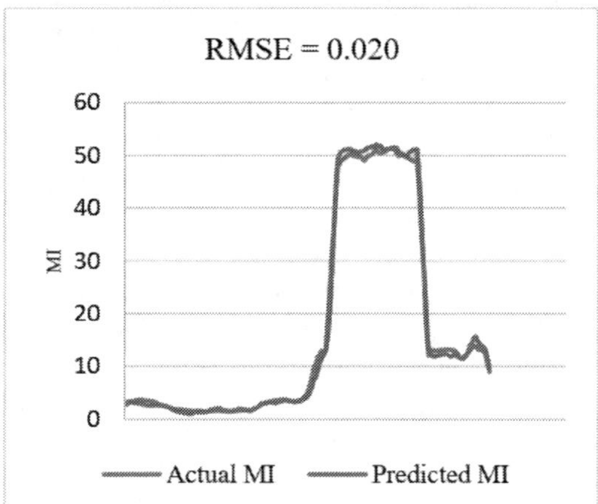

Figure 6. Predicted and Actual MI (SNN with 16 hidden nodes).

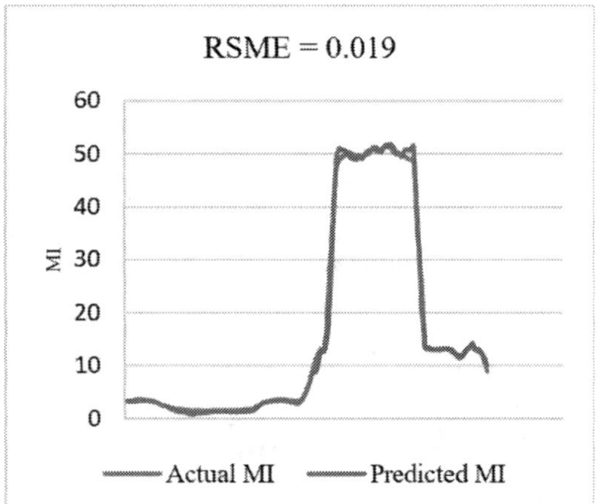

Figure 7. Predicted and Actual MI (Deep neural networks with 13 hidden nodes in 1st and 2nd hidden layer).

Based on Table 2, deep neural networks model performs the best prediction with the lowest RMSE value, followed by SNN and ANN.

ANN model can deal with complicated and poorly understood behaviour. ANN is able to understand the non-linearity of the process and provide the input-output relationship. SNN and deep neural networks are the models that are developed from ANN, which SNN targets to improve the generalisation capability of ANN by combining several networks to improve model predictions on unknown data [8]. Deep neural networks model aims to improvise the ANN to produce a better performance model by adding the number of hidden layers to increase the complexity and abstraction of the NN [9].

Based on the results, all NN models can predict the MI with high accuracy, with the deep neural networks model performing the best in estimation capabilities, data fitting and generalization, especially at the MI range 45 – 55. Besides the powerful predictive ability of deep neural networks, one more reason the model can obtain low RMSE and good correlation is that the training input variables were properly selected with large sample size. With this, it is proven that the variable chosen in Table 1 is significant and suitable to be used for MI prediction.

The research shows that the NN can accurately forecast PP MI using extrusion process parameters for various grades of PP. This technology can provide the PP manufacturer with a good tool to produce high-quality on-spec PP products and satisfies the customer needs. It can also help the manufacturer reduce the production of sub-prime products and indirectly reduce the cost of production.

5. Conclusion

In this work, the ANN, SNN and deep neural networks models were successfully developed to predict MI for the PP extruder. The results show that the three types of models are able to predict the MI of the polymer to a high degree of accuracy within a broad range of MI (1 – 55 g/10mins).

Based on the results, deep neural networks can perform the highest accuracy prediction with the lowest RMSE, followed by SNN and ANN. The RMSE are 0.0189, 0.0204 and 0.0240 respectively. All modellings proved that NN can perform non-linear function approximation for polymer extrusion. This indicates that the NN model is an effective analytical tool that can be adopted as a supporting MI prediction tool in the polymer industry.

Acknowledgments
The grant support from Universiti Teknologi Malaysia High Impact Research Grant, Vote number: Q.J130000.2409.08G86 is highly appreciated.

References

[1] Xu S and Liu X 2014 Melt index prediction by fuzzy functions with dynamic fuzzy neural networks *Neurocomputing* **142** 291-298
[2] Jumari N F and Mohd-Yusof K 2016 Comparison of product quality estimation of propylene polymerization in loop reactors using artificial neural network models *Jurnal Teknologi* **78** 6-13
[3] Li J and Liu X 2011 Melt index prediction by RBF neural network optimized with an MPSO-SA hybrid algorithm *Neurocomputing* **74** 735-740
[4] Abeykoon C, Martin P J, Kelly A L and Brown E C 2012 A review and evaluation of melt temperature sensors for polymer extrusion *Sensors and Actuators A: Physical* **182** 16-27
[5] Eitzlmayr A and Khinast J 2015 Co-rotating twin-screw extruders: Detailed analysis of conveying elements based on smoothed particle hydrodynamics. Part 2: Mixing *Chemical Engineering Science* **134** 880-886
[6] Funes E, Allouche Y, Beltrán G and Jiménez A 2015 A review: artificial neural networks as tool for control food industry process *Journal of Sensor Technology* **5** 28
[7] Hagan M T, Demuth H B, Beale M H and De Jesús O 1996 *Neural network design* (Boston: Pws Pub)
[8] Mukherjee A and Zhang J 2008 A reliable multi-objective control strategy for batch processes based on bootstrap aggregated neural network models *Journal of Process Control* **18** 720-734
[9] Debus B, Parastar H, Harrington P and Kirsanov D 2021 Deep learning in analytical chemistry *TrAC Trends in Analytical Chemistry* **145** 116459

ANN based Multi Model Predictive Control for pH-Control

ZY Pua[1], AW Hermansson[1] and CH Lim[1]

[1] School of Engineering and Physical Sciences, Heriot-Watt University Malaysia, 1, Jalan Venna P5/2, Precinct 5, 62200 Putrajaya, Wilayah Persekutuan Putrajaya, Malaysia.

a.hermansson@hw.ac.uk

Abstract. Artificial neural network (ANN) has many uses when non-linear behaviour is modelled. Here we are training a feedforward ANN that will mimic the behaviour of a Robust Model Predictive Controller (RMPC) for use in pH control. The training dataset were generated from running multiple tests on RMPC for different requirements and cases of pH-control. The training data focused on the control-inputs relating to the other process inputs. The training algorithm used in this neural network is Levenberg-Marquardt algorithm which is the most widely use algorithm in current machine learning industry. This neural network was trained by using the deep learning toolbox in Matlab®. Eight different cases is presented: four is for deploying neural network purpose, while the other four is for verification purpose. The result shows good control as long as the ANN-controller has been given a similar input and there are no multiplicity in the process input data.

1. Introduction

The emergence of the Industrial Revolution 4.0 in the last decade, has seen the process industries moving towards more automated control systems due to the increasing amount of data available and the complexity of process system and also more stringent requirement in product specification [3]. Robust automatic control is becoming the essential trend in the industry [11]. One of the candidates getting attention in the industry is model-based predictive control (MPC) which is able to handle high complexity and non-linear industrial systems. The main concept of the MPC is the prediction of the process behaviour and the controlled variable based on the past performance and current values, thus the MPC will optimize the manipulated variables to minimize the target requirements [9]. Due to these characteristics, MPC have become an attractive solution in pH control system with its non-linear behaviour. The studies of application of MPC to non-linear system have been done is the past decades, some of the recent review and improvement studies have also been done [4], [7], [13].

The summary is that when using MPC for systems where pH control is important the control engineer face the following problems: 1) MPC runs through an optimization process at each sampling time this makes it slow in generating the controller output; 2) MPC requires an accurate model of the system, which is nearly impossible to generate over a wide enough range do determine a suitable pH system model due to its complexity and uncertainty underlying in its dynamic behavior.

The first uses of MPC and ANN was implementation of ANN as a black-box model (Draeger, et al., 1995). ANN was implemented to improve the control performance of MPC in power converters system

(Wang, et al., 2022). This paper proposed an application of ANN based Model Predictive Control (ANN-MPC) for pH system, based on a discrete setup.

This paper is organized as follow. In Section 2, the brief introduction to the pH model and theoretical background about model-based predictive control (MPC), artificial neural network (ANN) and some application of ANN-MPC are presented. Section 3 will explain the research gap by providing problem statement and objectives of this paper. Methodology of Robust MPC algorithm is briefly described in Section 4 with the application of artificial neural network (ANN) as prediction model. Section 5 will be showing the results and critical analysis. The final section will conclude this paper with some suggestion for future work.

2. Literature Review

2.1 Conventional Model Predictive Control

MPC have a very strong capability to work within the constraints of the system through the online (real-time) optimization method, but the drawback for this model is the expensive computational cost when dealing non-linear system. MPC can also generalized to two major optimization group which is prediction-based and model-based [8]. We will be focusing on prediction-based model also known as black box modelling due to the further feature of ANN. In the study from [6], it shows robust offset-free control for pH by MPC relying on multiple models including Multi-Model Predictive Control, Min-Max MPC and Robust MPC. A comprehensive study on the pH control by different approaches of MPC has been reviewed in [4].

2.2 Artificial Neural Network (ANN) Based MPC

The world is undergoing the revolution towards Industry 4.0 with increasing automation and digitalization. The big leverage for machine learning in process control and modelling is ANNs capability of dealing with non-linear complex problems [16]. This section will briefly introduce the ANN and explain how it can be applied to MPC for pH control.

2.2.1 Artificial Neural Network

ANN is a subset of machine learning and is the heart for deep learning algorithms. It is inspired by human brain which mimic the biological neuron to transmit signal to each other. The basic components in ANN are the input layer, hidden layer and output layer combined through weights and biases and activation functions. With those components, ANN can be built up in various types of architectures. In this paper, the basic architecture structures will be briefly explained. Readers are referred to [2], [14], [12], [17] for further information.

The simplest architecture of ANN, in term of structure and complexity, is Feedforward Neural Network (FFNN). As neural network is arranged in "layer" structure, FFNN can be categorised into two groups which are single layer feedforward neural network (figure 1) and multi-layer feedforward neural network (figure 2). The single layer structure with direct connection between input and output with only one computational performing layer on the output layer. This makes the neural network easier to understand and making the computational time shorter with a single connecting layer between the input and output layer via series of weights, but one of the drawbacks is the accuracy of the data since the iteration process are greatly decreases. In figure 2, it shows that the additional of one hidden layer in the structure. This kind of structure are more likely to be used in the industry control. The advantages are the building complexity is low and a higher accuracy compared to single layer, thus possible to use for more complex problems. The disadvantage is this network required large amount of training set with input and target data to have a well-trained network.

There are multiple studies of chemical process control are using FFNN. Draeger et al did a study in 1995 [3] about the reactor pH control on laboratory-scale by using feedforward neural network-based model predictive control (MPC). The result is compared with the conventional PI-controller and the overall final result show that the FFNN based MPC have a very great advantage in control non-linear system, while in contrast the time for prediction and training are obviously longer. The study has provided few of the possible improvement can be done 1) improve model to multivariable (MIMO) system; 2) improve the training period by online time-varying dynamic model training.

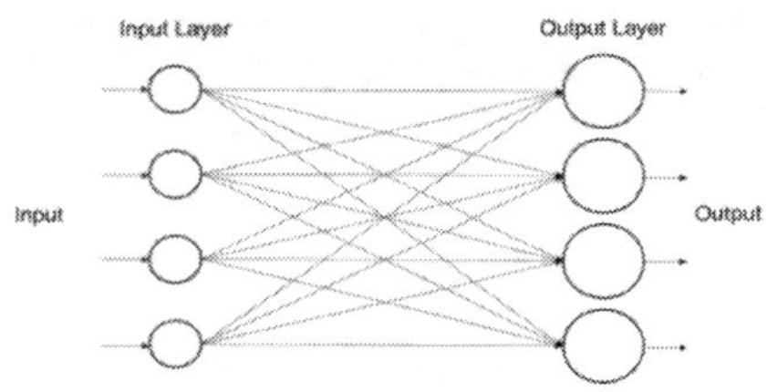

Figure 1: Single Layer FFNN

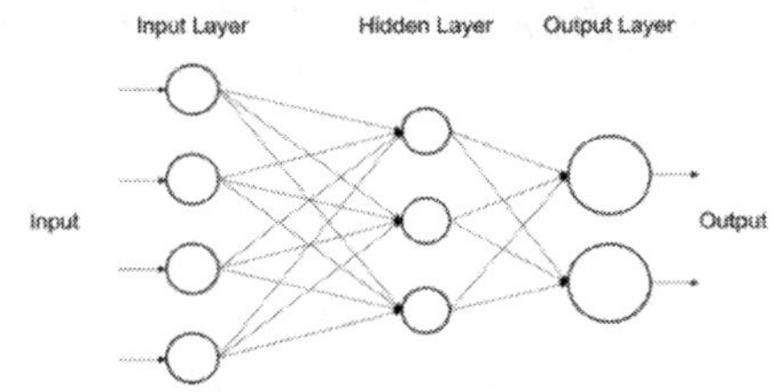

Figure 2: Multi-Layer FFNN.

The second improvement ideology have been done by [5] with the study of control strategies in fed-batch crystallization process. Online optimization type of FFNN based MPC has been used to force the crystallization reaction follow the optimal supersaturation trajectory. The outcome of this study shows there is an improvement on training period, but the constraint of this study is it did not include multiple input variables, thus it causes shorter training period.

The arbitrary structure of ANN controller was investigated by [1]. In this study, two case scenarios have been discussed 1) controlling pH in neutralisation reaction as non-linear system; 2) controlling temperature in a multivariable non-isothermal CSTR. The ANN structure used are FFNN based on explicit MPC (eMPC), where evaluation of the control law is faster. But one of the disadvantages is computationally expensive when dealing out of certain complexity of network.

3. Problem Statement and Objectives

Control of pH in chemical or biotechnology process is very important but may be extremely difficult due to its nonlinear behaviours. In this paper, the neutralization pH control will be used a FFNN with sigmoid activation function as it is suitable for large datasets. The generation of the input/output data will be done by controlling the pH model with the use of the RMPC presented in [6]. With the help of a proper model of ANN based MPC, controlling the pH can be easier implemented and automated to the system. Its ability to solve highly complex non-linear system with least prior knowledge is very attractive for detailed study. The possibilities of an ANN based MPC controller are:

- Reduce the computational time, so it is more practical to implement to fast systems.
- Handle multivariable process and unmeasured data (like disturbance).
- Obtain non-linear black box model for MPC by training ANN from RMPC simulation data.

4. Methodology

4.1 Training of Artificial Neural Network

As one of the goals is to obtain a non-linear black box model by train the ANN with measured data from RMPC simulation that created by Hermansson in 2019 [6]. So, the training set will be generated by the RMPC by having one-step pH setpoint changes. In the training of ANN, the process can be separate to several steps:

1. Generation of training data.
2. Train ANN via deep learning toolbox in Matlab®.
3. Analyze the training performance of the ANN.
4. Export the ANN model.

IOP Conf. Series: Materials Science and Engineering 1257 (2022) 012035 doi:10.1088/1757-899X/1257/1/012035

4.1.1 Generation of Training Data.
For the generation of training data, in this paper a pH setpoint change pattern 3-4-5-6-7-6-5-4-3 was selected. It covers the full range of pH set-points and have both upward and downward changes, so it was expected to be a suitable training dataset for the ANN. This training dataset will set at simulation length of 1200 minutes. Two basic training datasets will be simulated, one set is without model disturbance, and one including a 10% input disturbance. This is to test the ability of disturbance rejection of the ANN. An additional dataset will be generated by combining both basic training datasets mentioned before. The training dataset will consist of two measured time series data for controlled input, U, and pH value as Y. The set-point, y_{reff}, will also added in the training data.

4.1.2 ANN Training.
After the training datasets were generated, the dataset was input into Neural Network Fitting Tool in the Deep Learning Toolbox in Matlab®. Setpoint, y_{reff} and pH value, Y are input data as Predictors, while controlled input, U are the Responses in the Neural Network Fitting (NNF). In this NNF, feedforward neural network is used with a sigmoid activation function. The ANN topology selected will be consist of 2 neurons in the input layer, 10 neurons in the hidden layer and 1 neuron in output layer as shown in figure 3. The performance of ANN will be analysed and summarise in result and discussion section.

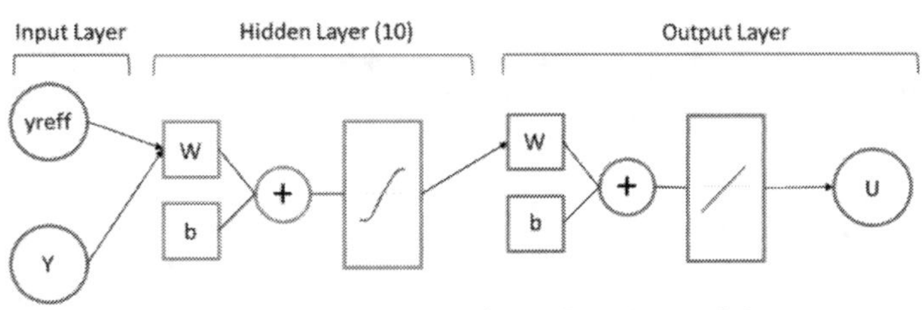

Figure 3: ANN Topology for pH model.

5. Results and Discussions

5.1 Case 1: Setpoint tracking for ANN-MPC (Without Model Disturbance)
To test the control ability for ANN(1), pH 3-4-5-6-7-6-5-4-3 have been created without disturbance. Figure 4 will show the response diagram. ANN-MPC achieve good control for setpoint tracking, while the only difference can be saw is during pH 5 to pH 6 and pH 7 to pH 6, it clearly shows that ANN-MPC will have or might need longer time to control due to the pH behaviour at this point is quite sensitive. Although ANN-MPC does not have significant fast settling time compared to Robust-MPC, the ANN-MPC is helping to reducing the average maximum overshoot but extending the rise time by around 45%. ANN-MPC showing a less oscillatory behaviour compared to Robust-MPC.

5.2 Case 2: Setpoint tracking for ANN-MPC (With Model Disturbance)
To test the control ability for ANN(2), pH 3-4-5-6-7-6-5-4-3 have been created with 10% model disturbance. Figure 5 will show the response diagram. As one the of significant different is during the pH control at pH 3. ANN-MPC seem does not have well data to control the pH=3 when there is a model disturbance happen. Although compared to Robust-MPC, the ANN-MPC have need longer rise time (68% increment), but it does help to lowering the overshoot for 26% difference. The rise time for Case 2 is actually short, but when the point pH 4 to pH3, ANN-MPC drag the rise time for 83 minutes, thus it causes the rise time average longer.

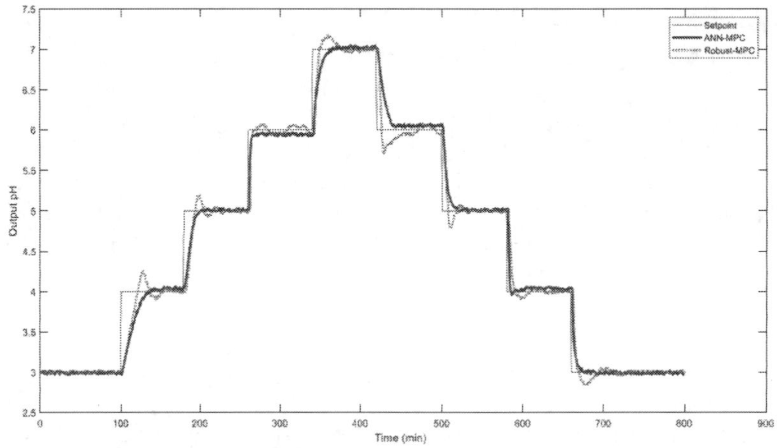

Figure 4: Response Diagram for setpoint tracking Case 1. Data for ANN-MPC and Robust MPC.

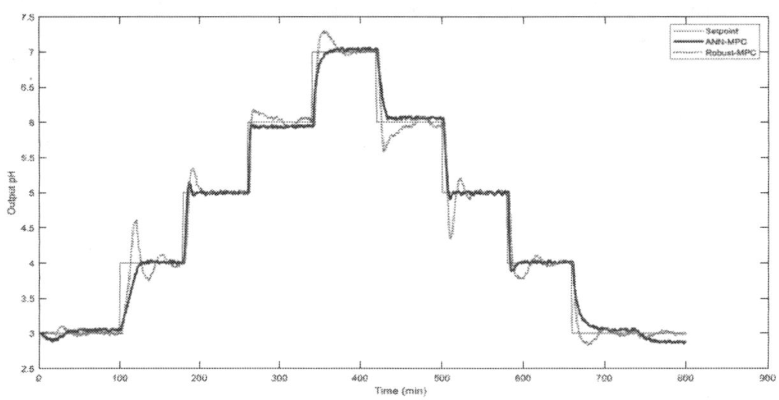

Figure 5: Response Diagram for setpoint tracking Case 4. Data for ANN-MPC and Robust MPC.

5.3 Computational Time
As one of the objectives is regarding to the computational time for ANN-MPC vs Robust-MPC, so the following analysis have been done. From the following table the, the result show that ANN-MPC have improved the computational time by reduce 60.2% from the Robust-MPC model.

Table 1: Computational Time in Seconds for each ANN.

Controller	ANN-MPC	Robust-MPC	Reduce
ANN(1)	746	1546	51.7%
ANN(2)	684	1327	48.5%
ANN(3)	730	2549	71.3%
Mean	**720**	**1807**	**60.2%**

6. Conclusions
In this paper, a non-linear black box artificial neural network-based model predictive control (ANN-MPC) is presented. Generally, the ANN-MPC can control the pH but limited to specific cases. But the application of ANN does reduce the computational time for MPC. Due to the black box behaviours, it is hard to defined what happen inside the ANN. Thus, the major limitation for this ANN-MPC in this paper is it only applicable to control the pH when it is training with the correct dataset. However, this should not happen if apply to industry control. Overall, the application of neural network base MPC

are still need further research and studies. But for sure that application of neural network will contribute a great result in current industry revolution 4.0. So, few of the recommendation for future work are listed:

1. Get high standard of understanding for neural network.
2. Improve the neural network by:
    a) Adjusting the number of neurons in hidden layer.
    b) Change a training algorithm.
    c) Use larger training dataset, but not overfitting.

References

[1] Akesson, B, & Toivonen, H 2006. A neural network model predictive controller. *Journal of Process Control, 16*, 937-946.

[2] AlDhaifallah, M, Alsabbah, Shebal, & Mujtaba, I 2016. A Predictive Neural Network-Based Cascade Control for pH Reactors. *Mathematical Problems in Engineering*, 7-15.

[3] Draeger, A, Engell, S, & Ranke, H. 1995. Model Predictive Control Using Neural Network. *IEEE Control Systems*, 61-66.

[4] Estofanero, L, Edwin, R, & Claudio, G 2019. Predictive Controller Applied to a pH Neutralization Process. *IFAC-PapersOnLine, 52*(1), 202-206.

[5] Georgieve, P, & Feyo de Azevedo, A 2006. Neural Network-Based Control Strategies Applied to a Fed-Batch Crystallization Process. *International Journal of Computational Intelligence, 3*(3).

[6] Hermansson, A 2019. ROBUST OFFSET-FREE CONTROL OF NONLINEAR SYSTEMS USING MODEL PREDICTIVE CONTROL AND INTEGRAL ACTION. *UPM Thesis*.

[7] Hermansson, A, & Syafiie, S 2015. Model predictive control of pH neutralization processes: A review. *Control Engineering Practice, 45*, 98-109.

[8] Huang, Y, Wang, H, Khajepour, A, He, H, & Ji, J 2016. Model predictive control power management strategies for hybrid electric vehicles: A review. *Journal of Power Sources*, 1-13.

[9] Kiš, K, & Klaučo, M 2019. Neural network based explicit MPC for chemical reactor control. *Acta Chimica Slovaca, 12*(2), 218—223.

[10] McAvoy, T 1972. Dynamic Models for pH and Other Fast Equilibrium Systems. *Industrial & Engineering Chemistry Process Design and Development, 11*(4), 630-631.

[11] Moumouh, H, Langlois, N, & Haddad, M 2019. A Novel Tuning approach for MPC parameters. *2019 IEEE 15th International Conference on Control and Automation (ICCA)*, 1638-1643.

[12] Pottman, M, & Seborg, D 1992. Identification of non-linear processes using reciprocal multiquadric functions. *Journal of Process Control, 2*(4), 189-203.

[13] Sha'Aban, Y, Tahir, F, Masding, P, Mark, J, & Lennox, B 2016. Control Improvement using MPC: A Case Study of pH Control for Brine Dechlorination. *IEEE Access, 4*, 1-12.

[14] Villaneuva Perales, A, Gutiérrez Ortiz, F, Vidal Barrero, F, & Ollero, P 2010. Using Neural Networks to Address Nonlinear pH Control in Wet Limestone Flue Gas Desulfurization Plants. *Industrial & Engineering Chemistry Research, 49*, 2263-2272.

[15] Wang, D, Shen, Z, Xin Yin, Tang, S, Liu, X, Zhang, C., . . . Norambuena, M 2022. Model Predictive Control Using Artificial Neural Network for Power Converters. *IEEE TRANSACTIONS ON INDUSTRIAL ELECTRONICS, 69*(4), 3689-3699.

[16] Wysotzki. F 1992. Machine Learning and its Application to Process Control. *Operations Research '91* (pp. 571–574). Physica-Verlag HD.

[17] Zarzycki, K, & Ławrynczuk, M 2021. LSTM and GRU Neural Networks as Models of Dynamical Processes Used in Predictive Control: A Comparison of Models Developed for Two Chemical Reactors. *MDPI Sensors, 21*, 5622.

IOP Conf. Series: Materials Science and Engineering 1257 (2022) 012036 doi:10.1088/1757-899X/1257/1/012036

Model Predictive Control – Sliding Mode Control of a pH system

W X Ooi, A W Hermansson and C H Lim

School of Engineering and Physical Sciences, Heriot-Watt University Malaysia Campus, 62200, Wilayah Persekutuan Putrajaya, Malaysia

a.hermansson@hw.ac.uk

Abstract. This paper studies the feasibility of using discrete sliding mode controller (SMC) to achieve offset-free control of nonlinear processes in the presence of disturbances. The performance of the SMC is compared to a multiple model predictive controller (MMPC) studying the ability of set-point tracking using the pH system as a case study. The results presented from the comparison show that SMC can perform offset-free control of a pH system, with the major drawback being slow response as well as oscillation at some pH values. Finally, a design of a combination between MMPC and SMC (MMPC-SMC) is proposed, with MMPC carrying out basic control response while the SMC fulfils the role of eliminating the offset. However, the inaccurate reading in the MATLAB simulation model does not generate the expected results on the pH control. Therefore, the modification on the MATLAB models is required to achieve the improved control system on the offset-free behaviour for the set-point tracking.

1. Introduction

In general, the processes involved in different industrial applications are exhibiting non-linear behaviours with different levels of severity, such as pH systems and distillation. The presence of nonlinearities in the process poses the challenges in the regulation of process variables around the target steady state as the closed-loop stability is difficult to achieve within non-linear systems [1]. The factors for the non-linear behaviour of the pH system includes the high sensitivity around the equilibrium point of neutralisation and transient behaviours related to the disturbances affecting the concentration of inlets [2]. Therefore, the control output is subject to fluctuation in linear controls and faces difficulties in convergence towards the desired pH setpoint adjustments. The following pH control problems illustrate the importance of development of non-linear models to improve the performance of stability and robustness of the control system while minimizing the computational complexity. Therefore, the use of advanced control techniques in dealing with complex non-linear systems is essential to achieve desirable control performance, such as Model Predictive Control (MPC) and Sliding Mode Control (SMC).

1.1. Model predictive control

The MPC is characterized by prediction the future process behaviour over a time interval and minimization a cost function. The optimization do penalization of the deviations between model and set-point over a pre-defined time horizon [3]. The optimal input sequence obtained is then repeated at each time step with the first input applied in the control sequence, allowing the cycles to recede into the future. One of the implementations of MPC is forming a multiple model predictive control (MMPC) with combination of multiple piecewise linear models (PWL) and the use of Bayesian weight calculator as optimization [4]. Referring to the research study by Hermansson and Syafiie, an adaptive integral controller is also augmented into the MMPC model (MMPC-I), leading to ability to reduce the tracking error formed from the external disturbances, contributing towards offset-free control [5]. However, there are several limitations to the current offset-free tracking approach, such as extra tuning effort, unguaranteed offset removal and increased computational load of the setup. In this context, the application of SMC for the integral action of the controller to remove the offset of the process is studied as the alternative methodology to control the non-linear processes.

1.2. Sliding mode control

The SMC is known as one of the variable structure controls where high-speed switching action is carried out to eliminate the initial errors for excellent robustness in achieving an offset-free tracking approach [6]. From the study of SMC by Ma, Zhao and Spurgeon, the SMC algorithms in discrete form is discovered to reduce the chattering effect caused by the infinite switching frequency [7]. Furthermore, the elimination of the external disturbance effects is also considered by adding a disturbance observer to measure the control input required to mitigate its deviation. The advantages of the SMC technique applied in multivariable pH control processes are described as the robustness and adaptive ability of the controller to handle the system with multiple uncertainties and disturbances [8]. In comparison, the performance output of the MPC and SMC in the pH control system are analysed to evaluate their abilities in achieving offset-free tracking and control response time taken. Other than that, Rubagotti, Raimondo, Ferrara and Magni also discovered the method of combining the MPC with SMC to utilize the features provided by both of the controllers to secure the robust stability of a continuous system in the presence of constraints and uncertainties [9]. In the following research study, a discrete SMC is also augmented into an MPC, but the MMPC control system instead to obtain the improvement in the controlling performance.

1.3. Problem statement

Although the implementation of integral control action in MMPC provides effective operation of disturbance rejection, the key concern of extra tuning effort is induced as different integral setting adjustments are required for each operating point in the multiple models. The occurrences of overshoot at the steep gradient change in the pH value are also observed from the offset tracking performance [10]. Furthermore, the robust stability in the MMPC is insufficient regarding the oscillation effect during the removal of offset. Therefore, the possible alternative to cope with the drawbacks is considered by applying the SMC technique. Discrete SMC model augmented with disturbance observer is modelled in the pH neutralization case study to compare the performance results. Last but not least, the design of MPC combined with discrete SMC control can be included in the research work to identify the feasibility of utilizing the MPC models in solving non-linear systems while SMC actions in reducing the uncertainties.

The primary aim of the research project is to study the implementation of control techniques to efficiently achieve the offset-free performance in non-linear system with reduced computational cost and time. The following objectives contributed to the project purpose included:

- Implementing Discrete SMC for a pH control system
- Comparing the performance output of pH control between offset-free MMPC and SMC
- Designing and implementing the model of combined MMPC-SMC in pH case study

2. Methodology

2.1. pH simulation

The main contribution of the research is demonstrating the performance of the advanced controllers through the simulation study of the pH behaviour by involving the neutralisation reaction between the acid and alkali in a continuous stirred tank reactor system (CSTR). As referred to the research conducted by Hermansson and Syafiie, the case study of neutralisation process is carried out in the CSTR with constant volume and constant flowrate of the feed stream [10]. The phosphoric acid solution (H_3PO_4) acts as the feed stream of the process while aqueous calcium hydroxide ($Ca(OH)_2$) is used as the control stream to manipulate the pH of the neutralisation reactions. The feed stream is installed into the CSTR system with constant flow rate but varying concentration while the control stream has constant concentration while the flowrate can be varied. The system includes further nonlinearities in the form of the mass transfers through precipitation or dissolution of the salt solids such as calcium and phosphates, $CaHPO_4$ and $Ca_3(PO_4)_2$ are modelled into the control system. The following pH case study is performed in MATLAB and applied for the non-linear control modelling by MMPC.

The MMPC is designed through the linearization around different base points (pH = 3, 4, 5, 6 and 7) using multiple PWL models and weighted by Bayesian weighting for switching between models. The general equation for a discrete PWL models is described below:

$$\tilde{x}(k+1) = A\tilde{x}(k) + B\tilde{u}(k-1) \quad (1)$$

$$\tilde{y}(k+1) = C\tilde{x}(k) \quad (2)$$

Which the $\tilde{x}, \tilde{u}, \tilde{y}$ represent the deviation variables for state input, control input and pH output respectively while A, B and C are constant matrices for the MMPC modelling. The computation of MMPC involves the minimization process of the cost function with the built-in tool *quadprog* for obtaining the optimal control output at each time step under the inclusion of limiting constraints ($0 < u < 0.012$, $\Delta u = 0.0025$). The integral controller action is also implemented into the MATLAB codes to show the effect of offset-free tracking. After finishing the setup, a series of simulation tests are conducted by tuning the pH operating setpoint in the sequence of 3-4-5-6-7-6-5-4-3 to observe the tracking abilities and disturbance rejection of the MMPC approach.

2.2. SMC

In comparison to the control performance of MMPC-I system, the SMC is analysed as the alternative control method using the similar MATLAB simulation of pH neutralisation in the CSTR system. In the following case, discrete sliding mode control is selected as the control approach to stabilize the pH control performance. The presence of external disturbances on the closed loop system is removed by adding the design of a disturbance observer. The discrete simulation model for SMC is designed based on the multiple discrete models available for different pH cases. These models are included with the Bayesian weighting by referring to the PWL models as shown in equation (1) and (2). Then, the control law input of SMC is applied to the models to simulate the pH control system. The control input of SMC is manipulated by sliding variable constant C_s and disturbance observer measurement constant, μ to adjust the control law of the SMC design. Subsequently, the following control input is simulated into the pH regulation system as similar to the case study in MMPC to compare their results on the tracking performance of the controller output in the presence of disturbances.

Besides, the possibility of combination between MMPC and SMC into a single control system is also considered in the research study. The controller model is created by adding the discrete sliding model control law as substitution to the adaptive integral action in the non-linear MMPC system. Thus, the control scheme of the MMPC-SMC controller handling the disturbances are presented to describe the stability properties of the control strategies and the comparison between the performance of offset-free tracking between the integral control and SMC action.

3. Results and discussion

3.1. Multi model predictive control implementation

The tracking response on the pH control simulation by MMPC with the addition of adaptive integral controller is presented in figure 1 below. The performance of MMPC-I model in the behaviour of setpoint tracking and disturbance rejection is used as the basis for the comparison study with the SMC.

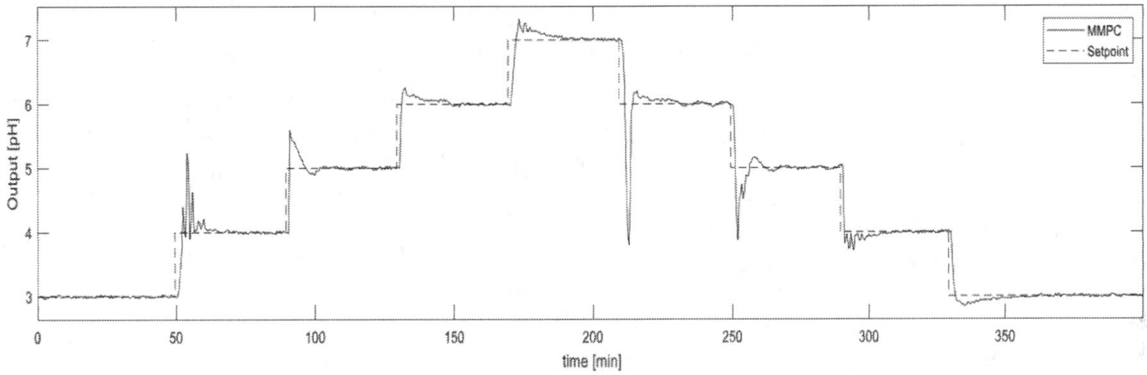

Figure 1. MMPC-I Response Diagram.

From the results obtained in MMPC-I, it is observed that the offset-free control is able to deliver a control output that eventually converges to the setpoint at different points of pH. However, the major drawbacks of the MMPC-I are displayed through the oscillation effects in the controller output and the overshooting behaviours at the initial stage of the step change in pH value. Hence, the control operation by SMC is studied to learn if any improvement is obtained from the control performance.

3.2. Sliding mode control implementation

In contrast to the operation of MMPC-I, the control input in SMC is obtained by feedback loop method where the deviation in the state input is measured at the sliding mode and the disturbance observer to generate the response required. Therefore, SMC model does not require the measurement of predicted output but instead finding the tracking response of the sliding manifold variable and the disturbance observer output.

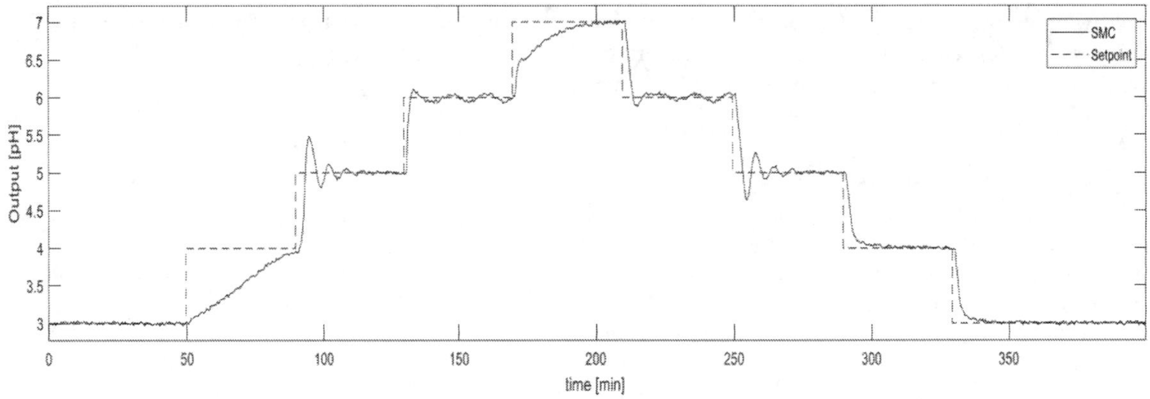

Figure 2. SMC Response Diagram.

The output response in SMC model also exhibits the achievement of the controller in setpoint tracking ability including the offset-free control performance. It is shown that most of the time the overshooting is less pronounced as compared to the response in MMPC-I. However, the rising time and settling time for different pH cases are observed to be significantly greater than the MMPC-I performance, especially in the case of pH = 4 and 7. The control behaviour in SMC is less favourable due to the slow response at the change in pH operating sequence. Moreover, the sliding mode controller are not sufficiently effective in compensating the oscillations of the control output, potentially due to the increased time constant at the switching of pH models. The limitations of using individual multiple model SMC in controlling the pH changes lead to the research attempt of combining both the MMPC and SMC.

3.3. Implementing MMPC-SMC

As progressed from the section 3.2, the combination of MMPC-SMC is generated based on reducing the load for SMC to work on the disturbance rejection as substituted for the role of adaptive integral controller in the MMPC. The change in control input from the SMC is implemented into the optimal control sequence of MMPC, alternating the change in control input for MMPC. The results obtained for the pH output are compiled in figure 3.

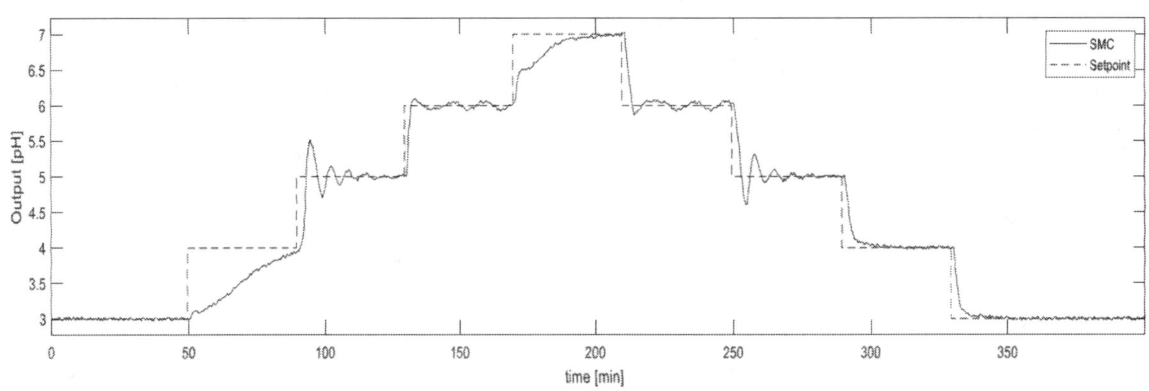

Figure 3. MMPC-SMC Response Diagram.

The simulation results provide that the offset between the control output and setpoint in MMPC control setup drastically increase after the addition of control input by SMC. The possible source of the deviation is the inaccuracy reading of the parameters in the SMC as the state input and control input variables are referred to the setpoint of the model instead of being the model output from MMPC. Therefore, this outcome is recorded as the research gap for this assignment which should be developed further by modifying the MATLAB codes for the improved referencing of the SMC variables.

4. Summary and conclusion

The discrete multiple model SMC is shown to achieve the offset-free tracking properties However, it is concluded to have a slower behaviour than the MMPC when dealing with pH control system. On top of that, the use of discrete SMC in the MMPC model does not show accurate results due to the error in modelling the variables of the controller setup. However, the potential in the combination of MMPC-SMC to be applied in non-linear control system has been shown to work for continuous cases, thus needing further work and should be possible as a research outcome. The future work should include the modification of the SMC-MMPC model to make a discrete combination to work. Alternatively, we could also turn towards further work of the multi model SMC, particularly increasing the switching speed through the reduced effect of chattering or oscillations and efficient offset removal.

Acknowledgements
A very great appreciation towards the supervision by Dr. Ayman for providing guidance and supports which leads to the completion of this research project. The assignment is only managed to be completed with the assistance and advices from both supervisor and co-supervisor, Dr. Lim.

References

[1] Camacho E F and Bordons C 2007 *Model Predictive Control* (London: Springer) pp 249-88
[2] Hermansson A W and Syafiie S 2015 Model predictive control of pH neutralization processes: a review *Control Engineering Practice* **45**
[3] Hermansson A W and Syafiie S 2020 Offset-free control of a pH system using multiple model predictive control *IOP Conf. Ser. on Mater. Sci. Eng.* vol 778 012072
[4] Hermansson A W, Syafiie S and Noor S M 2010 Multiple model predictive control of nonlinear pH neutralization system *Proc. on IEEE*
[5] Hermansson A W and Syafiie S 2014 Control of pH neutralization system using nonlinear model predictive control with I-controller *Proc. on IEEE*
[6] Zenteno-Torres J, Cieslak J, Davila J and Henry D 2021 Sliding mode control with application to fault-tolerant control: assessment and open problems *Automation* **2**
[7] Ma L, Dongya Z and Spurgeon S K 2018 Disturbance observer based discrete time sliding mode control for a continuous stirred tank reactor *15th Int. Workshop on Variable Structure Systems (VSS)*
[8] Chen J, Peng Y, Han W and Guo M 2011 Adaptive fuzzy sliding mode control in pH neutralization process *Procedia Engineering* **15**
[9] Rubagotti M, Raimondo D M, Ferrara A and Magni L 2011 Robust model predictive control with integral sliding mode in continuous-time sampled-data nonlinear systems *IEEE Trans. on Automatic Control* vol 56
[10] Hermansson A W & Syafiie S 2019 An offset-free MPC formulation for nonlinear systems using adaptive integral controller *ISA Transactions* **91**

IOP Conf. Series: Materials Science and Engineering 1257 (2022) 012037 doi:10.1088/1757-899X/1257/1/012037

A new way of working: Intelligent remote engineering via hybrid first principle modelling

A A M K Hing[1], A R Norman[1] and A F H A Mubin[2], M H Mansor[2]

[1]Group Technical Solutions, PETRONAS Technical Services Sdn. Bhd. (PTSSB) Malaysia

[2]Gas Transmission and Regasification, PETRONAS Gas Berhad (PGB) Malaysia

*e-mail: azleen.khairil@petronas.com, rahim.norman@petronas.com, ahmadfatthulhadi.ab@petronas.com, mohdhafeez@petronas.com

Abstract. The digital transformation era in the oil and gas industry is changing the way we operate our assets. A new way of working introduces automation of processes and a wider application of digital solutions. These improve remote engineering capabilities to maximise usage and applications of process advanced analytics to allow faster identification and troubleshooting of equipment and process issues. The foundation of remote diagnostic engineering is the hybrid of plant data and first principle simulation model. One of the main features of the tool is harnessing a process simulation model's strength in producing first-principle verified data and analysis, as example, a predictive composition properties model and virtual flow analysers. This soft sensor approach fills in the gaps of actual plant data availability and quality. Another main feature is the enablement of performance monitoring and diagnostics which includes equipment such as compressors, pumps, and heat exchangers. Adaptation of the new ways of working concepts will capture the intended outcomes namely reduced HSSE risks, improved profitability, higher capital efficiency and pandemic proofing. In the future, a wider application of the intelligent remote engineering tool is recommended to introduce prescriptive analysis, enriching insights for improved operational excellence.

1. Introduction

The digital transformation era in the oil and gas industry is creating a new way of working in the operations of our assets. Automation of processes, greater adaptation of Internet of Things technologies and wider application of digital solutions have direct positive impact towards remote engineering capabilities to eliminate human error and capitalize process advance analytics [1]. These has improved our asset operations by allowing faster identification and troubleshooting of equipment and process issues. Subsequently, this creates a more reliable asset operation creating opportunities for consistent plant optimisation. This paper will focus on the implementation of intelligent remote engineering via hybrid first principle modelling to address specific pain points faced by the operations of an asset. This implementation supports the new way of working for plant engineers to perform monitoring and reporting by leveraging on the strength of both first principle models and existing data infrastructure.

2. Methodology

The foundation of the remote diagnostic intelligent engineering tool that was developed is the hybrid of plant and calculated first principle data. The workflow starts with live data from sensors being captured into a historian system. Trusted and reconciled parameters captured from actual plant data were selected to be utilised as input to a first principle model to be run online and in near-real time.

From the first principle simulation, important calculated data was derived and sent to an asset framework solution which gives context to data and organises the information to align with asset needs. The depository of both plant and derived data is then utilized for further predictive analysis and monitoring via online dashboards. *Figure 1* shows a summary of typical workflow for the remote diagnostic intelligent engineering tool developed.

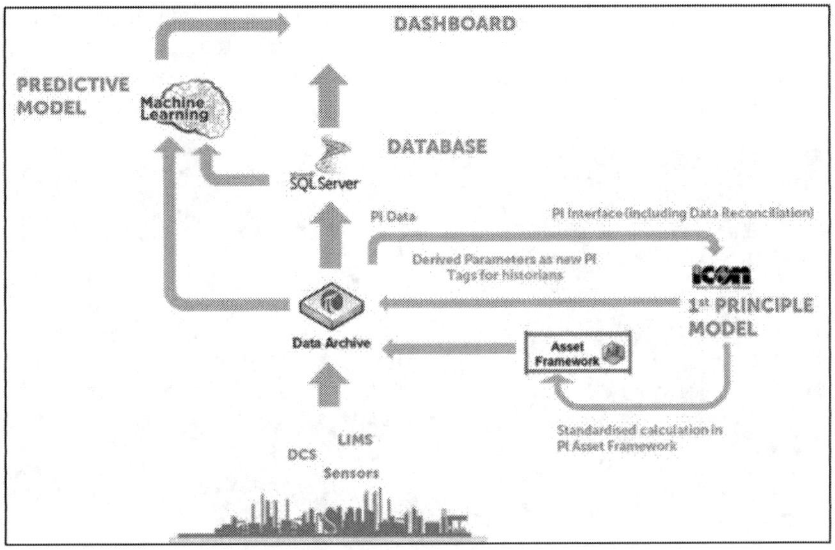

Figure 1. Remote diagnostic intelligent engineering tool workflow.

3. Case Studies

The implementation of intelligent remote engineering was carried out for an LNG regasification process. The regasification plant is operated to receive and store Liquefied Natural Gas (LNG) in LNG storage units, regasify the LNG for sales gas consumers and reload LNG into carriers for other alternative LNG customers. A digital transformation program was initiated to optimize operating cost and reduce health, safety, security, and environment exposure of the plant. Several pain points were identified to be addressed through intelligent remote engineering implementation to support the digital transformation program.

The first pain point identified is the unavailability of hard sensors of some critical parameters namely composition, flowrate and temperature readings at several important locations. This ultimately, reduces the accuracy in mass and energy balance calculations accounting and the estimation of LNG supply and sales gas fluid properties which impacted supply and sales balance.

The second pain point identified is having offline or manual calculations consisting multiple non-automated spreadsheet calculations for equipment and process monitoring which results in ineffective time spent by engineers for reporting and increases possibility of human errors. Calculated data or parameters reside in engineer's local machines which are not available online for interactions with other digital initiatives for more accurate plant performance insights [2].

The last pain point is the frequent change in plant's mode of operation. The plant switches between LNG unloading, LNG bunkering and holding modes are totally depending on the demand requirements of the customers. This requires the model to be dynamic to be able to generate accurate equipment and asset performance diagnostic results.

4. Results and Discussions

4.1. Soft Sensors

One of the main features of the tool is harnessing a process simulation model's strength in producing first-principle verified data and analysis, as example, a predictive composition properties model and virtual flow analysers. This soft sensor approach fills in the gaps of actual plant data availability and quality. Figure 2 and Figure 3 show the derived LNG composition and physical properties from the first principle mass and energy balance. The availability of more information derived from process simulation model using the virtual analysers such as calculated fluid composition, heating values and equilibrium flashing increased the accuracy of production planning activities.

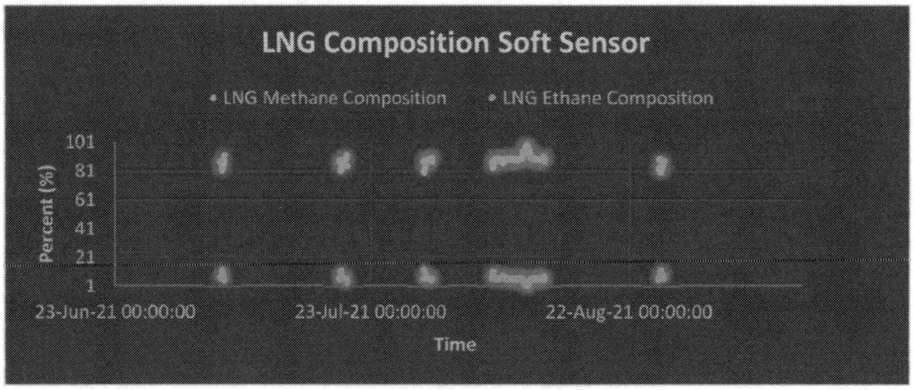

Figure 2. LNG composition derived data.

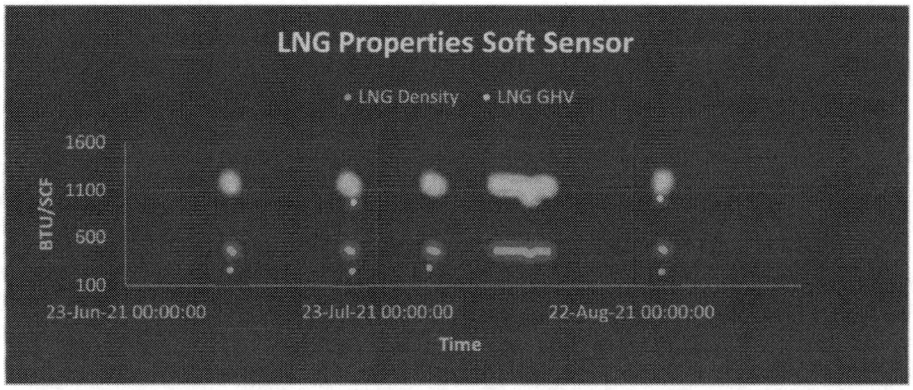

Figure 3. LNG properties derived data.

The derived data from first principle model are sent to an asset framework solution which gives context to data and organize the information to align with asset needs. The Asset Framework (AF)

IOP Conf. Series: Materials Science and Engineering 1257 (2022) 012037 doi:10.1088/1757-899X/1257/1/012037

supplements the Data Archive architecture by providing a hierarchal structure for all data in the organization [3]. The depository of both plant and derived data can then utilized for further predictive analysis and monitoring via online dashboards.

Some of the calculations performed in the Asset Framework Solution include:

- Mass balance reconciliation and accounting
- Reconciliation of plant and derived calculation results - removal of outliers, out of bound values
- Flowrate regression from pump current/flow lookup table
- Offgas compressor and LNG train running status

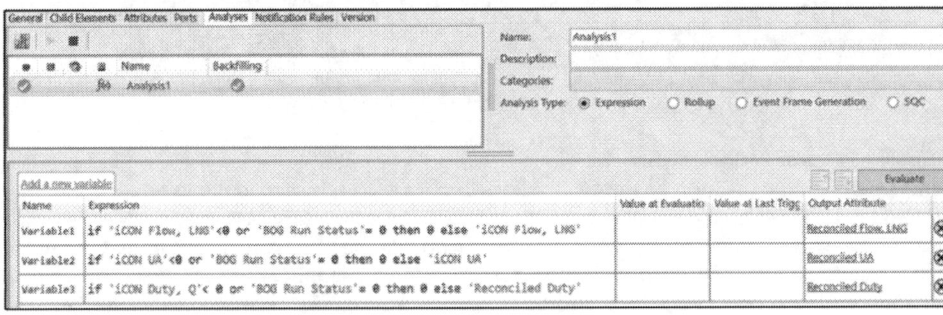

Figure 4. Snapshot of Asset Framework calculation.

The application of advanced analytics coupled with actual plant information automated the detection of operation modes, thus, making the hybrid first principle model dynamic to actual situations.

A main feature of the remote engineering tool developed is the enablement of performance monitoring and diagnostics which includes equipment such as compressors, pumps, and heat exchangers. Monitoring critical equipment performance indicators with higher accuracy allowed more timely troubleshooting and better maintenance planning. **Table 1** shows the typical operating parameters monitored.

Table 1. Typical equipment performance parameters monitored.

Equipment Type	Operating Parameters
Heat Exchanger	Duty, fouling factor, heat transfer coefficient
Rotating Equipment	Polytropic head, efficiency
Expander	Differential head, efficiency
Distillation Column	Maximum column flooding factor, vapor and liquid tray loading

An example of heat exchanger duty trending against design is shown in Figure 5.

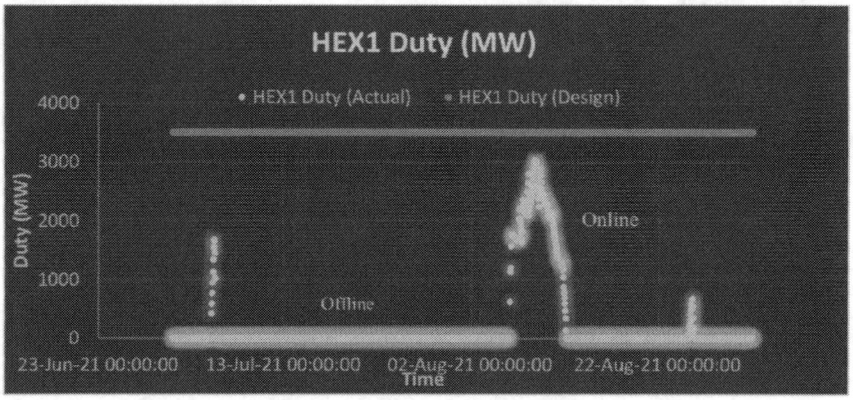

Figure 5. HEX1 Duty trending.

5. Conclusions and Future Work

The implementation of remote engineering via hybrid first principle model can address existing asset operational pain points creating a new way of working. The adaptation of the new way of working will capture the intended outcomes namely reduced HSSE risks, improved profitability, higher capital efficiency and pandemic proofing. In the future, a wider application of the intelligent remote engineering tool is recommended to introduce prescriptive analysis, enriching insights for improved operational excellence. The wider application can extend beyond asset operations to include financial and emissions information to support the oil and gas industry's aspiration of reaching Net Zero Carbon Emissions by 2050.

Acknowledgement

This work was supported by PTSSB and PGB engineering and digital teams.

References

[1] Norman A. *Remote equipment surveillance leveraging on 1st principle data.* 2022 2

[2] Hussein N. *Smart furnace analytics - delivering operational excellence and sustainability through domain and data* 2021 1

[3] OSISoft. *PI Server sp3 user guide, pi asset framework and pi system explorer.* 2018 6

IOP Conf. Series: Materials Science and Engineering 1257 (2022) 012038 doi:10.1088/1757-899X/1257/1/012038

Integration of neural network-based prediction for enhanced process controllability: application in hydrogen iodide decomposer for hydrogen production via water splitting process

SA Sirajul Adly[1] and N Mohd[1*]

[1]School of Energy and Chemical Engineering, Xiamen University Malaysia, 43900, Sepang, Selangor, Malaysia

*e-mail: noraini.mohd@xmu.edu.my

Abstract. Sulfur-Iodine Thermochemical cycle (SITC) process can be broken down into 3 sections. Based on the literature, there is limited study on the third section which is hydrogen iodide (HI) decomposition. In this work, a study to develop a model and controller of the HI decomposition is carried out. The goal of this work is to address this important gap, and more specifically to focus on the controllability of the HI decomposition section. Before the control and simulation study can be performed, a dynamic model of a HI decomposer is first established to obtain the baseline for the necessary data to be fed into the Artificial Neural Network for training to predict the correct outcome accordingly. The proposed controller is a Multi-Scale Control (MSC) integrated into an Artificial Neural Network (ANN) model (ANN-MSC-based-PID). It is worth highlighting that; the proposed model-based control strategy has proven to effectively control the HI decomposition reactor.

1. Introduction

The Sulfur-Iodine Thermochemical cycle (SITC) process is identified as a promising environmental friendly route to produce hydrogen fuel via the water-splitting method, commercially. A recent review reported that SITC is proven to produce the highest energy and exergy efficiency by using the lowest cost and no need for electricity when compared to other types of thermochemical cycles [1]. Besides, it can deal with massive production due to the ability of coupling with high-temperature solar [2]. A few factors presented by [3] increase the possibilities for scaling up the SITC process to an industrial scale. The factors presented by Zhang and co-workers are as follows [3]:

a) The SITC process is a purely thermal, the industrial scale is estimated to be very economical,
b) The SITC process is proven to achieve a thermal efficiency of 69%. Therefore, this is a good indicator for large-scale hydrogen production,
c) The SITC process is an all fluid state process, which makes it easier to be scaled up consequently realizing continuous operation.

Three SITC sections include Bunsen Reaction Section, Sulfuric Acid Decomposition Section and Hydrogen Iodide (HI) Decomposition Section which is seen in the chemical equation (1) to (3). The feedstocks are water, iodine and sulfur. The recyclable materials are iodine and sulfur.

Bunsen Reaction (Section 1)

$$SO_2 + I_2 + 2H_2O \leftrightharpoons (I_2 + H_2O + 2HI)_{LP} + (H_2SO_4 + H_2O)_{HP},\ \Delta H_1 = -75 \pm 15 \frac{kJ}{mol} \qquad (1)$$

Sulphuric Acid Decomposition Reaction (Section 2)

$$H_2SO_4 \rightarrow H_2O(l) + SO_2(g) + \frac{1}{2}O_2(g), \Delta H_2 = 186 \pm 3 \frac{kJ}{mol} \quad (2)$$

Hydrogen Iodide Decomposition Reaction (Section 3)

$$2HI \rightarrow H_2 + I_2, \Delta H_3 = 12 \frac{kJ}{mol} \quad (3)$$

Based on section 1 of SITC, the energy requirement for reversible reaction to occur was studied at 393K (120°C) [4]. In section 2 of SITC, the required temperature for the sulfuric acid decomposition is between 750°C to 900°C or higher and a catalyst should be presented in the reaction itself to obtain the desired product efficiently. While in section 3 of SITC, the hydrogen iodide decomposition reactor requires a temperature of 500°C with a catalyst presented inside the reactor for the reaction to occur.

One of the important studies in the HI decomposition section is on the catalyst effect on its productivity. A team of researchers from Korea [5] presented that using the platinum with a catalyst medium of gamma-alumina can produce hydrogen at a steady rate with a condition that the preparation of such catalyst shall be consistent. This catalyst is also known as a monometallic catalyst where a single metal element is present on the catalyst medium. Another team of a researcher in India [6] demonstrated that the performance of the catalyst can be enhanced by adding a non-noble metal into the catalyst preparation. The non-noble metal is nickel while the noble metal is platinum. This catalyst is known as the bimetallic catalyst.

Besides, process control plays an important role in the chemical process which contributed to optimizing the process, safety and health and reducing the environmental impact. HI decomposer temperature is an important variable that requires extensive control to maintain the desired value where there is scarce work found on its controllability. In this work, Artificial Neural Network, ANN model will be developed and trained to capture the SITC behaviour. Before ANN model development, a first principle (FP) dynamic model of HI decomposer is developed and the FP model will be validated in this work. The reason ANN is employed is due to its ability to mimic the actual process behaviour as presented by FP dynamic model.

2. Materials and Methods

The first principle (FP) model for the HI decomposer is developed using MATLAB software. The HI decomposition reactor is designed with a co-current configuration as shown in figure 1. Data and kinetics are collected from [7]. The configuration of the reactor uses a counter-current heating medium with a solid catalyst present in the reactor. The notation of p_{HI}^0 is the partial pressure of HI solution, $p_{I_2}^0$partial pressure of Iodine solution and $x_E$is the equilibrium conversion of HI. The rate of reaction is presented in equations (4) and (5):

$$r_{HI} = \frac{dx}{dt} = k(1-x) - k'\frac{x}{2}\left(\left(1+\sqrt{K_{I_2}}\sqrt{p_{I_2}^0 + \frac{p_{HI}^0 x}{2}}\right)^2\right)^{-1} \quad (4)$$

$$k' = 2k(1-x_E)(x_E)^{-1} \quad (5)$$

Equation 4 can be seen become a linear expression which is shown in equation 6.

$$y = p_1x_1 + p_2x_2 \quad (6)$$

$$y = \left(\frac{\left[(1-x)-\frac{(1-x_E)x}{x_E}\right]}{r_{HI}}\right)^{\frac{1}{2}}, x_1 = 1, x_2 = \left(p_{I_2}^0 + \frac{p_{HI}^0 x}{2}\right)^{\frac{1}{2}} \quad (7)$$

$$p_1 = \frac{1}{(k)^{\frac{1}{2}}}, p_2 = \left(\frac{K_{I_2}}{k}\right)^{\frac{1}{2}} \quad (8)$$

$$k = 3.13 \times 10^1 e^{\frac{-37.1}{RT}} \quad (9)$$

$$K_{I_2} = 1.80 \times 10^{-9} e^{\frac{75.1}{RT}} \quad (10)$$

The value of the x_2 is calculated from the measured data at each parameter while the $p_1$and p_2 obtain from the linear regression method. The rate of adsorption of the K_{I_2} and I_2 are estimated based on the value of using $p_1$and p_2 are then expressed as equations 9 and 10. The catalyst being used is a Platinum/Gamma-Alumina at wt% of 1.0 [7].

Simulation of the HI decomposition reactor in MATLAB Simulink involves 22 sections of the reactor calculation. The mole balance of the hydrogen iodide reactor input and output are represented in equation (11) to (13). The notation are y_A is the molar fraction, zone given *j* = *1,2,....n*, F_p is process stream flowrate and W_{cat} is the weight of catalyst.

$$\frac{dn_{A_j}}{dt} = F_p\left(y_{A_{j-1}} - y_{A_j}\right) - r_{A_j}W_{cat} \tag{11}$$

$$\frac{dn_{B_j}}{dt} = F_p\left(y_{B_{j-1}} - y_{B_j}\right) + \frac{1}{2}\left(r_{A_j}W_{cat}\right) \tag{12}$$

$$\frac{dn_{C_j}}{dt} = F_p\left(y_{C_{j-1}} - y_{C_j}\right) + \frac{1}{2}\left(r_{A_j}W_{cat}\right) \tag{13}$$

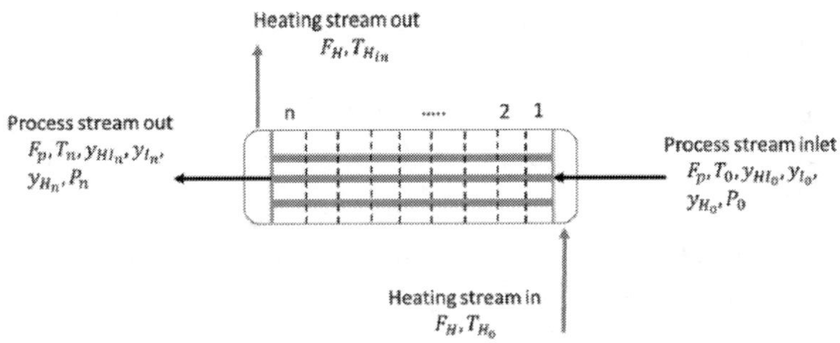

Figure 1. Schematic diagram of the co-current configuration HI decomposition reactor.

Component sub notation:
A = *HI*, B = H_2, C = I_2

The energy balances are shown in equations (14) and (15), where the notation of i and j refer to the number section. C_{p_j} denotes the heat capacity of the process fluid, C_{h_j}denotes the specific heat capacity of Helium gas, T_{p_j} is the process fluid temperature, T_{h_j} is the heating medium temperature, ΔH_r refers to the heat of reaction, $\rho_{h_j}$is the density of helium, $N_{t_j}$is the total number of shell per zone, A_s is the heat transfer area per zone, V_s is the volume of reactor per zone and $Q_{S_j}$is the heat being supplied from the heating medium.

$$\frac{d(\bar{C}_{p_j}N_{t_j}dT_{p_j})}{dt} = F_p\bar{C}_{p_j}\left(T_{p_{j-1}} - T_{p_j}\right) - \Delta H_r r_{A_j}W_{cat} + Q_{S_j} \quad \textit{(Catalyst Area)} \tag{14}$$

$$V_S\left[\frac{d\left(C_{p_j}\rho_{h_j}T_{h_j}\right)}{dt}\right] = F_h\left(C_{h_{j-1}}T_{h_{j-1}} - C_{h_j}T_{h_j}\right) - Q_{S_j} \quad \textit{(Shell and Tube)} \tag{15}$$

$$Q_{S_j} = UA_S\left(T_{h_j} - T_{p_j}\right) \tag{16}$$

3. Simulation Studies

3.1 First Principle (FP) Model Validation

For model validation, equations (4) to (16) were solved based on the numerical differentiation formulas (NDFs), and ode45s function. The parameter values used are given in table 1. Once the simulation process reached an equilibrium state, the result is taken to compare with the results of the laboratory reported in the open literature. Table 2 tabulates the parameters and the comparison results at two different temperatures, 400°C and 500°C, between the experimental model from the literature and the developed simulation model in the current work. It is showing that the developed model is comparable with the experimental data from literature and can be used to conduct further studies such as process controller design. Upon FP model validation, two different catalysts are introduced into the MATLAB Simulink system and simulated in an open-loop simulation. The monometallic catalyst is reported in [8] using platinum and gamma-alumina as the medium activation energy is at 37.1 J/mol. The bimetallic catalyst activation energy was reported at 23.9 J/mol using Platinum and nickel with activated carbon as the catalyst medium [9]. Both catalysts for monometallic and bimetallic were observed and it shows that the superiority of the catalyst favours the bimetallic.

3.2 Artificial Neural Network (ANN) Development

ANN have the potential to substitute the first principle (FP) model for chemical processes that have limited kinetics and experimental data in the literature. In this work, a custom code of Artificial Neural Network (ANN) was developed in MATLAB. The input-output data consisted of between 7700 to 17270 samples generated from the validated first principle model that was used for the development of the ANN model.

Table 1. Parameter for Validation of Bimetallic Catalyst.

Parameters	**Value**
Pressure	101.3 kPa
Temperature	400 to 550°C
Inner Diameter	0.016 m
Tube	1
Length of Tube	1 m
Flowrate of HI	2.8×10^{-8} kmol/s
The molar fraction of HI	0.55

Table 2. Validation of Bimetallic Implement into Simulation.

Catalyst	Ni(2.5)-Pt(2.5)	Ni(2.5)-Pt(2.5)
Medium	Activated Carbon	Activated Carbon
Temperature	400°C	550°C
Conversion of HI (Simulation)	19.09%	23.14%
Conversion of HI (Laboratory)	18.1%	23.9%
Deviation percentage	5.19%	3.18%

The ANN input use 5 parameters that were obtained from the validated first principle process model. The ANN consist of 4 hidden layers with the training algorithm using Levenberg-Marquardt optimization. The ANN required to large data set to achieve the highest accuracy which can be obtained from the open-loop test at different temperature conditions. The development of ANN uses 5 inputs which are the temperature of helium, the flowrate of helium, molar flowrate of HI, molar flowrate of iodine and molar flow rate of hydrogen. The equation of the ANN used in this work is shown in equation (17). The activation function used is the sigmoid function between 0 and 1 where 0 is the lowest probability and 1 is the highest probability of becoming the output. The output layer consists of the temperature of helium, the temperature of HI and the conversion rate. The notation Y_1 is the output where the subscript is the number of output 1, 2 and more. The notation W_1 is the weight of the input. While the X_1 is the inputs that are mentioned above. The subscript is determined by the number of input and output which in this case is 5 input and 3 output.

$$Y_1 = Activation\ (W_1X_1 + W_2X_2 + W_3X_3) \tag{17}$$

3.3 ANN Model Identification

Once the ANN has been trained successfully, the ANN model undergoes an open loop test for model identification to observe the behavior of ANN FP model which can be seen in figure 2. It is shown the ANN fit well with FP model.

4. Controllability Study

A linear analysis was conducted for both the monometallic and bimetallic catalyst FP model. The linearized FP model is then generated using the linear analysis function in MATLAB. Equation (18)

and equation (19) show the linearized transfer functions for monometallic and bimetallic respectively.
Monometallic

$$G_p = \frac{0.4e^{-0.119s}}{1.65s+1} \quad (18)$$

Bimetallic

$$G_p = \frac{0.11e^{-15.64s}}{23.46s+1} \quad (19)$$

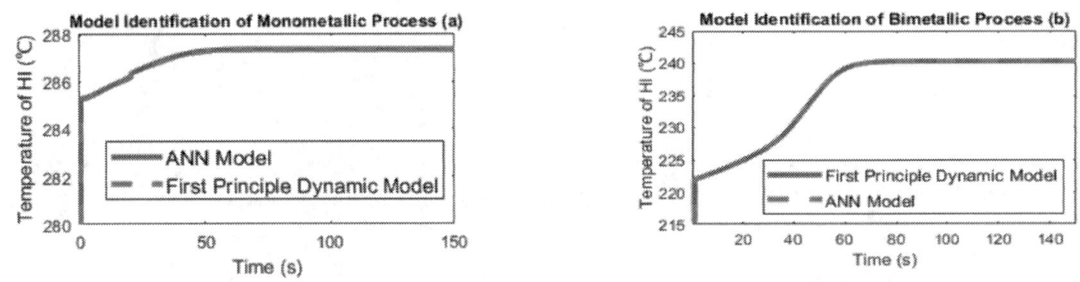

Figure 2. Model identification for Monometallic (a) and Bimetallic (b) with the ANN model.

4.1 ANN-MSC-based-PID Controller

After undergoing training, tested and validated, the ANN is linked with the MSC-based-PID controller in the MATLAB Simulink environment. The details of work on MSC-based-PID can be found in [10]. Figure 3 and 4 shows the response of MSC-based-PID (MSC-based-PID) in controlling the monometallic and bimetallic HI decomposition process which is represented by a trained ANN model respectively. The performance is compared to the linearized model controlled by IMC-based-PID (IMC-based-PID). Regardless of the catalyst type, the ANN model paired with MSC-based-PID controller configurations produced a smaller IAE value (2.175 monometallic and 10.93 bimetallic respectively) as compared to the linearized model paired with IMC-based-PID controller (7.989 for monometallic and 11.01 for bimetallic respectively). The control produces a satisfactory control response with fast settling time and low overshoot which can be seen in table 3. Noted that the MSC-based-PID is developed based on slow mode and fast mode transfer function of HI decomposition. Its ability to control a different model from its reference design shows that it is a robust controller. Overall the performance favours the ANN model paired with the MSC-based-PID controller.

Table 3. Overall characteristics of IMC-based-Monometallic, MSC Monometallic, IMC-based-PID Bimetallic and MSC Bimetallic.

Performance Criteria	Linearized FP Model-IMC-based-PID Monometallic	ANN-MSC-based PID Monometallic	Linearized FP Model-IMC-based-PID Bimetallic	ANN-MSC-based-PID Bimetallic
Rise Time	582.76	429.38	25.10	22.60
Settling Time	1.2095×10^3	1.5101×10^3	1.2335×10^3	1.7788×10^3
Overshoot	0.8347	11.9590	3.7211	5.3472

5. Conclusion

In this work, two different first principle models namely monometallic and bimetallic based catalyst models for HI decomposition reactor were developed and validated. Both catalysts for monometallic and bimetallic were observed and it shows that the superiority of the catalyst favours the bimetallic. The HI conversion rate using bimetallic improved by 16.98% at the temperature of 550°C. Controllability studies were carried out where IMC-based-PID shows efficiency in controlling the linearized models. ANN-MSC based-PID controller for the HI decomposition reactor is also simulated. Overall, ANN-MSC-based-PID has shown good performance, enhancing the controllability of the HI decomposition reactor. Noted that the dead time poses a limitation to the controller's capability. An improvement on the custom algorithm tailored of ANN to the specific requirement of the dynamic model shall be recommended. In addition, setpoint tracking and disturbance rejection tests shall also be performed in the future study.

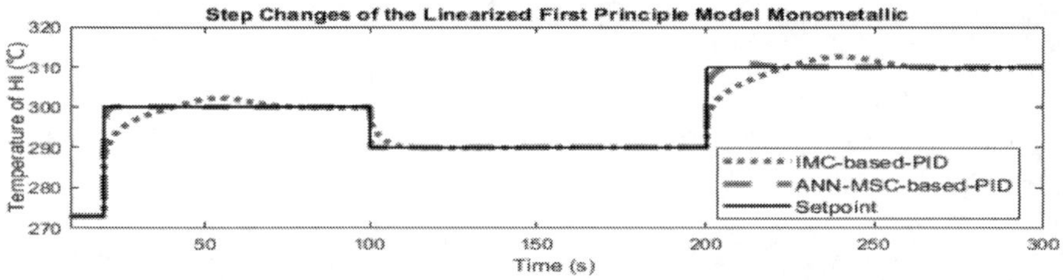

Figure 3. Step Response of Linearized FP Model-IMC-based-PID (IMC-based-PID) Controller and ANN-MSC-based-PID (MSC-based-PID) on controlling monometallic process.

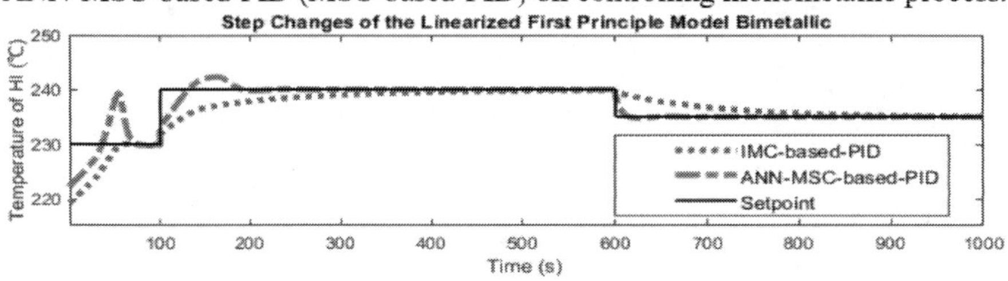

Figure 4. Step Response of Linearized FP Model-IMC-based-PID (IMC-based-PID) Controller and ANN-MSC-based-PID (MSC-based-PID) on controlling bimetallic process.

Acknowledgement
The financial support from XMUMRF grant (XMUMRF/2021-C7/IENG/0035) of Xiamen University Malaysia is greatly acknowledged.

References

[1] Mehrpooya, M., & Habibi, R 2020 *A review on hydrogen production thermochemical water-splitting cycles* **275** 31.

[2] C. Sattler, M. Roeb, C. Agrafiotis, and D. Thomey 2017. *Solar hydrogen production via sulphur based thermochemical water-splitting* **56** 30 – 47.

[3] P. Zhang, S. Z. Chen, L. J. Wang, T. Y. Yao, and J. M. Xu 2010. *Study on a lab-scale hydrogen production by closed-cycle thermo-chemical iodine-sulfur process* **35** 10166 – 172.

[4] L. C. Brown, G. E. Besenbruch, R. D. Lentsch, K. R. Schultz, J. F. Funk, and P. S. Pickard 2003. *High Efficiency Generation of Hydrogen Fuels Using Nuclear Power* **1** 1 – 343.

[5] L. Wang, K. Murata, and M. Inaba 2003. *Development of novel highly active and sulphur tolerant catalysts for steam reforming of liquid hydrocarbons to produce hydrogen* **257** 43 – 47.

[6] A. Alshammari, V. N. Kalevaru, and A. Martin 2016. *Bimetallic catalysts containing gold and palladium for environmentally important reactions* **6** 1 – 25.

[7] Nguyen, T. D. B., Gho, Y. K., Cho, W. C., Kang, K. S., Jeong, S. U., Kim, C. H., Park, C. S., & Bae, K. K 2014. *Kinetics and modeling of hydrogen iodide decomposition for a bench-scale sulfur-iodine cycle* **115** 531 – 39.

[8] G. Saracco, H. W. J. P. Neomagus, G. F. Versteeg, and W. P. M. Van Swaaij 1999. *High-temperature membrane reactors: Potential and problems* **54** 1997 – 2017.

[9] Q. H. Seer and J. Nandong 2017. *Stabilising PID tuning for a class of fourth-order integrating non-minimum-phase systems* **1** 1 – 17.

IOP Conf. Series: Materials Science and Engineering 1257 (2022) 012039 doi:10.1088/1757-899X/1257/1/012039

Life Cycle Optimisation for Bio-energy Systems with Carbon Capture and Storage in Electricity Generation

X H Cheng[1], Denny K S Ng[1], L Y Ng[2] and V A Murugappan[3]

[1] School of Engineering and Physical Sciences, Heriot-Watt University Malaysia, 62200, Putrajaya, Wilayah Persekutuan Putrajaya, Malaysia

[2] Department of Chemical Engineering & Petroleum Engineering, Faculty of Engineering, Technology and Built Environment, UCSI University (Kuala Lumpur Campus). USCI Heights, 1, Jalan Puncak Menara Gading, Taman Connaught, 56000 Cheras, Kuala Lumpur, Malaysia.

[3] Faculty of Engineering, Computing and Science, Swinburne University of Technology, Jalan Simpang Tiga, 93350, Kuching, Sarawak, Malaysia.

Corresponding Author: vmurugappan@swinburne.edu.my

Abstract. Bioenergy with carbon capture and storage (BECCS) has been identified as the most viable and cost-effective technology to achieving the 1.5°C targets set down in the 2015 Paris Agreement. It is essential to understand the impact of BECCS on carbon dioxide removal at a commercial scale. As such, this research presents an extensive life cycle analysis and optimisation of BECCS. To ensure the proposed methodology is appropriate for this research, case study of three cases is performed. In the case study, an existing coal-fired power plant is used as a base case for comparison between 2 different BECCS configurations. A regional assessment of the BECCS configurations from various critical performance aspects is then performed. This consists of assessing the system effectiveness, environmental impacts, energy efficiency, and cost optimisation from a life cycle perspective.

1. Introduction

Human activities have resulted in increased emission and accumulation of greenhouse gases (GHG) in the atmosphere. The largest source of anthropogenic GHG comes primarily from carbon dioxide (CO_2) emission through the combustion of fossil fuels for energy systems use [9]. The U.S. Energy Information Administration has estimated an increase global energy demand by at least 47% where fossil fuel remains the dominant source of energy [6]. This carbon-intensive economy has resulted in a 1°C average temperature rises since pre-industrial times and is having cascading detrimental effects on the environment [8]. Therefore, a shift towards sustainable sources of energy through legislation and international agreements is highly anticipated. In 2015, 195 nations have agreed on the Paris Agreement to limit the global temperature rise by 2100 to 1.5°C or 2°C [5]. In fact, leaders of many countries have recently met at the COP26 and reach for an agreement to phase out unabated coal [7]. To achieve the 1.5°C targets set by the Paris Agreement, emissions will need to be at net-zero levels to balance anthropogenic emissions by sources and removals of GHG by sinks. Thus, many climate policymakers and researchers have directed their attention to bioenergy carbon capture and storage (BECCS) to achieve the net-zero emission targets through energy system decarbonisation.

Research works have been directed towards evaluating the practicability of BECCS. For instance, Fajardy and Mac Dowell [3] have developed a modelling tool to evaluate the carbon-negative potential of BECCS power plant over its lifetime of 50 years. In terms of the technical feasibility of BECCS, Creutzig *et al.* [2] have performed in-depth research on the compatibility of BECCS with intermittent energy systems which consider direct and indirect land-use changes as one of the assessment criteria. Although there is increasing literature on the technical feasibility of BECCS, a holistic understanding of the overall life cycle environmental impacts of BECCS is still scarce. For example, an explicit framework was developed to assess the environmental benefits of combustion-based electricity generation systems with CCS through life cycle analysis (LCA) [12]. Apart from that, a life cycle meta-analysis was performed to explore the potential to retrofit existing coal power plants with BECCS [1]. In this study [1], the life cycle environmental impacts of eight carbon capture technologies were compared with sensitivity analysis performed by calculating the energy return on investment.

Most of the works discussed above focus on the environmental impacts or benefits of BECCS systems. However, life cycle cost analysis is crucial too. Moreover, there is no joint consideration on the practicability of BECCS in terms of environmental, cost and energy aspects from a life cycle perspective. Aside from that, previous LCA studies from the literature discussed are static assessments. They do not account for optimisation of different negative emissions pathways to determine the best solution in terms of environmental impacts, cost, and system effectiveness.

Therefore, to address the research gap mentioned, this research presents a comprehensive optimisation model of BECCS. A regional assessment of BECCS from various critical performance aspects is performed. This consists of assessing the system and cost effectiveness, environmental impacts and energy efficiency from a life cycle perspective. The resulting BECCS portfolios will then provide a new benchmark for comparing and choosing an optimal carbon removal approach to solve the climate emergency. This research aims to define the optimal negative emissions pathway within the existing power generation system while simultaneously minimising impacts on the environment, resources, and ecosystems. In the next section, the research methodology is demonstrated.

2. Methodology

The proposed methodology for this research is life cycle optimisation (LCO). LCO allows static LCA and techno-economic analysis to be performed synchronously to optimise and screen process alternatives [4]. In LCO, LCA is integrated with optimisation where an optimal solution will be selected with trade-offs between several components. On one hand, LCA assesses and quantifies the environmental impacts within all stages of the system's life cycle by adopting a cradle to gate approach. On the other, optimisation tools automate the generation of solutions that leads to environmental and economic improvements. Therefore, additional steps are required in LCO as compared to the conventional mathematical optimisation.

This research assessed the entire network represented in figure 1 by adopting a cradle-to-gate approach. The technologies included in this research were selected by defining the goal and scope of the study. Appropriate equations were then formulated based on the understanding of the working principle to mathematically model the behavior of each energy generation and carbon capture technologies pathway. The developed equations were then coded on a commercial mathematical optimisation software i.e., LINGO version 19, to develop an optimisation model for this research. Several equations that determine the key indicators used in this research were also formulated and discussed below. This includes total life cycle carbon emissions (TLCCE) and CO_2 offset.

The TLCCE for the overall system is computed using equation 1. TLCCE is calculated based on the summation of total CO_2 emissions from the system per MWh of electricity generated. CO_2 emissions can be classified as anthropogenic and biogenic. Anthropogenic CO_2 emissions refer to the emissions coming from fossil fuels. It can be generated from power generation, feedstock or product transportation, coal mining, and fertiliser production in biomass cultivation. On the other hand, biogenic CO_2 emissions refer to the CO_2 emissions originating from green resources. In this research, total CO_2 emissions from the system are denoted by the summation of both anthropogenic and biogenic emissions. This includes CO_2 emissions coming from feedstock production: fertilizer production, biomass mill, coal mining and power generation from feedstock combustion (Em_{ijp}), transport facility: biomass and coal transportation from production site (Em_{ij}), and carbon capture technology (Em_c). Meanwhile, the total energy produced from the system (F_p) is converted into MWh by multiplying with the system's total operating hour as given by TOH.

$$TLCCE = \frac{\sum_{i=1}^{I}\sum_{j=1}^{J}\sum_{p=1}^{P} Em_{ijp} + \sum_{i=1}^{I}\sum_{j=1}^{J} Em_{ij} + \sum_{c=1}^{C} Em_c}{\sum_{p=1}^{P} F_p \times \text{TOH}} \tag{1}$$

Then, CO_2 offset is calculated by using equation 2. As shown, CO_2 offset is denoted by *Offset* and is determined based on the resultant CO_2 emitted to the atmosphere. In the case where CO_2 offset is positive, it indicates net carbon emission. When CO_2 offset shows a negative value, it indicates negative net carbon emissions where the amount of CO_2 sequestered is higher than being released into the atmosphere. The total CO_2 emitted to the atmosphere is represented by the summation of Em_{ijp}, Em_{ij}, and Em_c whereas the total CO_2 captured by the system is denoted by Em_c^{Captured}. In this research, one of the objectives is to evaluate and optimise the environmental performance of different BECCS configurations. Therefore, CO_2 offset is used as the objective function that is desired to be minimised. This is to ensure that the selected negative emissions pathway is having the least amount of negative environmental impacts caused by CO_2 emitted to the atmosphere.

$$Offset = \sum_{i=1}^{I}\sum_{j=1}^{J}\sum_{p=1}^{P} Em_{ijp} + \sum_{i=1}^{I}\sum_{j=1}^{J} Em_{ij} + \sum_{c=1}^{C}\left(Em_c - Em_c^{\text{Captured}}\right) \quad \forall i, \forall j, \forall p, \forall c \tag{2}$$

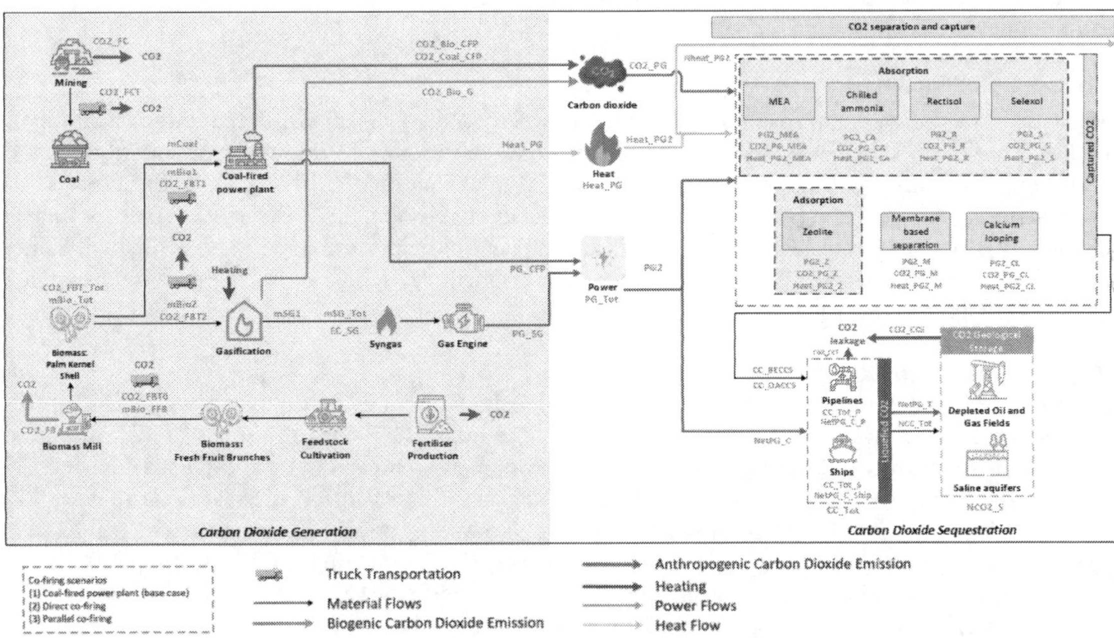

Figure 1. General superstructure.

Following this, the methodology described above is demonstrated based on the case study discussed in the following section.

3. Case Study

In this case study, three scenarios with different BECCS configurations are deliberately evaluated in achieving the research objectives mentioned above. A detailed discussion of all scenarios evaluated is provided in the following sub-sections.

3.1. Scenario 1: Coal-fired power generation plant without CCS (Base Case)

The first scenario is used to serve as a base case for comparison with BECCS retrofit. In this scenario, a detailed and transparent analysis of a coal-fired power plant is presented based on a reference coal-fired power plant located in Mukah, Sarawak. It is noteworthy that an electricity demand of 600MW is fixed for scenario 2 and 3. This is to ensure the consistent supply of energy to the end-user without being affected by the integration and/or implementation of BECCS. Moreover, a regional study for other case scenarios is performed where the selection of feedstock, the technology used, and the transport method are determined based on the conditions surrounding the Mukah region in Sarawak.

3.2. Scenario 2 & 3: Coal-fired power generation with biomass co-firing integrated with BECCS

BECCS is a type of negative emissions technologies that utilises biomass in power generation. The combustion of biomass in power generation will emit biogenic CO_2 instead of anthropogenic CO_2 from fossil fuels. According to the EU legislation, the carbon emission from the combustion of biomass is considered carbon neutral [10]. Then, the biogenic carbon emitted is captured and sequestered underground, thereby achieving negative carbon emissions.

This study assesses the GHG emission reduction capability by adopting biomass co-firing integrating with the BECCS system. Direct and parallel co-firing were considered and represented by case 2 and 3 respectively. The main objective of this scenario is to improve the usage of coal-fired power facilities while increasing domestic energy security using renewable energy sources. This can be achieved by optimising the CO_2 emission reduction capability of the selected technology by using the model developed.

4. Results and Analysis

In this section, life cycle optimisation is performed based on the CO_2 emission reduction capability of BECCS. This study is effective in identifying the main contributors to the environmental burdens of the system by performing LCO. LCO is also beneficial in evaluating the life cycle performance of the system whereby an optimal BECCS configuration is then selected. In a life cycle analysis, not all processes involved in the system pose a significant impact on the performance to be evaluated. Hence, in terms of economic perspectives, the comparison was made by focusing on the cost-intensive sub-processes involved in each scenario.

4.1. Life cycle optimisation of CO_2 emissions

Life cycle carbon emissions for the network presented in figure 1 were analysed with the objective function designed to minimise the total system cost. The total life cycle CO_2 emission distribution of all 3 scenarios is shown in figure 2 below. As shown, total CO_2 emissions from the base case (case 1) is around 11.7 $kgCO_2eq/MWh$ while power generation contributes to the highest emissions of 11.6 $kgCO_2eq/MWh$. In case 1, coal is used as the energy source in power generation, which releases a huge amount of anthropogenic CO_2 into the atmosphere. The remaining 0.07 $kgCO_2eq/MWh$ is the anthropogenic CO_2 emissions originating from feedstock production and supply chain. Likewise, case 2 and 3 show a TLCCE of 9.12 $kgCO_2eq/MWh$ and 1.28 $kgCO_2eq/MWh$ respectively. In case 2 and 3, although biomass is used as one of the energy sources in co-firing, the system's life cycle is not considered carbon neutral as there are still anthropogenic CO_2 emissions from coal combustion and feedstock supply chain which can negatively impact the environment.

Comparing case 2 and 3, case 2 shows a higher TLCCE. The biggest contributor to the higher TLCCE in scenario 2 is the coal transport. This can be explained as direct co-firing (case 2) has restrictions on the co-firing ratio, where a maximum of 30% is usually expected [11]. Likewise, parallel co-firing (case 3) does not have any restrictions on the co-firing ratio and allows a higher amount of biomass to be used. This reduces the CO_2 emissions from coal supply chain as the maximum capacity of coal transportation truck is 40% lower than that of biomass. Thus, case 3 exhibits an 86% lower TLCCE relative to case 2. Moreover, case 2 and 3 share the same biggest emitter of CO_2 in the system, which is power generation process. A life cycle carbon emissions distribution of 10.5 kgCO_2eq/MWh and 7.5 kgCO_2eq/MWh in power generation are given by case 2 and 3 respectively. As compared with the base case, case 2 and 3 shows lower life cycle carbon emissions in power generation as biomass is a low-carbon fuel with a lower CO_2 emission factor as compared to coal.

However, carbon capture is performed in case 2 where a total CO_2 of 9.7 kgCO_2eq/MWh is captured, resulting in a positive CO_2 offset of 9.1 kgCO_2eq/MWh TLCCE by the system. Likewise, in case 3, a life cycle CO_2 captured of 6.9 kgCO_2eq/MWh is observed which gives an overall TLCCE of 1.28 kgCO_2eq/MWh. Also, scenario 2 does not significantly reduce the CO_2 emissions as compared to scenario 1. Thus, in terms of carbon emissions, BECCS with parallel co-firing serves as a better option for BECCS retrofit due to the lower resultant CO_2 emissions to the atmosphere.

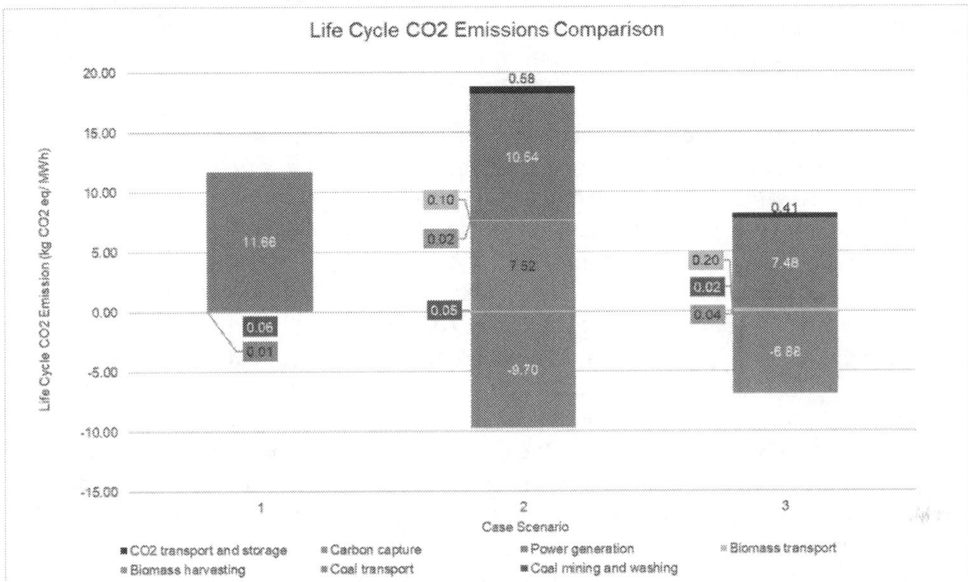

Figure 2. Life cycle CO_2 Emissions comparison for 3 different configurations.

Aside from this, life cycle carbon emissions analysis is also performed for the objective function designed to minimise the CO_2 offset from the system. The same set of results are obtained which can be explained as during system cost minimisation, the model aims to achieve the energy demand required by selecting the most cost-effective pathway. This can be done by minimising the number of equipment required and thus minimising the annual capital cost required. Furthermore, to minimise the total system cost, the model will also aim to achieve the electricity demand without producing excessive power, as an additional cost will be required for the latter. Meanwhile, when the objective function is designed to minimise the CO_2 offset from the system, the model will select the BECCS configuration pathway where the least CO_2 is generated. This can be achieved by minimising the power generated due to parasitic energy losses. This is because when a larger amount of power is required to support the carbon capture system, the system will require increasing fuel combustion which eventually leads to a higher CO_2 emission from the power generation system. Hence, it can be observed that both objective functions are indirectly targeted to minimise energy production which eventually leads to similar results being obtained.

5. Conclusion

In this research, the life cycle optimisation (LCO) of BECCS configurations to deliver a fixed amount of electricity to the consumer is performed. The objective functions were defined to minimise the total system cost and CO_2 offsets. The optimisation computed the total life cycle carbon emissions (TLCCE) of the selected negative emissions pathway. The results showed that although case 3 is effective in reducing the CO_2 emissions which gives the lowest TLCCE of 6.9 $kgCO_2eq/MWh$. In addition, the results also show that the objective function of minimum CO_2 emissions can be met in the minimum cost scenario as same results were obtained when the objective function is designed to minimise total system cost and CO_2 offset. However, this research looks at the overall CO_2 emissions from the system without considering the source of CO_2 emissions. In reality, CO_2 can be categorised as biogenic and anthropogenic CO_2 emissions as aforementioned. If only biogenic CO_2 emission is observed, the system's life cycle can be considered as carbon neutral and by implementing carbon capture, negative emissions is achievable.

Acknowledgements
The financial support from the School of Engineering and Physical Sciences (EPS) of Heriot-Watt University Malaysia is gratefully acknowledged. Moreover, the author also acknowledges LINDO SYSTEMS INC. for providing technical support.

References

[1] Bennett J A *et al.* 2021 Life cycle meta-analysis of carbon capture pathways in power plants: implications for bioenergy with carbon capture and storage. *International Journal of Greenhouse Gas Control.* **111** 103468. doi: 10.1016/j.ijggc.2021.103468.
[2] Creutzig F et al. 2019 The mutual dependence of negative emission technologies and energy systems. *Energy and Environmental Science.* **12(6)** 1805 – 17. doi: 10.1039/c8ee03682a.
[3] Fajardy M and Mac Dowell N 2017 Can BECCS deliver sustainable and resource efficient negative emissions? *Energy and Environmental Science.* **10(6)** 1389–1426. doi: 10.1039/c7ee00465f.
[4] Gong, J and You F 2017 Consequential life cycle optimization: general conceptual framework and application to algal renewable diesel production. *ACS Sustainable Chemistry and Engineering* **5(7)** 5887–5911. doi: 10.1021/acssuschemeng.7b00631.
[5] IPCC 2018 Global warming of 1.5°C. An IPCC special report on the impacts of global warming of 1.5°C above pre-industrial levels and related global greenhouse gas emission pathways, in the context of strengthening the global response to the threat of climate change. *Ipcc* - Sr15, 2(October) 17–20. Available at: www.environmentalgraphiti.org.
[6] Meghan Gordon M W 2021 Global energy demand to grow 47% by 2050, with oil still top source: US EIA. *S&P Global Platts.*
[7] Palmer I 2021 Coal is out at COP26 - except for countries where it's still in. *Forbes*. Available at: https://www.forbes.com/sites/ianpalmer/2021/11/13/coal-is-out-at-cop26--except-for-countries-where-its-still-in/?sh=2cd303a514f3.
[8] Rueda O et al. 2021 Negative-emissions technology portfolios to meet the 1.5 °C target. *Global Environmental Change.* 67(February). doi: 10.1016/j.gloenvcha.2021.102238.
[9] Sources of Greenhouse Gas Emissions 2021 EPA, United States Environmental Protection Agency. Available at: https://www.epa.gov/ghgemissions/sources-greenhouse-gas-emissions.
[10] Vandermel J 2020 Preventing Climate Change with BECCS: Bioenergy Carbon Capture and Storage. Available at: https://psci.princeton.edu/tips/2020/11/15/preventing-climate-change-with-beccs-bioenergy-with-carbon-capture-and-storage.
[11] Wang X et al. 2021 Experimental study and design of biomass co-firing in a full-scale coal-fired furnace with storage pulverizing system. Agronomy, 11(4) 1–11. doi: 10.3390/AGRONOMY11040810.
[12] Wang Y et al. 2021 Life cycle assessment of combustion-based electricity generation technologies integrated with carbon capture and storage: A review. Environmental Research. (October) 112219. doi: 10.1016/j.envres.2021.112219.

Synthesis of zinc oxide derived from wastewater of glove manufacturing via adsorption-desorption-precipitation path

C Y Pang[1], I Gulnaziya[1], Y L Pang[1] and M C Wong[1]

[1]Universiti Tunku Abdul Rahman, Jalan Sungai Long, Bandar Sungai Long, Malaysia

Gulnaziya@utar.edu.my

Abstract. The demand for ZnO significantly increased over the past decades. Extraction and refinement of zinc ores results in the release of hazardous wastes to the environment. In particular, the tremendous demand for disposable nitrile gloves during the Covid-19 pandemic has spiked superior demand for ZnO used as a crosslinking agent in the manufacturing process and leads to the generation of zinc containing wastewater traditionally removed via chemical precipitation. This paper focuses on the application of adsorption for zinc removal which opens an opportunity to recycle zinc to synthesize secondary zinc oxide. This study evaluated the feasibility of zinc oxide synthesis via the adsorption-desorption-chemical precipitation pathway. Palm shell activated carbon (PSAC), a low-cost adsorbent, was used for the removal of zinc from the synthetic and industrial wastewaters. Subsequently, zinc desorption was carried out using 0.3 M HCl. Then ZnO was synthesized from the desorption solutions via chemical precipitation using potassium hydroxide. Average zinc conversion rate of 98 % was achieved in this study. The synthesized ZnO exhibited high surface area of 97.4 m^2/g after calcination (400 °C, 3 hr), high purity and crystallinity. The results confirmed the feasibility of zinc ions recovery for recycling to produce secondary good quality zinc oxide.

1. Introduction

Glove manufacturing market was valued at 6.9 billion US dollars in 2019 and is one of the biggest markets worldwide due to its broad application range and the protection they provide [1]. With the recent worldwide Covid-19 pandemic, the demand for gloves has surged significantly. Crosslinking is a step during the production of nitrile and rubber gloves to strengthen the physical properties of the gloves. Zinc oxide is a very important chemical that is used during the crosslinking step in glove manufacturing as an activator and accelerator. Therefore, in correlation to the increased demand for disposable gloves, the demand for zinc oxide also significantly increased. Commercial zinc oxide is synthesized via the metallurgical methods using the direct or indirect method. These processes require large amount of zinc ores or zinc ingot as raw materials for the synthesis of zinc oxide and high energy input to raise furnace temperatures up to 1400 °C to vaporize the zinc from the raw materials [2]. Raw materials for these processes are usually obtained from mining activities and zinc refineries. Mining activities is known to heavily pollute the environment due to the discharge of acidic mining effluents [3] that may contain high amount of various heavy metals such as arsenic, lead, copper, and iron [3, 4, 5]. Zinc refinement can be done via different routes such as using blast furnaces or electrolysis [2]. Malaysia has one of the biggest glove manufacturing industries around the world. The use of zinc oxide during the glove manufacturing process produces wastewater discharge with zinc concentration higher than Malaysia environmental standards of 2.0 mg/L [6] and is traditionally treated treated via chemical precipitation due to its simplicity and effectiveness. Alternatively, application of adsorption using palm shell activated carbon (PSAC) for the removal of zinc can open up an opportunity for the recovery and recycling of spent zinc ions. PSAC is a low-cost adsorbent produced from palm shell kernels obtained easily from Malaysia's many palm oil industries, have shown good zinc ions removal [7]. As rubber and nitrile gloves manufacturing wastewater mainly contains zinc as the metal ions [8, 9], it can be effectively removed from wastewater via adsorption allowing the recovery of zinc ions. The adsorbent can also be regenerated using desorption agents such as inorganic acids [10] allowing the adsorbent to be reused while also producing a desorption solution containing zinc ions. Chemical precipitation can then be conducted to precipitate and synthesize zinc oxide from the precipitate allowing the recycling of zinc. Therefore, the objective of this work is to study the feasibility of zinc oxide synthesis via the adsorption-desorption-chemical precipitation pathway as a more environmentally friendly, sustainable, and quick method for zinc oxide synthesis.

2. Methodology

2.1. Materials

Palm shell activated carbon was purchased from Afza Maju Sdn. Bhd., grinded to 0.3 – 0.85 mm via pastel and mortar, washed and dried overnight in an oven at 80 °C overnight. Zinc nitrate hexahydrate ($Zn(NO_3)_2 \cdot 7H_2O$) (Sigma-Aldrich) was used to make synthetic wastewater for zinc adsorption as it is the most affordable amongst zinc precursors and easily available, hydrochloric acid was used as a desorption agent, and potassium hydroxide was used for chemical precipitation (All reagents used were of analytic grade).

2.2. Adsorption Parameters Study

Adsorption parameter studies were carried out in batch mode. Different zinc concentrations (5 – 200 mg/L) in aqueous solution was prepared by dissolving 4.550 g of zinc nitrate hexahydrate in 1 L of deionized water and further dilution was by adding deionized water to reach the required initial concentration. The adsorption experiments were carried out in a 250 mL conical flask where 50 mL of the zinc solution with a predetermined concentration was added into and was agitated at 200 rpm. A hot plate magnetic stirrer was used for the agitation and temperature control of the samples during the adsorption process. 1 g of PSAC adsorbent was weighted using an analytical balance and added to the conical flask. To determine the zinc removal efficiency and zinc adsorption capacity, equation (1) and (2) was used respectively.

$$R(\%) = \left(\frac{C_o - C_e}{C_e}\right) \times 100\% \quad (1)$$

$$q_e = \frac{(C_o - C_e)V}{m} \quad (2)$$

Where, R is the removal efficiency, C_o is the initial zinc concentration, C_e is the equilibrium zinc concentration, m is the adsorbent dosage, and V is adsorption volume.

2.3. Zinc Desorption

Zinc desorption studies were carried out in batch mode. 0.3 M hydrochloric acid was prepared by diluting the concentrated acid with deionized water. Desorption experiments were carried out in a 250 mL conical flask where 100 mL of the acid solution was added. 10 g of zinc loaded adsorbents with a known zinc adsorption uptake was added into the conical flask and was stirred at 200 rpm using an orbital shaker for 2 hours. To determine the zinc desorption efficiency equation (3) was used.

$$desorption\ efficiency = \frac{C_e(V)}{q_e(m)} \times 100\ \% \quad (3)$$

where, q_e is the predetermined zinc adsorption uptake, V is the desorption volume, m is the zinc loaded adsorbent dosage, and C_e is the zinc concentration in the desorption solution at equilibrium.

IOP Conf. Series: Materials Science and Engineering 1257 (2022) 012040 doi:10.1088/1757-899X/1257/1/012040

2.4. Zinc oxide synthesis via chemical precipitation

Zinc oxide synthesis was conducted in a batch process. Saturated potassium hydroxide was prepared by adding the potassium hydroxide granules to deionized water. Zinc oxide synthesis was carried out in a 500 mL beaker where 150 mL of zinc desorption solution with a known initial zinc concentrated was added to the beaker. To the solution, potassium hydroxide was added and then stirred at 800 rpm using a magnetic stirrer for 1 hour. After 1 hour, the solution was transferred to a centrifuge tube and was centrifuge at 10 000 rpm for 5 mins. After 5 mins, the zinc oxide was washed with deionized water and was centrifuge again at similar conditions and the cycle was repeated 3 times. The zinc oxide was then washed with deionised water and then with isopropyl alcohol and dried overnight in an oven at 80 °C. The dried sample was then calcinated in a tube furnace in air at 400 °C for 3 hours (10 °C/min rate).

2.5. Characterization of synthesized zinc oxide

Surface area of the prepared zinc oxide was determined using the Brunauer Emmett Teller (BET) method by N_2 adsorption (77 K) using Thermo Finnigan Sorptomatic 1990 series surface analyzer (ThermoFisher Scientific, USA) with a degassing condition of 200 °C for 6 hours. The surface morphology and elemental composition of the synthesized zinc oxide were identified using scanning electron microscopy coupled with energy dispersive x-ray (SEM-EDX) and energy dispersive X-ray spectroscopy (Ametek EDAX). The crystalline structure of the synthesized zinc oxide was identified using X-ray diffraction (XRD) (Shimadzu). Functional groups of the synthesized zinc oxide were identified using fourier transform infrared (Perkin-Elmer Inc., USA). Sample decomposition studies was conducted using thermal gravimetric analysis.

3. Results and discussion

3.1. Adsorption-desorption-chemical precipitation synthesis of zinc oxide

Adsorption parameters studies were conducted as a separate study and the optimal adsorption parameters were determined to be as follows: adsorption pH of pH 7, initial zinc concentration of 150 mg/L, adsorbent dosage of 40 g/L, and adsorption contact time of 2 hours. These optimal zinc adsorption parameters were used for further adsorption of zinc onto palm shell activated carbon (PSAC). Desorption of zinc ions was conducted using 0.3 M of hydrochloric acid (HCl) with a ratio of 0.133 g adsorbent/ mL HCl. 0.3 M HCl showed high zinc desorption efficiency achieving an average desorption efficiency of 91.5 %. HCl is a good desorbing agent for the desorption of zinc oxide due to its low acid disassociation constant, pKa of -6.3, meaning it can easily disassociate in H^+ ions. High concentration of H^+ ions will result in the protonation of the adsorbent surface allowing the removal of adsorbed zinc. After zinc was desorbed from PSAC, it can be reused for the adsorption of zinc while the zinc ions in desorption solution can be used to synthesize zinc oxide.

Zinc oxide was synthesized via chemical precipitation using saturated potassium hydroxide. Upon the end-point of the synthesis process, an average pH 13.3 was observed suggesting that most of zinc ions present in the solution have precipitated as zinc hydroxide or zinc oxide. Through mass balance using zinc concentration in the desorption solution before and after chemical precipitation, it was determined that the precipitation reaction achieved a high zinc conversion efficiency of 98 %. Furthermore, an overall zinc recovery of 39 % was calculated using the amount of zinc oxide recovered from an initial 1000 mL of zinc adsorption solution with concentration of 150 mg/L. The high zinc conversion efficiency and overall zinc recovery of the process suggests the reaction pathway is applicable for the recovery and synthesis of zinc oxide.

3.2. Characterization of synthesized zinc oxide particles

Thermal gravimetric analysis (TGA) was performed on the as-prepared zinc oxide (ZnO) sample. Figure 1 shows an approximate weight of 23 % mainly attributed to the evaporation of water molecules and decomposition of zinc hydroxide intermediates to ZnO up to 400 °C. As the weight of ZnO stabilized at 400 °C, this temperature was used for calcination to prevent sintering and agglomeration of the ZnO particles.

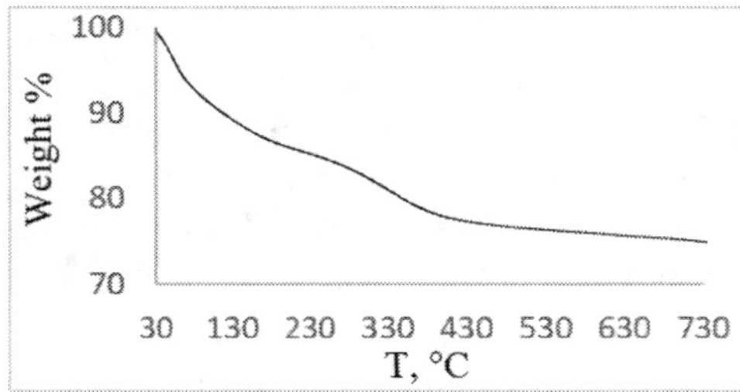

Figure 1. TGA analysis for uncalcined ZnO.

FTIR spectra for both calcined and uncalcined ZnO is shown in figure 2. The broad band corresponding to the stretching vibration of O-H bonds can be identified in both uncalcined and calcined ZnO samples at 3421 cm^{-1} and 3388 cm^{-1} respectively attributed to the presence of moisture and residue O-H from KOH used during chemical precipitation or undecomposed zinc hydroxide complexes. The band located at around 1488 cm^{-1} for uncalcined ZnO can be attributed to the asymmetric stretching of C=O while the band at 1414 cm^{-1} can be attributed to the symmetric stretching of C-O can be identified in both ZnO samples. Absorbance at 996 cm^{-1} is due to the presence of C-O-C. For the uncalcined ZnO sample an absorbance band related to the stretching of C-Cl was observed at 544 cm^{-1}. The absorbance identified at 427 cm^{-1} and 422 cm^{-1} for uncalcined and calcined ZnO can be identified as ZnO groups [11]. Significant change can be identified in the FTIR spectra for the uncalcined and calcined ZnO samples. After calcination, the absorbance intensity of the O-H stretching at 3421 cm^{-1} decreased attributed to the vaporization of moisture as well as the decomposition of unreacted zinc hydroxide complexes. Asymmetric stretching of C=O bond at 1488 cm^{-1} was not identified in the FTIR spectra for the calcined ZnO and the symmetric stretching of C-O at 1414 cm^{-1} was decreased. The absorbance band attributed to C-Cl stretching was also not identified after calcination.

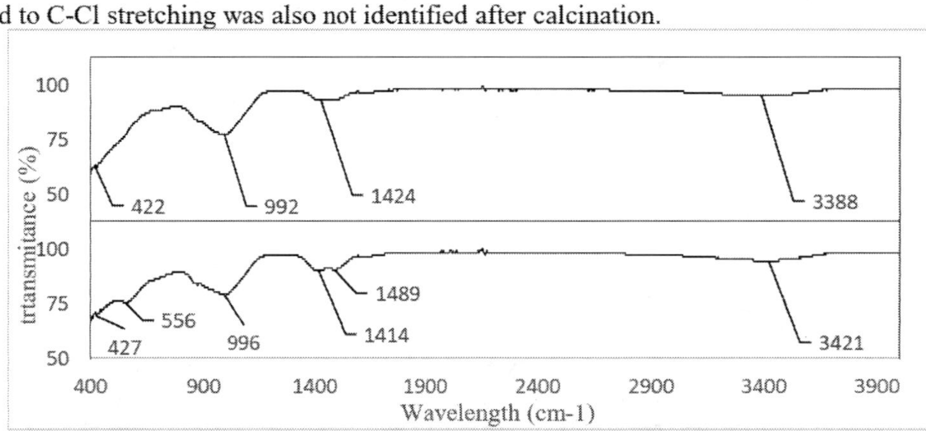

Figure 2. FTIR spectra for uncalcined ZnO (top) and calcined ZnO (bottom).

Figure 3 shows the XRD patterns of the uncalcined and calcined (400 °C and 3 hours) zinc oxide samples prepared from performing chemical precipitation using zinc desorption solution. The lack of any crystalline peaks observed in the uncalcined ZnO sample indicates an amorphous structure. On the other hand, high intensity and sharp peaks were observed for the calcined ZnO samples indicates high crystallinity. The 2θ peaks observed for the calcined sample indicate a hexagonal wurtzite structure attributed to the peaks at 31.90°, 34.58°, 36.38°, 47.67°, 56.72°, 63.01°, 68.09°, 77.0°, and 78.15° can be identified as the (100), (002), (101), (110), (103), (200), (212), and (201) crystalline planes respectively. Moreover, the peak observed at 29.6° for the calcined ZnO sample is likely due to the presence of potassium chloride or potassium perchlorate crystals [12, 13]. The effect of calcination is very obvious in the case of ZnO, where it changes from an amorphous nature to highly crystalline as the high calcination temperature provide sufficient energy to overcome the activation energy for crystal growth as well as removing various impurities.

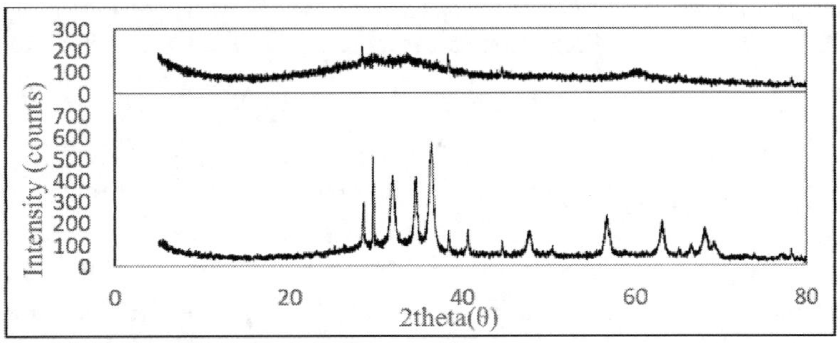

Figure 3. XRD spectra for uncalcined ZnO (top) and calcined ZnO (bottom).

Crystallite size of the ZnO samples was calculated via the Debye Scherrer equation (4) and is presented in table 1. It was observed that after calcination, ZnO crystallite size increases from 13.16 nm to 16.91 nm. The increase in crystallite size is likely attributed to the sintering phenomena during calcination that causes the deformation ZnO particles and increasing the crystallite size.

$$D = \frac{k\lambda}{\beta\cos(\theta)} \tag{4}$$

Table 1. XRD and BET analysis for Prepared ZnO Samples.

ZnO Samples	Crystalline plane (101)			BET specific surface area (m^2/g)	Average pore size (nm)	Total pore volume (cm^3/g)
	2Theta (θ)	FWHM (θ)	Crystallite size (nm)			
Uncalcined ZnO	36.1600	0.840	13.16	140.36	13.37	0.4691
Calcined ZnO	36.3816	0.594	16.91	97.38	15.56	0.3789

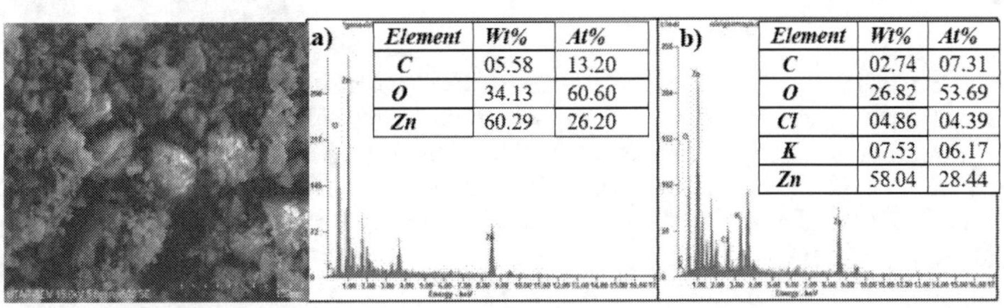

Figure 4. SEM image of ZnO (x500) and EDX spectra for a) uncalcined ZnO and b) calcined ZnO.

BET specific surface area, pore diameter and pore volume for uncalcined and calcined ZnO are also tabulated in table 1. The specific surface area for uncalcined and calcined ZnO are 140.36 m^2/g and 97.38 m^2/g respectively. Uncalcined ZnO also showed higher pore diameter and total pore volume of 13.37 nm and 0.4691 m^3/g compared to calcined ZnO with a pore diameter and total pore volume 15.56 nm and 0.3789 m^3/g respectively. The decrease in all three parameters after calcination can be attributed to the agglomeration and sintering effect that occurs during calcination. Adsorption-desorption isotherm plot for both ZnO samples shows a type 4 adsorption-desorption isotherm signifying a mesoporous structure (pore diameter 2 – 50 nm).

SEM images of the uncalcined ZnO (figure 4) shows an agglomerated spherical particles surface morphology with an estimated particle size of around 300 nm. The EDX spectra (figure 4a, b) indicates that samples contain high purity of zinc oxide in the synthesized powder with some potassium chloride or potassium perchlorate as impurities supporting the observations during XRD analysis. The elemental analysis data in table 2 shows an atomic ratio of zinc to oxygen of 0.43 before calcination and 0.53 after calcination. The increase in zinc atomic ratio shows a higher purity of zinc oxide after calcination, suggesting high quality and crystallinity zinc oxide

Table 2. EDX elemental and atomic composition for uncalcined and calcined ZnO.

Samples	Element	Wt(%)	At(%)	Zn/O ratio
Uncalcined ZnO	Zn	60.29	26.2	0.43
	O	34.13	60.6	
Calcined ZnO	Zn	58.04	28.44	0.53
	O	26.82	53.69	

4. Conclusion

Zinc oxide was successfully recovered and synthesized via the adsorption-desorption-chemical precipitation reaction pathway. The synthesis pathway was an effective method for the synthesis of zinc oxide from desorption solution achieving a high zinc conversion efficiency of 98 % for chemical precipitation and an overall zinc recover of 39 %. XRD analysis confirms the high crystallinity for synthesized zinc oxide while EDAX analysis confirms the high purity of zinc oxide in the product. FTIR analysis confirms the presence of ZnO bonding in the structure and the conversion of zinc hydroxide complexes to zinc oxide after calcination. Zinc oxide synthesized via this pathway also exhibit high surface area of 97.38 m^2/g after calcination. BET analysis shows the prepared samples follow a type 4 adsorption desorption isotherm suggesting a mesoporous structure. Overall, this synthesis pathway was found to be an efficient, quick, simple and environmentally friendly method for the production of zinc oxide from the recovered zinc ions.

Acknowledgements

We express great appreciation to the Centre for Photonics and Advanced Materials Research (CPAMR), Universiti Tunku Abdul Rahman (UTAR), Malaysia for the financial support under the UTAR Research Fund IPSR/RMC/UTARRF/2019-C2/G05 that allowed us to carry out this research.

References

[1] Anil C, Pratik M and Eswara P *Industrial Glove Market Outlook-2027* [Internet] 2020 May [2022, Mar, 28]. Available from: https://www.alliedmarketresearch.com/industrial-gloves-market
[2] Moezzi A, Andrew M M and Micheal B C 2012 *Chem. Eng. J.* **185** 1 – 12
[3] Magdalena J C, Katarzyna N 2016 *Environ. Monit. Access* **188** 262
[4] Amanda N and Moersidik S S 2019 *J. Phys. Conf. (Pekanbaru)* vol 1351 (United Kingdom IOP Publishing) p 12113
[5] Saba S, Aarif J, Christina E and Akhilesh K Y 2021 *Appl. Sci.* **11** 3910–19
[6] Department of Environment Malaysia *Environmental Requirements: A guide for investors* [Internet] 2010 Oct [2022, Mar, 28]. Available from: https://www.doe.gov.my/wp-content/uploads/2021/10/A-Guide-For-Investors1.pdf
[7] Baby R, Saifullah B and Hussein M Z 2019 *Sci. Rep.* **9** 18955
[8] Mohamadi M, Hasfalina C M, Hassan M A and Phang L Y 2010 *Afican J. Biotechnol.* **9** 6233
[9] Abu Bakar A, Rakmi A R, Mohamad A B, Kadhum A and Abdullah S 2009 *J. Sustain. Dev.* **1** 31
[10] Zhang X T, Hao Y N, Wang X M and Chen Z J 2017 *Materials* **10** 1002
[11] Wahab R, Kim Y S and Shin H S 2009 *Mater. Trans.* **50** 2092
[12] Agarwal M, Garg, S, Kandasami A and Kanjilal P 2017 *RCS Adv.* **7** 13836
[13] Samavat F, Haji Ali E, Solgi S and Taravati Ahmad 2012 *J. Phys. Chem* **2** 185

IOP Conf. Series: Materials Science and Engineering 1257 (2022) 012041 doi:10.1088/1757-899X/1257/1/012041

Investigation of glyphosate removal from aqueous solution using calcium peroxide

F Li[1], T S Y Choong[1,2*], L C Abdullah[1,2] and S N A Md Jamil[3,4]
[1] Sustainable Process Engineering Research Center (SPERC), Department of Chemical and Environmental Engineering, Faculty of Engineering, Universiti Putra Malaysia, 43400 UPM Serdang, Selangor, Malaysia.
[2] Institute of Tropical Forestry and Forest Product (INTROP), Universiti Putra Malaysia, 43400 UPM Serdang, Selangor, Malaysia.
[3] Department of Chemistry, Faculty of Science, Universiti Putra Malaysia, 43400 UPM Serdang, Selangor, Malaysia.
[4] Centre of Foundation Studies for Agricultural Science, Universiti Putra Malaysia, 43400 UPM Serdang, Selangor, Malaysia.
*Corresponding author's e-mail: csthomas@upm.edu.my

Abstract. Glyphosate, an organophosphate herbicide containing carboxyl group, is widely used in agriculture. It can be found as a contaminant in water, and procedures are required for its removal. The present study investigated the efficiency of glyphosate removal from aqueous solutions using a Fenton-like system. Calcium peroxide (CaO_2) was used as a source of hydroxyl free radicals with Fe^{2+} as the catalyst to trigger Fenton reaction. Fourier transform infrared spectroscopy (FTIR), X-ray diffraction (XRD) and potassium permanganate titration were carried out for characterization of calcium peroxide. The effect of operating parameters such as pH, molar ratio of Ca^{2+}: Fe^{2+}, and initial dosage of calcium peroxide on the total phosphorus (TP) removal efficiency was studied. The maximum TP removal was 94.50% under optimal conditions (pH = 3.0, initial glyphosate concentration = 50 ppm, initial CaO_2 dosage = 0.5 g, molar ratio of Ca^{2+}: Fe^{2+}= 6). This study demonstrated that calcium peroxide could be considered as an effective oxidant for glyphosate wastewater treatment.

1. Introduction

Water contamination, originated by diverse activities such as industrial, agricultural and domestic wastewater, is one of the primary confronted issues nowadays, which can result in environmental, health, social, economic concerns. Under this circumstance, agricultural wastewater caused by the extensive use of chemical fertilizers, pesticides and herbicides should be stressed [1]. Owing to the features that it is of high concentration, high salinity, high total phosphorus content and high COD value, agricultural wastewater must be effectively treated before discharged into the environment [2]. Glyphosate (N- (phosphonomethyl) glycine, $C_3H_8NO_5P$, PMG) is the active ingredient of Roundup, marketed as a non-selective, post-emergence, broad-spectrum organophosphate herbicides used widely globally in agriculture, forestry and cities. Considering the harm of excessive glyphosate to animals, plants, it is of great importance to remove glyphosate effectively, that is, to decrease the concentration of phosphorus. A range of conventional methods, such as adsorption and biological treatments, have been investigated for glyphosate removal. Advanced oxidation processes (AOPs) have also been developed as alternative treatment technologies for glyphosate removal. Adequate evidence exists to support that Fenton system including combination coupling Fenton system (i.e., electro-Fenton, adsorption-Fenton, photo-Fenton processes, and Fenton-like system) is significantly efficient. This is due to shorter contact time and fewer by-products on glyphosate removal under Fenton system [3]. Thus far, the major problem has received attention that it is quite challenging and inconvenient to store and transport liquid hydrogen peroxide (H_2O_2) [4]. To resolve this concern, calcium peroxide (CaO_2), known as the solid form of hydrogen peroxide, is an emerging oxidant in Fenton-based system to removal various types of contaminants. Due to the properties that it can decompose to H_2O_2 in water then release hydroxyl radicals (·OH), it is worthy studying on this material. In the present work, calcium peroxide, a solid form of hydrogen peroxide, was investigated for the removal glyphosate from aqueous solution. Fourier transform infrared spectroscopy (FTIR), X-ray diffraction (XRD) and potassium permanganate titration were conducted for characterization of calcium peroxide.

2. Materials and methodology

2.1. Chemical Reagents and Instruments

A 1000 ppm glyphosate stock solution was diluted from Roundup® (commercial grade, containing 360 g/l glyphosate). The glyphosate solutions with a natural pH of 4.9-5.4 were then prepared by diluting the stock solution with distilled water. Calcium peroxide (65%, Alfa Aesar), ferrous sulfate ($FeSO_4$), sodium sulfite (Na_2SO_3, 2.0 M), sodium hydroxide, sulfuric acid and other chemicals employed in the present study were of analytical grade chemicals. The pH of the solution was measured by the digital pH meter (Model: Sartorius PB-10). The desired pH of the solution was adjusted by the addition of standardized 0.1 M sulfuric acid and 0.1 M sodium hydroxide solutions. Ultraviolet-visible (UV) spectrophotometer (Model: Dynamica, HALO DB-20) was used to monitor the degradation process.

2.2. Methodology

Fourier transform infrared spectroscopy (FTIR), X-ray diffraction (XRD) and potassium permanganate titration [5] were conducted respectively to characterize calcium peroxide. Standard titration was run in triplicate and average value was calculated using equation (1).

$$Purity(\%) = \frac{5 \times C(KMnO_4) \times V(KMnO_4) \times 72.08}{2 \times m(CaO_2) \times 1000} \times 100\% \quad (1)$$

where $C(KMnO_4) = 0.02$ M.

50 ppm glyphosate containing solution was prepared in a volumetric flask with distilled water from the stock solution (1000 ppm). 100 mL 50 ppm glyphosate solution was taken into a 250 mL glass conical flask with the desired amount of CaO_2 and $FeSO_4$ added. After pH was subsequently adjusted, all experiments were carried out in equipped with a water bath shaker at a speed of 150 rpm under the controlled condition (initial CaO_2 dosage = 0.5 g, molar ratio of Ca^{2+}: Fe^{2+}= 6, 100 mL 50 ppm glyphosate solution, 150 rpm, RT= 25 °C, contact time= 120 minutes). The effect of pH on the TP removal was investigated at 2.0, 2.5, 3.0, 3.5 and 4.0. The effects of molar ratio of Ca^{2+}: Fe^{2+} (1, 2, 4, 6, 8 and 10) and the initial dosage of calcium peroxide (0.1 g, 0.3 g, 0.5 g, 0.7 g and 0.9 g) were also investigated. To quench the Fenton reaction, excess Na_2SO_3 (2.0 M) was added to the samples, ensuring the removal of the remaining H_2O_2 [6]. Ammonium Molybdate Spectrophotometry Method was applied to detect total phosphorus in aqueous solution. The absorbance of each sample was determined at λ= 880 nm and the calibration curve of glyphosate (R^2=0.9938) was prepared. The total removal efficiency was calculated by equation (2):

$$\text{Total Removal Percentage} = \left(1 - \frac{C_t}{C_0}\right) \times 100\% \quad (2)$$

where C_0 and C_t are the concentration of total phosphorus at initial and a given contact time t, respectively.

The following parameters were studied in the present study:

2.2.1 Effect of pH. In the Fenton process, the H^+ concentration greatly affects the formation of H_2O_2, which in turn controls the generation of the hydroxyl radicals. Thus, pH is one of the most important aspects to investigate. The Fenton reagent only works under acidic settings because the form of Fe in the solution is limited by pH. [7]. The effect of pH on TP removal was evaluated at the pH value of 2.0, 2.5, 3.0, 3.5 and 4.0. In this set, experiments were conducted under the basic set-up (initial CaO_2 dosage = 0.5 g, molar ratio of Ca^{2+}: Fe^{2+}= 6, 100 mL 50 ppm glyphosate solution, 150 rpm, RT= 25 °C, contact time= 120 minutes).

2.2.2 Effect of molar ratio of Ca^{2+}: Fe^{2+}. Another essential parameter for Fenton process is the molar ratio of H_2O_2: Fe^{2+}, CaO_2 for the present work, which impacts the generation of the hydroxyl radicals. Under the optimal pH of 3.0 that was obtained in the first parameter, other conditions were fixed (initial CaO_2 dosage = 0.5 g, 100 mL 50 ppm glyphosate solution, 150 rpm, RT=25 °C, contact time= 120 minutes). Fenton-based experiments were carried out to investigate the effects of molar ratio of Ca^{2+}: Fe^{2+} on glyphosate removal ranging from 1, 2, 4, 6, 8 to 10. Since the initial CaO_2 dosage was fixed at 0.5 g, the dosage of Fe^{2+} was varied based on the ratio required.

2.2.3 Effect of initial dosage of calcium peroxide. After the optimal molar ratio of Ca^{2+}: Fe^{2+} was determined, initial calcium peroxide dosage was a significant parameter to study. Under the optimal pH of 3.0 and the optimal molar ratio of Ca^{2+}: Fe^{2+} = 6 that were obtained from the first two parameters, other conditions were fixed (100 mL 50 ppm glyphosate solution, 150 rpm, RT=25 °C, contact time= 120 minutes). Fenton-based experiments were carried out to investigate the effects of initial CaO_2 dosage on glyphosate removal ranging from 0.1 g, 0.3 g, 0.5 g, 0.7 g to 0.9 g. The dosage of Fe^{2+} was changed with the dosage of CaO_2 because the optimal molar ratio of Ca^{2+}: Fe^{2+} was fixed at 6.

3. Results and discussion

3.1. Characterization

Figure 1 depicted the FTIR spectra of CaO_2 to identify the functional groups present in sample. According to the spectra, the peaks around 855 and 875 cm^{-1} corresponded to the O–O bridge of CaO_2. Similarly, peak at 590 cm^{-1} and the broad peak at 1487 cm^{-1} of CaO_2 were attributed to O–Ca–O vibration [8]. Moreover, the sharp peak around 3640 cm^{-1} representing O–H stretching band was also observed, which suggested the impurity of moisture in sample. According to the XRD image (figure 2), the sample showed the dominant peaks of 2θ at 30.1, 35.6, 47.3, 57.4 and 60.7, and were in agreement with standard data (JCPDS card no. 03-0865). The prominent peak suggests the tetragonal phase of CaO_2. Table 1 showed the purity of CaO_2 measured using the standard potassium permanganate titration method. The average purity is 63.30%, which is very close to the purity stated by the supplier.

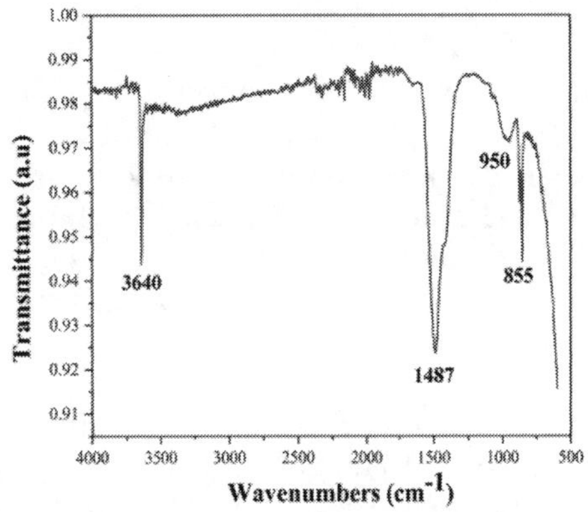

Figure 1. FTIR spectra of calcium peroxide.

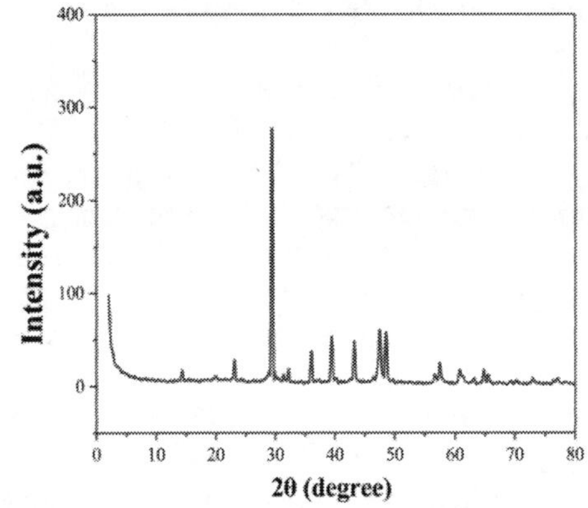

Figure 2. XRD patterns of calcium peroxide.

Table 1. Purity of calcium peroxide.

Sample	m(CaO_2)/g	V($KMnO_4$) / mL	Purity/%	Mean
1	0.0571	10.10	63.75	
2	0.0587	10.25	62.95	63.30 %
3	0.0545	9.55	63.15	

3.2. Effect of pH

As the Fenton process is substantially affected by pH, the performance of CaO_2/Fe^{2+}/glyphosate degradation was examined at various pH values. It could be observed from Figure 4 that pH affects the degradation efficiency to a great extent. Concisely, the total removal efficiency within 120 min could be sequenced: pH= 3 > pH= 3.5 > pH= 2.5 > pH= 2.0 > pH= 4.0. It increased considerably from 83.10% to 94.50% and peaked at pH= 3.0. The removal percentage declined slightly at pH= 3.5, where 92.32% TP removal was achieved. It was also can be noticed that only 69.69% TP removal was achieved when the pH value was increased to 4.0. At lower pH, due to the scavenging action by H^+ ions and reaction between $[Fe(H_2O)]^{2+}$ ion and H_2O_2 at a slower rate, lower pH values resulted in a significant decrease in the number of ·OH radicals. Another cause is stability of H_2O_2 at pH below 3 hindered by H_3O^{2+}, which prevents the generation of ·OH radicals [9].

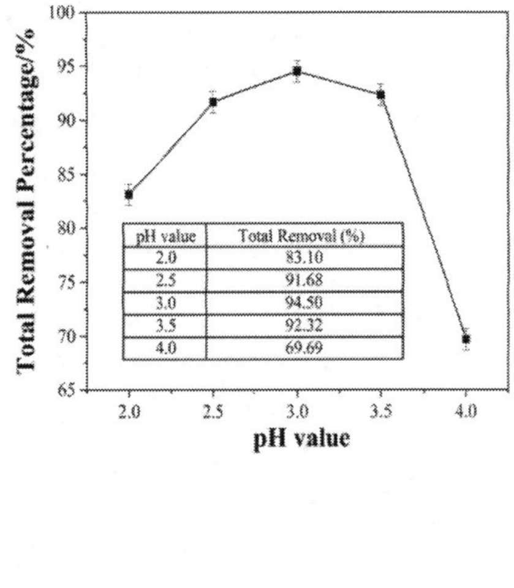

Figure 3. Effect of different pH on total removal percentage (initial CaO_2 dosage = 0.5 g, molar ratio of Ca^{2+}: Fe^{2+}= 6, 100 mL 50 ppm glyphosate solution, 150 rpm, RT= 25 °C, contact time= 120 minutes).

3.3. Effect of molar ratio of Ca^{2+}: Fe^{2+}

The ratio of Fenton reagent directly impacts the treatment cost and efficiency. The glyphosate removal corresponding to different molar ratios of Ca^{2+}: Fe^{2+} were measured and revealed in figure 4. TP removal climbed up steadily from 82.99% at ratio= 1 and peaked at 94.50% when molar ratio= 6. However, when the ratio was increased from 6 to 10, the TP removal dropped drastically to 66.92%. The reason behind this is that with a higher amount of Fe^{2+} (ratio=1 and 2), excessive Fe^{2+} reacts with·OH and results in an ineffective decomposition of H_2O_2 to release O_2. It can be concluded that the molar ratio of Ca^{2+}: Fe^{2+} = 6 was the most effective for the system within 120 min.

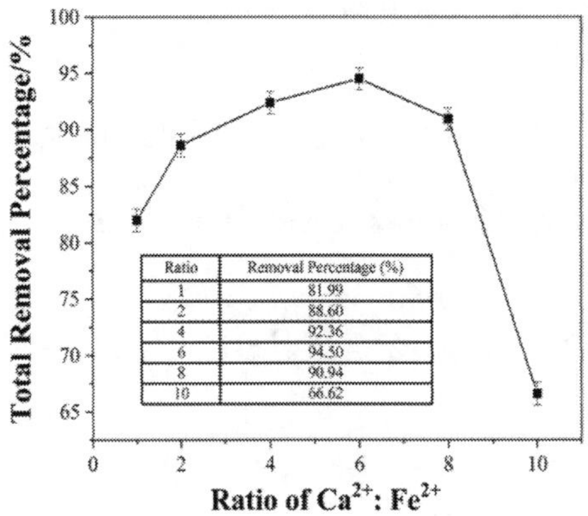

Ratio	Removal Percentage (%)
1	81.99
2	88.60
4	92.36
6	94.50
8	90.94
10	66.62

Figure 4. Effect of molar ratio of Ca^{2+}: Fe^{2+} on total removal percentage (initial CaO_2 dosage = 0.5 g, 100 mL 50 ppm glyphosate solution, 150 rpm, pH= 3.0, RT=25 °C, contact time= 120 minutes).

3.4. Effect of initial dosage of calcium peroxide

The effect of initial CaO2 dosage (0.1 g, 0.3 g, 0.5 g, 0.7 g, 0.9 g) was investigated and shown in figure 5. A dramatic increase in TP removal appeared from 55.08.% (0.1 g CaO_2) to 94.50% (0.5 g CaO_2). Nonetheless, as the initial CaO_2 dosage increased to 0.9 g, TP removal percentage dropped to 85.15%, which implied that an excess of CaO_2 negatively affected degradation. This is because excessive CaO_2 can emit an excessive amount of H_2O_2 which can scavenger the Fenton reaction. Therefore, the initial CaO_2 dosage= 0.5 g was evaluated as the most effective.

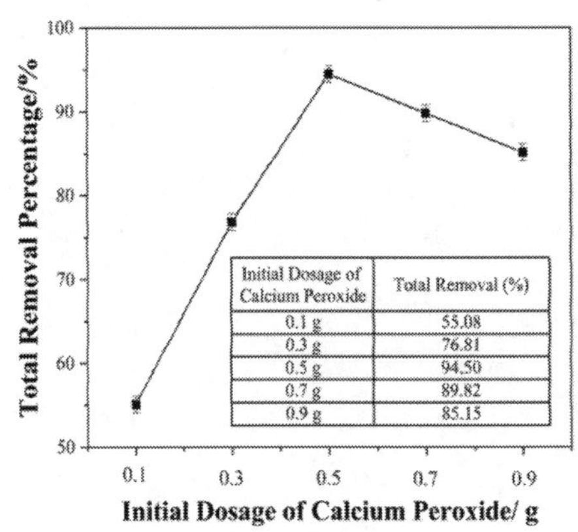

Initial Dosage of Calcium Peroxide	Total Removal (%)
0.1 g	55.08
0.3 g	76.81
0.5 g	94.50
0.7 g	89.82
0.9 g	85.15

Figure 5. Effect of initial CaO_2 dosage on total removal percentage (ratio (Ca^{2+}: Fe^{2+}) =6, 100 mL 50 ppm glyphosate solution, 150 rpm, pH= 3.0, RT=25 °C, contact time= 120 minutes).

4. Conclusion

The present work investigated the performance of CaO_2 using a Fenton-based system comprised of CaO_2 with the presence of Fe^{2+} catalyst for glyphosate removal. The optimal operating conditions were identified as: pH = 3.0, initial glyphosate concentration = 50 ppm, initial CaO_2 dosage = 0.5 g and molar ratio of Ca^{2+}: Fe^{2+}= 6. Under the given conditions, up to 94.50% TP removal was achieved within 120 min. The obtained results indicated that CaO_2 was an effective oxidant for glyphosate removal.

Acknowledgments

This project is funded by the Ministry of Higher Education (KPT), Malaysia under the Fundamental Research Grant Scheme FRGS /1/2020/TK0/UPM/01/2 (03 01 20 2250FR).

References

[1] Landrigan P J, Fuller R, Fisher S, Suk W A, Sly P, Chiles T C and Stephan B O R 2019 *Sci Total Environ* **650** 2389–94
[2] Iqbal M, Nauman S, Ghafari M, Parnianifard A, Gomes A and Gomes C 2022 *Biointerface Res Appl Chem* **12** 6336–60
[3] Jia D, Liu M, Xia J and Li C 2020 *J.Chem Technol Biotechnol* **95** 196–204
[4] Fan Y, Xuemei L, Zhiying L, Yanhua X, Lina S and Yekun J 2017 *Industrial Water Treatment* **5** 30-34
[5] Domaradzki M, Kaniewska J and Korpal W 2012 *Chemik* **5** 464–6
[6] Souza D R D, Trovõ A G, Filho N R A, Silva M A A and Machado A E H 2013 *J. Braz Chem Soc* **24** 1451–60
[7] Tran M H, Nguyen H C, Le T S, Dang V A D, Cao T H and Le C K 2021 *Environ Technol (United Kingdom)* **42** 1155–64
[8] Madan S S and Wasewar K L 2018 *Energy and Environment* 69–76
[9] Khamaruddin P F, Bustam M A and Omar A 2011 *Int Conf Environ Ind Innov IPCBEE* **12** 12–17

IOP Conf. Series: Materials Science and Engineering 1257 (2022) 012042 doi:10.1088/1757-899X/1257/1/012042

Optimization of aniline removal using graphite assisted by response surface methodology and box-behnken design

C X Tan[1], V-L Wong[1] and S P Yeap[2,3]

[1] School of Engineering ang Physical Sciences, Heriot-Watt University Malaysia Campus, 62200, Wilayah Persekutuan Putrajaya, Malaysia.

[2] Department of Chemical and Petroleum Engineering, Faculty of Engineering, Technology & Built Environment, UCSI University, Kuala Lumpur, Malaysia.

[3] UCSI-Cheras Low Carbon Innovation Hub Research Consortium, Kuala Lumpur, Malaysia.

Email: v.wong@hw.ac.uk

Abstract. The effectiveness of graphite as adsorbent in the removal of aniline by adsorption was investigated in batch mode based on four manipulated variables: pH of aniline solution (3, 7, 11), temperature of the incubator shaker (30°C, 40°C, 50°C), initial aniline concentration (5 ppm, 15 ppm, 25 ppm) and graphite dosage (0.5 g, 1.5 g, 2.5 g). The four manipulated variables were studied systematically through responses surface methodology (RSM) with 4-level of Box-Behnken design. A total number of 27 sets of designs with 3 centre points were created. The peak wavelengths of aniline working samples were measured at 230 nm and 280 nm using spectrophotometric method. Optimal condition of aniline adsorption revealed 15 ppm of aniline concentration with pH 3 at temperature 40°C using 1.5 g of graphite. The surface morphologies and atomic percentage of fresh and used graphite were examined by FESEM-EDX. Ultimately, graphite is proven to be an alternative in aniline removal, and more importantly, it is accessible easily as they require less processing and are abundant in nature.

1. Introduction

Aniline is an organic chemical raw material that is commonly used in the pharmaceutical industry, agrochemical manufacture and material science [1]. With the rapid development of the chemical industry in recent years, the global consumption of aniline is growing at an average annual rate of 4.8%, reaching more than 6.7 million metric tons in the year of 2019 [2]. This leads to not only existence of aniline compound in the wastewater effluent from the chemical industry, but also in the surface and ground water due to the excessive usage of aniline. Aniline is classified as a highly toxic chemical that is threatening to human health especially the upper respiratory tract. Moreover, due to the presence of benzene ring in the aniline, it is relatively stable and easily adsorbed by the sediments and became a permanent pollutant in aquatic environment. Therefore, the aniline removal methods were studied to minimize the negative effect of aniline to human and the environment [3].

The development of efficient aniline removal methods has gained much attention from the researchers due to its significant environmental concern. Moreover, the publications trend of articles related to aniline removal increases tremendously over the years and reached a total of 172 publications by the year 2019 according to statistics from Web of Science. Biological treatments [4], oxidation processes [5], absorption [6] and adsorption [7] are the recommended treatment methods to remove the aniline from the wastewater effluents. The adsorption process is regarded as an auspicious technique for aniline removal in the case of low concentrations because of its low cost and high efficiency [1]. Consequently, a number of adsorbents with unique characteristics have been investigated and suggested by researchers [1, 8, 9]. Selection of suitable adsorbent is important as different adsorbents demonstrate different removal efficiency.

In present work, graphite powder is selected as the new adsorbent to study the effectiveness of aniline removal. Graphite, a 3D layered structure of graphene sheets stacked together that displayed excellent chemical properties, thermal stability and mechanical strength is considered as an excellent material in aiding waste removal process [9]. Therefore, it is very likely that graphite possessed the similar adsorption performance with other excellent adsorbents such as activated carbon [8] and graphene oxide [1]. On top of that, the impact of ultrasonication on the aniline removal percentage of graphite is studied as well. Furthermore, the surface morphology of graphite before and after adsorption process was analyzed through FESEM-EDX. Additionally, the parameters to the adsorptive removal of aniline were optimized from the aspect of pH, incubator shaker's temperature, initial aniline concentration, and graphite dosage using response surface methodology (RSM).

2. Materials and Methods

2.1 Materials

Graphite powder (< 44μm; 99%), aniline (99%, MW: 93.13 g/mol), sodium hydroxide (99%, MW: 39.99 g/mol) and hydrochloric acid (99%, MW: 36.46 g/mol) were purchased from Sigma-Aldrich, Germany.

2.2 Batch Adsorption Process

10 mg/L of aniline solution was prepared in a 100 mL conical flask from the 100 mg/L aniline stock solution. The initial absorbance value of the aniline solution was measured using the UV-Vis spectrophotometer (HACH DR6000, USA) at 230 nm and 280 nm. 0.5 g of graphite powder was measured and added into the conical flask. The conical flask containing the graphite powder was inserted into the incubator shaker at 200 rpm for 6 hours at 30 °C. On top of that, the abovementioned procedures were repeated for ultrasonicated graphite powder with similar experimental condition. Total three (3) sets of samples were prepared to obtain the average absorbance value. 3 ml of the sample was collected from each conical flask and sent to centrifuge (Eppendorf Centrifuge 5430) for 3 minutes at 10000 rpm before the absorbance value of the solution was measured every hour. The removal percentage (R%) [10] of graphite and ultrasonicated graphite was calculated using equation (1) and the adsorption capacity was calculated using equation (2).

$$Removal\ Percentage\ (R\%) = \frac{C_o - C_t}{C_o}\ x\ 100\% \quad (1)$$

$$Adsorption\ Capacity, Q_e = \frac{(C_o - C_t)\,V}{m} \quad (2)$$

where R% is aniline removal percentage (%), Q_e is the adsorption capacity of graphite powder (mg/g), Co is initial concentration of aniline solution (ppm), Ct is concentration of aniline solution at t time (ppm), V is total volume of aniline solution (L) and m is the mass of graphite (g). The R% of fresh graphite and ultrasonicated graphite were compared and the adsorbent with better adsorptive performance was selected for further optimization in subsequent sections.

2.3 Response Surface Methodology (RSM)

A mathematical model was generated by RSM to predict the outcome of a system with any new conditions provided. In present study, Box-Behnken, a four independent variables fractional factorial design was chosen due to the fewer experimental sequences of the variables to evaluate potential complex response functions [11]. The pH value of aniline solution, temperature of incubator shaker, initial aniline concentration and dosage of graphite were the four variables studied in the optimisation experiments. A total of 27 sets experiments were required to justify the optimum conditions. The absorbance values were measured every hour using UV-Vis spectrophotometer for each set of experimental condition. The R% was calculated using equation (1).

3. Results and Discussions

At the preliminary stage, the adsorption performance of ultrasonicated pre-treated graphite and untreated fresh graphite were investigated. Based on figure 3.1, it shows that at λ_{max} = 230 nm, the ultrasonicated graphite and untreated graphite exhibited similar R% which ranged from 10% to 20% (Q_e = 0.15 mg/g to 0.31 mg/g) at pH 7. For λ_{max} = 280 nm, the R% for untreated graphite was closed to 40% (Q_e = 0.55 mg/g) compared to ultrasonicated graphite, which only had maximum removal of 15% (Q_e = 0.22 mg/g) at pH 7. Hence, the results did not demonstrate distinct advantage of ultrasonication in term of R%, therefore, the untreated fresh graphite was selected as the effective adsorbent for aniline removal in the subsequent optimisation studies. The results also revealed the R% of graphite was the highest at acidic condition (pH 4) for two wavelengths (230 nm and 280 nm) as the point of zero charge for graphite was pH 6 [12] and leads to the protonation of aniline to anilinium ions. The surface of graphite will be positively charged when the pH was below 6 and the anilinium ions displayed high affinity to graphite, removed from the solution by adsorbing on the surface of graphite [13]. Therefore, it concluded that aniline removal using graphite has the maximum R% at acidic condition.

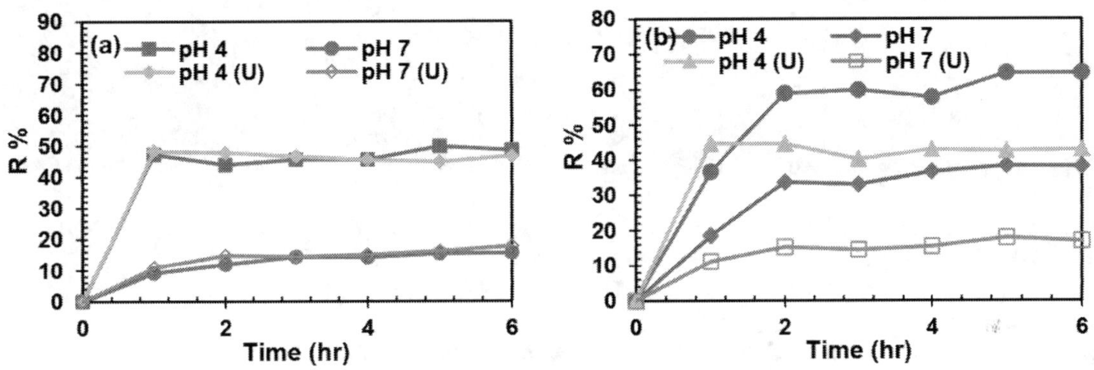

Figure 3.1: R% for ultrasonicated and fresh graphite at (a) 230 nm and (b) 280 nm.

Based on figure 3.2 (a), the FESEM-EDX image of fresh graphite displayed platelet-like crystalline carbon [14]. The similar surface morphology was observed on the used graphite after adsorption as shown in figure 3.2 (b). Hence, it revealed that aniline adsorption has minimal impact on the surface morphology of graphite before and after the adsorption process. Based on the FESEM-EDX elemental mapping as attached as figure 3.2 (c), it revealed that carbon, nitrogen, and oxygen were present in the graphite sample after adsorption. Trace amount of nitrogen was present in the sample due to aniline was originally a nitrogen-containing hydrocarbon and small quantity of nitrogen was released during the adsorption process [15]. 27 sets of experiment generated from RSM were conducted and the results were tabulated in table 3.1. Based on the results obtained, set 4, 14, 16 and 24 has the highest R% which ranged from 85% to 100% (Q_e = 0.11 mg/g to 0.52 mg/g) for both wavelength and the optimal aniline removal condition was determined at 15 ppm of aniline concentration with pH 3 at temperature 40°C using 1.5 g of graphite.

IOP Conf. Series: Materials Science and Engineering 1257 (2022) 012042 doi:10.1088/1757-899X/1257/1/012042

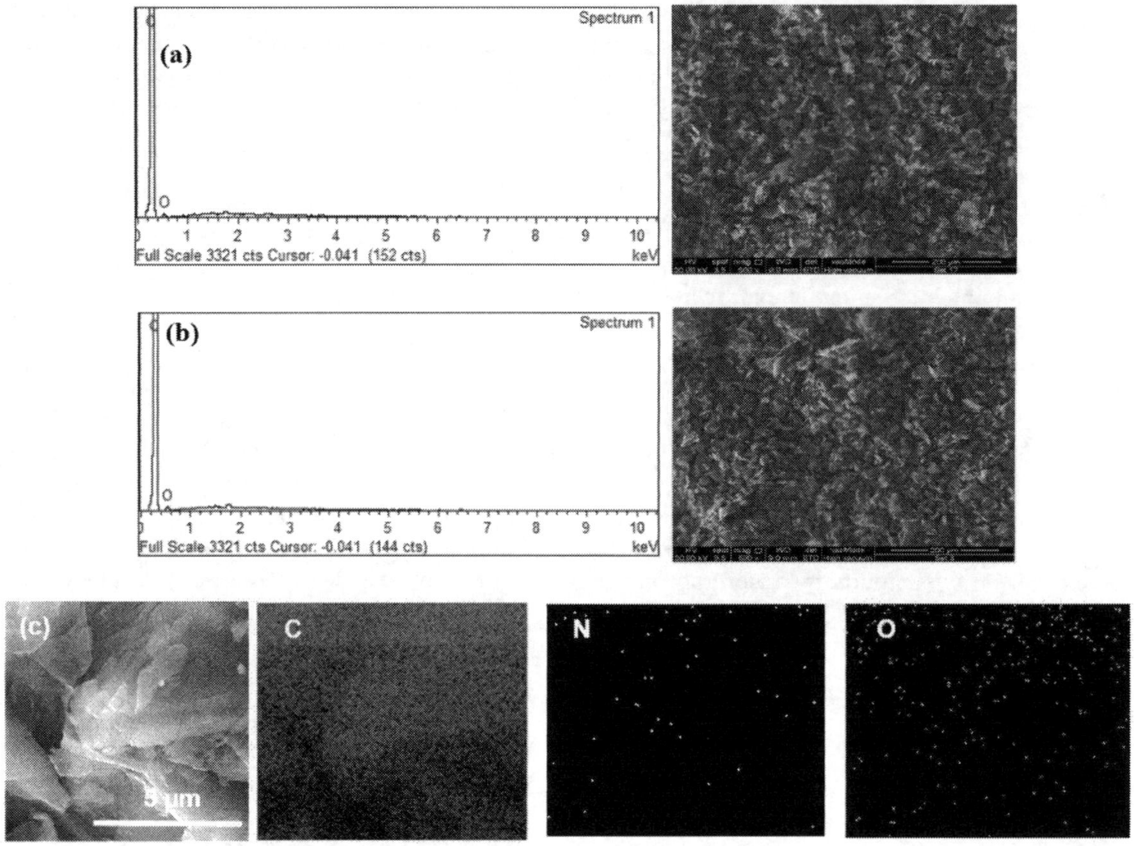

Figure 3.2: FESEM-EDX results for (a) fresh graphite, (b) used graphite after adsorption and (c) FESEM-EDX elemental mapping for used graphite after adsorption (pH: 7, Temperature: 50°C, Initial Concentration: 15 ppm and Dosage: 2.5 g).

Table 3.1: RSM results based on 4 experimental conditions.

Set	pH	Temperature (°C)	Initial Concentration (ppm)	Graphite Dosage (g)	R% (230 nm)	R% (280 nm)
1	11	30	15	1.5	16.57	13.90
2	7	40	15	1.5	38.75	41.05
3	7	50	15	2.5	58.28	64.63
4	3	40	15	2.5	85.10	87.16
5	7	40	15	1.5	35.39	37.21
6	11	40	5	1.5	27.13	0.00
7	7	40	5	2.5	60.80	77.03
8	7	50	15	0.5	44.04	51.46
9	11	40	15	0.5	29.36	28.78
10	7	40	5	0.5	32.93	38.96
11	7	50	5	1.5	65.23	74.51
12	7	50	25	1.5	46.54	52.00
13	7	40	25	2.5	32.34	26.90
14	3	40	5	1.5	96.54	100.00
15	7	30	25	1.5	18.01	11.68
16	3	30	15	1.5	93.70	85.37

Table 3.1: (cont.)

Set	pH	Temperature (°C)	Initial Concentration (ppm)	Graphite Dosage (g)	R% (230 nm)	R% (280 nm)
17	11	40	15	2.5	37.05	22.97
18	7	30	5	1.5	30.19	23.75
19	7	30	15	0.5	16.58	18.87
20	3	40	25	1.5	80.30	73.48
21	7	30	15	2.5	32.63	28.21
22	7	40	25	0.5	33.03	25.77
23	3	40	15	0.5	85.34	74.13
24	3	50	15	1.5	91.17	92.46
25	11	50	15	1.5	50.39	56.22
26	7	40	15	1.5	36.53	24.86
27	11	40	25	1.5	21.60	15.93

ANOVA analysis was also conducted to determine the significance of each parameter towards the adsorption of aniline. Based on figure 3.3, it revealed that the most significance parameter was the pH value of the aniline solution with contribution percentage of 60.74 % and 55.39% at respective wavelength.

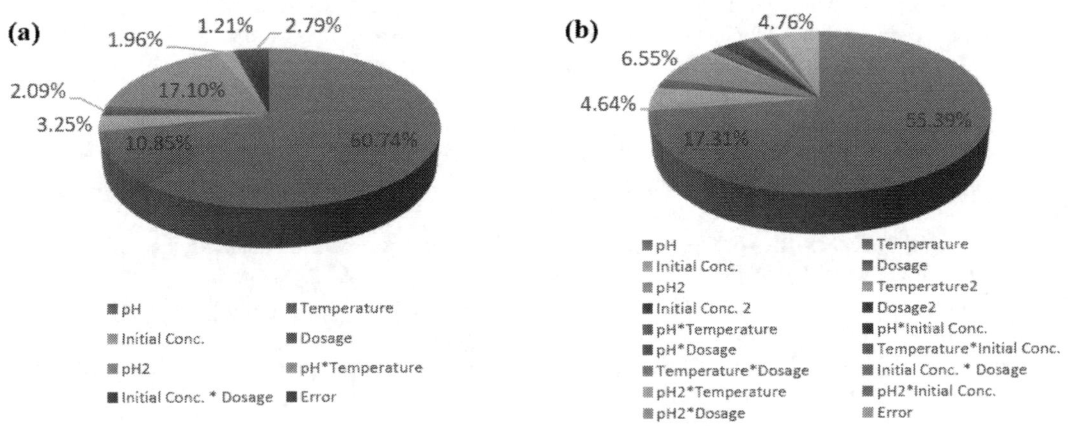

Figure 3.3: Contribution percentage of each parameter at (a) 230 nm and (b) 280 nm.

4. Conclusion

The effectiveness of ultrasonicated graphite and untreated fresh graphite were studied and the results revealed that pre-ultrasonicated graphite powder has similar aniline adsorption performance with the ultrasonicated graphite which was 10% to 20% (Q_e = 0.15 mg/g to 0.31 mg/g) at pH 7. Hence, untreated fresh graphite powder was applied in the following experiments. The morphology and surface characterisation of graphite was conducted through FESEM-EDX and based on the FESEM images obtained, it revealed that aniline adsorption has little impact on the morphology of the graphite powder. Based on the response surface methodology approach, the optimum conditions for aniline R% were determined at: pH 3, 15 ppm initial aniline concentration, 1.5 g of graphite and temperature of 40°C. ANOVA analysis was generated from Minitab 19 and revealed that pH of aniline solution was the most significant parameter (60.74% for 230 nm and 55.39% for 280 nm). Graphite can be used an alternative adsorbent for aniline adsorption in acidic condition and temperature of shaker was the less significant parameter affecting the adsorption performance of graphite.

IOP Conf. Series: Materials Science and Engineering 1257 (2022) 012042 doi:10.1088/1757-899X/1257/1/012042

References

[1] Deng Q, Chen C and Lei Q, 2018 Adsorption of aniline from aqueous solution using graphene oxide-modified attapulgite composites. *RSC Advance,* 8, pp. 23382-23389
[2] IHS Markit, 2020. Nitrobenzene and Aniline. *Chemical Economics Handbook*
[3] Li X, Xu H and Yan W, 2016 Electrochemical oxidation of aniline by a novel Ti/TiOxHy/Sb-SnO2 electrode. *Chinese Journal of Catalysis,* 37(11), pp. 1860-1870
[4] Jiang Y, Wang H, Yu S and Yang K, 2016 Simultaneous removal of aniline, nitrogen and phosphorus in aniline-containing wastewaster treatment by using sequencing batch reactor. *Bioresource Technology,* 207, pp. 422-429.
[5] Mohamed A A-R, Mohamed F S, Safinaz H E-D. and Ahmed M E-N, 2020 Simulated kinetics of the atmospheric removal of aniline during daytime. *Chemosphere,* 255
[6] Esin Poyrazoglu C, Haci Halil B and Ugur Can S, 2020 Use of nanocellulose and microcellulose for anline blue dye removal. *Fresenius Environmental Bulletin,* 29(11), pp. 9542-9549
[7] Xiao W, Zhou P, Mao X and Wang D, 2015 Ultrahigh aniline-removal capacity of hierarchically structured layered manganese oxides: trapping aniline between interlayers. *Journal of Materials Chemistry A,* 3(16), pp. 8676-8682
[8] Liu Q, Zhang L and Huang R, 2015 Removal of aniline from aqueous solutions by activated carbon coated by chitosan. *Journal of Water Reuse and Desalination,* 5(4), pp. 610-618
[9] Verma A, Thakur S, Mamba G and Prateek, 2020 Graphite modified sodium alginate hydrogel composite for efficient removal of malachite green dye. *International Journal of Biological Macromolecules,* 148, pp. 1130-1139
[10] Naga Babu A, Krishna Mohan G V, Kalpana K and Ravindhranath K, 2017 Removal of Lead from Water Using Calcium Alginate Beads Doped with Hydrazine Sulphate-Activated Red Mud as Adsorbent. *Journal of Analytical Methods in Chemistry,* 2017, p. 13
[11] Annadurai G & Sheeja R, 1998. Use of Box-Behnken Design of Experiments for the Adsorption of Verofix Red using Biopolymer. *Bioprocess Engineering 18,* pp. 463-466
[12] Jiendra U, Nagabhushana B & Bilehal D, 2019 Comparative adsorptive and kinetic study on the removal of Malachite Green in aqueous solution using titanium coated graphite and titanium coated graphite with CNT-ABS nanocomposite. *Desalination and Water Treatment,* 209(2021), pp. 392-401
[13] Dutta A K, 1958 Characteristics of free electrons in graphite from a study of its magnetic and other properties. *Physica,* 24(1-5), pp. 343-346
[14] Yau X H, Low F W, Khe C S and Lai C W, 2020 An investigation of the stirring duration effect on synthesized graphene oxide for dye-sensitized solar cells. *PLoS One,* 15(2)
[15] O'Brien J, Curtin T & O'Dwyer T, 2008 Removal of Aniline from Waste Streams using a Combined Adsorption and Catalytic Oxidation Approach. *Adsorption Science & Technology,* 26(5)

Foaming Tendency in Anaerobic Digestion of Wastewater Inoculated with Anaerobic Sludge and Cow Manure: Effect of Protein Concentration

S N Syed Sabeer Ali[1], R Omar[1,*], H Che Man[2], A I Md Idris[3], H A Tajuddin[4]

[1]Department of Chemical and Environmental Engineering, Faculty of Engineering, University Putra Malaysia, Serdang, Selangor, Malaysia.
[2]Department of Biological and Agricultural Engineering, Faculty of Engineering, University Putra Malaysia, Serdang, Selangor, Malaysia.
[3]Department of Chemical Engineering, Faculty of Engineering and the Built Environment, SEGi University, Kota Damansara, Selangor, Malaysia.
[4]Department of Chemistry, Faculty of Science, University of Malaya, Kuala Lumpur, Malaysia.
[*]Corresponding Author: Rozita Omar, Department of Chemical and Environmental Engineering, Faculty of Engineering, University Putra Malaysia, Serdang, Selangor

[*]Email: rozitaom@upm.edu.my

Abstract. Foaming has impacted the digestion process causing severe process failures. The basis of foaming is often unclear due to multiple foaming compounds in a single system. This study aims to elucidate the effect of protein concentration on foaming in anaerobic sludge and cow manure (CM) digesters inoculated with synthetic wastewater as a substrate. Gelatine was used to alter the protein concentration in the digesters. The foaming tendency, biogas production, protein, and chemical oxygen demand (COD) removal were measured to comprehend the findings. The results show that the foaming tendency increased with protein concentration in anaerobic sludge digesters. It also had the highest biogas production. There was no evident process inhibition observed in the anaerobic sludge digester, even at higher protein and COD concentrations. For CM digester, the foaming tendency and biogas production increased to 9 g/L; however, the biogas production decreased with increased substrate protein concentration, indicating process inhibition. The results indicate that the biogas production, foaming tendency, COD and protein removal increased steadily for both digesters. However, the trend decreased for CM digester at increased protein concentrations. This study also concluded that protein creates foaming when inoculated with anaerobic sludge and CM even at low substrate protein concentration.

Keywords: Anaerobic Digestion, Cow Manure, Anaerobic Sludge, Foaming, Protein

1. Introduction

Anaerobic digestion (AD) has gained interest as a cost-effective approach and a sustainable treatment technology over the years. The low energy requirement and low capital investment make AD an easy alternative for various industries and wastewater treatment plants worldwide. There is a considerable weightage of implementing AD in various industries, such as producing gas and developing environmentally friendly energy conversion technologies[1]. Even though AD offers a great advantage to may industries, there is plenty of room for process improvement. Several factors contribute to the deteriorating performance of AD. Foaming is regularly reported as one of the primary causes of unstable operating conditions leading to disturbance in microbial activity and biogas production [2]. Besides that, persistent foaming also decreases the digester working volume and cause overflow of reactors leading to economic loses [3]. Moreover, foaming causes adverse effects on the environment due to the overflowing of digesters and loss of methane in the liquid phase from oversaturation[4].

Amino groups in protein have the tendency to create foam, higher than that of lipid and carbohydrates[5]. Unlike carbohydrates, proteins can commonly be found in all organic substrates such as slaughterhouse waste, pig and chicken CM, food waste and domestic wastewaters[1] Protein molecules have a complex structure that can form a covalent and non-covalent bond leading to the formation of films on the interface. Protein has a hydrophobic and hydrophilic structure, resulting in the accumulation of proteins in the liquid foam interface, thus stabilizing foam[2]. Typically, proteins are introduced into the system through substrates. Proteins are present from 20 to 40% of chemical oxygen demand (COD) in domestic wastewater and 60-90% in sewage and food wastewaters[6]. The microorganisms also produce protein during enzymatic degradation forming extracellular polymeric substances (EPS), resulting in reduced surface tension and enhanced foaming [7].

The identification of foaming causes in a complex process is hard with the presence of multiple foaming compounds. Moreover, very few studies investigated foaming tendencies in CM and anaerobic sludge digesters treating protein-based wastewaters. Hence, this study aims to investigate the effect of protein concentrations on foaming in anaerobic digestion of synthetic wastewater with different inoculums. In order to understand the correlation between foaming and the performance of the digesters, several analytical methods were measured.

2. Materials and Methods

2.1 Anaerobic sewage sludge and Cow Manure (Inoculums)

The anaerobic (AN) sludge was collected from an anaerobic sludge tank of a municipal sewage treatment plant in Kuala Lumpur. Cow dung manure (CM) was collected from a waste collection tank at the UPM farm in Serdang, Selangor. The CM was filtered using a sieve to remove solid content upon collection and kept at a temperature of 2-4°C[8]. Table 1 tabulates the inoculums characteristics.

Table 1. Characteristics of AN sludge and CM

	Unit	AN sludge	CM	Methods
pH		7-7.5	6.5-7	pH Meter (HM Digital)
Biological Oxygen Demand (BOD_5)	mg/L	18,500	55,450	APHA Method (5210B)
Chemical Oxygen Demand (COD)	mg/L	27,360	86,250	APHA Method (5220C)
Ammonia	mg/L	227.08	63.0	APHA Method (4500C)
Alkalinity		1720	2620	APHA Method (2330B)
Total Solid (TS)	g/L	23.075	62.8	APHA Method (2540B)
Volatile Solid (VS)	g/L	16.2	56.05	APHA Method (2540E)
Protein	g/L	5-10	30-35	Bradford Assay Protocol
Oil & Grease (O&G)	mg/L	0	0	APHA Method (5520B)

2.2 Synthetic food wastewater (Substrate)

The synthetic food wastewater (SFW) based on food wastewater was prepared with yeast extract (Marmite) and meat extract (Bovril) as they are good micronutrient nutrient source for bacteria. The meat and yeast extracts at 1:1 ratio were mixed with distilled water as stock solution (COD of 30,000 mg/L). The blank solution (substrate) was prepared from the stock solution (3 g/L protein, 4608 mg/L COD). Gelatin was used to adjust the protein concentration as it is widely applied in food industries and is characterized as a strong foaming agent in a complex mixture[4]

2.3 Experimental procedure for the effect of protein on foaming tendency
Inoculums (AN sludge and CM) were first acclimatized by adding distilled water at 1:1 ratio and left at mesophilic temperature for bacteria digestion of the previous supernatant. After a few days, the inoculum was fed with SFW at gradual COD concentrations (1500, 2500, 3500 and 4500 mg/L) for bacterial adaptation. 90% COD removal achievement denoted SFW concentration increment. The foaming tendency experiment utilized batch process at digester working volume of 0.8 L. The protein concentration investigated was 3, 9, 15, and 21 g/L[4]. The temperature of the system was maintained at mesophilic using a water bath. pH of SFW was set at pH 7 by adding 3 N NaOH. The substrate to inoculum ratio used in this study was 1:1, which recorded the highest COD, BOD, TS, and VS removal in a previous study [9]. The experiment was conducted in triplicate for an accurate outcome.

2.4 Analytical methods
The COD and pH analysis was measured according to methods as in Table 1. Foaming tendency [12] was tested using the aeration method [5].

3. Results and discussion

3.1 Protein and COD removal efficiency
The protein removal efficiency for the reactor performance was investigated for AN sludge and CM as inoculum at different protein concentrations. More than 50% of the protein was depleted during the anaerobic treatment with AN sludge as presented in figure 3.1a. The 3 g/L protein addition into the SFW was converted at a rate 2-4 g/L per day for 4 days with the highest removal rate at 88.2% in AN sludge digester. However, for the subsequent substrate concentrations (9, 15, 21 g/L) the initial protein concentration in the system was higher at 11, 16 and 21 g/L, respectively, but the conversion was lower at 1-3 g/L per day. It is deduced that higher organic loading rates of protein in the digester reduced the protein degradation rates, consistent with a study [13] that proved that higher organic loading rates reduce process performance[10].

The highest removal in CM digestion was at a protein concentration of 3 g/L at 42.6%, followed by 31.1% and 29.9% for 9 g/L and 15 g/L, respectively (Figure 3.1 b). Higher initial protein concentrations in the SFW inoculated with CM resulted in a lower protein removal compared to AN digester with similar concentrations. Higher initial protein concentration in the CM inoculum as noted in the characterization (Table 1) may have contributed to this observation. The bacteria may not utilize high refractory proteins during a shorter time [11] Protein has a complex structure and configuration; hence, the protein hydrolysis process becomes a rate-limiting step delaying the degradation [13].

The COD removal reflects the consumption of organic matter within the digester[12]. The highest COD removal for AN sludge digester was 56% for 3 g/L protein and 61% for 9 g/L protein in CM digester. At 3 g/L protein concentration, COD removal was the highest for AN sludge digester but lowest for CM digester. Three times higher initial COD concentration in CM digester compared to AN sludge may be responsible for this observation. At the highest protein concentration (21 g/L), the COD removal was low for both digesters indicating unstable operation due to stress on the microorganism[13]. Over time, high organic loading leads to an increase in the population of acidogenesis bacteria that produces acids. However, the methanogenic bacteria would not be able to consume the acids as quickly, resulting in the accumulation of acids causing a large proportion of organic matter to remain in the digestate after the treatment [14].

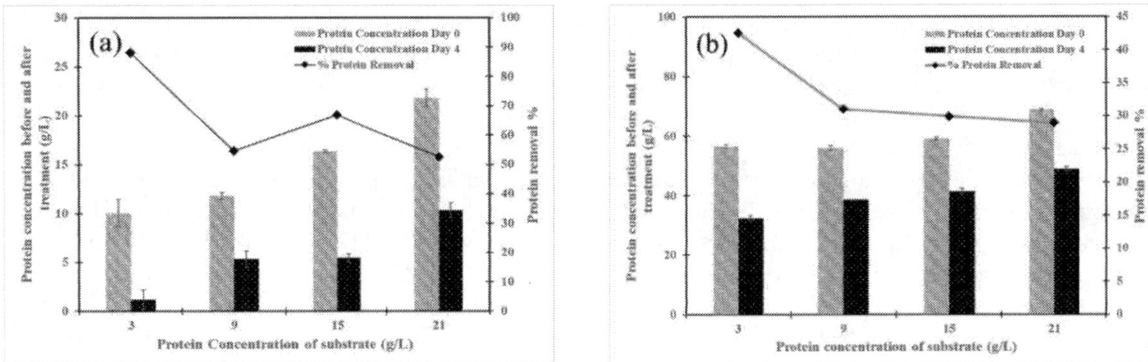

Figure 3.1 (a) Protein removal in AN sludge and (b) cow manure digesters

3.2 Foaming Tendency

The foaming tendency of the sample treated anaerobically using AN sludge and CM is plotted in figures 3.2 a and b, respectively. The foaming tendency for AN sludge digester was the highest at 1.0 mL-foam/(mL-air·min) at 21 g/L protein on day 3 followed by 9 g/L and 15 g/L at 0.77 and 0.73 mL-foam/(mL-air·min). Meanwhile, for the CM digester, the foaming tendency was the highest at 1.16 mL-foam/(mL-air·min) for 15 g/L protein on day 0 followed by 9 and 21 g/L on day 3. There was visible foaming observed in the CM digester at 24 hours after feeding. The foaming tendency at 21 g/L was relatively high for the AN sludge digester. It was suggested that the inefficient conversion of organic matter in the system cause the accumulation of hydrophobic materials at the interface, inducing foaming[15]. High organic compounds in the digesters denoted by high COD concentration (7987 mg/L) led to the accumulation of excess surface-active agents on the surface, triggering foaming[2]. For CM digester, the foaming tendency is seemingly higher, but the higher concentration of protein resulted in a lesser foaming tendency. The decrease in foaming tendency in CM digester could be due to the toxic effect of a high concentration of ammonia in CM digester, which was the digestion product from high initial protein in the digester. This observation is clear in which on Day 2 and Day 3, the foaming tendency of CM digester was lower than AN sludge digester at a substrate concentration of 21 g/L. The ammonia counteracts foam formation in the digester resulting in fewer gas bubbles[4]. Another possible explanation is that, since both reactors were fed with the same feed composition, it could be due to the influence of different microbial community in the digesters[16].

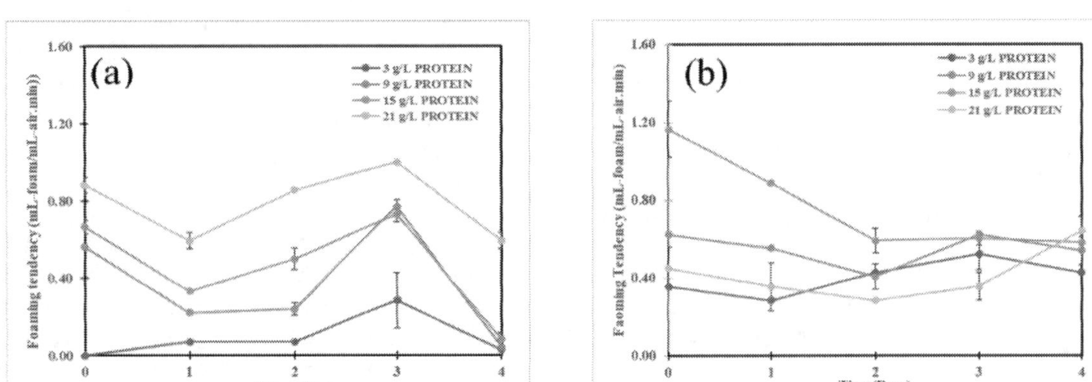

Figure 3.2 (a) Foaming tendency in AN sludge and (b) cow manure digesters

3.3 Biogas Accumulation

The biogas production in the AN sludge digester increases at higher protein concentrations (figure 3.3 a). The highest volume recorded was 1433 mL (21 g/L), whilst the lowest was recorded at 3g/L protein due to organic matter depletion. There was no evidence of inhibition of biogas production in the AN sludge digester at higher protein concentrations. Methanogenesis could have still occurred in the presence of acetoclastic methanogens that can adapt to high ammonium concentrations[17]. The presence of these genus bacteria is familiar in mesophilic digesters. Interestingly, the bacteria were observed to have higher resistance in digesters with high loading rates[18]. In CM digester, the lowest protein concentrations (3 g/l) resulted in lower biogas production at 175mL as shown in figure 3.3b. This could be related to the slow degradation rate of CM digestion at the initial phase of the treatment for bacterial adaptation[19] The highest observed biogas production for CM digester was 840 mL for 9 g/L protein. After that, an increase in protein concentration resulted in a decrease in biogas production. The typical amino groups in proteins have amine, ammonium ions, and potent methanogenic bacteria inhibitors are produced during the microbial degradation of proteins, causing biogas production restrictions. This reinstates that a higher ammonia concentration causes process instability and system failure[18]. CM used in this research had energy content at 89%, indicating a high potential to produce methane; however, extensive foaming observed in the digester led efficient gas-liquid transfer, thus decreasing biogas production[2].

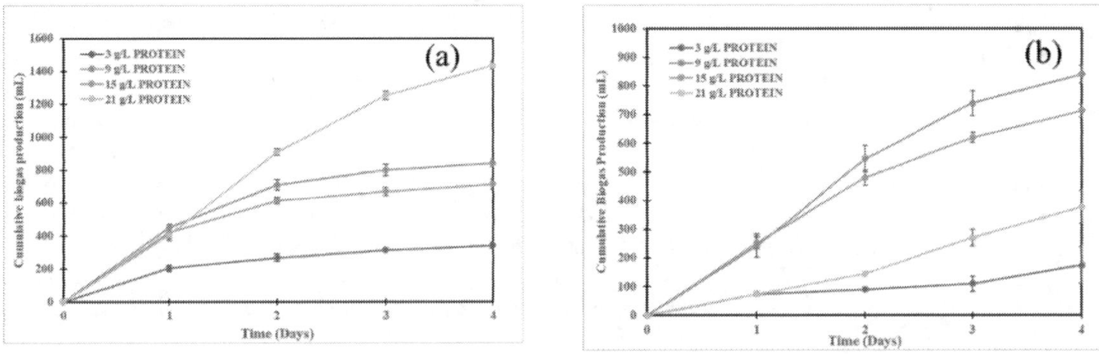

Figure 3.3 (a) Cumulative biogas production in AN sludge and (b) cow manure digesters

4. Conclusion

From this study, it can be concluded that when the protein concentration increases, the foaming tendency increases for AN sludge digester. At 21 g/L, the foaming tendency was highest at 1.0 mL-foam/(mL-air·min). However, the foaming tendency decreased for CM digester at higher protein concentrations. This indicated that there was a strong correlation between protein concentration and foaming tendency in AN sludge and CM digesters. Higher protein concentrations resulted in digester organic overloading, as shown in this study. The COD removal was the lowest for both reactors at 21 g/L indicating a slow degradation rate due to environmental changes such as pH fluctuations in the digesters. Nevertheless, higher protein concentration increased the foaming tendency for AN digester, but the trend for CM digesters reduced beyond 9 g/L protein. This could be a reflection of the bacteria diversity and environmental adaptions in both digesters.

References

[1] Chen X, Zhou W, Li G, Song Q, Ismail M, Wang Y, Ren L and Cheng C 2021 Anaerobic biodegradation of soybean-process wastewater: Operation strategy and sludge bed characteristics of a high-performance Spiral Symmetric Stream Anaerobic Bioreactor, *Water Research*, **197** 1-14

[2] Yang P, Peng Y, Tan H, Liu H, Wu D, Wang X, Li L and Peng X 2021 Foaming mechanisms and

control strategies during the anaerobic digestion of organic waste: A critical review, *Science of the Total Environment* **779** 1-15

[3] Kong X, Liu J, Yue X, Li Y and Wang H 2019 Fe0 inhibits bio-foam generating in anaerobic digestion reactor under conditions of organic shock loading and re-startup, *Waste Management* **92** 107-114

[4] Kougias P G, Boe K, and Angelidaki I 2013 Effect of organic loading rate and feedstock composition on foaming in CM-based biogas reactors, *Bioresource Technology* **144** 1–7

[5] Boe K, Kougias P G, Pacheco F, O-Thong S and Angelidaki I 2012 Effect of substrates and intermediate compounds on foaming in CM digestion systems, Water Science and Technology 66 10 2146–2154.

[6] Mahat S B, Omar R, Che Man H, Mohamad Idris A I, Mustapa Kamal S M, Idris A, Shreehivadasan C, Jamali and N S, Abdullah L C 2021 Performance of dynamic anaerobic membrane bioreactor (DAnMBR) with phase separation in treating high strength food processing wastewater Journal of Environmental Chemical Engineering **9** 1-15

[7] Moeller L, Zehnsdorf A, Pokorná D and Zábranská J 2018 Foam Formation in Anaerobic Digesters in *Advances in Bioenergy*, **3** 1–42

[8] Umar S, Abubakar I, and Ismail N 2012 Anaerobic digestion of cow dung for biogas production *Journal of Engineering and Applied Science* **7** 169-172

[9] Mahat S B, Omar R, Che Man H, Mohd Idris A I, Mustapa Kamal S M, Idris A and Anuar N K 2020 Influence of substrate to inoculum ratio (S/i) on the treatment performance of food processing wastewater containing high oil and grease (O&G) in batch mode *Desalination and Water Treatment* **203** 267–278

[10] Li J, Lu S, Wu S, Zhang W, Hua M and Pan B 2022 The breakdown of protein hydrogen bonding networks facilitates biotransformation of protein wastewaters during anaerobic digestion methanogenesis: Focus on protein structure and conformation *Environmental Research* **208** 1-8

[11] Yang G, Zhang P, Zhang G, Wang Y, and Yang A 2015 Degradation properties of protein and carbohydrate during sludge anaerobic digestion *Bioresource Technology* **192** 126–130

[12] Meegoda J N, Li B, Patel K and Wang L B 2018 A review of the processes, parameters, and optimization of anaerobic digestion *International Journal of Environmental Research and Public Health* **15** 1-16

[13] Jiang M, Qiao W, Wang Y, Zhou T, Lin M and Dong R 2022 Balancing acidogenesis and methanogenesis metabolism in thermophilic anaerobic digestion of food waste under a high loading rate *Science of the Total Environment* **824** 1-8

[14] Ekstrand E M, Björn A, Karlsson A, Schnürer A, Kanders A, Yekta A A, Karlsson M and Moestedt J 2022 Identifying targets for increased biogas production through chemical and organic matter characterization of digestate from full-scale biogas plants: what remains and why? *Biotechnology for Biofuels and Bioproducts* **15** 1-22

[15] Yang P, Peng Y, Liu H, Wu D, Yuan R, Wang X, Li L and Peng X 2022 Multi-scale analysis of the foaming mechanism in anaerobic digestion of food waste: From physicochemical parameter, microbial community to metabolite response *Water Research* **218** 1-12

[16] Kougias P G, Boe K, O-Thong S, Kristensen L A and Angelidaki I 2014 Anaerobic digestion foaming in full-scale biogas plants: A survey on causes and solutions *Water Science and Technology* **69** 889–895

[17] Westerholm M, Moestedt J and Schnürer A 2016 Biogas production through syntrophic acetate oxidation and deliberate operating strategies for improved digester performance *Applied Energy* **179** 124–135

[18] Vítěz T, Novák D, Lochman J and Vítězová M 2020 Methanogens diversity during anaerobic sewage sludge stabilization and the effect of temperature *Processes* **8** 1-12

[19] Zamanzadeh M, Hagen L H, Svensson K, Linjordet R and Horn S J 2017 Biogas production from food waste via co-digestion and digestion- effects on performance and microbial ecology *Scientific Reports* **7** 1-12

Investigation of fluid capillary rise using 3D printed microstructure for solar desalination

R B Ng[1], C L Hoo[1] and V L Wong[1]

School of Engineering & Physical Sciences, Heriot-Watt University Malaysia, Putrajaya, Malaysia.

rn34@hw.ac.uk, c.hoo@hw.ac.uk, v.wong@hw.ac.uk

Abstract. Current desalination technologies such as thermal desalination and reverse osmosis are a process of converting seawater to freshwater. This method brings a negative impact on the environment such as CO_2 emission. Solar desalination is one of the ways to resolve the current desalination challenges and more sustainable in terms of economic and environmental. Solar desalination uses solar energy as its energy consumption for evaporation. Solar desalination depends on the rate of evaporation specifically on the surface area. Hence, this study adopts the concept of cellular fluidics and 3D printing to perform capillaries rise and improve the liquid to the gas-surface area. The increase in strut diameter of the cell from 0.2 mm to 0.4 mm results in an increase of capillaries rise from 3.01 to 7.67 for Body Centered Cubic (BCC) cell geometrical shape. Different cell geometrical shape is also investigated with the strut diameter remaining constant. The Isotruss cell geometrical shape recorded a value of capillaries rise of 11.76 among the other geometrical shape due to Isotruss having more complexity in design and having internal structure.

1. Introduction

The mother planet 'Earth' consists of land and water. The majority of the water on Earth is from the ocean which contributed 97.5% to the global total water however seawater is not safe to drink due to high saline and high mineral. The remaining 2.5% is freshwater and two-thirds of it was trapped in either ice caps or glaciers. Another one-third of the freshwater also known as groundwater and lakes that were remained accessible [1, 2]. The scarcity of freshwater supply was the exponentially growing population of the world, by 2025 world's population will rise to nearly eight billion [1]. Besides, pollution affects the freshwater supply including agricultural fertiliser and industrial waste [1, 3]. The abundance of seawater remains unutilised on Earth, and scientists and engineers have introduced the desalination process. The desalination process uses seawater as the intake and removes the salts and minerals content in the process and ultimately produces freshwater that is suitable for human consumption and agriculture use, desalination remains a critical source of freshwater production in many water scarcity countries. The current desalination process comes with multiple methods such as multi-stage flash (MSF) distillation, multiple-effect distillation (MED), vapour compression, and reverse osmosis (RO) [4-6]. The current commercially viable desalination processes were dominated by MSF, MED and RO processes. MSF and MED processes were established as early as the 1930s in the Middle East in Saudi Arabia [6]. The MSF process used seawater as an intake to produce freshwater through evaporation by

IOP Conf. Series: Materials Science and Engineering 1257 (2022) 012044 doi:10.1088/1757-899X/1257/1/012044

reducing the pressure and increasing the temperature of the system [4]. Next, the MED process uses seawater as an intake and produces freshwater through multiple evaporators by reducing the pressure of the process [4]. Then, with the discovery of membrane technologies in the 1960s, the RO process started to gain popularity and slowly replace MSF and MED processes due to its space-efficient, scalable, and lower energy consumption [4-6]. RO process currently contributed 69% of the total volume of desalinated water produced, followed by MSF and MED both contributed 25%, and others by 4% [5]. To address the current desalination plant challenges high energy requirements, a sustainable solution with energy-efficient low-cost desalination and achieving zero carbon emission are critical for the future trend of desalination technology development. Some of the emerging desalination technologies in recent were forward osmosis (FO), membrane distillation (MD), and electrodialysis (ED) as they have a growing trend of research publication [6]. However, these emerging technologies still require high energy input to operate [6]. Besides, having unlimited energy from the Sun (solar energy) and an abundance of seawater on Earth, solar-powered desalination has gained popularity recently due to the growth of solar technologies and water demands. Hence, solar desalination is exciting to study and discover to achieve economically and environmental sustainability. Direct solar desalination systems such as solar still (SS) (**Figure 1**) in which using seawater as an intake and solar energy to heat the seawater in an enclosure, when evaporation takes place the water vapour then condenses on the glass wall and freshwater is produced [7]. SS process will have carbon net-zero emission to the environment due to it using renewable solar energy. However, solar still freshwater production was dependent on the passive solar energy and liquid to the gas-surface area for evaporation. The factor that affects the rate of evaporation were wind speed, humidity, temperature, and surface area. SS were dependent on the temperature from the solar radiation and liquid and gas surface area for evaporation. Research on solar-driven water evaporators to tackle high salinity desalination was conducted by using a bio-mimetic 3D structure with a freshwater collection rate of 1.72 kg/m^2/h with a promising 96% efficiency [8]. The concept of cellular fluidics was introduced to model a three-dimensional structure with a unit-cell design to promote the capillary rise in conjunction increase the liquid to gas surface area in an open-cell structure [9]. With the use of cellular fluidic, capillaries rise of the water will climb in the 3D-printed structure and promote more liquid to gas surface area, hence, improving the rate of evaporation of the SS process and ultimately more freshwater production.

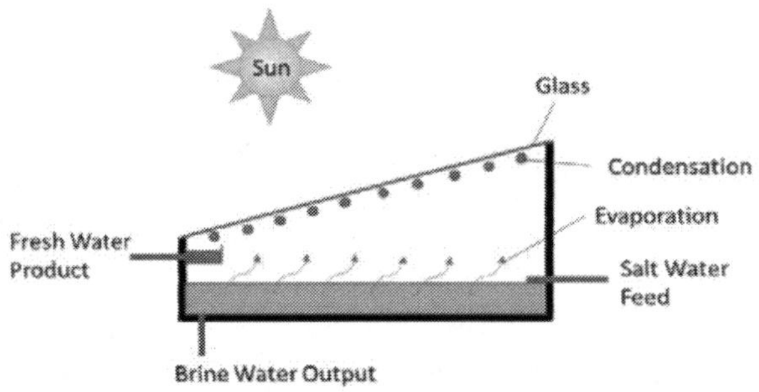

Figure 1. Solar Still Desalination Schematic

2. Theory of Capillary Rise

The liquid flow in the open-cell structure is driven by the cohesion forces, adhesion forces and gravity forces. The cohesion of the liquid was driven by the surface tension γ in liquid-vapour interfaces. Adhesion forces were driven by the contact angle of the liquid and solid surface in the solid-liquid interfaces. According to Newton's third law, when the fluid was in hydrostatic the adhesion forces will be countered act by the gravity forces [10]. Hence leads to the first equation where the adhesion forces are equal to the gravity forces (1). The adhesion force (2) and gravity force (3) is then converted into a dimensionless parameter to ease the comparison [9]. According to Dudukovic, to allow the capillaries to rise in action, the forces of equilibrium in the open cell structure should break unevenly. This can be done based on the adhesion force equation ((2) by enlarging the strut diameter (D), increasing the liquid to solid contact perimeter (s), and reducing the pore size of the cell (L). With changing of the parameters D, s, and L higher adhesion forces will be produced and will shift the equilibrium forces of the fluids resulting in higher capillaries rise. Then, the higher capillaries rise will result in a higher liquid height (h) hence gravity force equation ((3) will increase and eventually have equivalent forces to the adhesion force and capillaries rise will reach equilibrium at a higher level. Besides, the gravity force equation can be reduced with a lower volume of liquid in the open cell structure by reducing the cell strut diameter or changing different cell type designs [9].

$$F_{adhesion} = F_{gravity} \quad (1)$$

$$F^{*}_{adh} = \frac{\pi D^{*} \cos\theta \, s^{*}}{L^{*3}} \quad (2)$$

$$F^{*}_{gravity} = h^{*}(1-\emptyset) = h^{*}\varepsilon \quad (3)$$

3. Design Modelling

The cellular fluidic structure was modelled using nTopology (nTop) software with a built-in computer-aided-design (CAD) function. Using nTop was able to generate a lattice structure model with plenty of preset cell geometrical shapes such as Isotruss (ISO), Body Centered Cubic (BCC), and Face Centered Cubic (FCC) unit type that were presented and demonstrated by Dudukovic [9]. Besides, nTop has a function to determine the solid volume of the generated model, this function was convenient to effectively calculate the porosity of the model and predict the capillaries rise. Here, the design modelling methodology of a unit cell was presented. The modelling process started by creating a cube geometry with a dimension of 1.5 mm × 1.5 mm × 1.5 mm (L × W × H). Then, a lattice structure with a BCC unit type was imposed into the solid cube volume using the volume lattice function, hence generating a vector lattice framework in the geometry. After imposing the volume lattice, a trim lattice function was used to trim off excess lattices that overreach the desired geometry, this step is to ensure all lattices generated were within the geometry boundaries. After that, a thicken lattice function was used to thicken the vector lattice of the geometry, furthermore, the strut diameter will be controlled by the thickness of the lattices. Next, the volume of the thickened lattice was able to obtain using mass properties from body function. Afterwards, the mesh function was imposed into the thicken lattice followed by exporting the mesh in stereolithography (STL) file and ready for slicing and 3D printing. The investigation of the first and second objectives will be changing the strut diameter of the cell and using different cell geometrical shapes. Finally, the model used to verify the capillaries rise in this research was designed with a rectangular geometry with a dimension of 13.5 mm × 13.5 mm × 40.5 mm (L × W × H), each unit cell in the model was in 1.5 mm cube.

4. Fabrication Methodology

Commercial prototyping water washable resin S-900 (Jing Zhao Technology) was mixed with poly (ethylene glycol) diacrylate (PEGDA) with a weight ratio of 1:2 to produce a resin mixture for 3D printing. The purpose of mixing PEGDA in the commercial resin was to improve the hydrophilic properties of the material, hence improving the contact angle of the liquid to solid contact force and leading to higher capillaries rise [11]. Moreover, 1 wt % of photoinitiator powder was added to the resin mixture and stirred for 30 minutes by using a magnetic stirrer (IKA C-MAG HS 7) to ensure the full dissolution of the photoinitiator. After that, all cellular structures were printed using a masked stereolithography printer (MSLA, Phrozen Shuffle 2019). Before printing, the STL file generated from nTop software was imported to the Phrozen slicer software (PZSlice) to slice the STL 3D file to a 2D image. After that, the sliced file was sent to the 3D printer for printing via the USB port. After printing, all 3D printed parts were carefully removed from the build plate with a metal scraper. Then, immediately dipped into the distilled water for removing excessive resin on the model then sonicated for approximately 5 minutes under an ultrasonic cleaner (Elma Elmasonic EASY 10/H). Next, the sonicated 3D printed parts were removed and dried under natural ventilation for approximately 30 minutes. Then, the dried 3D printed parts were then exposed under a portable UV light with the illumination of 400 nm wavelength for 30 minutes to 1 hour to further strengthen the structure of the model by solidifying the resin. The UV light illumination was to enhance more cross-linking of the un-cure resin in the model. After that, the printed model can perform experimental testing.

5. Result and Discussion

The 3D printed model of BCC, FCC, and ISO with the strut diameter from 0.2 to 0.4 mm was printed and tested for the capillaries rise in a water tray. The capillaries rise on the BCC, FCC and ISO model was captured and the capillaries rise is measured using ImageJ (v2.3.0). Then, the result is illustrated in **Figure 2**. The investigation of changing strut diameter to improve capillaries was illustrated in **Figure 2**. The BCC cell geometrical shape (blue trend) with the increase of strut diameter (D) from 0.2 mm to 0.4 mm was performed and the capillaries rise (liquid height) increases from 3.01 to 7.67. Then, FCC cell geometrical shape (orange trend) with the increase of strut diameter from 0.2 mm to 0.4 mm was performed and the capillaries rise increased from 3.47 to 8.91. Next, ISO cell geometrical shape (black trend) with the increase of strut diameter from 0.2 mm to 0.4 mm was performed and the capillaries rise increased from 3.88 to 11.76. This observation proves the theory of capillaries rise mentioned earlier in equation (2), with the increase of strut diameter the adhesion force of the liquid to the solid surface will increase, meanwhile porosity (ε) of the cell size will be decreased, hence improving the capillaries rise into a higher level. **Figure 3** shows the result of using investigating different cell geometrical shapes (BCC, FCC, and ISO) and the effects of capillaries rising when the strut diameter is constant. This result is mapping back to the second objective. The result of cell geometrical shape used in the model in **Figure 3** has a sturt diameter of 0.4 mm for all three cells, giving a comparison of changing cell geometrical shape and effects of capillaries rise. In the Figure 3, Isotruss (ISO) cell geometrical shape shows it had the highest capillaries rise of 11.76 compared to the other cell geometrical shapes FCC at 8.91 and BCC at 7.67. This is due to the complexity of the Isotruss cell that has an internal structure from its design. Hence providing more liquid to solid contact perimeter (s*) will improve the adhesion force in equation (2) and result in higher capillaries rise. Besides, due to the complexity of the Isotruss structure, the porosity (ε) is lower comparing to BCC and FCC cell geometrical shape. Therefore, the higher capillaries rise was observed on Isotruss compared among the three different cell geometrical shapes. The limitation in this project lies on printing difficulty and complexity for higher strut diameter of the cell geometrical shape. As the higher strut diameter is modelled during 3D printing the porosity of the 3D printed model tends to trap the un-cure resin inside the small porous hole leading to a harder clean up during post-processing using alcohol. This will result in the model 3D printed as a whole solid instead

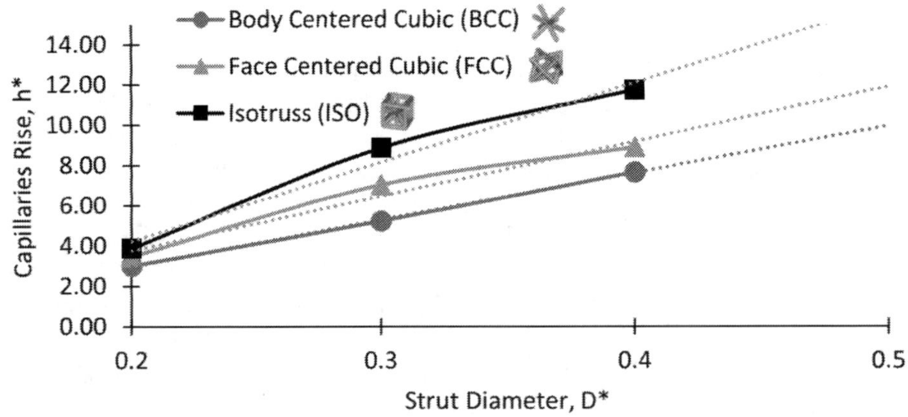

Figure 2. Capillaries Rise against Strut Diameter with different cell geometrical shape.

of a lattice structure. Leading to unsuccessful print in higher strut diameter which might be due to the selection of the resin materials. In order to print a higher strut diameter model, a higher precision resin material is required. Besides, some black spot was identified later in the 3D printer LCD screen, this black spot was unable to illuminate the UV light during the printing, and potentially limits the printing quality and consistency of the 3D printed model.

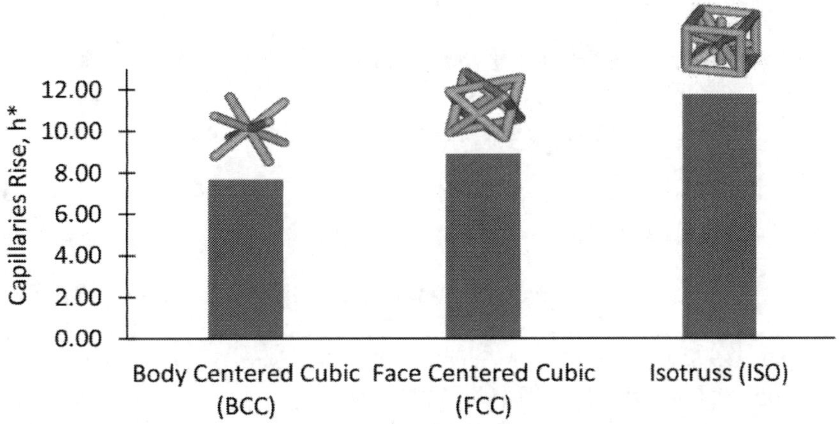

Figure 3. Capillaries Rise against different cell geometrical shape with strut diameter constant at 0.4 mm.

6. Future Plan

The future plan of the project is proposed to use a high precision or quality resin to achieve better print quality to achieve larger strut diameter printer and increase capillaries rise. Besides, materials selection of original hydrophilic resin or mixture of hydrophilic resin with PEGDA should be investigated to provide a better capillary rise. As proven in this study where the change of strut diameter will improve capillaries rise. Therefore, design modelling of different strut diameters from 0.5 mm onward shall be of great interest with different unit type cells in improve the capillaries rise further. The use of a higher precision printer will be able to achieve printing smaller cell sizes into 1 mm or smaller. Printing in a smaller size of the cell will be able to produce higher capillaries rise according to equation (2). Despite the fulfilment of the objectives in this project. A customisable cell geometrical shape other than BCC, FCC, and ISO is worth discovering to have a deeper understanding on how the fluid flow in the open cell structure. Investigating liquid contact angle and surface roughness in the future can be helpful to have a better understanding of the hydrophilic properties of the materials. This is because liquid contact angle will affect $\cos\theta$ in adhesion force in equation (2).

7. Conclusion

This study portrays a general view of the limited freshwater supply for the coming years and current challenges faced by desalination facilities, all of which are highly energy demand to run the plants and non-sustainable. Besides, there are also producing a lot of carbon emissions annually. Solar desalination is one of the ways to tackle the current challenges faced by desalination facilities. This study adopted the concept of cellular fluidics and capillaries rise is performed. The first objective investigated the increase of sturt diameter from 0.2 mm to 0.4 mm and observed the capillaries rise increase from 3.01 to 7.67 for BCC cell geometrical shape, 3.47 to 8.91 for FCC cell geometrical shape and 3.88 to 11.76 for ISO cell geometrical shape. The second objective, comparing different cell geometrical shapes with the strut diameter remains constant at 0.4 mm, the ISO cell geometrical shape recorded a value of 11.76 among the other two geometrical shapes BCC and FCC.

Acknowledgement
I would like to take this opportunity to thank nTopology, Inc for their support to provide the full version of nTopology software in this study that assisted us in the design modelling of the lattice structure generation.

References
[1] Patrick S M. *The Coming Global Water Crisis The Atlantic.* [Internet]. 2012 [cited 2022 Apr 09]. Available from: https://www.theatlantic.com/international/archive/2012/05/the-coming-global-water-crisis/256896/
[2] Perlman H. *Where is Earth's water?* [Internet]. 2013 [cited 2022 Apr 09]. Available from: https://web.archive.org/web/20131214091601/http://ga.water.usgs.gov/edu/earthwherewater.html
[3] WorldWildlifeFund *Water Scarcity.* [Internet]. 2021 [cited 2022 Apr 09]. Available from: https://www.worldwildlife.org/threats/water-scarcity
[4] Khawaji A D, Kutubkhanah I K, Wie J M 2008 *Advances in seawater desalination technologies. Desalination.* **221**(1-3) pp 47-69
[5] Jones E, Qadir M, van Vliet M, Smakhtin V, Kang S-m 2019 *The state of desalination and brine production: A global outlook. Science of The Total Environment.* **657**
[6] Ahmed F E, Khalil A, Hilal N 2021 *Emerging desalination technologies: Current status, challenges and future trends. Desalination.* **517**.
[7] Sharon H, Reddy K S 2015 *A review of solar energy driven desalination technologies. Renewable and Sustainable Energy Reviews.* **41** pp 1080-118
[8] Wu L, Dong Z, Cai Z, Ganapathy T, Fang N X, Li C, *et al* 2020 *Highly efficient three-dimensional solar evaporator for high salinity desalination by localized crystallization. Nature Communications.* **11**(1)
[9] Dudukovic N A, Fong E J, Gemeda H B, DeOtte J R, Cerón M R, Moran B D, *et al* 2021 *Cellular fluidics. Nature.* **595**(7865) pp 58-65
[10] Rodríguez-Valverde M A, Miranda M T 2011 *Derivation of Jurin's law revisited. European Journal of Physics.* **32**(1) pp 49-54
[11] Son K H, Lee J W 2016 *Synthesis and Characterization of Poly(Ethylene Glycol) Based Thermo-Responsive Hydrogels for Cell Sheet Engineering. Materials (Basel).* **9**(10) pp 854

IOP Conf. Series: Materials Science and Engineering 1257 (2022) 012045 doi:10.1088/1757-899X/1257/1/012045

Characteristics of alginate and soy protein isolates composite aerogels as oil absorbent

H S Roslan[1], A N Mustapa[1,2*], S H Hanipah[1], A B Alias[1], F M Fuad[1], Á Martín[3] and M J Cocero[3]

[1]School of Chemical Engineering, College of Engineering, Universiti Teknologi MARA, UiTM, 40450 Shah Alam, Selangor, Malaysia
[2]Integrated Separation Technology Research Group, School of Chemical Engineering, College of Engineering, Universiti Teknologi MARA, UiTM, 40450 Shah Alam, Selangor, Malaysia
[3]BioEcoUva, Research Institute on Bioeconomy, High Pressure Process Group Department of Chemical Engineering and Environmental Technology, University of Valladolid, Calle Doctor Mergelina s/n, 47011 Valladolid, Spain

*anajwa@uitm.edu.my

Abstract. Treating oil spill occurrence is crucial because of the toxic effects on marine life and the environment. Hence, absorption is used for a low-cost oil recovery process, simple mechanism, and sustainability to mitigate such occurrence. Synthetic absorbent is always associated with its poor degradability, which may cause water to turn toxic. Natural materials have been foreseen as excellent absorbent due to their cost-effectiveness, sustainability, and less environmental harm. In this work, alginate/soy protein isolates (SPI) aerogels are synthesized, undergo surface modification with methyltrimethoxysilane (MTMS) and tetraethyl orthosilane (TEOS) solution, and then dried by supercritical CO_2 at 110 ± 5 bar and 40 °C in 4h to produce aerogels absorbent. The synthesized absorbents were analyzed via FTIR, TGA, DSC, FESEM, contact angle and oil absorption analysis to determine their physical and chemical properties. Results demonstrated that the hydrophilic surface of the alginate/soy protein isolate could be tailored in a simple surface modification process. Its oil absorption capacity towards the engine oil can reach a range of 9.1-12.1 g/g, comparable to most reported natural absorbents. Thus, the findings in this study show that alginate/ SPI aerogels may have potential as an industry oil absorbent.

1. Introduction

Oil and gas are still relied upon as energy sources [1]; however, the oil and gas industries have been one of the primary sources of water pollution [2] until today. Oil spill adversely affect the environment and groundwater resources, endangering the well-being of humans and animals and indirectly causing global warming and climate change [2]. Therefore, every time an oil spill occurs, the main concern is to take the oil from the water as soon as possible [3]. Hence, the cleanup process must be done thoroughly. Unfortunately, most known cleanup processes have their limitations, and the following are some of the limitations: a) availability of trained workers, b) potential harm to the personnel's well-being, and c) the effect on the surrounding environment and marine life [4][5].

Interest in oil absorbents has grown remarkably over the past years due to their potential for low-cost oil recovery, simple mechanisms, and sustainability [6]. Aerogels have been reviewed as potential oil absorbents [5], and this is because aerogels are solid porous materials with a surface area of 2 x 105 to 1.5 x 106 m^2/kg, a density of 3 to 500 kg/m^3 and a refractive index of 1.007 to 1.24 [7]. Aerogels are ideal for a fast sorption rate and have good stability in floating on water [5]. There are three types of materials that can be used to prepare aerogels: a) inorganic materials, b) synthetic organic materials, and c) natural biomaterials. Alginate is a natural biopolymer polysaccharide with good chemical stability, pH sensitivity, form of gel barriers to water and gases, and biological functionality [8]. Another advantage of alginate, it is less harmful to the environment and is sustainable and cost-effective. However, biopolymer materials like polysaccharides are amphiphilic. Hence, surface modification of the sorbent is necessary. Prior research has shown that shrinkage of alginate gel still happens and eventually will lead to syneresis, the expulsion of liquid from the gel networks [9][10]. Therefore, adding additional biopolymers such as soy protein isolate (SPI) to produce composite alginate gels has been proposed to overcome such limitations.

Protein isolated from soy has been used in previous applications to strengthen alginate carriers for delivering drugs, plants extract, and biomedical applications [11]. SPI has showcased good advantages, such as being one of the cheapest proteins while having high hydrophilic capabilities and good processability [12]. Moreover, SPI are also promising in water-holding, oil binding and emulsifying properties. This work aims to synthesize hydrophobic alginate/soy protein isolates aerogels absorbent after undergoing surface modification and dried by supercritical CO_2 drying. The physical and chemical characteristics of the alginate/SPI aerogel were determined by the contact angle, oil absorption analysis, FTIR, DSC, TGA and DSC. The synthesized aerogel is foreseen could be an excellent candidate for oil absorbents application.

2. Materials and Method

2.1. Materials
Sodium alginate (SA), ammonium persulfate (APS), N, N-methylene bisacrylamide (MBAm), and glucono-δ-lactone (GDL) were purchased from Sigma Aldrich. SPI was purchased from Radiant Code Sdn. Bhd. Absolute ethanol and calcium carbonate ($CaCO_3$) were obtained from R&M Chemicals. Carbon dioxide (purity 99%) for the supercritical drying was supplied by SR Focus Trading. The agents prepared for the surface modification process were methyltrimethoxysilane (MTMS) and ethyl orthosilicate (TEOS), both purchased from Sigma Aldrich.

2.2. Preparation
2.2.1. Sodium alginate/SPI hydrogels. Three fractions of SA, 1, 2, and 3% w/w with an SPI fraction of 0.5% w/w, were synthesized in 100 mL distilled water. The alginate/SPI was prepared by internal gelation. According to the fraction set, distilled water and SA were stirred at 600 rpm at room temperature. Gradually during stirring, SPI was added to each beaker, and another additional beaker was prepared without SPI as blank. After obtaining a homogenized solution, MBAm and APS were added. The solution was then further stirred for 3h at 60 °C, followed by the addition of $CaCO_3$ to the solution. After the mixture was homogenized, GDL was added to initiate the chemical cross-linking and stirred for 3 minutes. The final solution was then transferred into moulds and refrigerated at 4 °C.

2.2.2. Surface modification of alginate/SPI hydrogels. The hydrogels undergo a stepwise solvent exchange sequentially every two hours; the following exchange was at 30, 50, 70, 90, and 100% v/v ethanol-water mixture. After the solvent exchanges step was completed, the alcogels proceeded with surface modification technique using TEOS and MTMS. The alginate/SPI alcogels were soaked for 24h at room temperature, followed by supercritical CO_2 drying. A portion of the alginate/SPI hydrogels was set aside without undergoing the surface modification and labelled as unmodified for analysis purposes.

2.2.3. Production of alginate/SPI hydrophobic superabsorbent aerogels. The alginate/SPI alcogels were dried using the supercritical CO_2 drying at 110 ± 5 bar and 40 °C in 4h at a continuous circulation of CO_2. The extracted liquid ethanol from the drying process was then removed every 1h, and the drying column was then depressurized at a rate of 2 bar/min. The production of hydrophobic superabsorbent aerogels was completed and then collected. The preparation was repeated three times using the same alginate/ SPI formulation.

2.3. *Characteristics*

The surface structure and morphology of the aerogels were determined using field emission scanning electron microscopy (FESEM) analysis by Jeol JSM-7600F. Information on the infrared spectrum of absorption was determined from Fourier transform (FTIR) analysis using Perkin Elmer Spectrum One. On the other hand, the aerogel's thermal stability and residue decomposition were measured using differential scanning calorimetry (DSC) Mettler Toledo DSC 1, and the analysis was carried out from 20 to 500 °C at a heating rate of 10 °C/min. The measurement is recorded over time as the temperature changes. Thermal gravimetric analysis (TGA) was used to measure the thermal characteristics of the aerogels, and the analysis was carried out from 20 to 400 °C at 10°C/min by using Mettler Toledo while the contact angle and its wettability were measured using the CVA3000 model equipment.

2.4. *Oil absorption*

Oil engine was selected as a model pollutant to analyse oil absorption of the alginate/SPI aerogel. The weight of the aerogel (1cm x 1cm x 1cm) was recorded before and then soaked inside in waste (engine oil) without water for 1 min to reach equilibrium. The soaked aerogel was then held in the air for 1 min before weighing. The absorption capacity, Q (g/g), was calculated using the equation (1) below:

$$Q = (M - M_o)/\, M_o \qquad (1)$$

where M is the weight after absorption and M_o is the weight after absorption (g).

3. Results and discussion

3.1. *Contact angle and oil absorption analysis.*

Figure 3.1(a) the condition of alginate/SPI; A (1:0.5), B (2:0.5), and C (3:0.5) after supercritical drying, (b) water dyed in methylene blue and oil droplets on A and C (c) water dyed in methylene blue and oil droplets on unmodified alginate/SPI aerogel (d) absorption of waste engine oil in water.

The contact angle of each aerogel (figure 3.1 (a)) is then analyzed, and the reading for A and B is greater than 90° indicating as hydrophobic; however, for the C, the reading registered it as hydrophilic. Further analysis of C shows that both water and oil droplets were absorbed (figure 3.1 (b)), suggesting no selective absorption similar to unmodified alginate/SPI aerogels shown in figure 3.1 (c). C's hydrophilicity may be due to the abundance of OH functional groups on its surface, rendering the ability to repel water [13]. While for A, the water droplets stay on top of the surface, and the oil droplets absorb entirely inside. A mini-experiment was done to showcase alginate/SPI's ability to absorb oil, as demonstrated in figure 3.1(d). The experiment shows that waste (oil engine) was removed when a piece of A was dropped on the oil layer.

Oil absorption capacity was measured when A and B were soaked in waste (engine oil). The absorption capacity of A and B were recorded at a range of 9.1-12.1 g/g. For comparison in work [14], the absorbent capacity for SA-Ca-Zr foam was in a similar range, while in work [15], the poly(lactic acid) materials have a significantly low absorption capacity. Both analyses were repeated three times, and the analyses' value was then taken at average. All results show that sodium/SPI aerogels (A and B) are hydrophobic and have good capability as an absorbent.

3.2. Fourier transform infrared spectroscopy (FTIR) analysis.

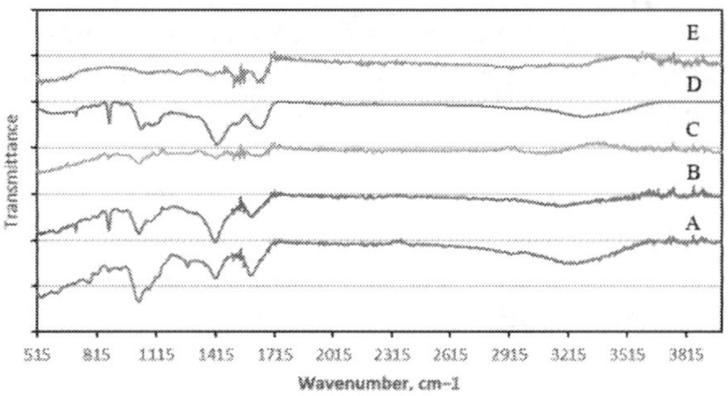

Figure 3.2 depicts the FTIR spectra for alginate/SPI with A) 1:0.5, B) 2:0.5, C) 3:0.5, D) blank alginate and E) powdered soy protein

The presence of alginic acids and soy protein can be identified from the spectra. The O-H stretching vibration appears as a broad peak at 3215 to 3515 cm−1, indicating alginate exists. The peaks appearing in the range of 815 to 1115 cm^{-1} are seen for the membranes corresponding to C-O and C-C and C-O-C linkage vibrations. Bands below 740 cm^{-1} indicate in alginate the C-O-C vibrations of glycosidic linkage and the deformation of pyranosyl rings [8]. The spectral region from 1200 to 1600 cm^{-1} indicates C-H and C-O-H deformation vibration, respectively. Based on [16], the C-H stretching vibration should be more visible after modification. However, the vibration appears to be subtle in C compared to A and B hence this explains the C sample present as hydrophilic meanwhile, the apparent C-H group has attributed to the A and B samples being hydrophobic. In addition, the OH groups have been reduced or replaced due to the surface modification process. Soy protein spectrum showed an amide I band, which can be found at the peak of 1589 cm^{-1}. The range between 1320 cm^{-1} and 1625 cm^{-1} shows C-N and N-H bands which signal protein, respectively [8].

3.3. Differential scanning calorimetry (DSC) thermal analysis and Thermogravimetric analysis (TGA) analysis.

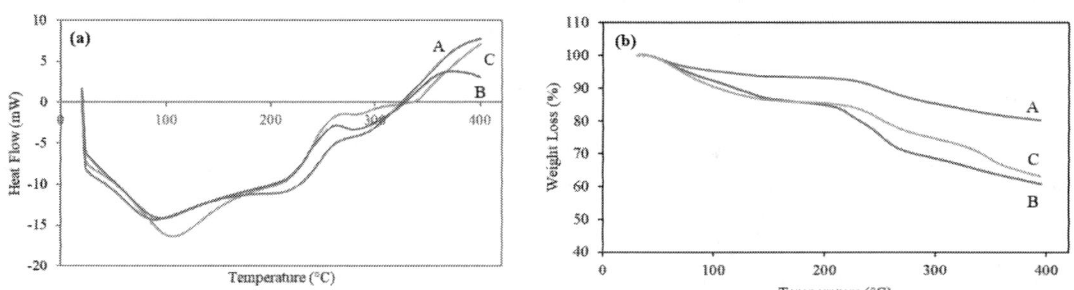

Figure 3.3(a) depicts the DSC analysis while **(b)** depicts the TGA analysis for alginate/SPI with A) 1:0.5, B) 2:0.5, and C) 3:0.5.

In the DSC analysis, the compatibility of polymer blends and the glass transition temperature (Tg) can be estimated. At around 65 °C, an endothermic peak can be seen (figure 3.3 (a)). Meanwhile, at around 350 °C and above, an exothermic peak can be seen, indicating alginate decomposition. From figure 3.3(a), the Tg values were recorded as 88.6, 96.26, and 107.67 °C for materials A, B, and C, respectively. The Tg value of the alginate/SPI blends increases with the increase of the alginate concentration, indicating the increased rigidity of the alginate/SPI chain. The water loss for A occurs at 85 °C, for B is at 96 °C, while for C is at 107 °C. The gradual shift to the right is because the hydrophobic groups require higher temperatures due to the surface modification process [16]. Figure 3.3(b) shows the TGA curves of all the blends, with a decomposition temperature in the range of 220-400 °C. The onset temperature ($T_{5\%}$) of the alginate/SPI increases as the alginate per cent decreases. The A blend appeared to be more stable compared to the B and C blends, where the A has a higher $T_{5\%}$ of 110.12 °C with a maximum decomposition temperature (T_{max}) of 256.87 °C.

On the other hand, the $T_{5\%}$ of B and C were 74.55 °C and 70.59 °C, respectively. However, towards the end, the C blend containing 3% w/w of alginate appeared to be more stable than B, which contains 2% w/w of alginate. The absolute weight loss observed was small at below 100 °C, corresponding to the samples' water loss. In the meantime, all samples showed a moderate loss between 100-200 °C. The decomposition of soy protein isolates occurred at around 210 °C to 250 °C with a sharp reduction of its weight. The decomposition of the carbonaceous matter started to occur at the temperature range from 280 to 400 °C. The slight weight loss in A indicates that the modified aerogel from this is thermally more stable than in B and C. Figures 3.3(b) shows that all blends exhibit a single decomposition temperature in the TGA curves, possibly indicating good compatibility.

3.4. Field Emission Scanning electron microscopy (FESEM) analysis.

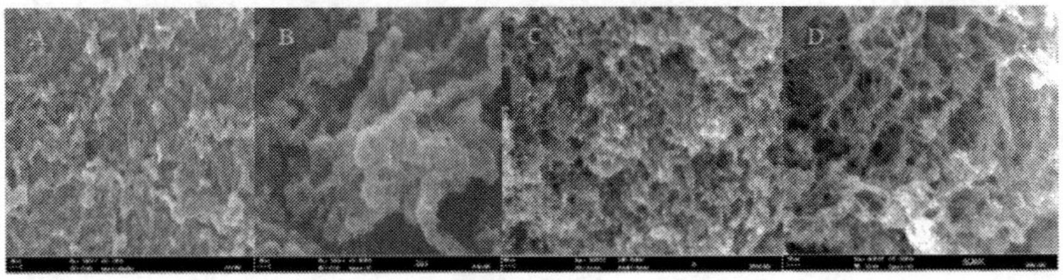

Figure 3.4 shows the FESEM image of the sample cross-section A) hydrophobic 1:0.5, B) hydrophobic 2:0.5, C) hydrophilic 3:0.5, and D) alginate blank.

All the images were taken at the same magnification of 50K on the cross-sectional area. Based on the figures (figure 3.4A to 3.4C), the textures of the alginate/SPI appear fibrous-like and became apparent as the alginate per cent increased from 1 to 3% w/w. The increasing alginate concentration has been attributed to the complexity of the fibrous structure. On the other hand, compared to figure D, the materials of A, B, and C containing soy protein show a decrease in porosity. This behaviour was similar to that of work reported by [17]. In general, the morphology of the A, B, and C samples display pleated surfaces nevertheless, it is more noticeable in the C. Indeed, the textural of the C depicts a rougher surface with more prominent micropores than the rest of the studied alginate/SI composition. These features of the C indicate its hydrophilic due to its higher alginate content, which may affect the material's morphological characteristics [8].

4. Conclusion

Alginate/SPI absorbent with alginate of 1 and 2% w/w was modified from hydrophilic to hydrophobic after soaking in the surface modification agent and possesses good oil absorption capacities as high as 12.1 g/g. In contrast, alginate prepared at 3% w/w present as hydrophilic and absorbed both oil and water. The C-H bonds appear more visible after modification for alginate concentration at 1 and 2% w/w, while the bonds appear subtle for alginate at 3% w/w. Alginate/SPI absorbent with an alginate concentration of 1% w/w are thermally more stable than alginate concentrations of 2 and 3% w/w. The morphology structure of the sample depends on the alginate concentration, as results demonstrated that the higher the concentration of the alginate, the stiffer the sample. This study uses a fixed amount of soy protein isolate of 0.5% w/w concentration. Hence further studies are envisaged to study the effects of different concentrations of soy protein isolates on the chemical and physical characteristics of the aerogels.

5. References

[1] Humbatova S I and Hajiyev N Q 2019 Oil Factor in Economic Development *Energies* **12** 1573
[2] Dorr B S, Mathewson P D, Hanson-dorr K C, Healy K A, Horak K E and Porter W F 2020

Landscape scale thermoregulatory costs from sublethal exposure to Deep Water Horizon oil in the double-crested cormorant *Mar. Pollut. Bull.* **152** 110915
[3] Ramseur J L 2010 Deepwater Horizon Oil Spill : The Fate of the Oil 1–20
[4] Yang Z, Chen Z, Lee K, Owens E, Boufadel M C, An C and Taylor E 2021 Decision support tools for oil spill response (OSR-DSTs): Approaches, challenges, and future research perspectives *Mar. Pollut. Bull.* **167** 112313
[5] Doshi B, Sillanpää M and Kalliola S 2018 A review of bio-based materials for oil spill treatment *Water Res.* **135** 262–77
[6] Yati I, Aydin G O and Sonmez H B 2016 Cross-linked poly (tetrahydrofuran) as promising sorbent for organic solvent / oil spill *J. Hazard. Mater.* **309** 210–8
[7] Hatami T, Viganó J, Innocentini Mei L H and Martínez J 2020 Production of alginate-based aerogel particles using supercritical drying: Experiment, comprehensive mathematical model, and optimization *J. Supercrit. Fluids* **160**
[8] Volić M, Pajić-Lijaković I, Djordjević V, Knežević-Jugović Z, Pećinar I, Stevanović-Dajić Z, Veljović Đ, Hadnadjev M and Bugarski B 2018 Alginate/soy protein system for essential oil encapsulation with intestinal delivery *Carbohydr. Polym.* **200** 15–24
[9] Crittenden R, Weerakkody R, Sanguansri L and Augustin M 2006 Synbiotic Microcapsules That Enhance Microbial Viability during Nonrefrigerated Storage and Gastrointestinal Transit **72** 2280–2
[10] Chandramouli V, Kailasapathy K, Peiris P and Jones M 2004 An improved method of microencapsulation and its evaluation to protect Lactobacillus spp . in simulated gastric conditions **56** 27–35
[11] Bel A, Karlovi S and Pavlovi V 2015 Food Hydrocolloids Protein-reinforced and chitosan-pectin coated alginate microparticles for delivery of fl avan-3-ol antioxidants and caffeine from green tea extract **51**
[12] Cuadri A A, Bengoechea C, Romero A and Guerrero A 2016 A natural-based polymeric hydrogel based on functionalized soy protein *Eur. Polym. J.* **85** 164–74
[13] Duan H, Lyu H, Shen B, Tian J, Pu X, Wang F and Wang X 2021 Science of the Total Environment Superhydrophobic-superoleophilic biochar-based foam for high-ef fi ciency and repeatable oil-water separation *Sci. Total Environ.* **780** 146517
[14] Wang Y, Feng Y and Yao J 2019 Construction of hydrophobic alginate-based foams induced by zirconium ions for oil and organic solvent cleanup *J. Colloid Interface Sci.* **533** 182–9
[15] Xue Z, Sun Z, Cao Y, Chen Y, Tao L, Li K, Feng L, Fu Q and Wei Y 2013 Superoleophilic and superhydrophobic biodegradable material with porous structures for oil absorption and oil–water separation *RSC Adv.* **3** 23432–7
[16] Meng C, Zhang H, Zhang S, Guo J and Zou X 2018 The preparation of hydrophobic alginate-based fibrous aerogel and its oil absorption property *J. Sol-Gel Sci. Technol.* **87** 704–12
[17] Silva R, Bulut B, Roether J A, Kaschta J, Schubert D W and Boccaccini A R 2014 Sonochemical processing and characterization of composite materials based on soy protein and alginate containing micron-sized bioactive glass particles *J. Mol. Struct.* **1073** 87–96

Acknowledgement
This work was supported by National Grant "Penyelidik Muda (YTR)." no. 600-IRMIS/YTR/5/3 (010/2020).

IOP Conf. Series: Materials Science and Engineering 1257 (2022) 012046 doi:10.1088/1757-899X/1257/1/012046

Synergistic adsorption-photocatalytic degradation of Rhodamine B dye using a 3D-printed photoreactor with g-CN homojunction-based thermosets coating: response surface methodology and adsorption modelling

EY Lee[1], SY Wong[1], VL Wong[1*], SJ Phang[1] and KH Cheah[2]

[1]School of Engineering and Physical Sciences, Heriot-Watt University Malaysia, Jalan Venna P5/2, Precinct 5, 62200 Putrajaya, Malaysia
[2]School of Aerospace, Faculty of Science and Engineering, University of Nottingham Ningbo China, 199 Taikang East Road, 315100 Ningbo, Zhejiang, PR China

E-mail: v.wong@hw.ac.uk

Abstract. In present work, graphitic carbon nitride (g-CN) homojunction were synthesized to investigate the synergistic effect of photocatalysis and adsorption on Rhodamine B (RhB) dye degradation. The photodegradation (DE%) and adsorption (R%) were evaluated under the irradiation of two 50 W LED lights and in dark for 6 hours, respectively. A three-level Box-Behnken design (BBD) were employed to study the effects of initial dye concentrations (5 to 20 ppm), recirculation flow rates (1 to 3 mL/min), and the number of photoreactors (2 to 6) with total 15 sets of experiment. A quadratic regression was developed to predict DE% with an error of less than 5.5% relative to the experimental data. Additionally, ANOVA analysis showed that the number of photoreactors is the most influential parameter ($p<0.05$); while the two-way interaction between variables are insignificant. For the adsorption studies, Langmuir isotherm showed the best-fit result ($R^2>0.98$) and the separation factor (R_L) has indicated the adsorption process is favourable ($R_L=0.05$). Additionally, the adsorption experimental data fit well with the pseudo-second-order kinetic model ($R^2>0.98$). Overall, the development of homojunction photocatalyst has provided an innovative and alternative approach for water remediation.

1. Introduction

The dyeing process in textile, printing and paper industries consumes massive amount of water and produces wastewater which contains dye and other contaminants. Due to the limitation of chemical and physical affinity between fibre and dye, more than 15% of the textile dyes is lost in the effluent during dyeing operation [1]. Efficient wastewater treatment is vital to maintain water quality and preserve clean water for life on the planet. Recently, photocatalysis treatment has gained attraction from researchers. The major challenge faced by photocatalysis treatment is the low usage of sunlight as photocatalytic reaction can only be excited under ultraviolet (UV) light irradiation at wavelength < 380 nm. Nonetheless, this issue can be countered by fabricating homojunction-based semiconductors [2], forming heterojunctions nanocomposites [3], and doping metal-ions to the photocatalyst [4]. These approaches have improved the photocatalytic activity considering that distinctive crystal phases, light adsorption ability, and electronic band structures were introduced to the original photocatalyst.

The photocatalytic performance can be influenced by various parameters, such as catalyst dosage, initial concentration of refractory organic contaminants, charge-trapping species, pH, temperature, and light intensity [5]. Response surface methodology (RSM) is one of the most popular experimental design methods to optimize the chemical processes based on several operating parameters. It is able to analyse the relative significance of several influence factors, thus determining the optimal condition for the experiment [6]. On top of that, the effect of interaction between different parameters can also be studied. RSM includes two designs, namely central composite design (CCD) and Box-Behnken design (BBD). BBD has been widely applied for the optimization of photocatalysis process, but the interactions between variables which affects photocatalysis efficiency remain a scarce and exploring state.

In present study, a 3D-printed photoreactor with graphitic carbon nitride (g-CN) homojunction-based thermoset coating was fabricated to investigate the synergistic effect of photocatalysis and adsorption on Rhodamine B (RhB) dye degradation. Three variables were alternated to study of the significance of each effect to dye degradation rate, including the recirculation flow rate of dye solution, initial concentration of dye solution, and the number of photoreactors (corresponding to the g-CN dosage). Analysis of variance (ANOVA) study was carried out to determine the interaction effects of each factor. In an overall view, this study is expected to help save study time and cost in industrial processes.

2. Methodology

2.1. Preparation of g-CN homojunction photocatalyst
Equal proportion (6 g) of urea and thiourea were dissolved in 30 mL of distilled water and stirred in a crucible. The resultant mixture was heated at 90°C for 24 hours to form nanocomposite. The slurry-like nanocomposite was then transferred to a furnace for calcination at 550°C for 2 hours with a ramping rate of 2°C/min. The yellow solid sample after the calcination was cooled, then grounded into fine powder. The fine g-CN powder was sealed in a zip lock bag and stored in a desiccator until further use.

2.2. Synthesis of gCN-based thermoset coatings and assembly of photoreactors
0.3 g of the synthesized g-CN was added into equal proportion (4.5 g) of EG and deionised water. Next, the mixture was sent for ultrasonication for 20 mins. HEMA (0.8 g) was then added into the mixture. The mixture was positioned between two 50W LED lights with gentle stirring for 3 hours. This step mainly initiated free radical polymerization between the HEMA and g-CN. 5wt% of citric acid was subsequently added to the viscous mixture for cross-linking formation. A single layer of g-CN mixture with the amount of 0.34 ± 0.02 g was applied evenly on the chamber of the 3D-printed photoreactor. The photoreactors with wet g-CN coating were then heated in an oven at 120°C for 3 hours to provide a complete cross-linking and evaporation of moisture. Approximately 8.34% of the photocatalyst retained on the photoreactor after thermal curing, which means the dried photocatalyst immobilized on the photoreactor is about 0.029 ± 0.02 g. In this case, the immobilized photocatalyst has a BET surface area of 16.74 m^2/g. For the assembly of the photoreactor, the lid was lined with a rubber ring, screwed, and sealed by UV resin to avoid leakage of the RhB solution. Figure 1 shows the graphical description of the fabrication of photoreactor.

2.3. Evaluation of photocatalytic activity
RhB dye is selected as a targeted pollutant, which is a cationic dye with two amino groups and one carboxylic group as its major functional group. All experiments were carried out in a dark box to avoid any disturbance of light wattage. Teflon tubes, which have better resistance to dye, were used to connect the 3D-printed photoreactors coated with gCN-based thermoset, a beaker containing 25 mL of pre-determined dye solution, and a peristaltic pump. Two 50W LED lights (IP66-5054) were placed 10 cm above the photoreactors as the power source. To examine the photocatalysis efficiency, the LED lamps were switched on, and the dye solution sample were analysed through UV-Vis spectrometer (Hach DR 6000) in a 1 hour time interval for 6 hours. The dye degradation efficiency was calculated using equation (1), where C_0 and C is the initial absorbance value and the absorbance value at time t, respectively.

$$Dye\ degradation\ efficiency\ (DE\%) = \frac{C_0 - C}{C_0} \times 100\% \quad (1)$$

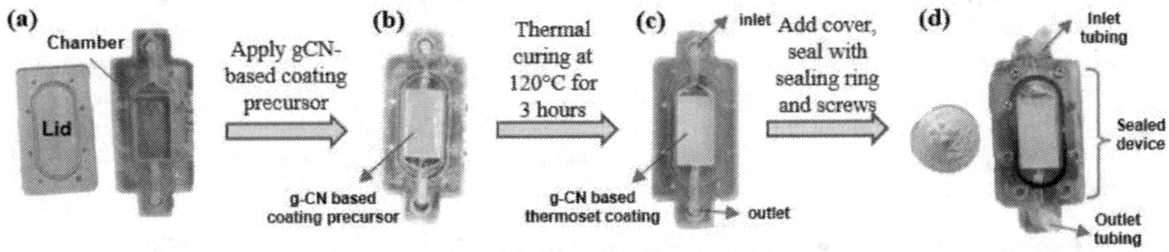

Figure 1. (a) Printed photoreactor cover and chamber, (b) monolayer g-CN coating before drying, (c) photoreactor with dried g-CN coating, and (d) assembled photoreactor.

2.4. *Experiment modelling using Box-Behnken Design (BBD) and statistical analysis*

The performance of the photocatalytic activity of g-CN can be optimized by a series of experimental runs. In this case, response surface methodology (RSM) was utilized to optimize the three parameters (dye solution flow rate, initial concentration of dye solution, and number of photoreactors). Equation (2) used to find out the BBD required experiment, where n is the number of variables and c is the number of centre points.

$$N = 2n(n-1) + c \quad (2)$$

Thus, 15 sets of experiments were carried out to study the combined effect of the three independent variables. The 3-factor levels of BBD can be interpreted as -1 (low), 0 (central point), and 1 (high). The manipulated variables in this study are given in table 1.

Table 1. Experimental design levels of chosen variables.

Variables	Levels in Box-Behnken design		
	Low (-1)	**Medium (0)**	**High (+1)**
Dye flow rate (X1, mL/min)	1	2	3
Dye concentration (X2, ppm)	5	10	15
Number of photoreactors (X3)	2	4	6

The Minitab Statistical Software (Version 19.1) was used for data processing and statistical analysis. Regression analysis was performed to evaluate the significance of the different linear, quadratic, and interaction terms, with the probability-value being the indicator. Analysis of variance (ANOVA) was also conducted to assess the model significance and adequacy, according to the correlation coefficient (R^2).

2.5. *Batch adsorption studies*

The control inputs, which are recirculation flow rates, initial dye concentration and dosage of adsorbents were altered to investigate the adsorption performance of g-CN-based thermoset coating. Batch adsorption experiments were conducted under a fixed volume (25mL) of RhB dye solution, with continuous stirring in dark for 6 hours. The R% and q_e of adsorbents were determined using equation (3) and (4):

$$R\% = \frac{C_0 - C_t}{C_0} \times 100\% \quad (3)$$

$$q_e = \frac{(C_0 - C_t)V}{m} \quad (4)$$

where q_e, C_0, C_t, V, m indicates adsorption capacity (mg/g), initial RhB concentration (ppm), RhB concentration at time t (ppm), volume of RhB dye solution (L) and mass of dried g-CN coating, respectively.

3. Results and Discussion

3.1. Field emission scanning electron microscopy (FESEM) characterisation

As seen in figure 2, a photocured HDT resin shows vertical and horizontal lines crossing on a homogeneous texture, which is a typical texture of a polymer resin [2]. Next, the g-CN based thermoset coating shows lamellar and dense structure with no specific shape. Similar surface morphologies were observed after the photocatalysis and adsorption process, which the surface becomes uneven and ragged. This indicates that organic particles were accumulated on the g-CN coating. This reduces the surface area, which affects the adsorption capacity.

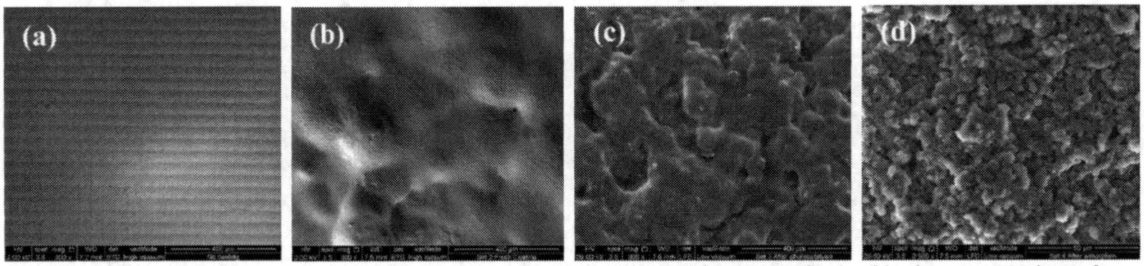

Figure 2. FESEM images of (a) photocured HDT resin (without coating), (b) clean g-CN based thermoset coating before reaction, (c) g-CN coating after photocatalysis, and (d) g-CN coating after adsorption.

3.2. Adsorption-photocatalysis studies

Batch adsorption experiments were performed with various initial dye concentration ranging from 5 to 20 ppm, and the number of photoreactor and flowrate are fixed at 4 and 1 mL/min, respectively. Without the presence of light source, the equilibrium adsorption data were analysed by Langmuir and Freundlich adsorption isotherm models. Langmuir model showed a best fitting with the highest corresponding value ($R^2 = 0.9822$) as compared to Freundlich model ($R^2 = 0.9596$). The separation factor of Langmuir model ($R_L = 0.05$) indicates that the adsorption of RhB onto the g-CN-based coating surface is the favourable process, which means there is no transmigration for adsorbate adsorbed on the surface with uniform monolayer energy. For the adsorption kinetic studies, pseudo-second order model shows a good fit compared to pseudo-first order model. It can be confirmed that as in the adsorption process, the concentration of the adsorbate is involved in rate-limiting step.

With the aid of LED light, the dye degradation efficiency increases from 63% to 81%, as shown in figure 3(c). The experimental conditions were also carried out at flow rate of 1 mL/min, initial dye concentration of 15 ppm and 4 photoreactors. Throughout the experiments, the photocatalysis kinetics fits the Langmuir-Hinshelwood kinetic model ($\ln C_0/C_t = kt$), which is a first-order rate equation. Figure 3(d) shows a linear plot between $\ln (C_0/C_t)$ and irradiation time (t) with high correlation coefficient ($R^2 > 0.99$) and the apparent first-order rate constant (k) of the photocatalytic reaction was found to be 0.284 hr^{-1}. Based on the batch adsorption experiment, the reaction rate constant was found to be 0.51 hr^{-1} using pseudo-first order model. Although the batch adsorption experiment showed the faster degradation rate, the degradation efficiency is relatively low (63%) compared to photocatalysis experiment (81%).

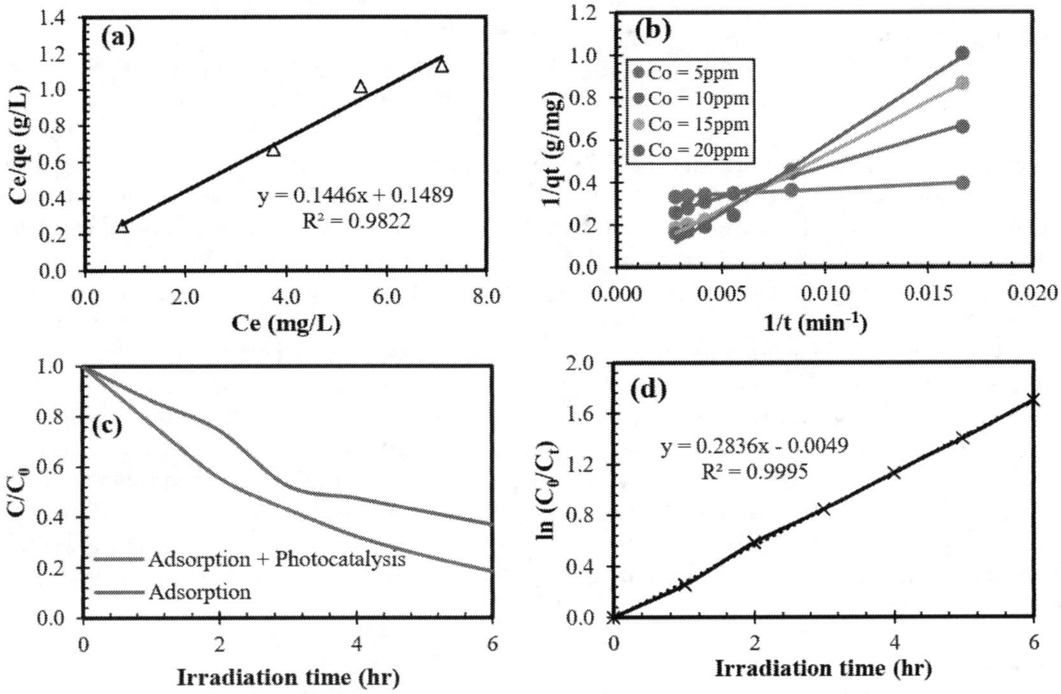

Figure 3. Plots of (a) Langmuir adsorption isotherm (b) Pseudo-second order adsorption kinetics (c) C/C_0 against time adsorption and/or photocatalysis process, and (d) Photocatalysis kinetics plot fitting Langmuir-Hinshelwood model.

3.3. Response surface methodology studies

The degradation of RhB on the g-CN thermoset coating is integrating adsorption and photocatalysis processes. Fifteen sets of experiment were carried out under light irradiation, and their respective dye degradation efficiencies were recorded.

Table 2. Box-Behnken design with experimental response and predicted response.

Set	Flow rate (mL/min)	Initial dye concentration (ppm)	No. of photoreactors	Experimental response	Predicted response	Percentage deviation
1	1	10	6	87.60%	87.77%	-0.20%
2	2	10	4	83.14%	84.90%	-2.12%
3	1	10	2	59.72%	58.82%	1.50%
4	2	5	2	58.89%	60.57%	-2.85%
5	2	10	4	83.83%	84.90%	-1.28%
6	1	5	4	85.31%	82.34%	3.49%
7	2	5	6	88.32%	89.51%	-1.35%
8	3	15	4	88.24%	87.46%	0.88%
9	2	10	4	83.48%	84.90%	-1.70%
10	3	10	6	91.11%	92.07%	-1.06%
11	1	15	4	81.05%	83.16%	-2.60%
12	3	5	4	89.31%	86.64%	2.98%
13	2	15	2	63.06%	61.39%	2.65%
14	2	15	6	92.73%	90.33%	2.59%
15	3	10	2	62.25%	63.13%	-1.41%

A second-order regression equation shown in equation (5) was established to predict the dye degradation efficiency:

$$Y = 0.1302 + 0.02154\,X_1 - 0.00082\,X_2 + 0.2614\,X_3 - 0.02363X_3^2 \quad (5)$$

By studying the probability value (p-value) of the different terms in the analysis of variance (ANOVA) study, it is found that the only the linear terms and the square term of X3*X3 are significant, while the other square terms plus all the 2-way interaction terms are insignificant. For example, the most significant factor, which is the number of photoreactors, has p-value lesser than 0.001, while all the 2-way interaction terms have p-value larger than 0.5. This model has high correlation coefficient (R^2) of 0.9796, and the R^2 for the predicted response is 0.95. It is also found that the percentage deviations of the experimental response with the predicted response are within 3.5%. Besides, the lack-of-fit also has p-value lower than 0.05, indicating that the contribution of the selected terms is adequately high for creating the whole model. The order or priority among the main effect of the variables was deduced, which is number of photoreactors (X3) > dye solution flow rate (X1) > initial concentration of dye (X2).

The model adequacy was verified by carrying out experiment with randomized parameters. Two replicative experiments were carried out with the experiment parameter of 2.5 mL/min flowrate, 6 ppm initial RhB concentration and three photoreactors. The experiment yields dye degradation efficiency of 80.38%, slightly higher than the predicted response of 76.05%. This can be presumed as there is a steeper change when the number of photoreactors change from two to three in experimental response compared to predicted response.

4. Conclusion

The immobilized g-CN thermoset coating used on the degradation of RhB dye molecules were studied. The characterization studies revealed that agglomeration of organic particles occurs as the surface morphology becomes clustered. Next, the response surface methodology was successfully applied, showing that the number of photoreactors is the most significant influence on the dye degradation efficiency. Besides, the interactions between parameters are insignificant. In this case, a second-order regression equation was developed to predict the photocatalysis performance, with good R^2 value of 0.95. Langmuir isotherm revealed that the process was favorable, and the adsorbates was limited in monolayer surface with uniform energy. On the other hand, pseudo-second order showed that the adsorption process was involved in rate-limiting step which is called chemisorption.

References

[1] Park H and Choi W 2003 Visible light and Fe(III)-mediated degradation of Acid Orange 7 in the absence of H2O2 *Journal of Photochemistry and Photobiology A: Chemistry,* **159** (3) 241 - 47

[2] Phang S J, Wong V L, Cheah K H and Tan L L 2021 3D-printed photoreactor with robust g-C3N4 homojunction based thermoset coating as a new and sustainable approach for photocatalytic wastewater treatment *Journal of Environmental Chemical Engineering* **9** (6)

[3] Chankhanittha T and Nanan S 2021 Visible-light-driven photocatalytic degradation of ofloxacin (OFL) antibiotic and Rhodamine B (RhB) dye by solvothermally grown ZnO/Bi2MoO6 heterojunction *J Colloid Interface Sci* **582** (A) 412 – 27

[4] Kasinathan K, Kennedy J, Elayaperumal M, Henini M and Malik M 2016 Photodegradation of organic pollutants RhB dye using UV simulated sunlight on ceria based TiO2 nanomaterials for antibacterial application. *Sci Rep* **6** 38064

[5] Molinari R, Argurio P, Bellardita M and Palmisano L 2017 Photocatalytic Processes in Membrane Reactors *Comprehensive Membrane Science and Engineering* 101 – 38

[6] Zhang J, Fu D, Xu Y and Liu C 2010 Optimization of parameters on photocatalytic degradation of chloramphenicol using TiO2 as photocatalyist by response surface methodology *Journal of Environmental Sciences* **22** (8) 1281 – 289

IOP Conf. Series: Materials Science and Engineering 1257 (2022) 012047 doi:10.1088/1757-899X/1257/1/012047

Comparative of alginate/graphene oxide and alginate/magnetite graphene oxide aerogel as adsorbents

A S Marzuki[1], A N Mustapa[1,2*], N H Othman[1], S H Hanipah[1], Á Martin[3] and M J Cocero[3]

[1]School of Chemical Engineering, College of Engineering, Universiti Teknologi MARA, 40450 Shah Alam, Malaysia
[2]Integrated Separation Technology Research Group, School of Chemical Engineering, College of Engineering, Universiti Teknologi MARA, 40450 Shah Alam, Malaysia
[3]BioEcoUva, Research Institute on Bioeconomy, High Pressure Processes Group Department of Chemical Engineering and Environmental Technology, University of Valladolid, Calle Doctor Mergelina s/n, 47011 Valladolid, Spain

*anajwa@uitm.edu.my

Abstract. Magnetite graphene oxide (MGO) has gained attention as a potential adsorbent in wastewater treatment. In this study, incorporation of MGO into biopolymer matrix and dried by supercritical CO_2 to improve the adsorption efficiency is presented. Alginate/MGO aerogel was prepared by blending the MGO with alginate and formed hydrogel by ionic crosslinking through external gelation. The blend ratio of MGO and alginate studied are comprised of 2 and 3 wt%, respectively. The hydrogel was dried by supercritical CO_2 at 110 ± 5 bar and 40 °C in 4h. The physical-chemical properties of alginate/MGO adsorbent and its adsorption efficiency via removal of methylene blue (MB) are studied. The sharp curve for 30% weight loss for both types of adsorbents has been shown around 200 °C in TGA correspond to the exothermic peak form in DSC analysis. Alginate/GO and alginate/MGO are stable over temperatures of 300 and 700 °C, respectively. Alginate/GO successfully adsorbed 90% of MB concentration meanwhile the alginate/MGO had removed up to 84% of MB. Another highlighted technique is supercritical CO_2 drying which considered as one of the most promising techniques to preserve the structure of aerogel with minimal shrinkage as low as 15% in comparison to ambient drying (50%).

1. Introduction

Graphene oxide (GO) has been accepted as one of the promising adsorbents for pollutants such as dyes and heavy metals due to its strong adsorption properties. However, its proclivity for forming aggregates is a major drawback because it prevents certain active sites from being used for pollutant adsorption. It is also a nanomaterial with a high dispersity in an aqueous solution, making it difficult to retrieve the material once it has been used. The sector would face high costs due to the time-consuming procedure of adsorbent renewal. To address this issue, Fe_3O_4 is added with GO to form magnetite graphene oxide (MGO), such that the adsorbent has magnetic characteristics, making the solid-liquid separation process easier [1]. Alginate has functional groups such as carboxyl and

IOP Conf. Series: Materials Science and Engineering 1257 (2022) 012047 doi:10.1088/1757-899X/1257/1/012047

hydroxyl that allow it to interact with metal ions and self-assemble a network structure [2]. By using the availability of free hydroxyl and carboxyl groups, which are highly reactive and sensitive to chemical modification, alginate has the potential to improve its properties in terms of solubility, hydrophobicity, and biological features [3]. The incorporation of GO or MGO into alginate may improve mechanical strength and adsorption capability [6] and reduce their toxicity while simultaneously assisting alginate in reducing pore size and increasing pore uniformity [4]. Furthermore, several works have demonstrated that combining an adsorbent with alginate can increase crosslinking density [5] and decrease nanoparticle leaching in a prospective application [6].

In this work, comparisons between alginate/GO and alginate/MGO are reported in terms of their characterization and their adsorption ability towards methylene blue (MB). The comparison aims to determine the adsorption performance of alginate/GO and alginate/MGO in removing MB from the solution. Both adsorbents were prepared in the form of thin gel layers with a ratio of 3:2 of alginate to MGO and GO via a sol-gel method and dried by supercritical CO_2 drying. The synthesized adsorbents were characterized by FESEM, XRD, FTIR, TGA, and DSC analyses. The adsorption performance of the synthesized aerogels is measured via analysis of MB removal from its solution.

2. Materials and methods

2.1. Materials

Graphite powder was purchased from Acros Organics B.V.B.A, whilst 35% hydrogen peroxide, 37% hydrochloric acid, potassium permanganate, 95.98% sulphuric acid, sodium nitrate, calcium chloride, ferric chloride, and ferrous chloride were obtained from R&M Chemicals. The 35% hydrogen peroxide and 37% hydrochloric acid were used to stop the reaction of $KMnO_4$ and to remove the excess of unreacted $KMnO_4$, respectively. Ferric chloride and ferrous chloride were used as a source of iron ions, whereas a 25% ammonia solution supplied by Merck was used to enable the precipitation of MGO. Sodium alginate was purchased from Sigma Aldrich. All the experiments were carried out using distilled water.

2.2. Preparation of graphene oxide (GO)

GO was synthesized from graphite by a modified Hummers method [7].

2.3. Preparation of magnetite graphene oxide (MGO)

Magnetite (Fe_3O_4) nanoparticles were incorporated with GO through co-precipitation with slight modification. GO was dispersed in 250 mL of water with sonication aid for an hour. Magnetite solution was prepared by mixing $FeCl_3$ and $FeCl_2$ with a 2:1 mole ratio in 200 mL of distilled water. Next, the solution was added slowly to the GO solution under nitrogen gas flow (40 mL/min) with rapid agitation. After an hour, 25% ammonia solution was added dropwise to obtain pH 10 and enable the precipitation of magnetite particles. The solution was regulated at 70 °C for three hours, then cooled down to room temperature. The resultant product was washed until neutral pH was achieved before drying in the oven at 60 °C.

2.4. Preparation of alginate/GO and alginate/MGO aerogel

GO was incorporated into alginate by an external gelation method. Sodium alginate and GO were added to 100 mL of water to obtain 3 and 2 wt%, respectively. The mixture was stirred for 24 hours for complete homogenization. Next, the solution was poured into a petri dish to form a thin layer and $CaCl_2$, a curing agent, was added to it. The mixture was left for 24 hours for complete gelation across the thin film. The hydrogel would undergo a series of solvent exchanges consisting of 10, 30, 50, 70, 90, and 100 v/v% ethanol prior to the supercritical CO_2 drying at 110 ± 5 bar for four hours. The process was repeated with MGO to produce alginate/MGO. Samples were replicated more than three times for the GO and MGO formulation with alginate.

IOP Conf. Series: Materials Science and Engineering 1257 (2022) 012047 doi:10.1088/1757-899X/1257/1/012047

2.5. Characterization of GO, MGO, alginate/GO, and alginate/MGO aerogel
The qualitative data on the physical structure and morphology of GO, MGO, alginate/GO, and alginate/MGO was obtained using Field emission scanning electron microscopy (FESEM, JSM-7600F, Jeol). X-ray diffraction (XRD, D/Max 2200V/PC, Rigaku) with operating tube voltage and current of 40 kV and 30 mA was used to evaluate the structural and compositional features of the materials. Data was acquired with a sample angle interval of 0.04° and diffraction angle (2θ) ranging from 5° to 90°. Fourier transform infrared spectroscopy (FTIR, Spectrum One, Perkin Elmer) examined the functional groups of the adsorbents. Thermogravimetric analysis (TGA, TGA851/LF/1600, Mettler Toledo) and differential scanning calorimetry (DSC, DSCI, Mettler Toledo) were performed with 8–10 mg samples under a nitrogen atmosphere. The experiments were carried out at a heating rate of 10 °C/min from 20 to 600 °C.

2.6. Adsorption experiment
Methylene blue (MB) adsorption studies were carried out with a magnetic stirrer (150 rpm) at room temperature and an uncontrolled pH (initial pH of MB solution). To begin, MB solutions were made by diluting a stock solution (100 v/v percent) to the concentrations desired. After that, 0.3 g of adsorbent was introduced to a beaker containing a target contaminant solution (100 ml). Ultraviolet-visible spectroscopy (UV-VIS, Lambda 750, Perkin Elmer) was used to evaluate samples collected at specific intervals. The experiment was conducted three times and reported as average data.

3. Results and discussions
Alginate/GO and alginate/MGO were synthesized in the form of thin film and dried by supercritical CO2 at 110 ± 5 bar and 40 °C in 4h. The volume shrinkage of the film was found as low as 15%. In comparison, similar samples were dried at the ambient condition for 24h, resulting in the shrinkage being more than 50% by volume. This shows that the supercritical drying method is reliable in preserving the pore structure and preventing them from collapsing.

3.1. Physical and chemical properties

3.1.1. GO and MGO. Figure 1(a) shows that the GO sheet had an uneven form with wrinkles and a wide surface area which confirmed the crinkled structure of GO [8] and allowed for multiple adsorption sites [9]. In comparison to the GO, the MGO surface was rougher as can be seen from the figure. The surface of the MGO was covered by magnetite nanoparticles [6].

(a) (b)
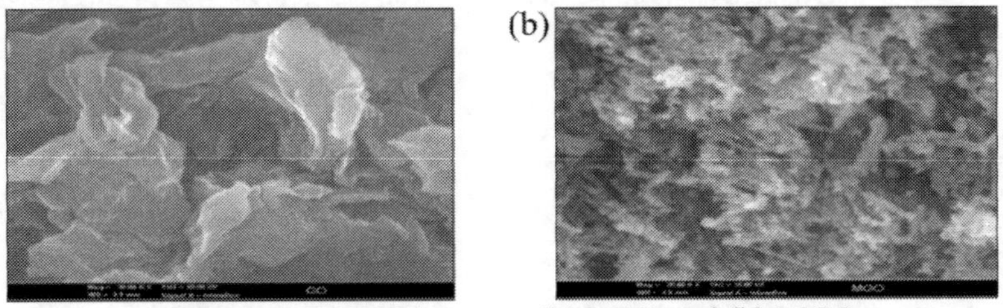

Figure 1. FESEM images of (a) GO and (b) MGO.

3.1.2. Alginate/GO and alginate/MGO. Due to the presence of alginate, the FESEM image of the alginate/MGO in Figure 2(c) shows more evident folding and wrinkling than MGO alone in Figure 1(b) due to the presence of alginate. Meanwhile, the active sites of alginate are adequately exposed between GO and MGO sheets, which can promote the interaction of alginate GO and alginate/MGO film adsorbents with target pollutants [10].

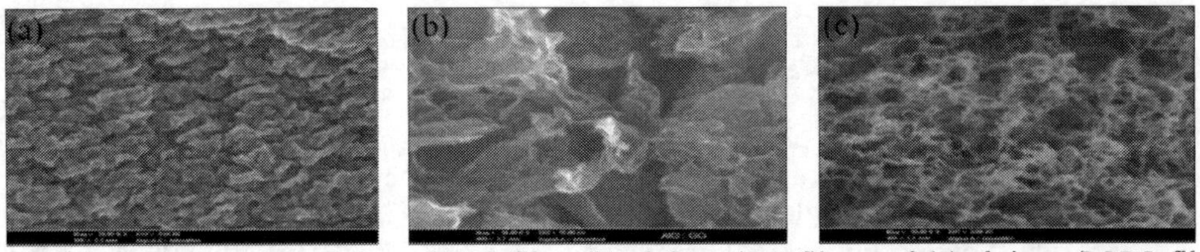

Figure 2. FESEM images of (a) alginate blank film, (b) alginate/GO film, and (c) alginate/MGO film.

Figure 3(a), IR frequencies for alginate blank at 3365 cm^{-1} attributed to hydroxyl group -OH in alginate matrix hydrogen bond contributed by the proton-donor and proton acceptor sites (stretching) [11], at 1600 cm^{-1} and 1419 cm^{-1} are for asymmetric and symmetric stretching vibrations of -COOH groups, respectively [12]. For alginate/GO and alginate/MGO, the broad peak shown at 3379 cm^{-1} resulted from the presence of hydroxyl group coming from the connected water, moisture, and/or intermolecular interactions between GO/MGO with alginate via hydrogen bonds [13]. Compared to the alginate blank, few changes were observed in the IR spectrum when GO/MGO were incorporated into it, such as the small shifted band that related to the stretching vibrations of carboxylate groups in metal-alginate interactions by ionic bonds [13]. FTIR spectra of alginate/MGO, which exhibit a peak at 589 cm^{-1} are caused by the vibration of the Fe-O bond in both octahedral and tetrahedral sites of the inverse spinel cubic of Fe_3O_4, respectively.

Figure 3(b) shows the spectrum of alginate/MGO, which is significantly matched to JCPDS file No. 19-0629. As previously reported by the [14], the spectrum obtained confirms the formation of magnetite nanoparticles, as indicated by the peaks at 2θ values of 30.05°, 35.3°, 43.95°, 53.3°, 58.20° and 63.45° that attributed to the crystal plan of magnetite at various planes (220), (311), (400), (422), (511), and (440). The appearance of diffraction peaks at less than 20° region elucidated the presence of alginate. In alginate/GO, the diffraction peak at 13° may correspond to the d-spacing of 0.680 nm in GO. The oxygen functional groups intercalated in the graphene interlayers and resulting in the increment of the d-spacing. Meanwhile, the weak diffraction peak that appears at 42.5° could be due to the incomplete oxidation. Both FTIR and XRD patterns results confirmed the successfulness of synthesizing alginate/GO and alginate/MGO.

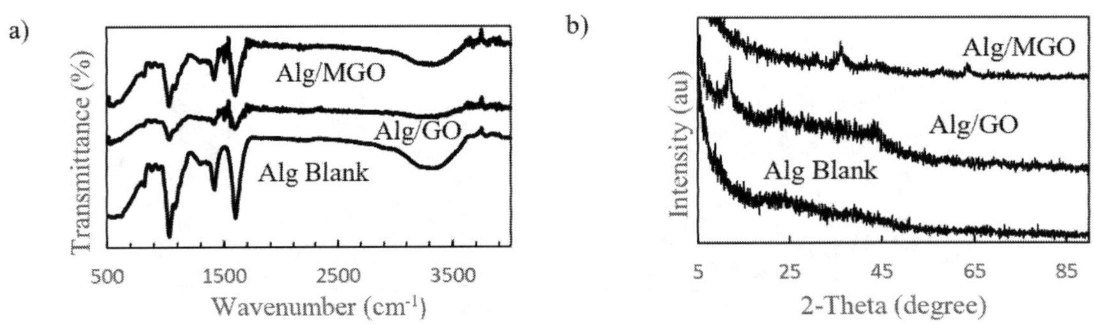

Figure 3. The FTIR spectrum (a) and XRD pattern (b) of alginate blank film, alginate/GO film, and alginate/MGO film.

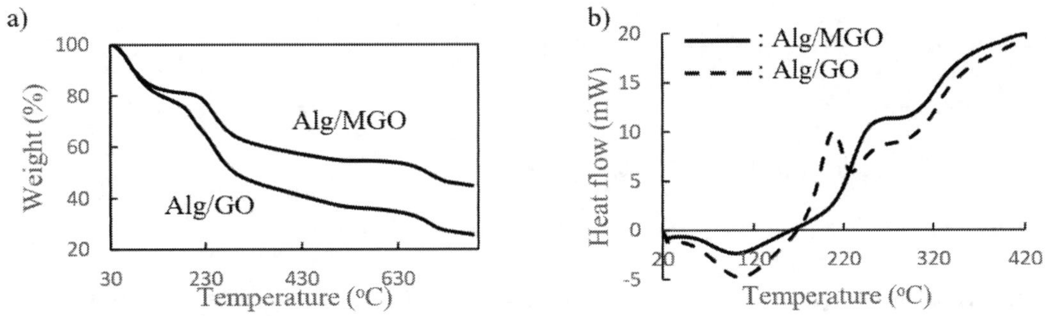

Figure 4. Alginate/GO and alginate/MGO curves for (a) TGA and (b) DSC.

The TGA and DSC curves are depicted in Fig. 4. For alginate/GO, the first mass loss between 35 and 133 °C (20%) corresponds to the endothermic peak at 110 °C, which was attributed to the dehydration process. On the other hand, about 50 wt% weight loss when the temperature increased from 147 to 337 °C. This happened due to the loss of the functional groups and degradation of the sample that gave an exothermic reaction at 210 °C, meanwhile the 50% of weight loss of the material occurred at 300 °C. As for alginate/MGO, the highlight is between 190 to 250 °C which gives the degradation of the sample in comparison with 240 °C of the exothermic reaction. The 50% weight loss of material occurred at 700 °C. After the temperature of 630 °C regions, the residue remains for the Alg/MGO is around 40% while the Alg/GO is around 20%. This shows that the Alg/MGO is more stable since the mass-loss rate for the adsorbent is much lower in comparison to the Alg/GO [15].

3.2. Alginate/GO and alginate/MGO effectiveness in the removal of methylene blue

To investigate the adsorption performance of alginate/MGO and alginate/GO adsorbents capacity, a continuous experiment was run for eight hours. The adsorption of MB started with 0.1 v/v% and the absorbance of MB decreased as the adsorbents had used up their capacity. Alginate/GO exhibited a good adsorption efficiency by removing a maximum of 90% of the MB concentration, while alginate/MGO removed 84% of the concentration. It is hypothesized that some of the alginate/MGO sites were occupied by Fe_3O_4 ion which makes the adsorption of the MB lesser compared to alginate/GO. Alginate also helps in inhibiting GO and MGO from aggregating, which improved adsorption efficiency.

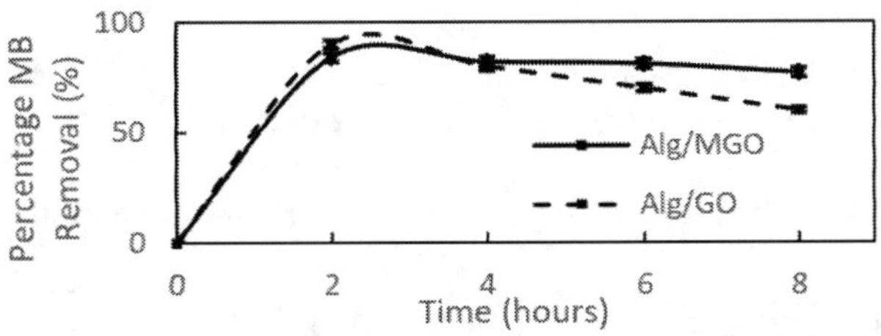

Figure 5. Effect of contact time of the MB dye solution adsorption onto alginate/GO and alginate/MGO adsorbents. Data represent an average of three experiments. Error Bar = ±SD.

4. Conclusion

The aggregation and leaching of the core material of GO and MGO have been prevented, which in turn improved its adsorption capacity. Although the percentage of removal using GO is higher, however, alginate/MGO is at an advantage due to its simpler and more efficient collection in post-waste treatment. The high temperature stability of alginate/GO and alginate/MGO over 300 °C as well as their excellent adsorption in the removal of MB indicates that both adsorbents appear as an excellent and versatile candidate for the water treatment.

Acknowledgements
This work was supported by Fundamental Research Grant Scheme (FRGS) from the Ministry of Education (MOE) no.600-IRMI/FRGS 5/3 (086/2019).

References

[1] Ninwiwek N, Hongsawat P, Punyapalakul P and Prarat P 2019 Removal of the antibiotic sulfamethoxazole from environmental water by mesoporous silica-magnetic graphene oxide nanocomposite technology: Adsorption characteristics, coadsorption and uptake mechanism *Colloids Surfaces A Physicochem. Eng. Asp.* **580** 123716
[2] Bo S, Luo J, An Q, Zhao X, Xiao Z, Zhai S and Li Z 2019 Circular utilization of Co(II) adsorbed composites for efficient organic pollutants degradation by transforming into Co/N-doped carbonaceous catalyst *J. Clean. Prod.* **236** 117630
[3] Maiti S and Kumari L 2016 Smart nanopolysaccharides for the delivery of bioactives *Nanoarchitectonics Smart Deliv. Drug Target.* 67–94
[4] Fei Y, Li Y, Han S and Ma J 2016 Adsorptive removal of ciprofloxacin by sodium alginate/graphene oxide composite beads from aqueous solution *J. Colloid Interface Sci.* **484** 196–204
[5] Yu Y, Shu Y and Ye L 2018 In situ crosslinking of poly (vinyl alcohol)/graphene oxide-glutamic acid nano-composite hydrogel as microbial carrier: Intercalation structure and its wastewater treatment performance *Chem. Eng. J.* **336** 306–14
[6] Vu H C, Dwivedi A D, Le T T, Seo S H, Kim E J and Chang Y S 2017 Magnetite graphene oxide encapsulated in alginate beads for enhanced adsorption of Cr(VI) and As(V) from aqueous solutions: Role of crosslinking metal cations in pH control *Chem. Eng. J.* **307** 220–9
[7] Othman N H, Alias N H, Shahruddin M Z, Abu Bakar N F, Nik Him N R and Lau W J 2018 Adsorption kinetics of methylene blue dyes onto magnetic graphene oxide *J. Environ. Chem. Eng.* **6** 2803–11
[8] Zhang B, Hu R, Sun D, Wu T and Li Y 2018 Fabrication of magnetite-graphene oxide/MgAl-layered double hydroxide composites for efficient removal of emulsified oils from various oil-in-water mulsions *J. Chem. Eng. Data* **63** 4689–702
[9] Shi P and Ye N 2014 Magnetite-graphene oxide composites as a magnetic solid-phase extraction adsorbent for the determination of trace sulfonamides in water samples *Anal. Methods* **6** 9725–30
[10] Zhang B, Hu R, Sun D, Wu T and Li Y 2018 Fabrication of chitosan/magnetite-graphene oxide composites as a novel bioadsorbent for adsorption and detoxification of Cr(VI) from aqueous solution *Sci. Rep.* **8** 1–12
[11] Ajeel S J, Beddai A A and Almohaisen A M N 2021 Preparation of alginate/graphene oxide composite for methylene blue removal *Mater. Today Proc.* **51** 289–97
[12] Hou Y, Zhong X, Ding Y, Zhang S, Shi F and Hu J 2020 Alginate-based aerogels with double catalytic activity sites and high mechanical strength *Carbohydr. Polym.* **245** 116490
[13] Jiao C, Xiong J, Tao J, Xu S, Zhang D, Lin H and Chen Y 2016 Sodium alginate/graphene oxide aerogel with enhanced strength–toughness and its heavy metal adsorption study *Int. J. Biol. Macromol.* **83** 133–41
[14] Ye N, Xie Y, Shi P, Gao T and Ma J 2014 Synthesis of magnetite/graphene oxide/chitosan composite and its application for protein adsorption *Mater. Sci. Eng. C* **45** 8–14
[15] Fei Y, Li Y, Han S and Ma J 2016 Adsorptive removal of ciprofloxacin by sodium alginate/graphene oxide composite beads from aqueous solution *J. Colloid Interface Sci.* **484** 196–204

Fluorine-free superhydrophobic cotton fabric for highly-efficient gravity-driven oil/water separation

WX Voo[1], WC Chong[1,2*], HC Teoh[1,2] and AW Mohammad[3]

[1] Department of Chemical Engineering, Lee Kong Chian Faculty of Engineering and Science, Universiti Tunku Abdul Rahman, Kajang 43000, Malaysia.
[2] Centre for Photonics and Advanced Materials Research, Universiti Tunku Abdul Rahman, Sungai Long Campus, Jalan Sungai Long, Cheras, 43000, Kajang, Selangor, Malaysia
[3] Chemical Engineering Programme, Faculty of Engineering and Built Environment, Universiti Kebangsaan Malaysia, Bangi 43600, Selangor, Malaysia.
*e-mail: chongwoonchan@gmail.com

Abstract. Rapid urbanisation and industrialisation have imposed great environmental impact due to the discharge of oily wastewater and oil spill incidents. The application of superhydrophobic and superoleophilic surfaces in oil/water separation (OWS) has attracted much attention due to their high selectivity of oil which leads to high separation efficiency in OWS. Herein, a fluorine-free superhydrophobic cotton fabric filter was prepared by loading SiO_2 onto the surface from the silica sol and followed by coating with stearic acid through facile dip coating method. The silica sol was prepared using tetraethyl orthosilicate (TEOS) where the loading was manipulated in order to study its effect on hydrophobicity. In the drying process, the effect of the drying temperature on the hydrophobicity was studied. All the pristine and modified cotton fabric samples were characterised by employing FTIR, SEM-EDX and water contact angle measurement. In the preparation process, the optimum TEOS loading was determined to be 4.5 mL. Additionally, the as-prepared superhydrophobic cotton fabric was used to perform OWS, under sole influence of gravity, by using different oil-water mixture solutions and the separation efficiency was above 99%. The results showed that higher drying temperature gives rise in hydrophobicity, and thus lead to more efficient OWS.

1. Introduction

Many industries, such as mining, textile, food, metal processing, oil and gas, are responsible in producing and discharging oily wastewater [1]. On the other hand, different phases involved in the oil industry, such as exploration, production, refinery, and transportation, have the risk of oil spill. Oil spills are also considered as a great waste of natural resources and can result in severe economic impacts. Currently, there are several remediation and treatment methods to tackle oily wastewater and oil spill, including gas flotation, coagulation, centrifugation, membrane filtration, adsorption, oil containment and skimming, and bioremediation [2,3]. Sadly, these conventional methods exhibit low selectivity and most importantly are time consuming. The porous materials, such as membrane and sorbent, that are commonly used to absorb oil in the event of emergency oil spill, will absorb both oil and water during deployment, thus led to poor separation efficiency [1]. The selectivity of the membrane or sorbent towards oil can be improved by altering their wetting properties, which can be done by modifying surface chemistry and roughness. The extreme non-wetting state of a solid surface by water is known as superhydrophobicity. This condition is also associated with water contact angle (WCA) > 150°, which requires high surface roughness and low surface energy [4]. Herein, SiO_2 particles were employed to increase the surface roughness whereas stearic acid was used to achieve surface energy. The use of stearic acid to alter the surface energy in the fabrication of superhydrophobic filter was reported by several researchers for its non-hazardous and environmentally benign properties [5–8]. The SiO_2 particles prepared from TEOS have been widely reported for their employment in enhancing surface roughness, however the effect of the TEOS loading on the hydrophobicity has not been reported. Although, it is also reported that temperature plays an important role in enhancing hydrophobicity [9]. The effect of temperature on the hydrophobicity provided by the surface coating of SiO_2 and stearic acid on cotton fabric was not reported before. Hence, the effect of TEOS loading and drying temperature on hydrophobicity of the filter were studied in this work. The OWS performance of as-prepared filter was also reported here.

2. Experimental

2.1. Materials
Cotton fabric (160g/m^2), tetraethyl orthosilicate (TEOS), stearic acid, ammonia solution (25%), ethanol (95%), acetone, oil red, chloroform, hexane. All chemicals were used without further purification.

2.2. Preparation of STA/SiO_2@cotton
The silica sols were prepared based on an improved Stober method reported by Zhao et al. [10]. Typically, 1 mL of TEOS was added into a mixture of ammonia solution (15 mL) and ethanol (150 mL). The solutions were transferred into a conical flask and stirred vigorously at room condition for 6 hours. Another three silica sols were prepared by mixing different loading of TEOS (4.5, 9 and 12 mL) into the ammonia-ethanol mixture.

A 7 cm × 7 cm cotton fabric was then immersed into the as-prepared silica sol for 2 hours with continuous stirring at 100 rpm. The cotton fabric was rinsed with acetone prior to the experiment. After that, the SiO_2-coated cotton fabric was dried at 65 °C for 2 hours in the oven. The sample is denoted as SiO_2@cotton. Meanwhile, 1 g of stearic acid was dissolved in 100ml ethanol by using ultrasonication for 10 minutes. The SiO_2@cotton was then dip coated in the 1wt% stearic acid solution at room condition for 2 hours with continuous stirring at 100 rpm. Next, the modified cotton fabric was dried at 200 °C for 2 hours. The final product is known as STA/SiO_2@cotton. Here, three more STA/SiO_2@cotton samples were prepared with different drying temperature (80, 120 and 160 °C).

2.3. Characterisation
The superhydrophobicity of the cotton fabric filters were assessed based on the WCA measured using the Drop Shape Analysis System DSA20S (Kruess, Germany) at room condition with a 10 μL droplet as the indicator. Functional groups on the filter surface were identified with FTIR using Thermo Scientific Nicolet iS10. Meanwhile, surface morphology of the cotton fabrics was studied based on the scanning electron microscopy images captured by Hitachi S-3400N in 15kV, whereas the elemental analysis was done with EDX.

2.4. Oil/water separation (OWS)
The OWS was performed solely under the influence of gravity. The oil/water mixture was prepared by mixing equal volume of oil and distilled water (30 mL each), where the oil was dyed red. Hexane and chloroform oil were used in OWS, which represented light and heavy oil, respectively. The performance of the filter in OWS was evaluated based on the separation efficiency, η, where it was calculated as equation (2.1), and the permeate flux, J, which was calculated based on equation (2.2).

$$\eta=\frac{M_f}{M_i}\times 100\% \tag{2.1}$$

$$J = \frac{V}{At} \tag{2.2}$$

where, M_i and M_f are the mass of oil (g) before and after separation, respectively. V is the permeate volume (L), A is the active area (m^2), and t is the recorded time of permeation (h).

3. Results and discussion

3.1. Preparation process

3.1.1. Effect of SiO_2 loading on hydrophobicity. The WCA for filters fabricated in silica sols of 1, 4.5, 9 and 12 mL TEOS loading were measured at 139, 151, 145 and 140°, respectively. The only sample that was able to achieve superhydrophobic state is the filter prepared from the silica sols of 4.5 mL TEOS. This phenomenon occurred due to the difference in the amount of the SiO_2 particles synthesised from the 4 different silica sols. The lower chemical loading would synthesise lesser amount of particles and hence the particles anchored on the filter surface would be insufficient in forming hierarchical structures that could provide high surface roughness. The increase of chemical loading would increase the amount of SiO_2 particles. giving rise in the aggregation and agglomeration of these particles. The particles in these states would greatly influence the hydrophobicity of the filter [11]. As the aggregated and agglomerated particles increase, a smoother surface would form instead of hierarchical structures, which is a combination of particles in the size of micrometer and nanometer, and hence resulted in less hydrophobic state [11,12]. The difference in particles amount can be observed from figure 1 for filters fabricated with TEOS loading of 1, 4.5, 9 and 12 mL.

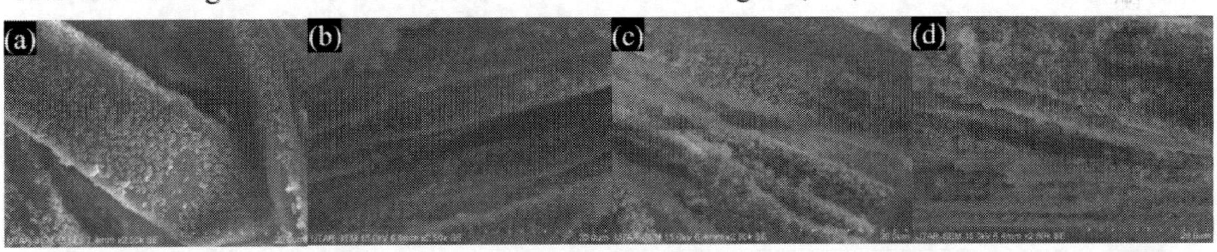

Figure 1. SEM images of filters fabricated with TEOS loading of (a) 1 mL, (b) 4.5 mL, (c) 9 mL and (d) 12 mL.

3.1.2. Effect of temperature on hydrophobicity. The WCA stated in table 1 shows that the filter can achieve better hydrophobicity at higher drying temperature. This was due to the disruption of hydrogen bonds present on the surface functional groups (i.e. -COOH), that often possess the highest hydrophilicity, at high temperature thermal treatment. [13,14]. As a result, the surface energy is lowered, and hydrophobicity was enhanced.

Table 1. Average WCA of cotton filters under different drying temperature.

TEOS loading (mL)	Drying temperature (°C)	WCA (°)
4.5	80	137
4.5	120	140
4.5	160	146
4.5	200	151

3.2. Characterisation

3.2.1. SEM-EDX analysis. Figure 2 depicts the SEM images captured on the surface of the pristine cotton fabric and STA/SiO2@cotton filter, as well as the EDX elemental mapping images for the STA/SiO_2@cotton. The deposition of SiO_2 particles on the cotton fibres as shown in figure 2b is apparent as compared to the smooth surface of the pristine cotton in figure 2a. The particles are the prominent contributor for the high surface roughness of the filter, which is essential in achieving superhydrophobic state [15]. Furthermore, the elemental mapping indicated that the SiO_2 particles were successfully anchored on the surface of the cotton fabric.

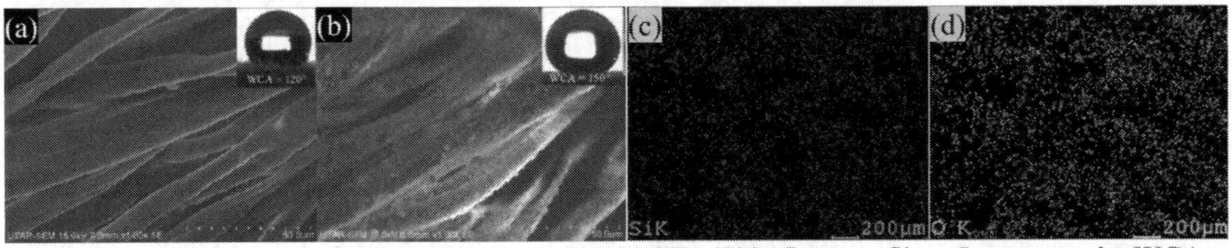

Figure 2. SEM images of (a) pristine cotton and (b) STA/SiO_2@cotton filter; Inserts are the WCA images of the respective sample; (c,d) EDX elemental mapping images of the STA/SiO_2@cotton filter.

3.2.2. FTIR analysis. Figure 3 illustrates the FTIR spectra of pristine cotton fabric, SiO_2-coated cotton fabric, STA/SiO_2-coated cotton fabric and SiO_2 powder. The broad adsorption band of 3332 cm^{-1} and 2899 cm^{-1} presented in spectrum of the pristine cotton fabric is dedicated to the stretching vibrations of –OH and –CH_2 groups, respectively [16]. On the other hand, the peak at 1160 cm^{-1} and 1103 cm^{-1} presented in the same spectrum are inferred correspondingly to the C–O and C–O–C group stretching vibration [17,18]. The absorption band at the wavelength of 3288 cm^{-1} reflects the hydrated hydroxyl groups (Si-OH) of SiO_2 on the surface of the modified cotton fabrics (see curve b and c) [19]. Meanwhile, the absorption bands at the wavelength of 2848 cm^{-1} and 2915 cm^{-1}, that are presented in the STA/SiO_2@cotton curve, are corresponded to the symmetrical and asymmetrical stretching of –CH_3 and –CH_2 groups, implied to the presence of the long alkyl chain of stearic acid [16,20]. Other than that, the peak at 1053 cm^{-1} in the spectra indicated the presence of Si-O-Si. Furthermore, the bending and symmetric stretching vibration of Si-OH are represented by the peaks at 795 cm^{-1} and 556 cm^{-1}, respectively [19].

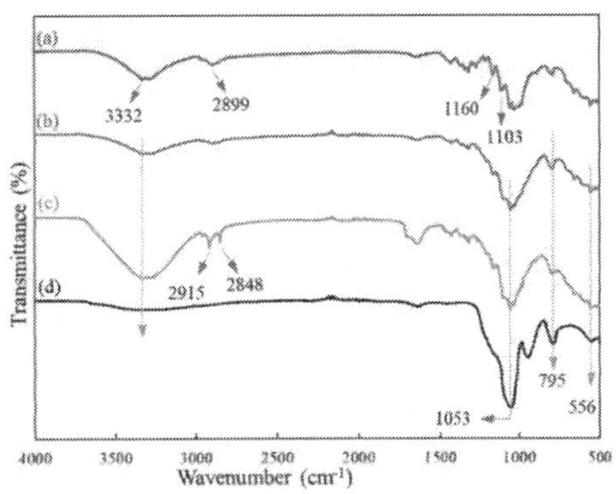

Figure 3. FT-IR spectra of (a) pristine cotton fabric, (b) SiO_2@cotton, (c) STA/SiO_2@cotton and (d) SiO_2 powder.

3.3. Oil/water separation (OWS)

Based on the WCA fabricated under different condition, the filter fabricated using 4.5mL TEOS loading and under drying temperature of 200 °C is chosen to perform OWS. In addition, the OWS for each type of oil was repeated 2 times and the average separation efficiency and permeate flux rate of the superhydrophobic filter was presented in figure 4. Figure 4 shows that the average separation efficiency as high as 99.7% can be achieved and the permeate fluxes range from 2000 to 3400 L h^{-1} m^{-2}. In comparison with the separation efficiency of OWS from other reported works, the as-prepared cotton filter in this work exhibited outstanding performance as shown in table 2.

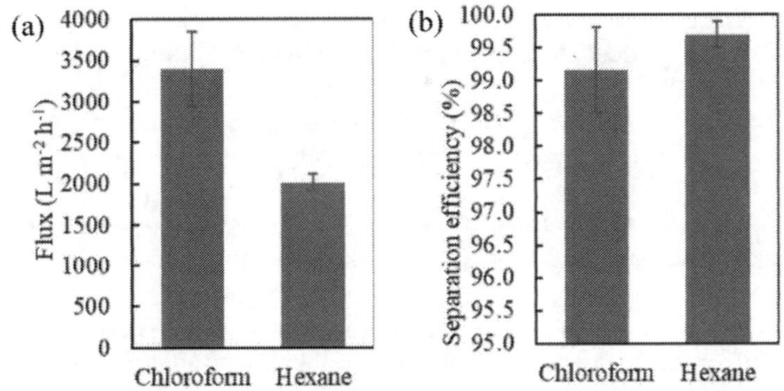

Figure 4. The performance of superhydrophobic STA/SiO_2@cotton filter in terms of (a) permeate flux (L m^{-2} h^{-1}) and (b) separation efficiency (%).

Table 2. Comparison of the separation efficiency of superhydrophobic cotton filters from some reported works.

Material(s)	Efficiency (%)	References
Eleostearic acid, SiO_2	99.5	[21]
SiO_2-boric acid-poly(vinyl alcohol) cross-linked polymer, polydimethylsiloxane (PDMS)	>90	[18]
SiO_2, siloxane	99.5	[22]
MnO_2, stearic acid	>90	[5]
PDMS, SiO_2	97-99	[23]
SiO_2, stearic acid	99.7	This work

4. Conclusions

To summarise, a fluorine-free superhydrophobic filter was prepared via a facile fabrication method using cotton fabric through silica sol immersion followed by dip coating in 1 wt% stearic acid solution. According to the characterisation, the optimum TEOS loading and drying temperature that can produce the most superhydrophobic filter are 4.5 mL and 200°C, respectively, where the WCA of the fabricated filter is 151°. Besides, the superhydrophobic STA/SiO_2@cotton filter exhibited excellent OWS ability with a separation efficiency of more than 99% in both the separation for hexane-water and chloroform-water mixtures.

Acknowledgement

The authors would like to express their gratitude towards the funding provided by Universiti Tunku Abdul Rahman Research fund (IPSR/RMC/UTARRF/2020-C2/T02).

References

[1] Gupta R K, Dunderdale G J, England M W and Hozumi A 2017 Oil/water separation techniques: A review of recent progresses and future directions *J. Mater. Chem. A* **5** 16025–58

[2] Dave D and Ghaly A E 2011 Remediation technologies for marine oil spills: A critical review and comparative analysis *Am. J. Environ. Sci.* **7** 424–40

[3] Yu L, Han M and He F 2017 A review of treating oily wastewater *Arab. J. Chem.* **10** S1913–22

[4] Sharma V, Goyat M S, Hooda A, Pandey J K, Kumar A, Gupta R, Upadhyay A K, Prakash R, Kirabira J B, Mandal P and Bhargav P K 2020 Recent progress in nano-oxides and CNTs based corrosion resistant superhydrophobic coatings: A critical review *Prog. Org. Coatings* **140** 105512

[5] Li D and Guo Z 2017 Stable and self-healing superhydrophobic MnO2@fabrics: Applications in self-cleaning, oil/water separation and wear resistance *J. Colloid Interface Sci.* **503** 124–30

[6] Hu C, Xie X and Ren K 2021 A facile method to prepare stearic acid-TiO2/zinc composite coating with multipronged robustness, self-cleaning property, and corrosion resistance *J. Alloys Compd.* **882** 160636

[7] Wang Z, Shen X, Yan Y, Qian T, Wang J, Sun Q and Jin C 2018 Facile fabrication of a PDMS @ stearic acid-Al(OH) 3 coating on lignocellulose composite with superhydrophobicity and flame retardancy *Appl. Surf. Sci.* **450** 387–95

[8] Zang D, Xun X, Gu Z, Dong J, Pan T and Liu M 2020 Fabrication of superhydrophobic self-cleaning manganese dioxide coatings on Mg alloys inspired by lotus flower *Ceram. Int.* **46** 20328–34

[9] Rasouli S, Rezaei N, Hamedi H, Zendehboudi S and Duan X 2021 Superhydrophobic and superoleophilic membranes for oil-water separation application: A comprehensive review *Mater. Des.* **204** 109599

[10] Zhao L, Du Z, Tai X and Ma Y 2021 One-step facile fabrication of hydrophobic SiO2 coated super-hydrophobic/super-oleophilic mesh via an improved Stöber method to efficient oil/water separation *Colloids Surfaces A Physicochem. Eng. Asp.* **623** 126404

[11] Ogihara H, Xie J, Okagaki J and Saji T 2012 Simple method for preparing superhydrophobic paper: Spray-deposited hydrophobic silica nanoparticle coatings exhibit high water-repellency and transparency *Langmuir* **28** 4605–8

[12] Silvia L, Hasanah I and Zainuri M 2019 Hydrophobic of polyethylene/SiO2 modified coating for self cleaning material *AIP Conf. Proc.* **2202**

[13] Drelich J, Chibowski E, Meng D D and Terpilowski K 2011 Hydrophilic and superhydrophilic surfaces and materials *Soft Matter* **7** 9804–28

[14] Kuo S W, Wu Y C, Wang C F and Jeong K U 2009 Preparing low-surface-energy polymer materials by minimizing intermolecular hydrogen-bonding interactions *J. Phys. Chem. C* **113** 20666–73

[15] Zhai G, Qi L, He W, Dai J, Xu Y, Zheng Y, Huang J and Sun D 2021 Durable super-hydrophobic PDMS@SiO2@WS2 sponge for efficient oil/water separation in complex marine environment *Environ. Pollut.* **269** 116118

[16] Chung C, Lee M and Choe E K 2004 Characterization of cotton fabric scouring by FT-IR ATR spectroscopy *Carbohydr. Polym.* **58** 417–20

[17] Jiang B, Chen Z, Sun Y, Yang H, Zhang H, Dou H and Zhang L 2018 Fabrication of superhydrophobic cotton fabrics using crosslinking polymerization method *Appl. Surf. Sci.* **441** 554–63

[18] Jannatun N, Taraqqi-A-Kamal A, Rehman R, Kuker J and Lahiri S K 2020 A facile cross-linking approach to fabricate durable and self-healing superhydrophobic coatings of SiO2-PVA@PDMS on cotton textile *Eur. Polym. J.* **134** 109836

[19] Wang X, Ding H, Wang C, Zhou R, Li Y, Li W and Ao W 2021 Self-healing superhydrophobic A-SiO2/N-TiO2@HDTMS coating with self-cleaning property *Appl. Surf. Sci.* **567** 150808

[20] Yu N, Xiao X and Pan G 2019 A stearic acidified-ZnO/methyl polysiloxane/PDMS superhydrophobic coating with good mechanical durability and physical repairability *J. Dispers. Sci. Technol.* **40** 1548–58

[21] Lin H, Chen K, Zheng S, Zeng R, Lin Y, Jian R, Bai W and Xu Y 2022 Facile fabrication of natural superhydrophobic eleostearic acid-SiO2@cotton fabric for efficient separation of oil/water mixtures and emulsions *Sustain. Mater. Technol.* e00418

[22] Yu X, Liu X, Shi X, Zhang Z, Wang H and Feng L 2019 SiO2 nanoparticle-based superhydrophobic spray and multi-functional surfaces by a facile and scalable method *Ceram. Int.* **45** 15741–4

[23] Talebizadehsardari P, Seyfi J, Hejazi I, Eyvazian A, Khodaie M, Seifi S, Davachi S M and Bahmanpour H 2020 Enhanced chemical and mechanical durability of superhydrophobic and superoleophilic nanocomposite coatings on cotton fabric for reusable oil/water separation applications *Colloids Surfaces A Physicochem. Eng. Asp.* **603**

IOP Conf. Series: Materials Science and Engineering 1257 (2022) 012049 doi:10.1088/1757-899X/1257/1/012049

Aspen adsorption simulation on biosorption between water hyacinth (*Eichhornia crassipes*) and Pb (II) ions in packed bed column

C Y Boon[1,2] and TVN Padmesh[1,3]

[1]Department of Chemical Engineering, Manipal International University, 71800 Nilai, Negeri Sembilan, Malaysia.
[2]vincent_boon@rocketmail.com
[3]padmesh.tnp@miu.edu.my

Abstract. Biosorption offers low capital and operating cost on removal of heavy metals from contaminated aqueous wastewater. This research study evaluated the effect of flow rates, initial concentration of Pb (II) ions, bed height and operating conditions of the continuous flow adsorption process on the breakthrough time of the bed column containing water hyacinth as biosorbent through simulation by using Aspen Adsorption. The simulation was conducted whereby the flow rates (Q) were varied between 0.2 L/s and 0.8 L/s, the initial concentration of Pb (II) ions (C_o) were varied between 20 mg/L and 200 mg/L, and the bed height (h) were varied between 0.25m and 1.00 m. The objectives are achieved whereby, when Q decrease, C_o decreases and the h increases, the breakthrough time increases. When Q is 0.2 L/s, C_o is 20 mg/L and h is 1.00 m, the maximum breakthrough times achieved for Study A, B and C, are 7.70 hours, 9.27 hours and 6.2 hours respectively. Study B has the longest maximum breakthrough time due to it having operating temperature of 35 °C as well as having the optimum water hyacinth biosorbent particle size which is 0.75×10^{-3} m.

1. Introduction

Adsorption, particularly biosorption, is becoming an attractive alternative for the removal of heavy metal from contaminated wastewaters (Dang et al, 2009) for its cost effectiveness and low energy consumption. Moreover, Aspen Adsorption (Adsim) software can be used to simulate dynamic adsorption, which provides models for the generation of breakthrough curves and estimation of performance parameters for continuous fixed bed. The main objective of this study is to simulate the continuous fixed-bed adsorption of Pb (II) onto water hyacinth (*Eichhornia crassipes)* using Aspen Adsorption®. This study will be able to provide information on biosorption of Pb (II) on water hyacinth in a continuous fixed-bed adsorption process, the effects of different parameters for an adsorption process as well as provide insight in designing and optimizing bed columns used by heavy metal adsorption facilities.

2. Materials and Methods

2.1. Materials & Apparatus

The removal of Pb (II) by water hyacinth has been simulated using Aspen Adsorption @V8.4.

2.2. *Methods*

2.2.1. Parameters. In this paper, the optimum condition for batch adsorption is used as the operating condition for the continuous flow adsorption of Pb (II) ion by water hyacinth as biosorbent in a bed column via simulation. The optimum conditions were obtained from 3 different studies. From each study, the optimum conditions were different in terms of the optimum temperature, maximum adsorption capacity, energy constant and size of adsorbent used.

As shown in Table 1, the three different optimum condition for batch adsorption were obtained from Yi et al. (2019), Murithi et al. (2012) and Mahamadi & Nharingo (2010). Each of these three optimum conditions for batch adsorption will be applied as the operating conditions for one continuous flow adsorption study. For example, the optimum condition in Yi et al (2019) will be applied as operating condition for Study A, the optimum condition in Murithi et al. (2012) will be applied as operating condition for Study B and the optimum condition in Mahamadi & Nharingo (2010) will be applied as operating condition for Study C.

Table 1. The optimum condition for batch adsorption from three different studies.

	Yi et al., 2019 (Study A)	Murithi et al., 2012 (Study B)	Mahamadi & Nharingo, 2010 (Study C)
Optimum Temperature (°C)	25	35	30
Particle Size (m)	0.18×10^{-3}	0.75×10^{-3}	2.5×10^{-3}
Maximum Adsorptive Capacity, q_m (mg/g)	75.44	163.9	26.32
Energy constant, k_L (L/mg)	0.0476	0.0234	0.61
IP_1 (kmol/kg)	0.0002278	0.0004949	0.00007947
IP_2 (m^3/kmol)	15765.12	7750.08	202032

As indicated in Equation 1, the equation on the left side is the common form of Langmuir isotherm and the equation on the right side is the form that Aspen Adsorption uses (Juela, 2020). To use the Langmuir isotherm in Aspen form, the maximum adsorption capacity, q_m, and energy constant, k_L, from Table 1 were converted to IP_1 and IP_2 respectively via Equation 2, whereby M is the molecular weight of lead (II) nitrate while q_e (or w_i) is the amount of solute adsorbed by an amount of adsorbent, in equilibrium with a solution of concentration C.

$$q_e = \frac{q_m k_L C}{1+k_L C};\ w_i = \frac{IP_1 IP_2 C}{1+IP_2 C} \quad (1)$$

$$IP_1 = \frac{q_m}{1000M};\ IP_2 = 1000 k_L M \quad (2)$$

Since water hyacinth is used as adsorbent, the properties of water hyacinth adsorbent are needed. These physical properties of water hyacinth in Table 2 were obtained from other papers due to the authors, which are Yi et al. (2019), Murithi et al. (2012) and Mahamadi & Nharingo (2010), in the batch adsorption papers did not report the physical properties of the water hyacinth adsorbent used in respective papers. The physical properties of water hyacinth as adsorbent are assumed to be same for Study A, B and C, and will be keyed into the simulation.

Table 2. The physical properties of water hyacinth.

Physical Properties	Values	Reference
Bulk density	0.031 g/cm^3	Soriano et al. 2016
Particle density	0.023 g/cm^3	Soriano et al. 2016
Pore space volume	2.2 ml/g	Soriano et al. 2016

Pore volume	13.75 cm^3/g	Soriano et al. 2016
Average pore diameter	5.22 nm	Rashwan & Girgis 2004
Particle Porosity	0.806	Mohammed & Davies, 2011
Pore diameter	5.22 nm	Rashwan & Girgis 2004
Interparticle voidage	0.0682 m^3/m^3	-

$$pore\ space\ volume \times bulk\ density = interparticle\ voidage \tag{3}$$

The interparticle voidage is needed in the bed configuration during the simulation in Aspen Adsorption and was calculated based on Equation 3 (Nieva et al., 2019). For Table 3, as water was used as solvent for all the studies, the association factor of solvent for water is 2.26 and the molecular weight of the water is 18.01 g/mol (Juela, 2020). The dynamic viscosity of the solvent is assumed to be the same as the dynamic viscosity of the solution and both are assumed to be the same as the dynamic viscosity of the water at respective temperature (Wright State University, 2022). The solution density is also assumed to have the same density as water at respective temperature. The solute molar volume at its normal boiling point, being that the solute is lead (II) nitrate, is 0.1194 cm^3/mol (Saeed, 2014). The temperatures used for respective studies were obtained from Table 1.

Table 3. Solvent and solution properties.

Properties	Study A	Study B	Study C	Units
	Values	Values	Values	
Association factor of solvent, α_A	2.26	2.26	2.26	-
Molecular weight of the solvent, M_s	18.01	18.01	18.01	g mol^{-1}
Dynamic viscosity of the solvent, η_s	0.000891	0.00072	0.000798	kg m^{-1} s^{-1}
Temperature, T	298	308	303	K
Solute molar volume at its normal boiling point, V_m	0.1194	0.1194	0.1194	cm^3 mol^{-1}
Solution density, ρ_{sol}	998.17	995.17	997.17	kg m^{-3}
Dynamic viscosity of the solution, η_{sol}	0.000891	0.00072	0.000798	kg m^{-1} s^{-1}

Using the raw data from Table 2 and 3 as well as Equation 4 to 10, obtained from (Juela, 2020), the mass transfer coefficient for different flow rate in Table 4 was calculated.

$$Re = \frac{d_p v_s \rho_{sol}}{\eta_{sol}}; \quad Sc = \frac{\eta_{sol}}{\rho_{sol} D_m}; \tag{4}$$

$$Sh = \frac{k_f d_p}{D_m} = 2 + 1.58 Re^{0.4} Sc^{1/3} \tag{5}$$

$$\tau_p = \varepsilon_p + 1.5(1 - \varepsilon_p) \tag{6}$$

$$D_k = 97 r_p \left(\frac{T}{M_s}\right)^{0.5} \tag{7}$$

$$D_m = 7.4 * 10^{-8} \frac{(\alpha_A M_s)^{0.5} T}{\eta_s V_m^{0.6}} \tag{8}$$

$$\frac{1}{D_p} = \tau_p \left(\frac{1}{D_m} + \frac{1}{D_k}\right) \tag{9}$$

$$\frac{1}{K_i} = \frac{R_P}{3 k_{fi}} + \frac{R_p^2}{15 \varepsilon_p D_p} \tag{10}$$

Table 4. The mass transfer coefficient at different flow rates for individual studies.

Flow rate, Q (L/s)	Study A	Study B	Study C
	Mass Transfer Coefficient (1/s)	Mass Transfer Coefficient (1/s)	Mass Transfer Coefficient (1/s)
0.2	0.7133	0.06131	0.005549
0.4	0.7795	0.06586	0.005849
0.6	0.8177	0.06835	0.006005
0.8	0.8442	0.07002	0.006108

2.2.2. Simulation. For the feed and product stream, the reversible model was chosen. The flow rate, temperature and pressure for both feed and product streams are the same, assuming that the pressure drop is zero. The concentration of Pb (II) ions for the product stream was set to zero to determine the breakthrough time (Juela, 2020). For bed configuration, under the general tab, discretization method selected is Upwind Difference Scheme 1 (UDS 1) and the number of nodes is set as 20 by default. Under material/momentum balance tab, convection with estimated dispersion is chosen (Soriano et al., 2016). Under the kinetic model tab, the solid film model was chosen for the film model assumption and the lumped resistance in linear function was chosen for kinetic model assumption (Soriano et al., 2016). Under isotherm tab, Langmuir 1 was chosen. Under energy balance tab, isothermal was chosen as energy balance assumption. Under the initials tab, the water concentration is always set to 1 kmol/m^3 and set to free.

To study the effect of Q on the breakthrough time of the bed column, the Q is manipulated between 0.2 L/s to 0.8 L/s with increments of 0.2 L/s and with parameters obtained for Study A, B, and C, while maintaining C_o at 80 mg/L and h at 0.5 m. To study the effect of C_o on the breakthrough time of the bed, C_o was manipulated between 20 mg/L to 200 mg/L with increments of 30 mg/L and with parameters obtained for Study A, B, and C, while maintaining Q at 0.2 L/s and h at 0.5 m. To study the effect of h on the breakthrough time of the bed column, h was manipulated between 0.25 m to 1.0 m with increments of 0.25m and with parameters obtained for Study A, B, and C, while C_o and Q were maintained at 20 mg/L and 0.2 L/s respectively. The bed diameter was kept constant at 0.5 m.

3. Results and Discussion

The breakthrough time is determined when the outlet concentration of Pb (II) ions of the bed column reaches 0.05 mg/L, which is same as the permitted concentration limit for Pb (II) ions in wastewater set by the United State Environmental Protection Agency (EPA) (Arbabi, 2015). The limit set by EPA is used since it is lesser than the limit set by the Malaysia's Environmental Law, which is 0.1 mg/L (Environmental Quality Act, 1974, 2015).

3.1. Effect of Flow Rate

As Q increases from 0.2 L/s to 0.8 L/s, the breakthrough time decrease for Study A, Study B and Study C as shown in Table 5. This is because the increase in Q, increases the Reynold's number and, subsequently, the mass transfer coefficient increases based on Equation between 4 and 10. When the mass transfer coefficient increases, the external film diffusion mass transfer resistance at the surface of biosorbent particles decreases (Zhong et al., 2018), making it easier for Pb (II) ions to diffuse into the pore of the adsorbent through the water medium and adsorbed onto the pore surface. Therefore, the amount of adsorbent being used up in adsorbing Pb (II) ions will increase and, subsequently, the bed column will exhaust at a faster rate, which results in the decrease in breakthrough time.

Table 5. The breakthrough time achieved for each individual study under different Q.

Flow rate, Q (L/s)	Study A	Study B	Study C
	Breakthrough time (h)	Breakthrough time (h)	Breakthrough time (h)
0.2	1.79	2.93	0.79
0.4	0.90	1.45	0.37
0.6	0.60	0.96	0.23
0.8	0.45	0.72	0.16

The other reason for the decrease in breakthrough time when Q increases might be the inadequate residence time of the solute in the packed bed column (Mishra et al., 2016). In higher Q, the Pb (II) ions have shorter residence time in the bed column, which results in shorter contact time with the water hyacinth biosorbent. This results in less opportunities for the Pb (II) ions to reach the meso and micropores of the adsorbent, resulting lesser chances for adsorption equilibrium between the Pb (II) ions and the surface of the pores to occur (Juela, 2020; Mishra et al., 2016; Zhong et al., 2019). Therefore, the maximum adsorption capacity of most biosorbent particle will not be reached resulting the bed column exhaust at an increase rate as well as the decrease in breakthrough time of the bed column.

3.2. *Effect of Initial Pb (II) Ions Concentration*

As C_o increases from 20 mg/L to 200 mg/L, the breakthrough time decreases for Study A, B & C as shown in Table 6. This is because, when C_o increases, the amount Pb (II) ions surrounding the biosorbent particles also increases. This creates an increased concentration gradient of Pb (II) ions between the surrounding biosorbent particle and the inside pores of the biosorbent particles.

Table 6. The breakthrough time achieved for each individual study under different C_o.

Initial Pb (II) ions concentration, C_o (mg/L)	Study A	Study B	Study C
	Breakthrough time (h)	Breakthrough time (h)	Breakthrough time (h)
20	3.84	4.60	3.05
50	2.47	3.63	1.26
80	1.79	2.93	0.79
110	1.40	2.43	0.57
140	1.15	2.07	0.44
170	0.98	1.80	0.35
200	0.85	1.59	0.30

The concentration gradient acts as a driving force that leads to the increased amount of Pb (II) ions diffusing from the surroundings of the biosorbent particles into the medium inside the pore of the biosorbent. Thus, increasing the number of Pb (II) ions interacting with and getting adsorbed onto the active sites or surface of the pores of the biosorbent, subsequently, increasing the rate of biosorbent particles being used up. This eventually leads to the decrease in breakthrough time of the bed column.

3.3. *Effect of Bed Height*

From Table 7 as the breakthrough time increases for Study A, B and C, when the h increases from 0.25 m to 1 m. This is because when the h increases, the number of biosorbent particles within the bed column increases, as well as the total surface area and the number of active sites available for adsorption of Pb

(II) ions also increases (Juela, 2020), which leads to the increased amount of Pb (II) ions being adsorbed. Since C_o ions is keep constant, the present of more biosorbent particles increases the breakthrough time of the bed column as shown in Figure 3.

Table 7. The breakthrough time achieved for each individual study under different h.

Bed height, h (m)	Study A	Study B	Study C
	Breakthrough time (h)	Breakthrough time (h)	Breakthrough time (h)
0.25	1.92	2.27	1.48
0.50	3.84	4.60	3.05
0.75	5.77	6.93	4.62
1.00	7.70	9.27	6.20

3.4. Effect of Applying Different Optimum Condition of Batch Adsorption Process as Operating Condition in Continuous Adsorption

As shown in Table 7, when C_o is 20 mg/L, Q is 0.2 L/s and h is 1.00 m, Study B produces the longest maximum breakthrough time (9.27 hours), followed by Study A (7.7 hours) and Study C (6.2 hours). One of the reasons is Study B has the highest operating temperature, followed by Study A and C, as shown in Table 7 and 1. The increase in operating temperature opens up additional sorption sites in the biosorbent, which increases each biosorbent's Pb (II) ions uptake capacity (Murithi et al., 2014), which ultimately increases the adsorption capacity of the bed. Another reason is that the biosorbent particle size used in Study B is bigger than that of in Study A and smaller than that of in Study C, as shown in Table 7 and 1. When the particle size increases, the adsorption capacity of each biosorbent particles also increases due to the increase in surface area for the adsorption of Pb (II) ions. Resulting in the maximum breakthrough time obtained in Study B longer than that of in Study A. However, if the particle size is too big, it will result in slower pore diffusion rate because the diffusion path along the pores is larger and the resistance to diffusion is higher (Mishra et al., 2016). As a result, most Pb (II) ions can only be adsorbed on the outer portion of the biosorbent particles which leads to a faster breakthrough time. This is evident whereby Study C used the largest biosorbent particles size among the other studies which resulted in the shortest maximum breakthrough time.

4. Conclusion

Through this experiment, it is evident that the breakthrough time increases when Q decreases, C_o decreases and h increases. In addition, the breakthrough time can be maximized by using the optimum operating temperature as well as optimum biosorbent particle size. For this paper, the optimum operating temperature was found to be 35 °C while the optimum water hyacinth biosorbent particle size was 0.75 $\times 10^{-3}$ m. The maximum breakthrough time obtained in this paper was in close approximation with the simulated breakthrough time in Soriano et al., 2016, where in that paper the longest simulated breakthrough time was around 6.67 hours, with conditions that are similar to the condition tested in this paper. This shows that continuous flow absorption simulation via Aspen adsorption using the data obtained from the batch adsorption is feasible, in which theoretical maximum breakthrough time of the bed column can be obtained without the need for actual experiment.

References

[1] Arbabi M Hemati S and Amiri M 2015 Removal of lead ions from industrial wastewater: a review of removal methods *International Journal of Epidemiologic Research* **2** 105-9.

[2] Dang, V. B. H., H. D. Doan, T. Dang-Vu and A. Lohi 2009 Equilibrium and kinetics of biosorption of cadmium (II) and copper (II) ions by wheat straw *Bioresour. Technol* **100** 211-219.

[3] Environmental Quality Act, *1974 Kerajaan Malaysia Official Portal of Department of*

Environment [Internet] 2015. Available from: http://www.doe.gov.my/portalv1/wp-content/uploads/2015/01/Environmental_Quality_Sewage_And_Industrial_Effluents_Regulations_1979_-_P.U.A_12-79.pdf

[4] Juela D M 2020 Comparison of the adsorption capacity of acetaminophen on sugarcane bagasse and corn cob by dynamic simulation Sustainable Environment Research **30**.

[5] Mahamadi C and Nharingo T 2010 Utilization of water hyacinth weed (*Eichhornia crassipes*) for the removal of Pb(II), Cd(II) and Zn(II) from aquatic environments: an adsorption isotherm study *Environmental Technology* **31** 1221–28.

[6] Mishra A Tripathi B D and Rai, A. K 2016 Packed-bed column biosorption of chromium (VI) and nickel (II) onto Fenton modified Hydrilla verticillata dried biomass *Ecotoxicology and Environmental Safety* **132** 420–428

[7] Mohammed U S and Davies R M 2011 Moisture-dependent engineering properties of water hyacinth parts *Singapore Journal of Scientific Research* **1** 253-263.

[8] Murithi G Onindo C O and Muthakia G K 2012 Kinetic and equilibrium study for the sorption of Pb(II) ions from aqueous phase by water hyacinth (*Eichhornia crassipes*) *Bulletin of the Chemical Society of Ethiopia* **26** 181-193.

[9] Murithi G Onindo C O Wambu E W and Muthakia G K 2014 Removal of cadmium(II) ions from water by adsorption using water hyacinth (*Eichhornia crassipes*) biomass *BioRes* **9** 3613-31.

[10] Nieva A D Garcia R C and Ped R M R 2019 Simulated biosorption of Cr^{6+} using peels of *Litchi chinensis sonn* by Aspen Adsorption® V8.4 *International Journal of Environmental Science and Development* **10**.

[11] Rashwan W E and Girgis B S 2004 Adsorption capacities of activated carbons derived from rice straw and water hyacinth in the removal of organic pollutants from water *Adsorption Science & Technology* **22** 181 – 194.

[12] Saeed W 2014 *Diffusion coefficients (diffusivities)* [Internet]. Available from: https://chemicalprojects.net/2014/07/04/diffusion-coefficients/.

[13] Wright State University *Property Tables And Charts (SI Units)* [Internet] 2022. Available from: http://cecs.wright.edu/people/faculty/sthomas/htappendix01.pdf

[14] Yi Z Liu J Liu X Zeng R and Cui Y 2019 Lead (II) removal from wastewater by water hyacinth *IOP Conference Series: Earth and Environmental Science* **310**.

[15] Zhong Q Q Xu X Shen L Han Y J and Hao Y C 2018 Breakthrough curve analysis of Enteromorpha prolifera packed fixed-bed column for the biosorption *Environmental Engineering Science* **35** 864–874.

IOP Conf. Series: Materials Science and Engineering 1257 (2022) 012050 doi:10.1088/1757-899X/1257/1/012050

EVALUATION OF THE STRUCTURE CHARACTERISATION OF HYDROXYPROPYL BISPALMITAMIDE MEA AND CERAMIDE 3

C H Chong[1]* and T-J Khoo[2]

[1]Department of Chemical and Environmental Engineering, Faculty of Science and Engineering, University of Nottingham, Jalan Broga, Semenyih 43500, Malaysia.

[2]School of Pharmacy, Faculty of Science and Engineering, University of Nottingham Malaysia, Jalan Broga, Semenyih 43500, Malaysia.

ChienHwa.Chong@nottingham.edu.my

Abstract. The skin is the largest organ in the body and the skin barrier is important to reduce moisture loss, prevent irritations and bacterial infections. The complex skin barrier consists of essential lipids such as ceramides together with cholesterol and free fatty acids. Hydroxypropyl Bispalmitamide MEA or PC-104 is claimed to be a compound created to mimic ceramides and claimed to have similar effectiveness as Ceramide 3 at moisturising the skin and restoring the skin barrier. The objective of this study is to evaluate the structure characterisation of Hydroxypropyl Bispalmitamide MEA and whether it can be categorised as a ceramide. Three different spectroscopy and physico chemical analysis conducted on the two samples revealed the structural formula and chemical composition of Hydroxypropyl Bispalmitamide MEA to be different from Ceramide 3. Therefore, the two compounds cannot be considered to contain the same chemical structural make up, based on spectral analysis generated by instrumentation. Hydroxypropyl Bispalmitamide MEA cannot be categorised as a ceramide because it is lacking the sphingosine moiety.

1. Introduction

Ceramides play a major role in maintaining the efficient barrier function of the stratum corneum (SC). It is important in structuring and preserving the water permeability barrier and creating a natural physiological environment for skin, together with cholesterol and fatty acids [1] [2]. Ceramides are used in cosmetics especially for dry skin conditions. A synthetic pseudo-ceramide called Hydroxypropyl Bispalmitamide MEA (PC-104) is a compound created to mimic the properties of natural skin-barrier building ceramides. It is claimed to be as effective as Ceramide 3 at moisturising the skin with lower melting point, and hence easier to formulate and is cheaper than real ceramides [3]. (www.bumtorg.ru). However, the presence of sphingosine base in ceramide is the major difference between the two [4]. Ceramides, classified as sphingolipids, are composed of a sphingosine base (18 carbons) and an amide-linked fatty acid (14–26 carbons). Sphingosine is the characteristic structure for the identification of ceramide (Figure 1. and table 1). The objective of this study is to evaluate the characteristics of Hydroxypropyl Bispalmitamide MEA and Ceramide 3.

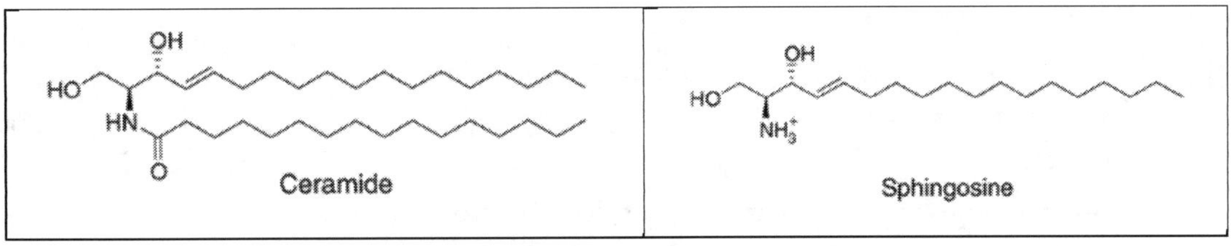

Figure 1. Ceramide chemical structure and sphingosine section.

Table 1. IUPAC name, chemical , chemical formula, molecular weight, theoretical m/z (mass divided by charge number) and CHNO elemental %. [5, 6]

	Ceramide 3	Hydroxypropyl Bispalmitamide MEA
IUPAC name	N-((2S,3S,4R)-1,3,4-trihydroxyoctadecan-2-yl)stearamide	1,3-Propanediamino, 2-hydroxy-, N,N'-bis(1-oxohexadecyl)-N,N'-bis(2-hydroxyethyl)-
Chemical Formula	$C_{36}H_{73}NO_4$	$C_{39}H_{78}N_2O_5$
Molecular Weight	583.98	655.06
Theoretical m/z	583.55 (100.0%), 584.56 (38.9%)	: 654.59 (100.0%), 655.59 (42.2%)
CHNO Elemental %	C: 74.04; H: 12.60; N: 2.40; O: 10.96	C: 71.51; H: 12.00; N: 4.28; O: 12.21

2. Research Methodology

Samples of Hydroxypropyl Bispalmitamide MEA and Ceramide 3 were submitted to an independent laboratory for analysis. Three different Spectroscopy and elemental analysis techniques were conducted using Infra-Red (IR) Spectroscopy, Liquid Chromatography coupled with Mass Spectrometry (LCMS) and Elemental (CHNOS) analysis.

2.1. Elemental analysis (C, H, N, O, S)

The experiments were carried out using a CHN628 Elemental Determinator. The instrument has been switched on for more than 12-hours prior to operation. All incoming gas has been switched Open. (O2, He, Compressed Air). The temperature of the Primary Furnace was at its optimum setting at 950°C and the Secondary Furnace was set at 850°C. Blank samples were used for calibration and this is followed by the sample analysis.

2.2. LCMS-M2Scan

The experiment adopted the phospholipid characterization or lipid omics from Shimadzu with some modification [7]. The Shimadzu method is well described in literature and documented as two stages method. Stage 1 describes the elimination of polar head group followed by method stage 2 where the fatty acid composition can be analysed according to target peaks and production ions (Figure 2). The fatty acid that makes up the long carbon chain can be identified and differentiated according to the number of hydrocarbon aliphatic carbon chains and the position of the double alkene bond. In the case of Ceramide 3, the main functional group is an amide as compared to ester in phospholipid. It is best to profile the long fatty acid chain using an MRM Library. Solvent used in this method must be of HPLC Grade. During the experiment, optimisation was performed to optimise the parameter that suits both samples. The parameters were then optimised such as LC program, Collision energy, flow rate gas and etc. until the stage of MS2Scan. The solvent used in this method was of HPLC Grade.

1. Main instrument: Triple Quadrupole LC-MS/MS
2. Method Stage 1: Polar head group elimination.
3. Method Stage 2: MRM Library to determine fatty acid composition analysis of target peaks and product ions.
4. Software library: MRM Library for phospholipids profiling
5. Analytical Column at oven temperature of 45 C: Phenomenex Kinetex C8 (150 mm L × 2.1 mm I.D., 2.6 μm)
6. Ionisation method: Electrospray Ionisation (ESI +/-)
7. Mobile Phase A: 20 mM ammonium formate and Mobile **Phase B: Acetonitrile/Isopropanol (1:1)**

Figure 2. The Shimadzu method, software library and mobile phases used in LCMS analysis.

2.3. FTIR analysis

A 0.1 g of samples were prepared and put on a plate. The samples were scanned and the process and based line were clicked and selected using a FTIR-Spectrum 400. The sample was scanned. Lastly, an automatic button was clicked to get a smooth curve and to label all peaks. The curve was expressed in percentage Transmittance versus wavelength (cm^{-1}). The experiments were conducted in duplicates to confirm the accuracy of the results.

3. Results and discussion

3.1 Spectroscopy and Elemental Analysis Report Detailing Comparison between Ceramide 3 versus Hydroxypropyl Bispalmitamide MEA.

3.1.1 Chemical Structure. Comparison and differentiation of Hydroxypropyl Bispalmitamide MEA versus Ceramide 3 starts at the molecular level. We first consider the chemical structure, functional groups and elemental composition of the two chemicals individually [5].

Ceramide 3 – Chemical Structure, IUPAC name, formula and weight.

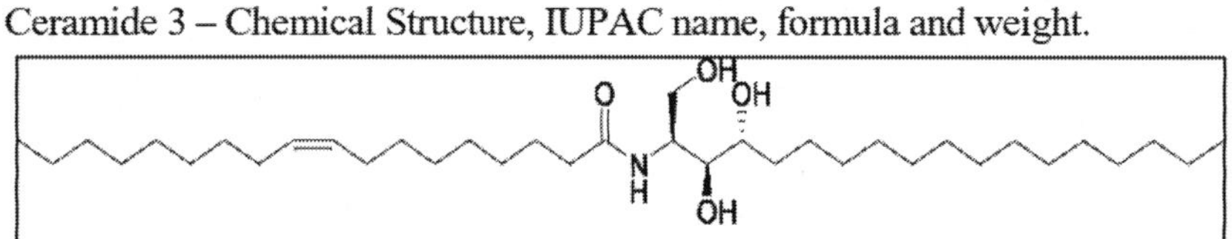

(Z)-N-[(2S,3S,4R)-1,3,4-trihydroxyoctadecan-2-yl]octadec-9-enamide

Figure 3. Ceramide 3 chemical structure.

Ceramide 3 has the following Chemical Formula: $C_{36}H_{71}NO_4$ with a molecular weight of 583.98. (Figure 3). The following are breakdown of theoretical composition in percentage C, 74.30; H, 12.30; N, 2.41; O, 11.00. Ceramide 3 is made of an amide functional group together with a three-alcohol functional group. The side chain are aliphatic hydrocarbons with an alkene.

In Hydroxypropyl Bispalmitamide monoethanolamide (MEA); -

Hydroxypropyl Bispalmitamide MEA – Chemical Structure, IUPAC name, formula and weight [6].

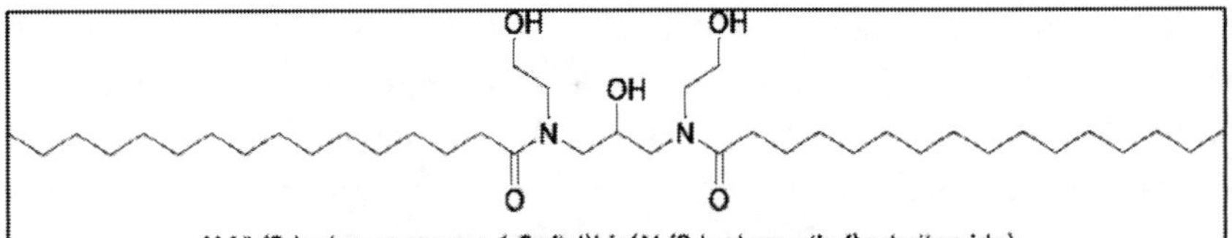

Figure 4. Hydroxypropyl Bispalmitamide MEA chemical structure

Hydroxypropyl Bispalmitamide MEA has the following Chemical Formula: $C_{39}H_{78}N_2O_5$ with a molecular weight of 655.06 (Figure 4). The following is the breakdown of theoretical composition, in percentage, for the element C, 71.51; H, 12.00; N, 4.28; O, 12.21. Two amide functional groups together with three alcohol made up the structure. A tertiary amine can be considered to be present if nitrogen is not part of the amide group. Side chain on both end of Hydroxypropyl Bispalmitamide MEA is made of aliphatic hydrocarbons the structure.

3.1.2 Spectroscopy and Elemental Analysis. Hydroxypropyl Bispalmitamide MEA chemical and structural composition can be differentiated from Ceramide 3 by conducting spectroscopy and elemental analysis. In determination of differences by % and types of elements present in the two compounds, the CHNOS elemental analysis was conducted. Predominant functional groups which give rise to the chemical properties and nomenclature were determined by InfraRed (IR) Spectroscopy. A detailed marker to differentiate the total make up of Hydroxypropyl Bispalmitamide MEA versus Ceramide 3 was analysed from the spectra result of Liquid Chromatography Mass Spectrometry.

3.1.3 CHNOS elemental analysis. Recorded elemental composition by percentage for sample Hydroxypropyl Bispalmitamide MEA indicates the percentage closest to (theoretical values) of C: 72.4 (71.51%); H: 11.3 (12.00%); N: 4.1 (4.28%) and O: 12.2 (12.21%). When compared to Ceramide 3, the recorded values are closest in (theoretical values) of C: 73.9 (74.30%); H: 11.3 (12.30%); N: 2.8 (2.41%) and O: 12.0 (11.00%). The percentage errors for Hydroxypropyl Bispalmitamide MEA ranged from 0.08% to 5.8% meanwhile the percentage errors for Ceramide 3 ranged from 0.54% to 16%. The recorded number when compared to theoretical value of N is a confirmative indication of the existence of sphingoid moiety in Ceramide 3. Theoretical values were calculated based on Molecular Weight 655.06 of Hydroxypropyl Bispalmitamide MEA with Chemical Formula $C_{39}H_{78}N_2O_5$ while Molecular Weight 581.54 of Ceramide 3 with Chemical Formula $C_{36}H_{71}NO_4$. In both samples, sulphur was not present.

3.1.4 IR spectrum analysis. Spectra of both samples were not superimposable indicating a general shift of wavenumbers and peaks representing the chemical functional groups. The effect of its differences in composition on vibrational and stretching energy deviates the wavenumbers at fingerprint region clearly indicates that Ceramide 3 composition is entirely different from Hydroxypropyl Bispalmitamide MEA. Therefore, they cannot be considered the same, similar or vice versa. Amide is represented by two peaks at 3385.49 cm^{-1} and 3306.48 cm^{-1} in the spectra of Hydroxypropyl Bispalmitamide MEA. In the spectra of Ceramide 3, indication of a single peak at 3339.19 cm^{-1} of one amide group is present. The wavelengths 938.71 cm^{-1} and 1074.48 cm^{-1} can be attributed to =C-H bending in Ceramide 3, otherwise not observed to be present in the spectra of Hydroxypropyl Bispalmitamide MEA. (Figures A1 and A2).

3.1.5 LCMS. Further evidence of the sphingoid moiety in Ceramide 3 is represented by the m/z fragment peak. Predominant peak m/z 579.100 in the ESI negative scan indicates the presence of ionised parent mass of Ceramide 3 while the ESI positive scan of Hydroxypropyl Bispalmitamide MEA with peak at m/z 655.500 shows a clear indication of parent mass. In the ESI Positive scan of Ceramide 3, there appears to be a peak at m/z 304.3, a confirmatory assignment of sphingoid moiety presence. This peak is not present in m/z of the ESI positive scan of Hydroxypropyl Bispalmitamide MEA. Both samples are of high purity with single peaks appearing at Rt 5.8 min Ceramide 3 and Rt 4.9 min Hydroxypropyl Bispalmitamide MEA, respectively. Tertiary amine in Hydroxypropyl Bispalmitamide MEA fragment containing amide bond can be assigned to m/z peak at 304.3. (Appendices A3(a), A3(b), A4(a) and A4(b).

Three different spectroscopy and physico chemical analysis conducted on the two samples reveals that the structural formula and chemical composition of Hydroxypropyl Bispalmitamide MEA to be different from Ceramide 3. Therefore, the two compounds cannot be considered to contain the same chemical structural make up, based on spectral analysis generated by instrumentation.

4. Conclusion

Based on the analytical results by cross comparing and interpreting the spectra of LCMS, IR Spectra and CHNOS elemental analysis of functional groups, we can characterize the individual chemical composition of Hydroxypropyl Bispalmitamide MEA and Ceramide 3. After cross comparison, the results point towards the lack of sphingosine moiety in Hydroxypropyl Bispalmitamide MEA and therefore this compound cannot be categorised as ceramide.

Acknowledgement

This work was supported by GOOD PHARMA DERMATOLOGY PTE. LTD under the research grant S0035.54.04.

References

[1] Vavrova K, Kovačik A and Opalka L 2017 *European Pharmaceutical Journal* **64** 28–35.

[2] Mutanu Jungersted J, Hellgren L I, Høgh J K, Drachmann T, Jemec G B E and Agner T 2010 *Acta Dermato-Venereologic* **90** 350–353.

[3] *What are ceramides for hair. Ceramides in cosmetics. What does ceramide deficiency lead to?* Bumtorg [Internet]. 2020 [cited 2020 Sep 22]; Available from: https://bumtorg.ru/en/aktyory/chto-takoe-ceramidy-dlya-volos-keramidy-v-kosmetike-k-chemu-privodit/

[4] Iwai I, Han H M, den Hollander L, Svensson S, Ofverstedt L, Anwar J, Brewer J, Bloksgaard M, Laloeuf A, Nosek D, Masich S, Bagatolli L A, Skoglun U and Norlén L 2012. *Journal of Investigative Dermatology* **132** 2215–2225.

[5] *PubChem Compound Summary for CID 74945544, Ceramide 3.* PubChem [Internet]. 2004 [cited 2022 July 5]. Available from: https://pubchem.ncbi.nlm.nih.gov/compound/Ceramide-3.

[6] *PubChem Compound Summary for CID 11193108, Hydroxypropyl bispalmitamide monoethanolamide.* PubChem [Internet]. 2004 [cited 2022 July 5]. Available from: https://pubchem.ncbi.nlm.nih.gov/compound/Hydroxypropyl-bispalmitamide-monoethanolamide.

[7] *Solution for Pharmaceutical Analysis.* Application Notebook [Internet]. 2015 [cited 2022 July 5];1–75. Available from: www.shimadzu.com/an/.

IOP Conf. Series: Materials Science and Engineering 1257 (2022) 012050 doi:10.1088/1757-899X/1257/1/012050

Appendices

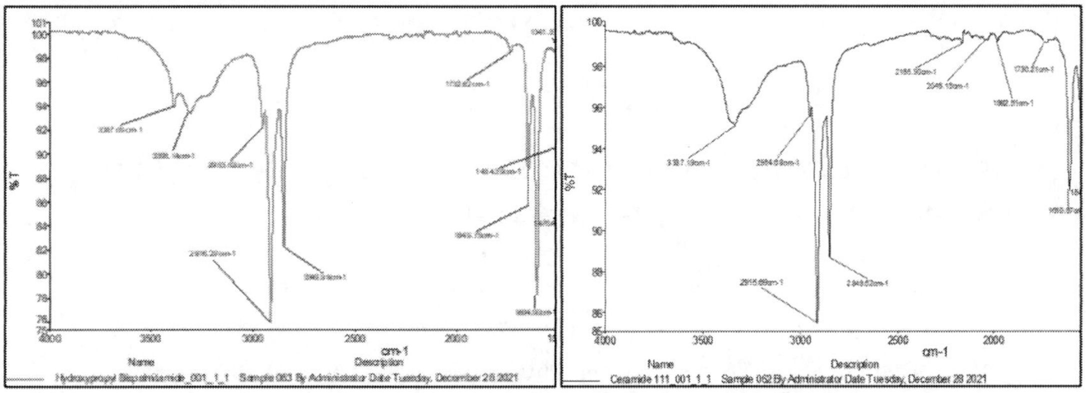

Figures. A1 and A2. FTIR analysis for Hydroxypropyl Bispalmitamide MEA and Ceramide 3.

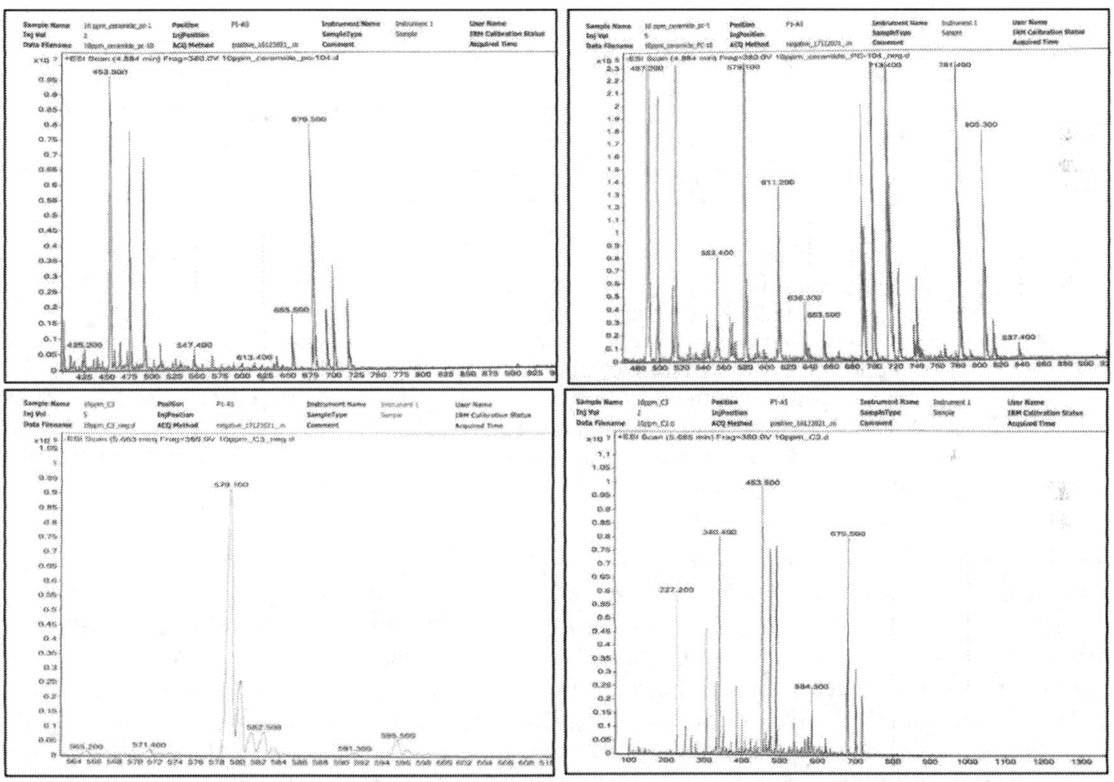

Figures A3(a) LCMS analysis positive scan for Hydroxypropyl Bispalmitamide MEA, **A3 (b)** LCMS analysis negative scan for Hydroxypropyl Bispalmitamide MEA, **A4(a)** LCMS Ceramide 3 negative scan and **A4(b)** Ceramide 3 Positive Scan. A peak at m/z 304.3, a confirmatory assignment of sphingoid moiety presence.

A Sensitivity Study of Salt Concentration in Brine to The Solubility of CO2 in Saline Aquifers

A S Setiawan[1], M A Hamid[1], E J Mackay[2].

1. Department of Engineering and Physical Science, Heriot-Watt University Malaysia, 62200 Putrajaya, WP. Putrajaya, Malaysia
2. School of Energy, Geoscience, Infrastructure and Society, Heriot-Watt University, Campus The Avenue, Edinburgh EH14 4AS, United Kingdom

e-mail: as2158@hw.ac.uk; m.ali@hw.ac.uk; E.J.Mackay@hw.ac.uk

Abstract. In CO_2 sequestration, four main trapping mechanisms are expected such as structural, residual, solubility, and mineral trapping. This paper aims to analyse the impact of salt concentration in brine on the CO_2 solubility trapping mechanism in the saline aquifer. The CO_2 solubility decreases with salinity. In this study, a sensitivity run will be done on an existing saline aquifer dataset to investigate the impact of salt concentration using the compositional mode simulation approach. Different types of salt will give different levels of CO_2 solubility in an Aqueous solution, for example, at isotherm conditions and at increasing pressure, the same amount of NaCl will be more soluble. At the same temperature condition, higher salinity will decrease CO_2 solubility with pressure increases and produce a salting-out phenomenon at the near injector area. The reference dataset was completed using the black-oil simulation with the gas-water system, where the CO_2 act as the gas phase and brine as the water phase. The black-oil simulator can only handle isotherm and a single salinity value, which is 200,000 ppm in this case. Thus, this study is expected to give more information and improve the solubility trapping mechanism in a saline aquifer.

1. Introduction

According to the global climate resolution from the UN Climate Change Conference 2015, through its Paris Agreement article 4, paragraph 1, that carbon sequestration (sink) plays a significant role in reducing carbon dioxide (CO_2) emissions to the atmosphere, and it is expected to achieve Paris Agreement's target by the second half of this century. The CO_2 sequestration can be performed under a depleted hydrocarbon reservoir, a coal bed methane reservoir, or a saline aquifer reservoir, as illustrated in *Figure-1*. Although CO_2 flooding has been around in the oil industry since the 60s, but there is still room for improvement, especially in CO_2 injection into a saline aquifer. In many cases, numerical simulation studies are required to perform proper injection planning.

There are four main CO_2 trapping mechanisms that generally happen in the sequestration site during and post CO_2 injection. Referring to the IPCC report in 2005, shown in *Figure-2*, there are (1) structural & stratigraphic trapping, (2) residual trapping, (3) solubility trapping, and (4) mineralisation. The first trapping, structural and stratigraphic, relies on the geological features of the storage site and happens instantaneously from the beginning of the injection. The remaining trapping mechanism will take effect longer, especially the mineralisation trapping. Another prolific trapping is CO_2 solubility, especially its solubility with the brine in the saline aquifer storage site. The parameter impacting CO_2 solubility in a brine solution is mainly the water chemistry which includes the type of salt, salt concentration, salt composition, in-situ pressure, and temperature.

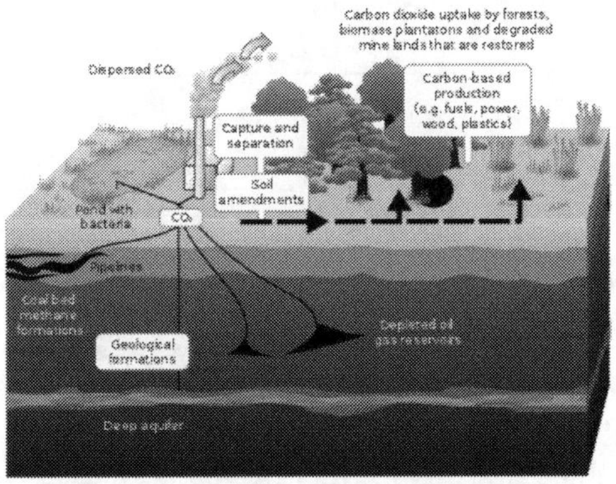

Figure-1: Illustration of CO_2 Sequestration (UKCCSRC, 2020).

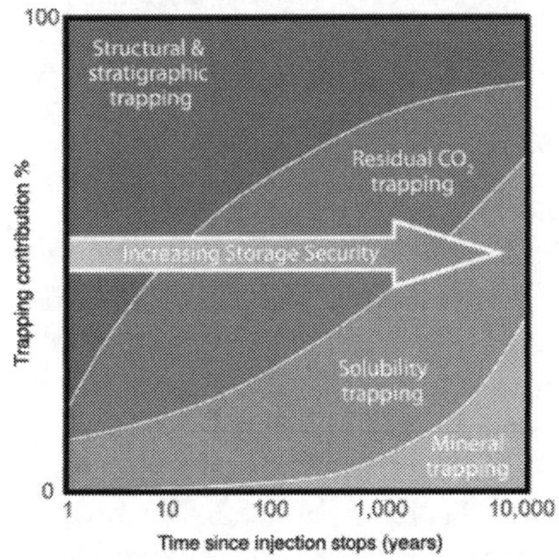

Figure-2: Trapping mechanism and contribution through time (IPCC, 2005).

This paper will share the result of numerical simulation evaluation on CO_2 solubility corresponding to a single salt type and its salinity sensitivity. The evaluation is done using a case study of a sequestration model from the United Kingdom performed by Pale Blue Dot Energy.

2. CO_2 Solubility in Brine

The CO_2 solubility increases with pressure but decreases with temperature, while the increase of salinity of the brine will reduce the CO_2 solubility. The interaction between CO_2 as the gaseous phase and brine (formation water) as the aqueous phase has been experimentally studied by many. The modelling accuracy of this interaction is essential for the whole CO_2 sequestration study, especially in the saline aquifer reservoir.

Li and Ngheim[5] shared the constants as functions of pressure, temperature and salinity that cover many gases, including CO_2, at the saturation pressure and temperature of H_2O. Henry's constants are expressed as follows:

$$\ln H_i = \ln H_i^* + \frac{1}{RT}\int_{p^*}^{p} \bar{v}_i \, dp \quad (1)$$

where,

Hi = Henry's constant at p and T
Hi* = Henry's constant at p* and T
R = gas constant
$\bar{v}_i$ = partial molar volume of component i in solution
p = pressure
R = ideal gas constant
T = temperature

The Henry's constant will then be used in the fugacity calculation of the aqueous phase. Fugacity of the gaseous and aqueous phase is modelled as in the equilibrium condition or expressed by:

$$f_{CO2,a} = f_{CO2,g} \quad (2)$$

where,

$f_{CO2,a}$ = aqueous fugacity of CO_2
$f_{CO2,g}$ = gaseous fugacity of CO_2

Fugacity is a partial pressure that substitutes the pressure of ideal gas in the computation of chemical equilibrium. In the equation (2), the gaseous phase's fugacity of CO_2 can be calculated using the cubic equation of state published by Peng and Robinson in 1976 (EOS-PR). While the aqueous phase's fugacity will be solved using Henry's law through the following equation:

$$f_{co_2,a} = H_{co_2,a} \cdot y_{co_2,a} \quad \text{........(3)}$$

where,
$H_{co_2,a}$ = Henry's constant for CO_2 solubility in brine
$y_{co_2,a}$ = mole fraction of CO_2 in brine

3. Salinity Impact on CO_2 Solubility

The formation water typically contains various salt and mainly is an NaCl type of salt. The presence of salt reduces solubility. Several experiments were done by Takenouchi & Kennedy (1964), and Yasunishi and Yoshida (1979), which have shown several correlations between CO_2 solubility with NaCl brine, as illustrated by *Figure-3*. The experiment produces interaction parameters between aqueous and non-aqueous in a mixture, k_{ij}^{AQ} and k_{ij}^{NAQ} respectively. These parameters are the main input to calculate the Equation of State (EOS) constants a.

$$k_{ij}^{AQ} = -0.31092(1 + 0.15587c_{sw}^{0.7505}) + 0.2358(1 + 0.17837c_{sw}^{0.979})T_{ri} - 21.2566 \exp(-6.7222T_{ri} - c_{sw}) \quad \text{..........(4)},$$

where c_{sw} is the salinity of the brine. While the non-aqueous parameter for CO_2 is 0.1896.

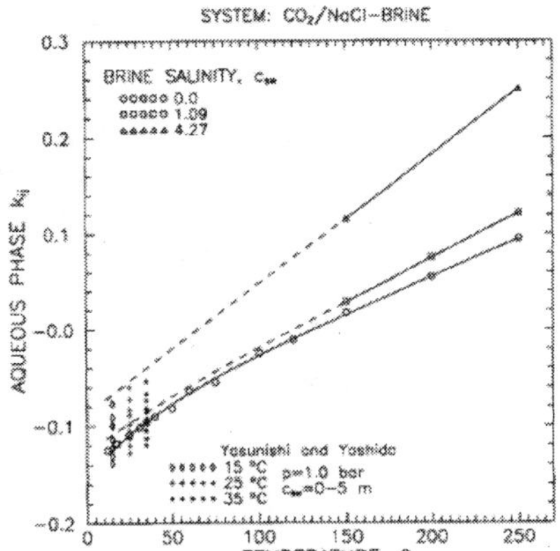

Figure-3: Aqueous phase interaction parameters for CO_2/NaCl brine binary systems as a function of temperature and brine salinity.

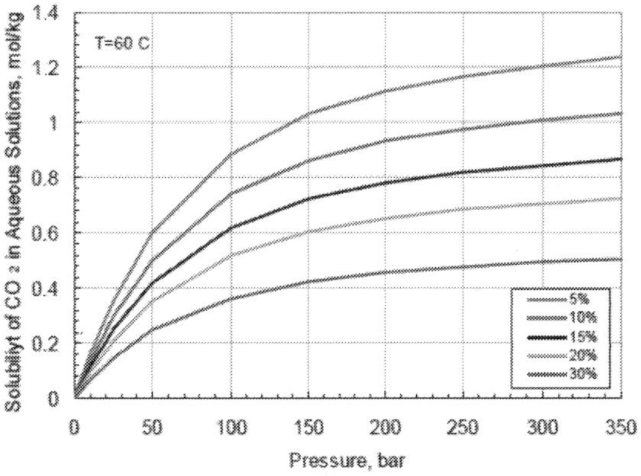

Figure-4: Solubility of CO_2 in aqueous of NaCl with varying total salinity at isotherm condition.

Further simulation works based on the above findings were done by Chang and Coats[2] based on data matching with Wiebe (1941). The Chang and Coats correlation was developed to estimate the properties of CO_2-saturated water for binary CO_2-water systems as a function of temperature, pressure, and salinity. This relationship is illustrated in *Figure-4*. The increase of CO_2 solubility will increase brine density then produces convection current to facilitate the mixing of CO_2 with unsaturated brine, therefore will increase solubility trapping (Nghiem et. Al, 2004).

4. Bunter-36 Case Study

To evaluate the impact of salinity on the CO_2 solubility, and then ultimately on the solubility trapping, a numerical model of CO_2 storage in the saline aquifer was used and established using a compositional simulation machine. The model that is being used for this evaluation is coming from the Bunter-36 field located in the United Kingdom. The model as illustrated in *Figure-5*, was originally done in a black-oil simulator with the assumption of a high salinity value taken from a nearby field.

4.1. *Model Description*

The Bunter-36 field was considered as one of the storage sites in the United Kingdom CCS project. The primary storage is an extensive sandstone water-bearing reservoir located within the Triassic age formation. The wells drilled within the area did not discover any hydrocarbon accumulation. A very minimum data acquisition has made the uncertainty becomes high in the reservoir characterization. Sensitivity will be done only on the salinity value of the model. The summary of Bunter-36 base case model properties are as follows:

Table 1: Summary of the model properties.

Parameter	Information	
Type	Saline Aquifer	
Active cells	603,394	
Grid size	200x200	m^2
Connected Aquifer Size	270×10^9	m^3
Avg NTG	0.85	fraction
Avg Poro	22	(%)
Avg Perm	210	(mD)
Supply	7 (40yrs)	(Mt/yr)
Storage	391	(Mt)
Init.Res-Press	124.4	barg
Init. Res-Temp	45	(°C)
Depth	1170	(mTVD-SS)
Injector	4	
Depo.	Fluvial, lacustrine	
Salinity	205K	(ppm)
Tcr	30.98	°C
Pcr	73.77	bara
Vcr	0.0939	m^3/kg.mole
Acentric factor	0.239	
MW	44.01	
SG	1.53	
Boiling Point	-78.45	°C

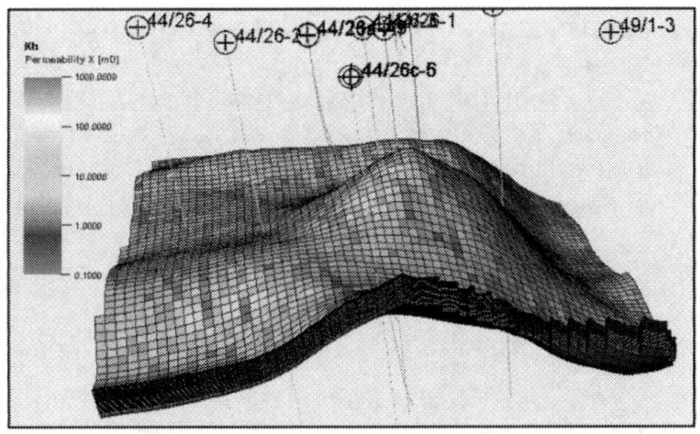

Figure-5: Bunter-36 in 3D view (Pale Blue Dot, UKCSS, ETI, 2016).

4.2. *Uncertainty Background*

The formation water salinity data of the model is not available because there was no direct measurement performed. The base estimate value was referring to "An assessment of carbon sequestration" [1]. The study covered several other Bunter Closures, excluding Bunter-36, in the area. In this study, the salinity value established is within a range of 130,000 ppm to 205,000 ppm. The best estimate value is also a mid-range value of the Caister gas fields study (Ritchie & Pratsides, 1993), with a high value at 250,000 ppm. The value range is also consistent with the data provided in Ketter's[4] study of the Esmond, Forbes, and Gordon fields (Ketter, 1991). Thus, it was decided no further sensitivity was done to cover the Bunter-36 area. However, this study is not to seek the best estimate and the range of salinity values, but more to demonstrate the impact of salinity on the CO_2 solubility and the solubility trapping by using a broader range of salinity values with a single type of salt.

4.3. *Simulation Description*

The numerical simulation in this evaluation is performed using a compositional simulator with CO_2 sequestration mode. The main component involved were CO_2 as injection fluid, freshwater as aqueous phase and NaCl as the solid or salt composition, the last two are the main ingredient of formation water (brine mixtures). The CO_2 sequestration mode functionality is based on a paper by Spycher & Pruess[6]

IOP Conf. Series: Materials Science and Engineering 1257 (2022) 012051 doi:10.1088/1757-899X/1257/1/012051

about the phase partitioning model for CO_2-Brine mixtures. A set of salinity values was utilized in the evaluation, which includes the minimum, best estimate, maximum (all explained in subsection 4.2), drinking water standard, freshwater formation, dead-sea, and other deliberate values.

5. Result and Discussion

CO_2 dissolves the brine mixture in the aquifer and forms a new ionic compound. These new ionic compounds are stable and may also precipitate salts or normally called a 'salting-out' process. The reaction will directly affect the density and viscosity of the formation brine. This reaction can be modelled using compositional simulation options that calculate mutual solubilities of CO_2 in the brine, and the brine in CO_2. A similar profile of solubilities should be matched with the laboratory analysis result.

The solubility trapping is dependent on the in-situ liquid composition and characteristics. In the saline aquifer, the solubility of CO_2 determines by salt composition and concentration. The external parameter including pressure and temperature will also affect the CO_2 solubility in the liquid phase. The solubility trapping itself theoretically can be illustrated by the following reactions:

$$CO_2 + H_2O \leftrightarrow H_2CO_3, \quad (5)$$

$$H_2CO_3 + H_2O \leftrightarrow HCO_3^- + H^+ + H_2O, \quad (6)$$

$$HCO_3^- + H_2O \leftrightarrow CO_3^{2-} + H^+ + H_2O, \quad (7)$$

All cases were run up to 500 years without stopping the injection to see the evolution of CO_2 solubility through time, and the simulated impact of solid precipitation nearby the injectors. CO_2 solubility decreases with the increase of salinity, this is demonstrated clearly in *Figure-6*. The freshest brine with 10,000 ppm, as we normally found in the fluvial dominated depositional environment, shows the preferable case of CO_2 injection. The increases of CO_2 solubility in the liquid phase, in this case, a NaCl mixed brine affect the storage capacity as demonstrated by *Figure-7*. Theoretically, using the same rate of 7Mt/year or the equivalent of 361 MMscfd, a maximum of about 24 Tcf (equivalent to 1200 MT) of CO_2 can be stored in the Bunter-36 sandstone.

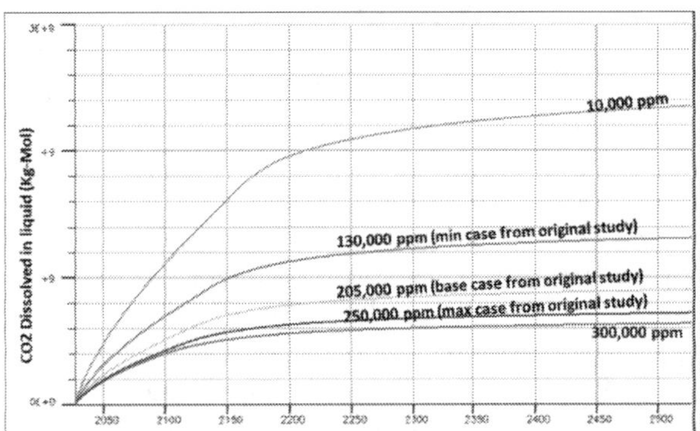

Figure-6: CO_2 dissolved in the liquid varies with the brine salinity in the aquifer through the injection time.

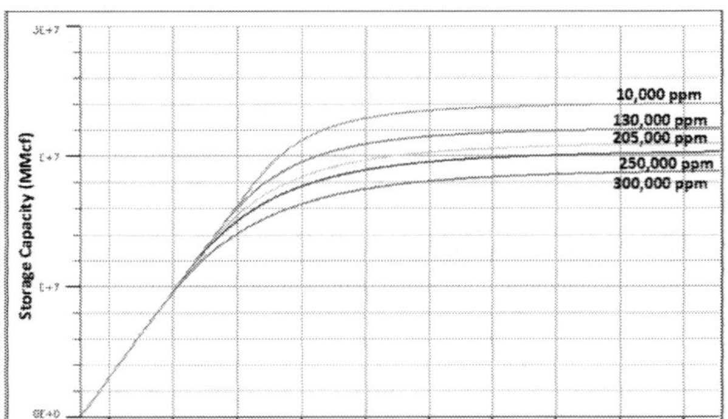

Figure-7: CO_2 storage capacity varies with the brine salinity in the aquifer.

Figure-8 shows the impact of solid precipitation on injectors and limited soluble CO_2 capacity in the storage site. The injectivity is decreasing earlier compared to the fresher aquifer. This is also reflected by *Figure-9*, which shows a rapid pressure build-up nearby the injection interval and in the reservoir for the case of highest salinity brine caused the least amount of CO_2 dissolved in the liquid. The simulation result also shows that solid precipitation is less likely to happen in a fresher aquifer brine compared to the most saline aquifer.

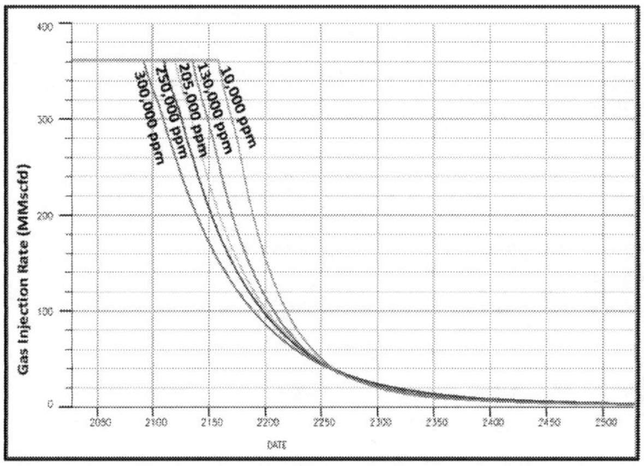

Figure-8: CO_2 injection rate varies with the brine salinity in the aquifer.

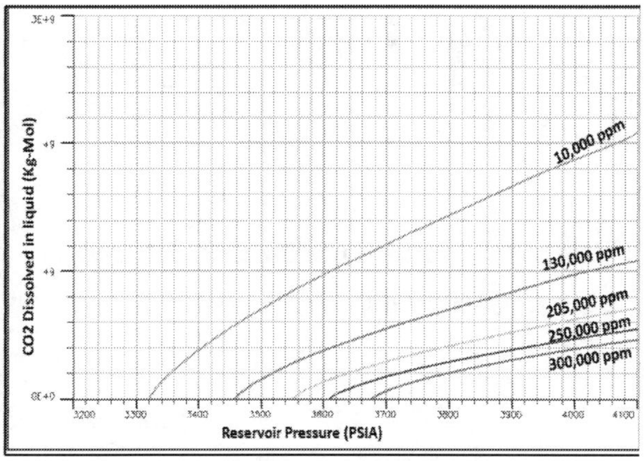

Figure-9: The profile of reservoir pressure build-up due to injection.

6. Conclusion and Way Forward

This paper demonstrates and describes the modelling of an important parameter affecting CO_2 storage in the saline aquifer, which is salinity concentration. The increases in brine salinity levels in the saline aquifer will reduce the CO_2 solubility hence the solubility trapping capacity. Additionally, solid precipitation is observed severely on the higher salinity brine based on this sensitivity study. The solid precipitation affects the injectivity index hence reducing the CO_2 capacity.

In a shortcoming future, a sensitivity study with more parameters such as pressure, temperature, total dissolved solids, and salt types is necessary to be performed to comprehend and be able to select the best storage site, especially in the saline aquifer.

Acknowledgement

This work is based on the information taken from the Strategic UK CCS Storage Appraisal Project[3], funded by DECC, commissioned by the ETI and delivered by Pale Blue Dot Energy, Axis Well Technology and Costain. The authors also greatly acknowledge Heriot-Watt University Malaysia for support and publication permission for this paper.

References

[1] Bentham M 2006. "*An assessment of carbon sequestration potential in the UK–Southern North Sea case study*." Nottingham: Tyndall Centre for Climate Change Research.

[2] Chang Y B, Coats B K and Nolen J S 1998 "*A Compositional Model for CO2 Floods Including CO2 Solubility in Water*" paper SPE 365164 first presented at the 1996 SPE Permian Basin Oil & Gas Recovery Conference, Midland, Texas.

[3] James A, Baines S and McCollough, S 2016. *Strategic UK CCS Storage Appraisal - WP5A - Bunter Storage Development Plan, ETI.* https://doi.org/10.5286/UKERC.EDC.000267

[4] Ketter F J 1991 "*The Esmond, Forbes and Gordon Fields, Blocks 43/8a, 43/13a, 43/15a, 43/20a, UK North Sea.*" Geological Society, London, Memoirs 14.1: 425-432.

[5] Li Y and Ngheim L X 1986 "*Phase Equilibria of Oil, Gas and Water/Brine Mixtures from a Cubic Equation of State and Henry's Law*" Canadian Journal of Chemical Eng. 64, 486– 496.

[6] Spycher N and Pruess K 2010 *Phase-Partitioning Model for CO2–Brine Mixtures at Elevated Temperatures and Pressures: Application to CO2-Enhanced Geothermal Systems Transport Porous Medium* 82:173–196. http://dx.doi.org/10.1007/s11242-009-9425-y.

Baking Kinetics and Characteristics of Cake Mixture Formulated with Mealworm Powder

C L Hii[1*], N A R S Kumaran[1], C H Cheah[1], K L Tan[1], B Mistry[1] and C L Chiang[2]

[1]Department of Chemical and Environmental Engineering, Faculty of Science and Engineering, University of Nottingham Malaysia, Jalan Broga, 43500 Semenyih, Selangor Darul Ehsan, Malaysia
[2]Foundation in Engineering, Faculty of Science and Engineering, University of Nottingham Malaysia, Jalan Broga, 43500 Semenyih, Selangor Darul Ehsan, Malaysia

*Corresponding email:Ching-Lik.Hii@nottingham.edu.my

Abstract. Yellow mealworm (*Tenebrio Molitor*) was reported containing 24.3 to 27.6% protein content. However, public acceptance is still relatively low as many tend to reject it as a source of food. Hence, studies were carried out to incorporate dried mealworm powder into a standard butter cake mixture formulation. Yellow mealworms were dried using an air ventilated oven at 80°C. The dried mealworms were then ground into powder and added in a standard butter cake formulation (0%, 5% and 10%). Baking was carried out in a convective oven at 140°C for 30 minutes. Moisture content, colour, texture and protein content were analysed. The effective diffusivity values of the cake samples varied from 4.46×10^{-11} m^2/s to 4.72×10^{-11} m^2/s. Significant observations were recorded for the L^* and b^* values obtained among the cake samples. The firmness of the cake increased while the springiness decreased as addition of the mealworm powder was increased. An increase in protein content was observed from 5.99% to 11.12% with the addition of mealworm powder in the cake formulation.

1. Introduction

The Food and Agriculture Organization of the United Nations predicts that human population will reach 9 billion people by 2050 [1] and this will require a greater demand for food. In some countries, the practice of eating insects or known as entomophagy have been practiced (e.g in Africa and Asia) in order to obtain protein source [2]. Yellow mealworm (*Tenebrio Molitor*) is known as a novel protein source and an average mealworm contains 24.3% to 27.6% protein [3]. Unfortunately, public acceptance towards consuming mealworms is relatively low as expected. However, by incorporating as an ingredient in food may have a greater acceptance e.g. as flavoring ingredient in chocolate [4].

The study was aimed to investigate the effect of incorporating yellow mealworm powder (0%, 5% and 10%) into a cake mixture formulation. Replacement or addition of ingredients into a cake formulation will affect not only the flavour but also the physical, nutritional and textural characteristics of the cake [5]. Specific objectives of the study are (i) to investigate the baking kinetics of the cake added with dried mealworm powder, (ii) to determine the colour attributes of the cake after baking, (iii) to determine the firmness and springiness of the cake after baking and (iv) to determine the protein content of the cake.

IOP Conf. Series: Materials Science and Engineering 1257 (2022) 012052 doi:10.1088/1757-899X/1257/1/012052

2. Materials and Methods

2.1. Sample Preparation

Mealworm larvae (about 4-5 cm) were purchased locally from Semenyih, Selangor (Figure 1). The larvae were starved for 48 hours and then refrigerated at -10°C (48 hours) and stored at 4°C. Six packets (500 gm each) of butter cake formulation were used. Before drying, the mealworms were first boiled in hot water and then sieved/wiped dry. About 500 g mealworm were put on a baking tray (6 hours) in an air-ventilated oven (Memmert, Germany) at 60°C. After drying, the mealworms were left to cool and blended into powdered form. The powder was stored at 4°C before use.

Figure 1. Mealworm larvae.

2.2. Baking Procedure

The experiments were conducted using a 500 g standard butter cake formulation (0% mealworm powder) with the addition of 275 g butter and five grade A eggs. The butter and eggs were whipped in a steel bowl followed by the addition of the cake formulation. Twelve cupcakes were baked in a convective oven (Samsung, Malaysia) at 140°C for 30 minutes. The experiments were then repeated by adding 25 g (5%) and 50 g (10%) of the mealworm powder into the cake formulation.

2.3. Moisture Content

During baking, the weight was measured every 10 minutes until completion (30 minutes). The moisture content (m_i) of the cake was determined using equation (1).

$$m_i = \frac{w_s - w_d}{w_d} \quad (1)$$

Where w_s and w_d represent the sample and bone-dry weights, respectively. The bone-dry weight was obtained by drying the cake sample for 24 hours in an oven (Memmert, Germany) at 105°C.

2.4. Effective Diffusivity

Due to the irregular shape, the cakes were assumed as spheres and the height (H), depth (E) and width (W) were measured. Equation (2) was used to calculate the approximate diameter (D_{cake}) [6].

$$D_{cake} = \sqrt[3]{H \times E \times W} \quad (2)$$

The one-dimensional unsteady state diffusion in a sphere is shown in equation (3) [7].

$$MR(\text{Moisture ratio}) = \frac{m_i - m_e}{m_o - m_e} = \frac{6}{\pi^2}\sum_{n=1}^{\infty}\frac{1}{n^2}\exp\left[-\frac{n^2\pi^2 D_{eff} t}{r^2}\right] \quad (3)$$

For long drying periods, the first term (n = 1) is sufficient to provide an estimate of the general solution. By linearizing the first term, the equation can be further simplified to equation (4).

$$\ln MR = \ln \frac{6}{\pi^2} - \frac{\pi^2 D_{eff} t}{r^2} \quad (4)$$

where r represents the radius (1/2 of D_{cake}). The effective diffusivity (D_{eff}) can be determined by plotting ln MR against time (t) to obtain a liner graph to estimate the D_{eff} from the gradient.

2.5. Colour Analysis

Colour analysis was conducted by using a colorimeter (Precision Colorimeter, China) by pointing the aperture directly at the cake surface for L^* (light-dark), a^* (green-red) and b^* (blue-yellow) [8] attributes. Total colour change (ΔE) was determined according to equation (5) [9].

$$\Delta E = \sqrt{(L^* - L_0{}^*)^2 + (a^* - a_0{}^*)^2 + (b^* - b_0{}^*)^2} \quad (5)$$

Where subscript 0 is referring to the reference sample without addition of mealworm powder (0%).

2.6. Texture Analysis

A texture analyzer (Stable Microsystems TA. XT TEE32, UK) utilizing a compression platen was used. The platen compressed the cake until reaching 25% of the product height and after 30 seconds, the platen was withdrawn from the sample. The firmness and springiness of the cupcake were determined from the force versus time curve generated using the built in software.

2.7. Protein Content

Crude protein content was analyzed using the Kjeldahl method. A digester (Buchi KjelDigester, USA) was used to digest 0.1 g of cake sample in the presence of a Kjeldahl tablet (copper catalyst) and concentrated sulphuric acid. Five digester test tubes were filled with concentrated sulphuric acid and Kjeldahl tablets, which were labelled as blank samples. The remaining 15 digester tubes were prepared similar to the blank samples with the addition of 0.15 g of cake sample of varying amount of mealworm powder. Each analysis was conducted in five replicates. The digestion procedure was carried out at 380°C for 5 hours until a greenish-blue solution was obtained. Protein nitrogen was liberated to form ammonium ions during the digestion procedure. Distillation of all the samples in the test tube were then conducted using 32% sodium hydroxide (NaOH) and 4% boric acid (H_3BO_3) solution of pH 4.5 (Buchi KjelFlex Distillation Unit, USA) to capture ammonium ions in a receiving conical flask. The conical flask containing the ammonium ions was then titrated using 0.1M hydrochloric acid. The end point of titration was determined when the solution reached a pH of 4.65. The nitrogen content (%N) in each sample was calculated using equation (6) [10]. The crude protein content was determined by multiplying a factor 6.25 to the nitrogen content (6.25 × %N).

$$\%\,N = \frac{(\text{Volume of standard acid (ml)} - \text{Volume of blank (ml)}) \times \text{Normality of acid} \times 1.4007}{\text{Weight of sample (g)}} \quad (6)$$

3. Results and Discussion

3.1. Baking Kinetics

The moisture content profiles of the cakes incorporated with 0%, 5% and 10% mealworm powder are as shown in Figure 2.

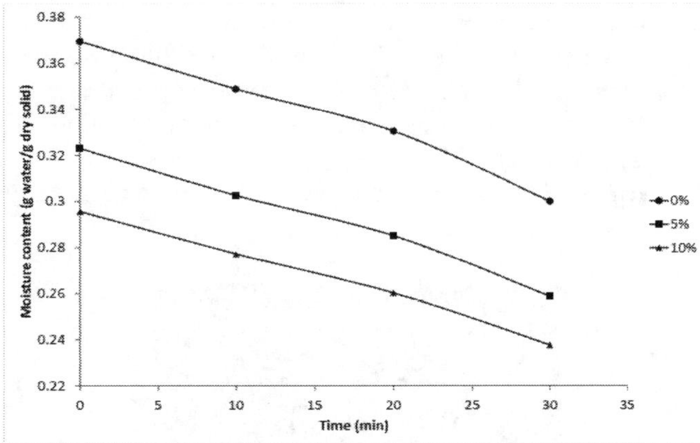

Figure 2. Moisture content profiles of the cakes upon addition of mealworm powder.

A slight reduction in initial moisture content was observed when the cake was added with the mealworm powder due to addition of the dry ingredients. Initial baking rates were determined at 0.0021, 0.002 and 0.0018 g water/g dry solid.min, respectively, for 0%, 5% and 10% addition of mealworm powder. Overall, baking rates were determined at 0.0023 g water/g dry solid.min, 0.0021 g water/g dry solid.min and 0.0019 g water/g dry solid.min for each respective percentage addition.

3.2. Effective Diffusivity

The effective diffusivities of the cakes containing 0%, 5% and 10% addition of mealworm powder were determined at 4.46×10^{-7} m^2/s, 4.72×10^{-7} m^2/s and 4.97×10^{-7} m^2/s respectively. These values are near to the order of magnitude reported for other cake products ($10^{-8} – 10^{-11}$ m^2/s) (Table 1).

Table 1. Comparison of effective diffusivities among baked products.

Material	Effective Diffusivity (m^2/s)	Reference
Butter cake	$4.46 \times 10^{-7} – 4.72 \times 10^{-7}$	Current work
White cake	$2.66 \times 10^{-8} – 1.67 \times 10^{-10}$	[11]
Cake	$3.8 \times 10^{-8} – 9.9 \times 10^{-11}$	[12]
Madeira cake	$1.43 \times 10^{-8} – 8.44 \times 10^{-8}$	[13]

3.3. Colour Analysis

The L^* and b^* values were observed to decrease as the amount of mealworm powder increased (Table 2). This showed that the cake samples turned darker and less yellowish upon addition of the mealworm powder, which could be due to the excessive browning that occurs during drying. However, no significant observation was recorded for the a^* values.

Table 2. L^*, a^*, b^* and ΔE values of the cake samples after baking.

Addition of mealworm powder in cake (%)	L^*	a^*	b^*	ΔE
0	71.09	7.17	37.04	-
5	63.28	7.60	28.34	11.71
10	56.17	7.30	22.96	20.52

It was observed that the colour of the cake samples incorporated with the mealworm powder differed significantly from the reference sample (Figure 3). The higher addition of mealworm powder (10%) resulted in greater total colour change (ΔE = 20.52) as compared to the lower amount (5%) (ΔE = 11.71). Further research should be carried out to determine the optimum level of mealworm powder addition that is acceptable by consumers. The drying conditions used in preparation of the mealworms should be optimized as well to produce an acceptable colour of the dried mealworm powder.

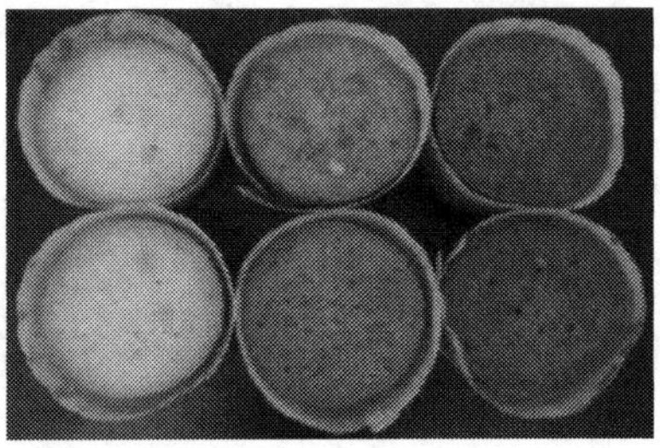

Figure 3. Baked cake samples (left to right: 0%, 5% and 10% mealworm powder).

3.4. Texture Analysis
Table 3 shows the firmness and springiness of the cake samples after baking. As the amount of mealworm powder increased, the firmness of the cake samples increased due to addition of more solid ingredients (mealworm powder). The increased values could be attributed to the increased fibre and protein content from the mealworm powder. These macromolecules (protein and fibre) form hydrogen bonds with water, reducing the amount of free water and overall mobility, causing an increased firmness as the addition of mealworm powder increased. Table 3 also showed that the addition of mealworm powder decreased the springiness after baking. It caused the product matrix to become denser, therefore the cakes were not able to recover immediately after deformation.

Table 3. Firmness and springiness cake samples after baking.

Addition of mealworm powder in cake (%)	Firmness (N)	Springiness (%)
0	15.71	61.27
5	18.36	59.88
10	21.85	55.24

3.5. Protein Content
The addition of mealworm powder increased the crude protein content as shown in Table 4.

Table 4. Protein content of cake samples after baking.

Addition of mealworm powder in cake (%)	Protein Content (%)
0	5.99
5	8.67
10	11.12

The crude protein content was determined at 8.67% and 11.12% for cake containing 5% and 10% addition of mealworm powder, respectively. The incorporation of mealworm powder in the cake formulation yielded higher protein content and showed great potential to be consumed as a protein-rich delicacy. The cake samples containing no addition of mealworm powder (0%) was determined to contain 5.99% of protein as compared to a typical butter cake with protein content of 6.50% [14]. This could be due to the different recipe and amount of ingredients used.

4. Conclusions

Current study has successfully incorporated mealworm powder in a butter cake formulation. D_{eff} values were near to the literature range (10^{-11} - 10^{-8} m^2/s). The cake samples showed obvious total colour change (*ΔE*) mainly due to the dark-brown colour of the mealworm powder. Firmness and springiness were found to increase and decrease, respectively, as the amount of mealworm powder increased. Protein content increased with the addition of mealworm powder. Further research should attempt to determine the optimum drying conditions and the amount of mealworm powder that can be incorporated. Factors such as dough beating and butter whipping might also affects quality. It is also recommended to do sensory evaluation and incorporate mealworm powder in other baked products.

Acknowledgements

The funding from the Department of Chemical and Environmental Engineering is acknowledged.

References

[1] Alexandaratos N 2012 *How to feed the world in 2050.* Available at: http://www.fao.org/3/a-ap106e.pdf.
[2] Vane-Wright R I 1991 Why not eat insects? *Bull. Entomol. Res.* **81** 1-4
[3] Ghaly A and Alkoaik F 2009 The yellow mealworm as a novel source of protein *Am J Agric Biol Sci.* **4** 319–31
[4] Caparros M R, Sablon L, Geuens M, Brostaux Y, Alabi T, Blecker C, Drugmand D, Haubruge É and Francis F 2014 Edible insects acceptance by Belgian consumers: Promising attitude for entomophagy development *J. Sens. Stud.* **29** 14-20
[5] Arora B, Kamal S and Sharma V P 2017 Sensory, nutritional and quality attributes of sponge cake supplemented with mushroom (*Agaricus bisporus*) powder *Nutri. Food Sci.* **47** 578-90
[6] Bart-Plange A and Baryeh E A 2003 The physical properties of category B cocoa beans *J Food Eng.* **60** 219-27
[7] Hii C L, Law C.L, Cloke M and Suzannah S 2009 Thin layer drying kinetics of cocoa and dried product quality *Biosys Eng.* **102** 153–61
[8] Ee C T, Hii C L, Ong S P, Law C L and Tan C H 2022. Hybridization of freeze drying and impacts on drying kinetics and dried product quality of kedondong fruits *Dry. Technol.* Inpress
[9] Raquel P F G and Maria J B 2012 Effect of drying treatments on texture and color of vegetables (pumpkin and green pepper). *Food Bioprod. Process.* **90** 58-63
[10] AOAC 2005 Official methods of analysis in *Official Methods of Analysis of AOAC International.*
[11] Sakin M, Kaymak-Ertekin F and Ilicali C 2007 Modeling the moisture transfer during baking of white cake *J Food Eng.* **80** 822–31
[12] Baik O D and Marcotte M 2003 Modeling the moisture diffusivity in a baking cake *J Food Eng.* **56** 27–36
[13] Megahey E K, McMinn W A M and Magee T R A 2005 Experimental study of microwave baking of Madeira cake batter *Food Bioprod. Process.* **83** 277-87
[14] Begum F, Sarker A, Ahmed S, Mondal K and Anjum S 2016 Physical and nutritional quality of retails cakes of Dhaka city *Bangladesh J. Sci. Ind. Res.* **51** 35–40

AUTHOR INDEX

Abas, N.Z. ... 97
Abdul-Hakim, N A. ... 37
Abdullah, L C ... 241
Adly, SA Sirajul ... 223
Ahmad, Z ... 127
Ali, S N Syed Sabeer ... 253
Alias, A B ... 265
Alias, S ... 37
Alwi, H ... 19
Alwi, S R W ... 200
Andiappan, V ... 25, 163, 169
Asmadi, A Ahmad ... 49
Azizi, C Y ... 19
Azmi, I. SH ... 97
Bakar, N A Abu ... 73
Boon, C Y ... 289
Brendan, L K W ... 13
Chai, S Y W ... 139
Chai, Y H ... 145
Cheah, C H ... 308
Cheah, KH ... 271
Chee, A L K ... 91
Cheng, X H ... 229
Chew, J J ... 151
Chia, S Y ... 1
Chiang, C L ... 308
Chin, B L F ... 91, 145
Chong, C H ... 296
Chong, EHY ... 163
Chong, WC ... 283
Choong, T S Y ... 241
Chua, W S ... 157
Cocero, M J ... 265, 277
Djarum, D H ... 127
Farooqi, A S ... 194, 194
Fuad, F M ... 265
Goodman, A H ... 7
Gulnaziya, I ... 235
Hadj-Kali, MK ... 133
Hamid, M A ... 302
Hanipah, S H ... 265, 277
Hashim, H H ... 73
Hashim, MA ... 133
Hermansson, A W ... 212
Hermansson, AW ... 206
Hian, M Lim Moy ... 79
Hii, C L ... 308
Hing, A A M K ... 218
Hocking, R K ... 151
Hoo, C L ... 259
How, B S ... 61, 79
Hussein, N ... 194
Hwang, J Z H ... 25
Idris, A I Md ... 253
Ihsan, A Kamal Ariffin Mohd ... 108
Ismail, K I ... 182
Jalil, M.J ... 97
Jamil, S N A Md ... 241
Kalib, N S ... 121
Kalib, N Shieela ... 102, 108
Khaerudini, D S ... 151
Khoo, T-J ... 296
Kiew, P L ... 145
Kiwan, B ... 121
Koh, K S ... 175
Kong, K G H ... 61
Krishnan, S ... 67
Kumar, A N ... 91
Kumaran, N A R S ... 308
Lahuri, H M ... 7
Lai, M K ... 182
Lam, H L ... 79
Lam, M K ... 145
Lau, J S Y ... 145
Lee, EY ... 271
Li, F ... 241
Lim, C H ... 79, 169, 212
Lim, CH ... 206
Lim, C-H ... 55
Lim, J W ... 145
Lim, M W ... 1, 13, 157
Lim, S S ... 175
Lim, S ... 55
Lo, S L Y ... 61
Lock, S S M ... 194
Luthfi, AAI ... 133
Mackay, E J ... 302
Mahadi, M. B ... 97
Man, H Che ... 253
Manaf, N Abdul ... 49
Mansor, M H ... 218
Martín, Á ... 265
Martin, Á ... 277
Marzuki, A S ... 277
Mistry, B ... 308
Mohamed, A R ... 43

Mohammad, AW 283
Mohd, N 223
Mokhtar, M A Mohd 49
Mubin, A F H A 218
Muchtar, A 102, 108, 121
Murugappan, V A 229
Mustapa, A N 19, 265, 277
Naji, S Z 43
Narayanasamy, G 67
Ng, D K S 25, 169
Ng, Denny K S 229
Ng, DKS 163
Ng, L Y 169, 229
Ng, LY 163
Ng, R B 259
Ng, W P Q 79
Ngu, L H 139
Norhanifah, A H 85
Norliza, A R 85
Norman, A R 218
Nurhaslina, C R 19
Oh, P C 188
Omar, R 253
Ooi, W X 212
Othman, N H 277
Padmesh, TNP 67
Padmesh, TVN 289
Pang, C Y 235
Pang, R 182
Pang, Y L 235
Pang, Y-L 55
Phang, SJ 271
Pikri, M H Muhamad 73
Pua, F-L 55
Pua, G 55
Pua, ZY 206
Rafidah, J 85
Rahman, M.A. 97
Rahman, Mohammad Nurizat 114
Ramasamy, G 7
Ramli, R M 194
Rasib, I.M. 97
Rasid, N I Mohd 49
Ravi, R 102
Ren, Y 31
Rivero, R 31
Roslan, H S 265
Sabil, K M 7
Safira, M R 157
Saleh, S M 188
Salleh, MZM 133
Setiawan, A S 302
Shah, S S Md 7
Shahril, Norshakina 114
Shariff, Ismail 114
Shivaprasad, P 163
Somalu, M Rao 102, 108
Sunarso, J 61, 151
Suparmaniam, U 145
Tajuddin, H A 253
Talip, N A Abdul 73
Tan, C H 200
Tan, C X 247
Tan, I S 145
Tan, K L 308
Tang, M F 61
Teh, H Y 145
Teoh, HC 283
Tye, C T 43
Voo, WX 283
Wajahat, S M 194
Wong, J L 151
Wong, K W 169
Wong, M C 235
Wong, SY 271
Wong, V L 175, 259
Wong, V-L 247
Wong, VL 271
Wong, W-Y 55
Yap, M Y 13
Yap, T C 182
Yeap, S P 247
Yong, J 175
Yuan, V Zhen 108
Yusof, K M 200
Yusup, Suzana 114
Zakaria, A Wan 73
Zhang, J 127
Zulkifli, A S 188
Zuri, W N S Wan Mohamed 49